MAYA COSMOS

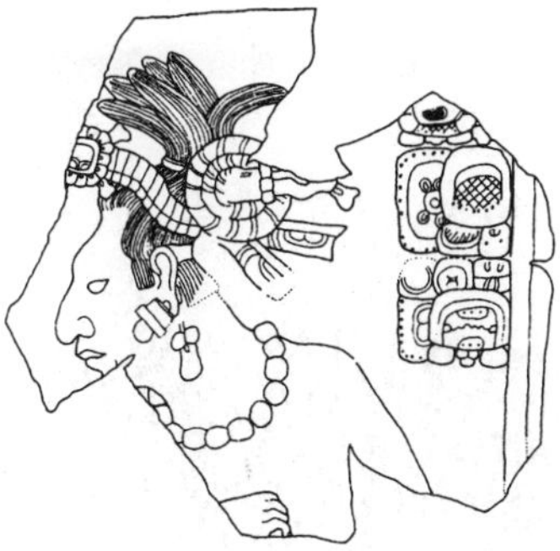

The fragment used on the jacket belongs to a shattered panel that archaeologists found among the rubble on the north end of the Palace at Palenque. The image shows Pakal, the king of Palenque, wearing the headdress of Itzamna, the first sorcerer and diviner.

ALSO BY LINDA SCHELE

Maya Glyphs: The Verbs (1982)

The Blood of Kings:
Dynasty and Ritual in Maya Art (1986)
with Mary Ellen Miller

ALSO BY LINDA SCHELE AND DAVID FREIDEL

A Forest of Kings:
The Untold Story of the
Ancient Maya (1990)

ALSO BY DAVID FREIDEL

Cozumel: Late Maya Settlement Systems (1984)
with Jeremy Sabloff

MAYA COSMOS

Three Thousand Years on the Shaman's Path

David Freidel,
Linda Schele
& Joy Parker

Photographs
by Justin Kerr &
MacDuff Everton

QUILL
WILLIAM MORROW
NEW YORK

It is the policy of WIlliam Morrow and Company, Inc., and its imprints and affiliates, recognizing the importance of preserving what has been written, to print the books we publish on acid-free paper, and we exert our best efforts to that end.

Library of Congress Cataloging-in-Publication Data

Freidel, David A.

Maya cosmos : three thousand years on the shaman's path / David Freidel, Linda Schele, and Joy Parker: photographs by Justin Kerr and MacDuff Everton.

p. cm.

ISBN 0-688-14069-6

1. Mayas : Religion and Mythology. 2. Mayas—Rites and ceremonies. 3. Mayas—Philosophy. 4. Shamanism—Mexico. 5. Shamanism—Central America. I. Schele, Linda. II. Parker, Joy, 1953—. IIII. Title.

F1435.3.R3F74 1993

299'.79281—dc20 93-2696

CIP

Printed in the United States of America

First Quill Edition

5 6 7 8 9 10

BOOK DESIGN BY CHARLOTTE STAUB

THIS BOOK IS DEDICATED TO

Evon Z. Vogt,

David Humiston Kelley, and

Helen Woods Parker

A Personal Note from the Authors

The sky has become important to us in a way we never anticipated. For each of us the way to the sky has been different. The first time David Freidel really fell into the night sky, he was a boy lying on a hill, in the open country of southeast Wyoming. The Milky Way arched wide and sparkling in the blackness above him, and he lost the hills and the trees as he flew upward into it. That summer he was a high school volunteer working on ancient Indian campsites. It was the first step on the journey that turned him into an archaeologist and led to this book.

For Linda Schele, it was different. She was a professional painter and studio teacher who happened to take a vacation in 1970. The place she went to visit was México and her life was forever changed by the experience. But the sky did not really come home to her in all its glory until the summer of 1992 when she finally found a place—Tixkakal Guardia in Quintana Roo, México—where it was dark enough and cloudless enough to see the crystalline glory of the Milky Way. In the pattern arching across the sky, she saw the face of Creation and the World Tree standing at the center of the Cosmos.

These exhilarating moments have taken on special meaning because of what we discovered by writing this book: that the patterns of stars in the night sky, and especially the Milky Way, are the primary symbols of Maya Creation. We thought their Creation myth was just a story—but before this book was finished we realized that this myth presented a breathtaking and literal map of the sky. Writing together, changing each other's words and ideas, striving for some way to describe what we had come to understand about the Maya world, we had stumbled into the Creator Shaman's attic. There we found the Maya sky, a painted book whose constellations unfold its picture story of genesis across the face of the cool darkness while the sun illuminates the Otherworld.

Over the last twenty-five years, we've spent a great deal more time looking down in the dirt or at dusty stone inscriptions than up at the sky. Working with teams of dedicated people, we've slashed through snake-infested undergrowth to lay down map trails, pried big rocks out of small holes, squinted with sweat-stung eyes to scribble notes in the blinding midday heat. Or we have patiently sat in contorted positions and traced the lines of an eroded inscription onto a paper damp with the sweat that dripped from our chins. In return for all this hard work, we've been privileged to stand and gaze at man-high painted masks of the Maya gods, released from their darkness for the first time in two thousand years. We've watched ruined cities emerge and seen fallen stone facades restored to comprehensibility. We've seen the walls of long-buried temples appear under the hands of men patiently chipping away at the stone-filled, concrete-hard dirt in the dark, hot tunnels under pyramids. We've been there when the names of powerful leaders dead and forgotten for more than a thousand years have been spoken again, not only by people from our world, but by their living descendants.

Beyond the sense of accomplishment we feel in our craft and the sheer joy of the work, we have felt a driving need for something more: not just to find the Indian behind the artifact—as the adage goes in archaeology—but to understand that person's way of seeing the world. This book is about the Maya way of seeing. At its heart is the experience of a spiritual and magical world. It is a world in which ancestors affect the fate of the living, where human beings can transform into their animal counterparts, where ritual transforms space and objects into powerful energy carriers. It is a world alive in all its parts.

Before writing this book, we studied the beliefs the ancient Maya recorded in their art and architecture as if they were part of a fascinating alternate reality that had little relevance to our personal lives. We did what scientists are supposed to do—keep an objective distance. Like the majority of our colleagues, we were secure in the superiority of our own worldview. Preparing this book has changed our attitude. We now come to this alien world with respect and awe, knowing that the worldview created by the Maya was, and is, just as powerful, meaningful, and viable as our own.

To write this book, we had to reach inside ourselves to find both empathy for their way of being and a method to communicate this worldview to others. Because their way is pervasively spiritual, we had to discover our real feelings about spirituality and faith in the shared wisdom that guides daily life and conduct. To write about the Maya cosmos, we had to accept the significance of its supernatural reality and learn to play by the rules of its own internal logic.

We haven't turned our backs on scientific knowledge or methods. On the contrary, in the course of describing the Maya cosmos we have systematically referred to concrete and material patterns of Maya things—to buildings, pots, sculptures, inscriptions, and to the contexts in which they were found. We have also included germane historical and contemporary eyewitness accounts of actual Maya behavior in many different situations. Yet the heart of this book—or perhaps its soul—is our personal encounter with the Maya, past and present, and our adventure of discovery as professed novices in the world as seen through their eyes, words, and deeds.

Far from presenting a detached point of view, we have integrated ourselves, our experiences, and our understanding into what we have to say. Because our book is about a body of knowledge created and perpetuated by the Maya, it must also be about how we ourselves have begun to absorb that knowledge, learning a different rationality just as one learns a new language. Contemporary scientific research has, in our view, taken the human witness out of our cosmology. Maya reality places human beings at the center of the cosmos and makes humanity responsible for creation as an ongoing, endless process. We have tried to take a page from their book by using our own experiences to open a window of understanding into what they have created.

When we first proposed this book to our editor, we thought it would be a book on shamanism in the religion of the ancient Maya. As we began to research the themes we had chosen, to read the work of ethnographers, and most of all, to participate in modern Maya ritual, the book transformed itself before our eyes. It became a book on the continuity of Maya reality from the ancient past into the present. None of us will ever again be able to walk the streets of Yukatan, Chiapas, or Guatemala and think of the Maya we see as merely biological descendants of the ancients, cut off from their past by the trauma of the Conquest. Our view of Maya history and the continuity of their cultural heritage has changed.

Yet even with this change, shamanism remains at the heart of our book. From his student days onward, David Freidel has been fascinated with shamanism and he has transferred that interest to Linda and Joy. One of his earliest encounters with this belief system was in the work of Mircea Eliade. *Archaic Ecstasy*, especially, convinced him that shamanism is a very old, coherent, and broadly diffused mental paradigm. It draws its powers of persistence from its capacity to organize knowledge about the world by way of a rather simple set of symbols and assumptions. The most important premise is that the spiritual force that every human being has personally experienced is ambient, of the world, of the cosmos and everything in it. Birth and death are just especially overt examples of the capacity of spirit to move through material forms. Johannes Wilbert, friend and adept in the shamanism of the Warao people of Venezuela, told us at lunch in Puerto Rico recently that there were two ways to regard the divine—as a transcendent Creator out there, or as an immanent Creation potentially manifest everywhere. We think he was referring to the same basic premise of shamanism we described above. The wonderful thing about this idea is that it nurtures a sense of belonging to the world and to the cosmos. We need not look up at the night sky or the sun and feel insignificant. Rather, shamanism teaches that those bright beings are part of a shared history and life-force that belong to all humanity.

We believe that shamanism can function as a guide in complex, civilized societies and that as a way of organizing knowledge about the world it spans the Americas today. We had the privilege of experiencing this truth at the 1992 Austin Workshop. A Native American shaman named Sakim, holy man for people of the Creek nation in Florida, attended the workshop at the behest of Kent Reilly, specialist in the shamanism of the Olmec civilization that preceded the Maya in ancient México. Linda had arranged for several of the Maya people who are learning to read and write in their ancestors' glyphs to attend the workshop and participate in the process of unlocking their past. At the traditional Saturday night party, which brought people together at Linda's house for a potluck dinner, David found himself in conversation with Martín, a Kaqchikel Maya from Guatemala, and Sakim, the Creek holy man. These men, speaking Spanish and English respectively while David provided a halting translation, began to explore the common dimensions of their understanding of the world. Martín spoke of maintaining balance between the forces of the world, from the center of heaven to the center of the earth. Sakim enthusiastically agreed and spoke of the responsibilities of people in this balance. They saw together that in the European languages, space had the connotation of emptiness, while in their words space is full and substantive. So balancing all the things embraced within it made sense. Martín said that many things are alive, things of stone and wood (and here he touched the coffee table in front of him with its stone top and wood side). Sakim said that anything with form, substance, purpose, and place is a being. Inspired, Sakim drew off his sacred shell pectoral and placed it on Martín. Martín, deeply moved, broke into a speech in Kaqchikel. By this time, David had recruited Dennis Tedlock to help in translating an increasingly complicated discussion. Dennis Tedlock is a trained K'iche' day-keeper. Linda stopped by to show Sakim a framed page of paper she treasured greatly. It was a thank-you note written by Martín in modern Kaqchikel using the old writing system of hieroglyphs. It was, to her knowledge, the first example, in at least three centuries, of a Maya using his native writing system to record something he wanted to say. Sakim, no longer able to contain himself, began to pray in Creek over the page. Surrounded now by three adepts speaking four languages, David felt as if he was witness to the first meeting of an American united nations. We firmly believe that the Maya vision celebrates a way of conceiving the world that lives in the hearts and minds of the firstcomers to this hemisphere, a vision that is the natural legacy of all the American peoples and that can help to unite them in purpose as we face the challenge of rethinking our relationship to the life system that spawned us all.

These ideas, new to Linda and Joy and old to David, have relevance to many more people than just us. The contemporary Maya were taught, or came to believe, that they had inherited nothing from the Precolumbian past. Some scholarly theories hold that modern Indian institutions and beliefs are a product of the Conquest itself and were created as a cultural entity by the Spanish conquerors. The ruins left by the ancient Maya, and the cultural legacy they represent, have been appropriated by the Ladinos and mestizos as the patrimony of modern nations, without reference to the Indian peoples who still live among them. We have learned a different way of understanding the modern Maya who have become our friends. We accept that they are not exactly the same as their ancient forebears, any more than we are the same as the people who fought the Vikings in the England of the eighth century. Yet the history written and preserved in images on hundreds of stone monuments and clay vessels, the architecture of their Precolumbian cities and residences, the fossilized remains of rituals in temples and hamlets, all hold a vision for the modern Maya that reflects their heritage because it still makes sense in their languages and practices. The modern Maya can claim their past because it lives in their present. Their pageants and beliefs shed light on the ancient arts of their forebears in ways that cannot come from our world or our science, and they offer us the gift of new knowledge about an ancient cosmos.

DAVID FREIDEL	LINDA SCHELE	JOY PARKER
Dallas	Austin	Los Angeles

FOREWORD

THE ORTHOGRAPHY

In Maya studies, spelling has always been a problem. When the Spanish came, they found people using consonants and vowels that did not exist in their own language, and just as bad, they found that consonants and vowels in their own language were unknown to the Maya. The early friars, usually using conventions developed within their respective orders, worked out standard ways of representing the sounds they tried to record, while other chroniclers used their own systems. The result was a hodgepodge of orthographies that has not been helped by the introduction of modern technical systems.

Glyphic studies traditionally use the colonial orthography imposed on Yukatek by the Spanish. This orthography is still used by scribes among the Cruzob Maya today to write their books of prophecy. It has served the Yukateks well and yielded an enormous literature, including numerous dictionaries, the Books of Chilam Balam, and the writing of modern h-men of the Cruzob Maya. It has also become the traditional orthography used in epigraphy: it was the calendric terms from Bishop Diego de Landa's *Relación de las Cosas de Yucatan* that early researchers used for their own writing.

Unfortunately, there are many problems caused by using these early orthographies, including the Yukatek one. Many Mayan languages have contrasting sounds that do not exist in Spanish; often Spanish chroniclers never heard them at all or simply ignored them. Furthermore, the Spanish settled on different and inconsistent systems for different languages. The orthography they developed for Yukatek, for example, cannot be used to write most of the Cholan languages and none of the highland

Maya languages. As we moved into the twentieth century, the situation only got worse, because a new set of orthographies was added to the alphabet soup, including practical orthographies developed by linguists on the one side and translators of the Bible on the other side. When the international phonetic alphabet was added to the already confused situation, we ended up with some systems using diacritics, some systems that didn't use them, and almost as many ways of spelling Maya words as there are Maya languages and researchers.

As a result, the same word is often spelled in more than one way within one document. One of the most confusing problems facing students of the Maya, especially those new to the field, is how to make sense of the many different ways the same word can be spelled. The word for "lord" can appear as *ahaw*, *ahau*, *ajau*, *ajaw*, or *axaw*, depending on which orthography is used. Having to cope with all these different ways of spelling is often the first problem to be faced, even before cognate sets[1] between different languages can be determined.

In earlier, more technical publications, we tried to use linguistic orthographies, but since early glyphic studies began using the system of colonial Yukatek, many terms simply looked funny when they were written in a technical orthography. We tried keeping established terms in the old orthography and changing any new ones into modern orthographies—of which several are used by different researchers. Unfortunately, we ended up with a combination system that was more confusing than simply keeping the old Yukatekan system. Chinese presents a similar problem with multiple orthographic conventions and the same resulting confusing plethora of different spellings for the same word.

In our last book, *A Forest of Kings*, we chose to retain the Yukatekan orthography that is traditional to the field of epigraphy, with a few alterations to accommodate particular difficulties. That decision worked adequately when we were working with the languages of the inscriptions—Yukatekan and Cholan. However, in this book we found ourselves facing a different problem, since we were including words from most of the extant Maya languages. In this context, the old Yukatekan orthography simply will not work.

Furthermore, since the publication of *A Forest of Kings*, the Maya of Guatemala have themselves adopted a uniform alphabet with which to record their own languages. The Maya of the Proyecto Lingüistico Francisco Marroquín and the Academia de Lenguas Mayas, in cooperation with writing and language groups all over the highlands of Guatemala, adapted the alphabet from the practical orthography developed at the PLFM during earlier research with professional linguists from the United States. In 1989, the Ministry of Culture and Sports of Guatemala adopted the alphabet and Margarita López Raquec's *Acerca de los alfabetos para escribir los idiomas Maya de Guatemala* as the official government publication of the alphabet and its history.

A very similar alphabet is now being used by the writers' cooperatives of Tzotzil and Tzeltal, and several organizations have adopted the alphabet developed by Barrera Vásquez for Yukatek. Since these very similar alphabets have been developed in cooperation with or by native speakers and adopted by them for writing their own languages, we have decided to adopt their uniform alphabet for *Maya Cosmos*, as a means of reducing the confusion of alternate spellings and, most of all, in recognition that the Maya have the right to decide how their own languages should be written.

Our decision brings with it both advantages and disadvantages. The major disadvantage is that words with traditional spellings, including place names, suddenly look very strange to someone used to seeing them in the old form. The advantage is that everything is spelled in the same way and that most of the traps that led English speakers into mispronouncing Mayan words are gone. Words now look the way they sound. *Cimi* is *kimi* and *ceh* is *keh*. The biggest problem is the traditional calendar terms and the place names that are fixed in the colonial spelling of Yukatek. At first, they look very odd in their new spellings. *Dzibilchaltun* becomes *Tz'ibilchaltun* and *Uaxactun* becomes *Waxaktun*. We debated what to do with these situations where there are traditional spellings and decided that we had to apply the new alphabet to all words that are Maya in origin, whether or not they have become Hispanicized as place names. We have decided to keep the traditional spellings when place or ethnic names derive from Nahualt, so that Olmec, Toltec, Aztec, and Zinacantan remain the same. Our decision may be confusing for some, an anathema to others, and appreciated, we hope, by the rest, but we believe our decision is the right one to make. Besides, we have found the adjustment quick and relatively painless.

The traditional orthography of Yukatek corresponds to this new uniform alphabet as follows:

a = a (pronounced like the *a* in f*a*ther)
b = b
c = k (pronounced like English *k*)

e	=	e	(pronounced like the *e* in set)
h	=	h	(pronounced like an English *h*)
i	=	i	(pronounced like the *ee* in *see*)
j	=	j	(pronounced like a hard *h* sound, as with the Spanish *j*)
k	=	k'	(pronounced like a *k*, but with the glottis closed)
l	=	l	
m	=	m	
n	=	n	
o	=	o	(pronounced like the *o* in h*o*ld)
p	=	p	
pp	=	p'	(this is a glottalized *p*, pronounced with the glottis closed)
q	=	q	(this post-velar *k* is pronounced deep in the throat and has no equivalent in a European language; neither is it present in the lowland languages)
q	=	q'	(glottalized version of the same consonant)
s	=	s	
t	=	t	
th	=	t'	(this is a glottalized *t*, pronounced with the glottis closed)
tz	=	tz	(this is another consonant that does not exist in English or Spanish)
dz	=	tz'	(to the confusion of both native and non-native speakers, the old Yukatek orthography had *dz* for the glottalized form of this consonant)
u	=	u	(pronounced like the *oo* in z*oo*)
u	=	w	(the Spanish used the letter *u* to write the consonant *w* so that many traditional spellings, such as *ahau* or *Uaxactun*, have a *u* in the place of the consonant)
x	=	x	(this is equivalent to the English *sh*)
y	=	y	
z	=	s	

There are other consonants, especially the retroflexives and an *n* pronounced like the final sound in *sing*, that are used by several Mayan languages, but because we do not use vocabulary from those languages, we have not included them in our list. However, there is one correspondence problem that should be noted, because we do refer to terms using them. *L* in all other Mayan languages is pronounced *r* in Chorti, but in Kaqchikel and K'iche', *r* corresponds to *y* in other Mayan languages. We will note these exceptional spellings when they occur in the text.

Finally, we will use the Yukatek Maya plural suffix *-ob* for Maya words. More than one Chak will be written Chakob, more than one sahal will be sahalob.

ACKNOWLEDGMENTS

We wish to acknowledge the many people who helped us with the ideas presented in *Maya Cosmos* and who contributed to its writing and its production. First and foremost is Maria Guarnaschelli, senior editor and vice-president of William Morrow and Company. She nursed us through the process, often with a blunt honesty that stimulated us to break out of the limitations of our imagination and find new ways of writing about our world. Her reading of the last draft penetrated to the heart of the matter with precision and amazing insights. She was encouraging, enthusiastic, and helpful when we needed it, and by expecting us to do more than we thought we could do, she made the book better than it would have been without her.

Our copy editor, Joan Amico, gave us able and meticulous editing with a very complicated text. Her skill and broad knowledge was a comfort: for we knew she would find what we had missed. And our designer, Charlotte Staub, brought her cooperation, skill, and commitment to a very complicated design problem. Because of her, we were able to create unusually complex illustrations and knew they would come out as we intended. Her work was essential.

Chas Edwards, junior editor at William Morrow, was vital in bringing all the pieces together. Thanks, too, to Deborah Weiss Geline in copyediting, Nick Mazzella, Harvey Hoffman, and Tom Nau in production, and Russell Gordon and Bob Aulicino in jacket art for their invaluable skills.

Our agent, Elaine Markson, encouraged us and helped us extend the deadlines when Linda became ill and the complication of the material overshot our estimates. She and her agency smoothed our path in many ways.

For those interested in the process of collaborative writing, the manuscript was written using *Nota Bene* by N. B. Informatics Incorporated as the primary word processor. This was the second book the three of us had written together, so we had worked out many of the kinks in the process and most of all built complete confidence in each other. Linda or David wrote the first drafts of individual chapters. The primary writer then sent a disk with the file to the other one. The second author "massaged" the text—as we came to call the process—by changing, adapting, rewriting, adding, and deleting until satisfied with the product. The chapter then went back to the primary author, who massaged it again.

This process sometimes went through one cycle, sometimes two if there was particular difficulty or disagreement. Finally, the text went to Joy, who massaged it in her own very important way. Her special responsibility was to make sure the text was readable to nonacademic people. We communicated a lot by phone. During the six months before the manuscript was sent to Morrow, major discoveries forced the reorganization of some chapters. A final pass through the hands of all three authors and several conference calls brought the changed manuscript to its final state.

Linda designed and prepared the illustrations using a 486 computer, a UMAX 1200 DPI scanner, and Aldus Pagemaker under Windows. To design the book, Charlotter Staub used proofs of the illustrations printed on a Hewlett-Packard Laserjet III. Linda made copy corrections on the original files and made design changes as requested. Final reproduction quality proofs were printed on a Lasermaster Unity 1200S Plain Paper Typesetter. This process allowed us to create very complicated labeling and designs at a low production cost.

Many of the ideas in this book come from years of interchange with friends, colleagues, collaborators, and our students. We wish to acknowledge in particular the contributions to this process made by (alphabetically) Ira Abrams, Timothy Albright, Anthony Aveni, Grace Bascope, Victoria Bricker, Michael Coe, Garrett Cook, Barbara Fash, William Fash, John Fox, Jerry and Judy Glick, Gary Gossen, Gillett Griffin, Nikolai Grube, David Kelley, Barbara and Justin Kerr, Ruth Krochock, Robert Laughlin, Matthew Looper, Floyd Lounsbury, Barbara MacLeod, Peter Mathews, Warner Nahm, Kent Reilly, Merle Robertson, Sakim, David Stuart, Charles Suhler, Karl Taube, Barbara and Dennis Tedlock, Evon

Vogt, Kristaan Villela, Johannes Wilbert, and the many participants in the Texas Meetings on Maya Hieroglyphic Writing. Most of all, Linda and David wish to thank their graduate students. The contributions that they have made to our growth and flexibility of mind are incalculable. In addition, we have sent chapters or materials to various colleagues who have offered suggestions and criticisms that have been invaluable. These people include Garrett Cook, Gary Gossen, Evon Vogt, John McGee, William and Barbara Fash, Kent Reilly, Barbara Tedlock, Dennis Tedlock, Anthony Aveni, Nikolai Grube, Karl Taube, Sam Edgeton, Walter "Chip" Morris, Johannes Wilbert, and Enrique Florescano.

We particularly wish to thank people who provided photographs for us. These include Ed Krupp, Duncan Earle, John McGee, and Gary Gossen. Very special thanks are due to MacDuff Everton, who offered us his extraordinary wraparound photographs and slides from twenty years of work with the Maya. Our friends Justin Kerr and Barbara Kerr gave us access to their photographic archives, including roll-outs of pottery without which we could not have done our work. They also provided us with bed and breakfast when we came to New York, as well as minds on which we could test our ideas.

We must also give a particularly heartfelt thanks to Nora England, who helped organize the hieroglyphic workshops with the Maya in Antigua. Our special friend, Pakal Balam José Obispo Rodríguez Guajan, shared his family and Imixche' with Linda. Martín Chacach Cutzal has been a special friend to Linda and was the first to ask her to give the workshops in Antigua. The members of Oxlajun Keej Maya' Ajtz'iib, a group of Maya writers who work with Nora, have opened their world to Linda in the last four years. Many of the anecdotes Linda tells recall adventures with these friends. They are Nikte' Juliana Sis Iboy, Waykan Ch'imil José Gonzalo Benito, Saqijix Candelaria López Ixcoy, and former member Saqch'en Ruperto Montejo Esteban. Steve Eliot and the people of CIRMA in Antigua, Guatemala, have always been supportive and immensely helpful, lending us a place to have the workshops, tables so that we could work, and occasionally a van so that we can visit distant places. In Yukatan, we are especially grateful to John Sosa for sharing his insight into Maya cosmology with us, to Don Emet, a true Maya leader, to Don Pablo, and our many friends in Yaxuna village. Joann Andrews is responsible for helping maintain David Freidel's sanity and purpose in his research; she is truly the First Mother of northern research. Thanks also to Mary Dell Lucas and Alfonso Escobedo for opportunities to travel in

Yukatan. Without the help and friendship of all these people, *Maya Cosmos* would not have the form it does.

A very special thanks is also owed to Duncan Earle, who made possible Linda's experience of the Chamula Festival of Games. He served as her guide and teacher. We also want to thank Alfred Bush, who lent his house to Linda for two nights, and Suzanna Ekholm, who took us in when Alfred's other house guests arrived. Norman Shaifer and his wife provided the car that got us to Romería. Carol Karasik graciously let Linda and Duncan disturb her privacy when we were in San Cristóbal.

Federico Fahsen, a good friend, has worked with Linda and Nikolai Grube in mini-conferences we have created in Guatemala for the last two summers. The three of us have spent many wonderful hours working together at his farm, his house, in Antigua, and in Guatemala. That lively debate and sharing between the Guatemala conferistas has contributed to many of the new discoveries we present in *Maya Cosmos*. When she was Minister of Culture of Guatemala, Marta Regina Fahsen provided Linda and her Maya friends a letter of introduction and cooperation for a visit to Tikal in 1990. It was an important visit for both Linda and for the Maya who were with her.

Research by Linda Schele, as it is presented in various chapters, has been supported by the John D. Murchison's Regents Professorship at the University of Texas. She has also worked with the Copan Acropolis Archaeological Project, which is under the direction of Dr. William Fash, and the Instituto Hondureño de Antropología e Historia. José María Casco, the current genente, has been very supportive and gave permission to use material from the Copan excavations. We would also like to thank Ricardo Agurcia, co-director of the CAAP, for his support. The wonderful hours of debate and exchange with Barbara Fash, William Fash, Robert Sharer, Will Andrews, David Sedat, Julie Miller, Alfonso Morales, and Rich Williamson have been central to the joy of working at Copan. Some of Linda's happiest hours have been spent in the tunnels and the acropolis courts trying to understand the history and meaning of that magnificent structure. And Linda extends a very special thanks to E. Wyllys Andrews V for his long years of friendship and support.

Dr. William Gutsch, director of the American Museum of Natural History's Hayden Planetarium, helped us check the astronomy of Creation in fast time. Von Del Chamberlain of the Hansen Planetarium in Salt Lake City spent many very early hours with Linda checking the astronomy and casting it back to 15000 B.C. We also sent early versions of the discovery to Anthony Aveni, Barbara and Dennis Tedlock, Johannes Wilbert, David Kelley, Floyd Lounsbury, and Nikolai Grube. All of them made important contributions as did Werner Nahm and Victoria Bricker.

Research at Cerros mentioned in various places was carried out under the auspices of the office of the Archaeological Commissioner of Belize. Joseph Palacio, Jaime Awe, Elizabeth Graham Pendergast, and Harriot Topsey served in that office and greatly facilitated our research. The Cerros work was supported by the National Science Foundation (BNS-77-07959, BNS-78-2470; BNS-78-15905; BNS-82-17620) and by private donations by citizens of Dallas to the Cerros Maya Foundation. Thanks to T. Tim Cullum and Richard Sandow for their support. Stanley Marcus—and through Mr. Marcus many other individuals—supported the Cerros work throughout its duration. Mr. Marcus has been a special mentor and friend to David Freidel throughout his career in Dallas. The research at Cerros was originally directed by Dr. Ira Abrams, and we thank him for that.

Research at Yaxuna mentioned periodically throughout the book is being carried out under the auspices of the Instituto Nacional de Antropología e Historia, México. Maestra Lorena Mirambel has been especially helpful in her capacity as head of the Archaeology Council of the INAH. The Directors of the INAH in Mérida, Ruben Maldonado and Alfredo Barrera, have greatly facilitated our work at Yaxuna. Dr. Fernando Robles, senior investigator of the INAH, and Dr. Anthony Andrews, first took David Freidel to Yaxuna and have strongly encouraged the work at the site. The Yaxuna research is supported by the National Endowment for the Humanities (RO-21699-88; RO-22349-91), the National Geographic Society, the Provost's Office of Southern Methodist University, and private benefactors in Dallas through Mr. Stanley Marcus. The current phase of research would not have been possible without the generous support of Bernard Selz and the Selz Foundation of New York.

CONTENTS

CHAPTER ONE

WORLDS APART, JOINED TOGETHER: The Road of Maya Reality

THE CH'A-CHAK CEREMONY

(as told by David Freidel)

The plains of Yaxuna in northern Yukatan are usually covered with a green sea of waving maize plants and waist-high grasses in the month of July. The temple-mountains of stone rubble rise skyward like the gray, forested islands of a landlocked archipelago, but in the summer of 1989, the sea had become a desert. Red dust jumped in small puffs as I stumbled, heat-stupid, through the stunted weeds and stillborn cornstalks from the June planting—the second failed planting of that year. Stripped of its life-sustaining greenery, the plain was a maze of low rock patterns, the homes and household lots of the ancient community's farmers, warriors, craftsmen, and merchants. The earth, torn up by burrowing iguanas, was littered with the broken pottery trash from two thousand years of habitation. The sherds glittered in the relentless glare of midmorning.

I climbed the steep broken stairway of a temple pyramid and saw two Maya farmers sweating over thirty-pound chunks of quarried stone at the bottom of the shallow square hole on the summit. Torn from the earth two thousand years ago by their ancestors, who bore them by tumpline from nearby quarries and piled them into platforms to raise the eyes and voices of kings and shamans to the horizon above the tangled forest, the stones were being moved once again by the muscle and will of Maya men, who were heaving them into neat piles by the side of the pit. At the bottom of the pit, the men had cleared away the rubble to reveal the summit of an earlier temple—a mountain inside a mountain. Its stairway of brown-plastered masonry disappeared under the rubble of the temple built on top of it. I jumped down and carefully brushed clean the surface of the floor so that I could stand on the ceremonial platform. Maya rulers had last stood there when the Maya civilization was new, centuries before the birth of Christ in an alien world far away. This was one of the first temple-mountains raised by the Maya in this northern country.

My exhilaration of discovery was tempered by worry as I struggled out of the pit and scanned the eastern horizon for the blue-black shadows of rain-yielding thunderheads. The cloud-borne Chakob were riding the wind, but they weren't coming our way. They were headed south toward the village of Santa María. No one around me had spoken of rain for some time; it had been days since the last sprinkle on the lands of Yaxuna.

"They held a Ch'a-Chak at Santa María yesterday. The shaman from K'ankabtz'onot came and they killed two deer for it," said a young villager.

Sometimes shamans perform Ch'a-Chak ceremonies for the tourists who throng to Yukatan to admire the temples. These rituals are supposed to call upon the ancient gods to bring rain. Here in Yaxuna, however, Ch'a-Chakob were a deadly serious business, a plea for relief from the drought that threatened the lives and well-being of my friends.

Drought was perfect working weather for my project, but it was catastrophe for the villagers who lived along the western edge of the ancient city around the ancient well of Yaxuna. Drought is a time when old tensions surface and chronic afflictions feel worse. It is a time to redress the balance between the people and their place in the world through communion with the unseen beings of the Otherworld. In the old days, said some of our workers, people could seek out wild honey in the high forest in times of drought and live on honey mixed with water to still the grumbling in their bellies and stretch their dwindling supplies of corn. Now the high forest is gone, cut down for lumber, cleared for farmland, and replaced by tangled young growth. Wild honey is hard to find.

The archaeologists working on the ancient site of Yaxuna were a part of that tension. We brought desperately needed wage work. Now that the stocks of seed corn were depleted, we brought money to buy new seed corn and money to buy corn to eat. We had divided up the work among as many families within the village as possible, scheduling the men in awkward three-day shifts during a field season when it would have been better for us to have the same men working longer stretches of days. We had brought the village some financial relief, but we had also dug into the temple-mountains, the dwellings of the old gods. Perhaps by upsetting them, we were bringing misfortune along with opportunity. We had a camp of thatched-roof native houses at the edge of the village where the ruins began, on land that belonged to everyone, lent to us by every family in Yaxuna. Because of all of these things, we were of the community—and we were part of its crisis.

Next door to us lived Don Pablo, the village *h-men* or "doer." He was joining in the common effort of the shamans from villages throughout the neighboring region by preparing a Ch'a-Chak ceremony to bring the rain gods to Yaxuna. Several nights earlier, he had come over to our household after dinner with Don Leocario, the mayor, to visit and to discuss the coming ceremony. But he had said nothing to us directly about this sacred work. I reminded him that we had offered to help and asked what we could contribute. Since Don Pablo lives and works with people of modest means, his request that night, out of courtesy, was for small quantities of inexpensive things—two bottles of the cheapest rum, ten candles, ten bags of incense. I suggested that we might supply chickens as well, and he agreed that we could bring two live chickens for the ceremony.

Since Ch'a-Chak ceremonies are a community effort, the work and the materials were all contributed voluntarily. Don Pablo brought his knowledge, his prayers, and his sacred stones. The other men built the altar and dug the nearby fire pit where the sacred breads would be cooked. Their altar looked simple and improvised, a shaky table of poles held together with vines. During the ceremony, however, it would become the center of the cosmos. All the participants brought the dough made with corn ground by their womenfolk for the sacred breads that are layered, like heaven and the underworld, on the altar, as well as the cooked meats, and the "wine" made from honey and "virgin water" from a deep natural **Plate 28**
well.

That day, while I was studying the ancient cityscape and presiding over our excavation in the temple-mountain, the ceremony was well under way in the woods at the far side of the village. Don Pablo and his helpers had set aside three days for the prayers and preparations. The following day, all the villagers and archaeologists together attended the climax of the Ch'a-Chak ritual and witnessed the legacy of thousands of years of Maya devotion and ritual knowledge. This is the *Ch'a-Chak*, the "Bring-Rain," ceremony.

FIGURE 1:1 Don Pablo censes the mesa in the Ch'a-Chak ceremony at Yaxuna in 1989

Once we were assembled, Don Pablo began. With quiet dignity, he crouched down beside the altar and picked up a tin can nailed to a stick. Holding the handle of this homemade brazier, he rummaged in his pocket, pulled out a little bag of white powder—*pom*, dried fragrant tree sap—and sprinkled it on the coals that were burning inside the can. As the sweet smoke rose in billows, he raised his voice in prayer and began moving counterclockwise around the leafy green arbor of bound saplings and baby corn plants (Fig. 1:1). The incense cloud undulated back and forth in the hot air.

Pitching his voice into a higher octave to show the proper respect, Don Pablo spoke softly to the Chak Lords, the rain gods. He circled around the tall, thin cross made of sticks that stood behind the altar and the four young men stationed at its four corners. These men were making the roaring sounds of thunder, *ruum-ruum-ruum*. They were clapping small wooden sabers and pistols together, and sprinkling fresh, clean water from little gourds onto the boys who crouched at their feet beside the corn plants of the arbor. The men embodied the Chakob, and the boys were the frogs of the rainstorm. The boys chirped and croaked, "whoa-whoa," imitating the many night sounds that are heard when the land and the crops are sated with the rain they were trying to bring with this ritual.

As Don Pablo passed in front of the altar, he paused to pray more loudly, shaking the arching branches of the arbor the way the thunder shakes the roof of a house. He pulled on one of the six vines that radiated outward from the center of the arbor. Entranced, his eyes half closed, he raised his face heavenward and summoned the gods to save the crops of the farmers who stood anxiously around the altar observing him. The late-afternoon air hung heavy, still, and expectant over the fallow cornfield with its tangled thicket of new-growth trees. Clouds passed by above the treetops and suddenly everyone heard a distant rumble of thunder. Perhaps it would not come that day, but soon, they knew, the Chakob would bring back the rain. They had heard the h-men's prayers.

Every one of us, the men and boys of the village and the motley crew of American and Mexican archaeologists working nearby, was caught up in that moment. That hopeful rumble of thunder broke the tension caused by our hesitant, sympathetic attempt to believe and the mild embarrassment of the villagers caught between their faith in ancient knowledge and their aspirations to become "modern" people. Being present at this ceremony was like something glimpsed from the corner of your eye, something you're not really sure you've seen. For a brief time Don Pablo had gone a little beyond our view, into a place represented by an altar set with cooked breads, gourd cups of corn gruel, and magical stones. He talked to god in that place and god listened to him. Don Pablo is an *h-men*, a "doer," the shaman of his town.

Shamans are specialists in ecstasy, a state of grace that allows them to move freely beyond the ordinary world—beyond death itself—to deal directly with gods, demons, ancestors, and other unseen but potent beings. Shamanic ecstasy can last moments, hours, or even days, but the amount of time spent in trance is less important than the knowledge of its existence. As the spell broke—and the villagers began joking, passing around drinks of honey wine, and doling out the feast of breads and chicken stew that had been sitting on the altar—we, the archaeologists, believed that we had at least tried to help our friends. By our presence, our goodwill, and our heartfelt desire to suspend our own disbelief, we had aided Don Pablo on his journey to ease the suffering of his village from the drought that burned their land.

On the surface Don Pablo appears an ordinary man, a farmer, robust and enormously energetic, with dark, intelligent eyes and a quick smile. Although he heard the calling rather late in life, he has walked the shaman's path for some years. Yet his journey did not begin with the dreams that called him to help his people. It began thousands of years ago, when human beings first conceived of a place beyond death inhabited by ancestors, spirits, and gods—the place between the worlds. To journey to that place in ecstasy and return alive is a very special talent, and shamanism is a special institution we humans have invented to harness that particular talent.

The idea we are asking you to entertain as you read this book is that this place, which the ancient Maya called *Xibalba*, the "place of awe," is real and palpable. It is real because millions of people over the millennia have believed it to be real, and have shaped their material environment to accommodate that reality. The Maya world we are entering is a world of living magic. There we will see the earth and the heavens through the eyes of educated Maya whose ceremonies reaffirm age-old ideas about how things work—ideas that have endured through a hundred generations to the present day. We come in search of the buried power and ruined knowledge of Don Pablo's predecessors, the sages and holy lords who ruled, divined, and taught in Don Pablo's world eons before the discovery of the Americas by Europeans.

Today it is not only the archaeologists who are searching to uncover the magic of the past. A scant few months before Don Pablo performed his rain ritual, and only a few miles north of his village, a strange scene had unfolded. On March 20, the road from the tawdry little tourist town of Piste had been filled with processions of pilgrims, 35,000 strong, all heading for the grand plaza of Chich'en Itza. To the sound of bronze gongs and chanting, through clouds of incense that rose above the crystal-wielding crowd, modern celebrants came to witness the diamond-shaped shadows snake down the serpent balustrades of the great pyramid, the Castillo, on the day of the spring equinox (Fig. 1:2). A thousand years after the Maya built this sacred place, it has once again become a haven for believers seeking a deeper truth beyond the chaos and confusion of their daily lives. In an age of industrial power beyond the imagination of *our* ancestors, Western people still come searching for miracles in the magical past of the Maya. Within the massive ruined pyramids that rise above the forests of Central America, they come seeking the mythical mountains of the mind.

A hundred and fifty years ago, the gleaming beige colonnades, grand stairways, and intricate carvings of downtown Chich'en Itza were buried deep inside mounds of dirt and coarse rubble under a carpet of bush and

(© E. C. Krupp, Griffith Observatory)

FIGURE 1:2 The Castillo at Chich'en Itza
Equinox observance photographed at 4:55 P.M. on March 21, 1987

scraggly forest. Were it not for the ancient fame of some extraordinary buildings like the Castillo and the Great Ballcourt, the city might have passed for an oddly bumpy stretch of terrain in the otherwise flat plain of northern Yukatan. Indeed, for the intrepid tourist willing to venture away from the excavated portion of Chich'en on small, unmarked paths, the typical scene of forested hillocks with odd bits of rock sticking out of them is still there. Cleared, consolidated, and partially restored by archaeologists, the ruins at the center of the city were recaptured from nature to stand as living testimony to the architectural brilliance and cultural sophistication of an ancient people. Preserved as a huge open-air museum, these ruins inspire the admiration and the imagination of thousands of modern pilgrims, even as the portraits of their ancient makers slowly dissolve back into the natural limestone from which they were brought forth by artisans a thousand years ago.

The ruins of Chich'en Itza are not the Chich'en Itza conceptualized by the ancient Maya. This city has become part of our contemporary experience. Because we modern pilgrims are ignorant of the intentions of the original builders, we impress our own meanings and aesthetic values on the Maya monuments, just as we always do when we contemplate masterworks of art from other cultures and other times. At Chich'en, all that is different is the scale—in this huge museum, visitors stand inside the artifact they have come to see. And our ignorance is convenient, for it allows free rein to the modern imagination. We see in these ruins what we want to see, be it affirmation of the romantic mysticism of springtime pilgrims, or the practical materialism of many modern scholars who devote lifetimes to studying the Maya. And when we begin to understand what the builders intended—through our decipherment of the Maya dedicatory texts, analysis of their images, and the study of their architecture and the artifacts they left littering the landscape, we learn that our imaginations have been constrained by our own cultural filters in tricky and deceptive ways. Though we have tried to interpret the intentions of the Maya within the context of what we see as "civilized behavior," somehow their culture just does not make a "fit" with the patterns of behavior laid down in our own world. We are forced to acknowledge that our perception of the past is always a prisoner of the present. Our reconstruction of the mountains inside the Maya mind is, like the pyramidal mountain restored by the careful archaeologist, an interpretation and not the true original.

We of the modern world come from a society in which mystical knowledge is sometimes admired and honored, but is more often regarded as lunatic and irrational. When the Europeans conquered the globe, they had to believe themselves superior in knowledge, insight, morality, and technology to the peoples they came to dominate. We assert, however, that what our nineteenth-century European and American ancestors took for global destiny—progress and enlightenment of the world through acquaintance with the Western perception of culture—was a historical happenstance. As the domination of European and North American society slowly dissipates in the twilight of the twentieth century, those still-remaining, vibrant cultures throughout the world are now reaching out to share the technological and economic power we thought was ours alone.

People in the West are learning that we live on a planet with many societies, each harboring different visions and ways of looking at things that are, in fact, alternative realities. Acknowledging the equality of these different realities is a matter of human justice. Tapping their strengths to foster cleaner, more humane, and more survival-oriented forms of material power is a matter of common sense. Learning to understand and respect these alternate cultural realities is a first step onto the road toward a better world. Behind us on this road are the byways of our several pasts; ahead is the inevitable coming together of our futures. The ability to successfully pass back and forth between alternate realities is a fundamental feature of mysticism in general, shamanism in particular, and anthropology in practice. A convergence of the spiritual and the material domains is perhaps disturbing to some scholars in their citadels of Western rationality, but we believe it is our best hope if we wish to create a future of tolerance and effective collaboration between peoples.

Our own culture's premises about the nature of civilized society and what it should achieve provide only partial and sometimes misleading clues to the understanding of the Castillo at Chich'en Itza, of the leaders who envisioned it, and of the many people who built it. Neither the romantic projection of our yearning for the mystery beyond our material existence, nor our dreams of the power it took to harness thousands of lives to the task of creating human-made mountains, can ultimately encompass the past we are seeking. There are rationales for power that are different from those in our world. Our rationales mostly honor power over people rather than power shared between them because we moderns identify power with sources outside the human spirit. In the scientifically envisioned world of energy and matter, people own power, they don't engender it. Once we have discovered the rationales of the Maya—through the words and actions of the Maya themselves—it is our heartfelt belief that we will also discover a central truth: that Don Pablo's rickety altar of saplings erected in the parched woods, the ruined temples of Yaxuna built at the dawn of the Maya civilization (Fig. 1:3), and the Castillo of Chich'en Itza raised in its final glory are essentially forms of the same thing. They are all symbols of the Creation of the cosmos. They are all instruments for accessing spiritual power from the creative act, and that power continues as a fundamentally human experience and responsibility. In this common purpose, they are signposts on the Maya road to reality stretching across the landscape of history.

Maya kings and lords of old were shamans, but contemporary Maya shamans are not kings. Maya society has changed profoundly over the last three thousand years, especially since the arrival of Columbus and those who followed him. The Spanish conquerors and their descendants worked hard to destroy all vestiges of indigenous government, and they tried even harder to warp Maya belief to fit their own expectations. Under this pressure, the cultural reality of the Maya changed and adapted—but it endured. Just how much that reality has changed is a matter of intense interest to many anthropologists working with contemporary Maya, and historians working with archival records from the last five hundred years.

FIGURE 1:3 Early Classic Mound at Yaxuna

The nature of these changes is even more important to a small but growing number of educated Maya, striving to understand their own historical roots.

Some scholars have presumed that the trauma of the Conquest destroyed most of indigenous Maya culture. From their point of view, the rituals performed and the beliefs held by Don Pablo and his people are products of the colonial experience of Maya peasant farmers during the five centuries since the Conquest. We prefer the view of other scholars who perceive the Maya as adaptable and capable of absorbing and synthesizing new ideas into their vision of the cosmos. We recognize, of course, that Maya culture and cosmology have changed over the centuries. The world around them has surely changed. The Maya have experienced alterations in their physical environment with the introduction of new domesticated animals, new crops, and new ways of growing them. They have also experienced modifications in the material conditions of their lives through the appropriation of their lands, slavery, the bondage of the haciendas, the exploitation of wage labor, and the maintenance of rural poverty. Nevertheless, we believe that changes in the world of their actual experience caused by the arrival of the Spanish have been accommodated by their capacity to transform their models of the cosmos without destroying the basic structures of the models themselves.

Perhaps the most dramatic example of this kind of transformation is embodied in the Maya's adaptation of the Cross of Christ, the central symbol of European domination. The Maya promptly appropriated and reinterpreted this most Christian of all symbols by merging it with the World Tree of the Center (Fig. 1:4), the *yax che'il kab,* as the Conquest period Yukatek Maya called it. The Christian cross became, quite literally, the pivot and pillar of their cosmos, just as the World Tree had been before. Anthropologist Evon Vogt often recalls with irony and amusement his discovery that the ostensible Christian piety that the present-day Zinacanteco Maya of Chiapas display toward their wooden crosses is, in fact, a declaration of cultural autonomy from their oppressors.[1] This, we believe, is how the Maya vision of the cosmos works. It is a dynamic model combining historical knowledge, myth, and the practical experience that is perpetually being re-created through ritual performance.

In many ways writing this book has set us on a journey of discovery. The effort has revealed to us, with dramatic poignancy, that a significant number of Precolumbian Maya ideas from the deep past have survived up into the present time. Maya cultures evince continuity particularly in their core ideas about the essential order of the cosmos, its patterns and purposes, and the place of human beings in it. Maya shamanism as a social institution has survived the last two and a half millennia because the shamans help their neighbors in their communities to re-create this view of reality over and over again—when they heal a sick child, or bless a new home, or renew the nurturing bonds between the inhabitants of this world and those of the Otherworld. Through participation in these rituals, the Maya, both exalted and ordinary, reaffirm their culture's

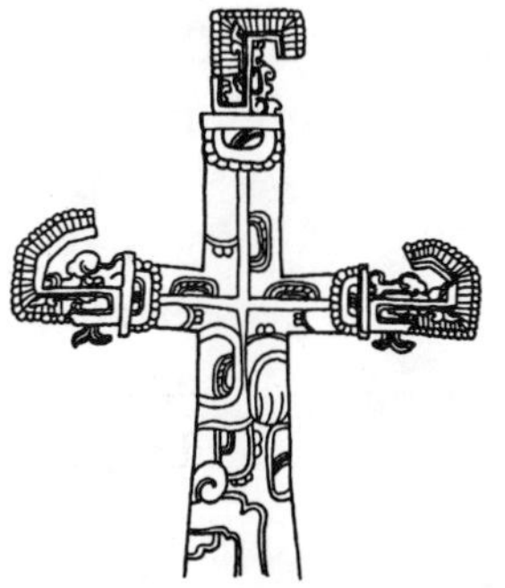

the cross symbol of the Classic period representing the tree at the center of the cosmos

FIGURE 1:4

Itzamna making a turkey offering in the New Year's Pages of the Dresden Codex

Itzamna calling up a crocodile tree

FIGURE 1:5 Itzamna

deepest truths. It is no wonder, then, that when the Spanish arrived, the principal god of the Yukatek cosmos was Itzamna, the greatest shaman of all and the patron of his human counterparts (Fig. 1:5).

Modern Maya live in a metaphysical, philosophical, and religious tradition bridging the ages, from the time of the Classic kings to the shamans and ritual leaders of today; from sovereign states of old with their huge royal capitals embracing tens of thousands to the villages and small towns that now encompass the lives of most Maya. The very persistence of these traditions implies that the Maya of antiquity were unified in their view of the world regardless of their station in life. A vision that embraces and accommodates social differences and inequalities has to be informed by a limited number of central ideas that everyone can comprehend. These ideas must be universal enough that all who adhere to the tradition conceive of themselves as belonging to one substance and nation through their shared understanding of the nature of reality. The great cultural and religious traditions of our contemporary world, among them Christianity, Judaism, Islam, Buddhism, Hinduism, and Confucianism, work in exactly this way. The Precolumbian world of Middle America gave birth to a great universalizing tradition just as successful as that of the European Old World, and the Maya cultivated one of the most eloquent of its variants.

People from the Western tradition have difficulty perceiving the holistic vision that undergirds contemporary Maya practices and their connection to the ancient past. Contemporary Maya culture seems spread out and fragmented; villages are enclosed within the borders of many different modern countries and separated from one another by great distances.

Looking back into antiquity, it is easier to see the fundamental unity of the Maya world. Fifteen hundred years ago, people in the great city of Tikal (in what is now Guatemala) in the southern lowlands (Fig. 1:6) wrote in Yukatekan,[2] an ancestor of the language spoken by Maya now resident throughout the northern lowlands (in what is now México). Scribes wrote Yukatekan at the vigorous Classic city of Caracol in central Belize and at Nah Tunich, a huge shrine cave in southern Belize. So the Precolumbian territory inhabited by speakers of this language was larger than any of the modern Central American states with Maya populations. We also know that in the eighth century the educated people who lived in the western city of Palenque, and their counterparts in Copan in the southeast, a thousand kilometers away, spoke the same language, used the same writing system, and participated in the same cosmology and political institutions. These two cities even established alliances through marriage between their royal families. The unity of ideas that illuminates the wonderful art, architecture, and written texts of the Precolumbian Maya reflects a common cultural vision that existed all across this vast region. While this unified worldview never manifested in a single overarching government or political state, it bound the multitude of Maya towns, cities, and kingdoms into a single world as decisively as the vision given to Muhammad and written down in the Koran binds together the many nations of Islam in a common culture, whatever their political contentions.

This Maya tradition of thought was systematically subverted during colonial times. Europeans tried to eradicate the ancient ways by educating Maya children in the ideas and language of the conqueror and through the mass destruction of Maya art and images ("idols"), the repression of ritual, the burning of books, and the methodical eradication of literacy in the old writing system. Where once Maya communities had been in intensive communication and interaction with one another over long distances, now they were fragmented by the Spanish and kept in deliberate isolation.[3] Contemporary Maya groups today remain isolated and alienated from each other by poverty, by policies that discourage aggregation and communication, and by national borders. They all share a history of local oppression and difficult encounters with the dominant cultures of the modern nation-states in which they find themselves. Today there is not a single Maya society, linguistically or culturally, but rather a mosaic of rural enclaves, each preserving branches of the original vision planted by their ancestors. These branches are changed by centuries of social

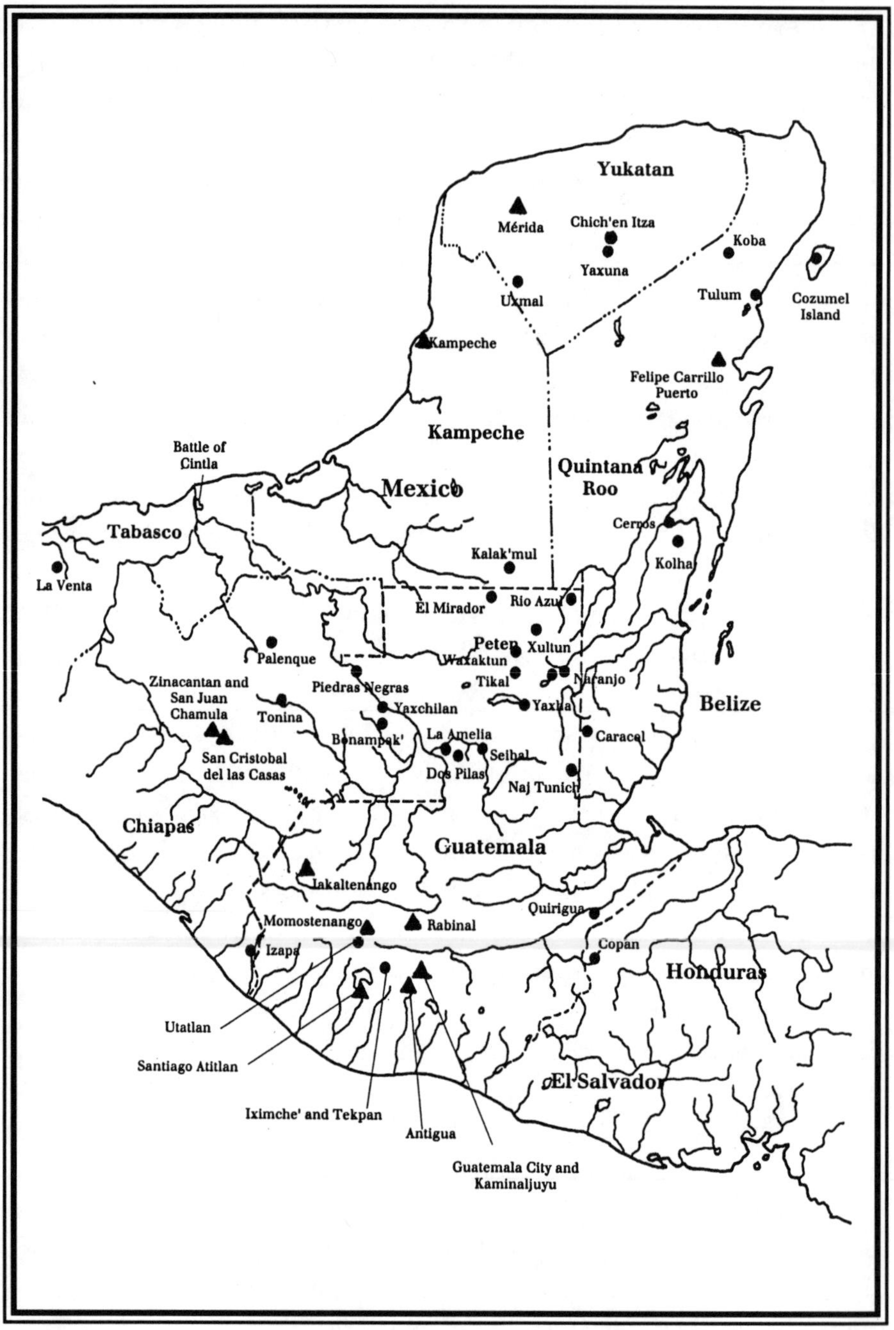

FIGURE 1:6 A Map of the Maya Area

pressure and evolution, by interaction with the Christian traditions and dominant national governments. Yet they have all sprouted from the original Creation tree of Maya culture, and in spite of everything, they all still bear the fruit of ancient truths.

We knew about the coherency and integrity of ancient Maya thought long before we began to work on this book. Our own studies and more than a century of research attest to the overall homogeneity of Maya art, writing, and symbolism throughout the entire Classic period. Nevertheless, we were surprised that every element of Maya cosmology, no matter where we started, drove us toward a few basic central themes: the creation of the cosmos; the ordering of the world of people, and of the gods and ancestors of the Otherworld; the triumph of the ancestral humans over the forces of death, decay, and disease through cunning and trickery; the miracle of true rebirth out of sacrifice; and the origins of maize as the substance of the Maya body and soul. All of these themes are expressed in the Popol Vuh, the Book of Council of the K'iche' Maya of highland Guatemala.[4] The genesis stories in the Popol Vuh are a redaction of the central myths celebrated by lowland Classic Maya as a fundamental expression of their own genesis. These stories, as our colleague Michael Coe puts it,[5] anchor Classic Maya thought in the same way that the Mahabharata and the Ramayana epics anchor popular Hindu experience and notions of royal power in Bali. Because of their great importance to any understanding of modern Maya life, we will be returning to these stories frequently in the coming pages.

The Popol Vuh is a touchstone and the closest thing to a Maya bible surviving to the present, but it is not the sole surviving expression of great Maya ideas. We have numerous other documents from the K'iche', a history called the Annals of the Kaqchikel, and even a play called the Rabinal Achi. There is also the Ritual of the Bakabs, a remarkable Yukatek Maya book of incantations. The Books of Chilam Balam, also from Yukatan, contain important references to Maya cosmology. Many of these references are still obscure, but others make sense in light of new understandings recovered from Classic-period records. By working back and forth from the ancient cosmology to the modern, we are becoming increasingly convinced of the integrity and continuity that Maya thought displays over thousands of years of history. Just a short distance below the surface of apparent differences, the ancient roots live on.

The road that led to our belief in the essential integrity of the Maya tradition has been rocky and filled with detours. Its source stretches back over a hundred years,[6] to an era when three important types of resources became available to the leading Mayanists of the day. The first was Brasseur de Bourbourg's[7] discovery and translation into French of the *Relación de las cosas de Yucatan* (published in 1864), a description of

Maya life written around 1566 by Diego de Landa, the first bishop of Yukatan. The publication of this book and other early source materials, especially the accounts of the various early expeditions into the Maya region and the histories compiled by Spanish eyewitnesses that were already available, enabled students of the Maya to have access to first-hand descriptions of Maya religion and rituals recorded in the years just after the Conquest.[8] Landa and other early chroniclers' accounts were, of course, biased, but in spite of the distortions caused by the cultural blinders they wore, their descriptions provided fundamental information about pre-Conquest culture.

This information was especially valuable when applied to the second great discovery of the age—three Maya books that came to be called the Dresden (Fig. 1:7), the Paris, and the Madrid codices, named after the cities where they were found. A fourth, known as the Grolier codex, was published in 1973 by Michael Coe in his book, *Maya Scribe and his World.* These accordionlike books, made of beaten-bark paper covered with a thin layer of plaster, are almanacs for the timing of ritual. They record, in pictures and writing, information about which gods and what acts were associated with each day in the calendar cycles. Astronomical tables for anticipating favorable cycles of Venus and eclipses of the sun are also included as books of learning and prognostication for priests

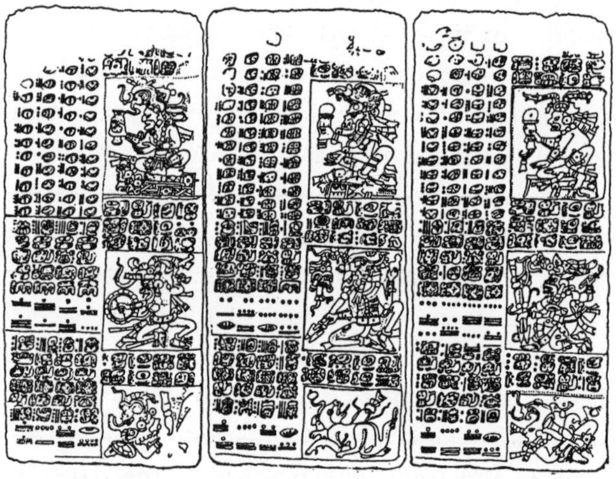

three pages from the Venus tables in the Dresden Codex (after Villacorta 1977)

FIGURE 1:7 A Maya Codex

specializing in the use of the calendar. Before the discovery of these codices, scholars had no Maya books. This circumstance was due to two factors: the widespread burning of Maya texts by the Spaniards and the destructive influence of the wet climate upon those books that were found by archaeologists in excavated tombs. Even to this day, we have only four codices.

The third important resource to become available to Mayanists in the early part of this century came from the efforts of a new breed of explorers and scholars. A series of resourceful and resilient people, including Alfred Maudslay, Teobert Maler, Sylvanus Morley, Franz Blom, and others,[9] traveled throughout the Maya world mapping, making casts, using the new art of photography, and creating the first accurate drawings of the writing and imagery left by the ancient Maya when they abandoned their cities. The work of Maudslay, Maler, and Morley made available to interested scholars, for the first time, a major selection of texts and images that could be studied in the comfort of their homes and libraries, and as a result, the understanding of Maya art and writing took a major leap forward.

At the same time another set of scholars[10] began to publish studies of the monuments, of the three Maya codices, of Maya pottery, and of the archaeological information that was being provided by organizations like Harvard's Peabody Museum of Archaeology and Ethnology, the Carnegie Institution of Washington, Tulane's Middle American Research Institute, and the Field Museum of Chicago. Still other scholars, led especially by the great Mayanists Alfred Tozzer, Ralph Roys, and Adrián Recinos, among many others,[11] preserved, translated, and analyzed documents from the colonial period and more recent times.

Alfred Tozzer was also one of the first people from our world to study the modern Maya, their beliefs and their way of life. The tradition he started has continued, with many great ethnographers[12] publishing studies of different Maya communities in México, Guatemala, and Belize. These studies have their own set of problems stemming from the biases of the researchers, but they also provide insights into the lives of these Maya communities from the first two decades of the twentieth century until the present. They offer us a chance to see how the Maya have changed, adapted, and preserved old ways through a century of experience. We will use information from many of these sources to present our own understanding of the Maya, ancient and modern.

By the middle of the twentieth century, these many avenues of research had come together to form a broad direction, and on hand to take advantage of the situation was an Englishman named J. Eric Thompson. As one of the most famous and influential of all Maya scholars, he shaped the modern vision of the ancient Maya with his persuasive and eloquent writing. Sir Eric produced a body of work that combined all the disparate investigations he inherited from his predecessors in a new and particularly effective way. He studied the modern Maya, their folklore and languages. He read historical records from the colonial period to reconstruct who, where, and what the Maya were at the Conquest. And he cast what he learned from both the modern and colonial Maya back to the Classic period.[13] He excavated ancient sites, and following the death of Sylvanus Morley in 1948,[14] he reigned as the leading expert on Maya glyphic writing until Tatiana Proskouriakoff supplanted him in 1960 with her spectacular studies of dynastic history.[15] He was a master of Maya religion, affirming the existence and nature of a pantheon of Maya deities through a series of important articles and books.

Toward the end of his career, Sir Eric proposed that the Classic Maya were developing a kind of monotheism centered on an important god called Itzamna (or God D according to Paul Schellhas's alphabetic nomenclature of Maya gods in the codices).[16] Several of Thompson's predecessors[17] had already associated the image of this particular god with the Itzamna of the Conquest-period Maya of Yukatan. Starting from this foundation, Thompson proposed that Itzamna means "reptile (iguana) house" and surmised that this "iguana-house" divinity had been as central to Classic Maya thought as the sources from the time of the Spanish Conquest had suggested it was for the Postclassic Maya.

As it happens, saurian imagery is everywhere in Maya art—and Thompson went on to identify a wide variety of these images, including the Cosmic Monster, Vision Serpents, and the Manikin Scepter (God K in the Schellhas system, known to the Maya as K'awil), as the god Itzamna.[18] These three categories of supernatural beings are terribly important to any discussion of Classic Maya cosmology. Scholars today, however, have a great advantage over Thompson in their investigations of symbolism: they are able to decipher the glyphic names of many of these supernaturals. The upshot of these decipherments has been the discovery that Itzamna, God D, *did* exist in the Classic period, but that he was not the "iguana house" fellow that Thompson identified. Itzamna is one of the creator gods—one of several. He is also one of the aged gods who oversee ritual scenes in the palaces we find on the elaborately painted clay vases of the Classic period. Along with God L and God N, Itzamna ruled the Otherworld of Maya cosmology as its principal shaman and doer, making him very important to the themes of this book.

This name glyph is an excellent object lesson in the way interpretations of Maya cosmology change when new kinds of evidence and new approaches are found—a process that will surely not cease with this book. Beyond this fundamental fact, Thompson's proposal of Classic-period monolatry, intensive focus on one god in a cosmology that includes many gods, raised some fundamental issues we must tackle. To begin, we'll let him speak for himself:

> The conception of Itzam Na is indeed a majestic one. One realizes why the Maya rulers came at one time to consider him the only great god, for it rather looks as though the Maya of the Classic period had developed a cult of Itzam Na into something close to monotheism, with all other beings, such as sun and moon, probably the Chakob, and so on, as Itzam Na's servants or his manifestations, expressed by setting their heads in his open jaws. We find the god retaining his iguana form, but also developing into two anthropomorphic manifestations, Gods D and K. Perhaps such ideas were too abstract to appeal to the Maya peasant, for, as we have seen, the cult of Itzam Na completely disappeared with the collapse of the old ruling class following the Spanish Conquest.
>
> With the end of the Classic period, the fragile cult of postulated near-monotheism was overcome. It was still remembered six centuries later as a golden age before "idolatry" was introduced. . . .
>
> The children of Israel found difficulty in adhering to monotheism notwithstanding the fulminations of their prophets; perhaps there was always greater difficulty in persuading the Maya peasant to abandon his down-to-earth Chakob for the monotheistic abstractions of the Itzam Na cult, but after the collapse of the old order and the end of the ceremonial centers there may have been no Maya prophets around to destroy the peasants' golden calves and altars to "Baal" and "Moloch."
>
> (Thompson 1970:233)

The implicit assumption buried in Thompson's interpretation is that the ancient Maya elite moved their religious thinking in the direction of monotheism in the course of becoming "civilized" during the Late Classic heyday of their hierarchical society. In his conception, however, the majority of people, the commoners, found the unification of the many spiritual forces of their beliefs into a single universal concept of one god unconvincing. Thompson believed the Classic-period commoners and villagers were involved in a "simpler" vision of the cosmos, explicated by their shamans and performed in ritual by their family leaders. This worldview he saw as perpetuated to the present day among the predominantly rural populations of Mayan-speaking people. To Thompson then, these practices of the peasants, both ancient and modern, were disconnected from the ancient religion of the elites.

For us, this is *the* fundamental issue. For if the ancient Maya elite believed in a different kind of universe than the people subject to their authority, then their ritual performances, art, and glyphic texts were meaningful only to themselves and offer little insight into the broader experience of the majority of the people. Moreover, if this spiritual chasm between elite and commoner existed, then the great artistic and intellectual achievements, the massive public construction efforts in the hundreds of ruined urban centers, must be regarded as the bitter fruit of the sustained oppression of the majority by the elite minority. Rather than being products of one of the greatest universal visions humanity has ever invented, these cities and their monuments merely presage the oppression of the Europeans and their descendants that has shaped Maya life for the last half millennium. Thompson believed that this great gulf did indeed exist, and that the great collapse of the ninth century was caused by peasants revolting against their elite oppressors.

Our experience and study have convinced us of the opposite—that a unified view of Maya ritual and cosmology has endured for at least two millennia. We have even found that in some situations the structures of belief have descended from the Olmec, with roots that are three thousand years old and perhaps even more ancient. For us, these persisting patterns refute the difference of religious vision that Thompson believed existed between the exalted and the humble Maya of antiquity. There is also a direct linkage between the rituals and beliefs of modern villagers and their ancient forebears. Since we accept the ninth-century collapse of the Classic kingdoms and the Spanish Conquest as cataclysmic catastrophes that shook the Maya cosmic vision to its core, our task is to explain how this worldview still managed to survive both events, disguised and transformed but essentially intact. For this effort, we draw inspiration from several quarters.

Ethnohistorian Nancy Farriss, in a brilliant and subtle study, has documented how the Yukatek Maya cosmos survived the Spanish Conquest. She ably confronts Thompson's arguments that monotheistic religions are "rational" evolutionary advancements over "primitive," magical folk religions. In the process, she exposes the cultural biases undergirding our fascination with monotheism.[19]

Farriss suggests that at the time of the Conquest, there was a three-part hierarchical division of Yukatek religious practice that matched analogous levels in the Spanish Catholicism of the times. She proposes interaction and syncretism between the religions at all three levels. At the base, she places the level of personal experience. This is the level of the local practitioner, the shaman.[20]

In the middle level,[21] Farriss documents the process by which the traditional Yukatek Maya elite, reduced by the Conquest to living on the same economic level as the commoner, oversaw the transformation of indigenous Maya worship of the pantheon of deities into a Christianized worship of a pantheon of saints. Feasting, the elaborate festivals of local saint cults, the devotion to and material adornment of saintly images, and the construction of churches to house them were all methods of transforming and disguising Maya indigenous beliefs so that their community life and culture could survive. The Maya simply replaced their "idols" and many spirits with "images" and many saints and proceeded to adapt Spanish Catholicism to their own worldview. We could not agree more with her interpretation of this process and will explore this syncretism further in our own journey into the Maya cosmos.

Farriss's concept of the upper level of Maya spiritual belief—which involves the more universal side of religion as opposed to the more personal side—corresponds to the Christian idea of God Almighty. She speculates[22] that the educated elite had no great difficulty absorbing such notions as the Trinity, since their divinities already had multiple (commonly fourfold) aspects. But like Thompson, she associates the concept of a high god, such as the cult of K'uk'ulkan or the idea of Itzamna, with the elite and assumes that this concept had no real meaning to the rural peasants. In other words, Maya farmers felt no more connection to the monotheistic Christian God of their new overlords than they had for the K'uk'ulkan or Itzamna of their old rulers. The disenfranchisement of the native elites who now had to live among the commoners in the small towns and villages did not help the matter, for they, like their former constituents, began to feel that the magical performance of village ritual had more relevance to their lives than the high god of the Church.

So, like Thompson, Farriss believes there was an uncrossable chasm between the cosmological perspectives of the elites and the commoners during Precolumbian times. However, unlike Thompson, Farriss does not regard this dichotomy as central to her argument, nor as a matter of great consequence in Maya social organization. Instead, her study focuses on the remarkable unity that the elite and common folk shared after the Conquest in their devotion to a pantheon of gods who became saints. As we mentioned above, the transformation of the old Maya gods into the Catholic saints preserved the core of the Yukatek Maya vision and enabled their collective cultural survival. Nevertheless, where Farriss joins Thompson in seeing a division between a single universal god embraced by the elites and the local divinities of the commoners, we must part company.[23] The worlds of the elite and the commoners of Maya antiquity may look different, but they could not have become so closely joined together after the Conquest without a shared worldview and cosmology that reached from the bottom of society to its very top.[24]

Magical and personal experiences are the elements that form the lowest level of Farriss's model, and this level incorporates the world of the village shaman—the cosmos of Don Pablo. Modern village shamans cure individuals of illness, assuage the afflictions of the household, and help neighbors find peace with the spirits of their maize fields. But their work is not exclusively confined to the personal, nor even to the local scene. Shamans are also responsible for community propitiation of the Chakob, the rain gods; of the Dueños or Yum Kaax, the spirits of the ruins and forests; of the Babatunob, the divinities who hold up the sky; of the Balamob, the Jaguar Protectors of the fields and towns; and of Halal Dios, God Almighty, who is the sun. Whatever the variations, these are not local divinities but rather broadly acknowledged ones.[25] Most important, these modern gods and beings have the same functions, and often the same names, as the major gods of the Precolumbian pantheon, as written down and pictured in Maya books at the time of the Conquest. Although the Spanish burned these books by the thousands, we know this from the four surviving Maya codices we have in our possession.

As we mentioned above, the triune aspect of the Spanish God would have made as much sense to the Maya farmer as to his king because all Yukatek Maya understood the fourfold nature of divinity.[26] This concept remains at the center of their religion today. Hunabk'u, the Oneness God of the Spanish, also made sense to Precolumbian village shamans and kings because all Yukatek Maya understood, and understand today, that at the center of the fourfold cosmos is the one. Like the purloined letter in Poe's story, the unitary ideas of the Maya cosmos, the prime elements of Farriss's highest and most abstract level, are not hidden behind a veil of secret knowledge.[27] They are out in plain sight, manifested openly in the household and field rituals of the modern Maya.[28]

When Don Pablo performs his rituals, he regenerates the order of the cosmos and rejoins the two separated worlds, the human world and the Otherworld, by creating a portal. Within this holy space, he calls forth and binds together the fourfold gods: the rain nurturers, the protectors, and the burden-bearers at the edges of the world. From the center, he brings the source of life to his people. Through the now-open portal to the Otherworld, he sends maize and other sweet and fresh things to the other side so that they may nurture and honor both the lesser gods and God Almighty. This sacred, universal space that he creates is the center of the heavens, and the center of the earth. He calls it *u hol gloriyah*, the "glory hole."[29] The ancient Maya had other words for it—the Black-Transformer, the mouth of the White-Bone-Snake, the *yol* ("the heart of"). Though called by a variety of names throughout the ages, the experience has not changed. Whether opened by the ancient Maya or Don Pablo, an inheritor of their understanding, the portal lies at the beginning of the path to the Otherworld. And those of us who learn to see it are on the Maya road to reality.

Itzamna, an ancient form of Almighty God, who was supposed to have been meaningful only to the elites of older times, can also be found in the contemporary Maya world. The *itz* of *yitz ka'an*, the blessed substance of the sky, which flows through the portal represented by the hanging sky platform on the shaman's altar (literally, "its blessed substance, the sky"), is the *itz* of *itzamna*. *Itzam* (literally, "one who does *itz*" or an "*itz*-er") is the term for shaman—the person who opens the portal to bring *itz* into the world. What is *itz*?[30] For the Maya it is many things: the milk of an animal or a human; the sap of a tree, especially copal, the resin used as incense; it is the sweat from a human body, tears from a human eye, the melted wax dripping down the side of a candle, the rust on metal. These substances are secreted from many kinds of objects. Many of them—like milk, tree resins, and candle wax—are considered precious substances that sustain the Gods. *Y-itz ka'anil* is the magic stuff the shaman brings through the portal from the Otherworld. As an *itz*-er, the shaman is the direct analog of Itzamna,[31] the greatest *itz*-er of them all. When the village shaman opens the portal from this side, Itzamna opens it from the other and sends the precious *itz* through to nourish and sustain humanity in all its diversity. Obviously, a reciprocity is at work here.

As we try to understand the bonds between the ancestral Maya world and the world of the living Maya, all five million of them, we are in good company. Many specialists who know a lot more about the contemporary Maya than we do, and who have spent years studying with shamans and other knowledgeable people, have been following this road for a long time. This is the same path walked by Tozzer and his compatriots who earlier in this century recorded the many ways the modern Maya propagate their reality. The work of all the specialists of the last hundred years tells us the Maya have not ceased changing, adapting, and transforming their modern experiences into meaningful patterns.

Our own contribution to this enterprise comes from our study of the ancient Maya. Until the decipherment began in earnest thirty years ago, students of the ancient Maya could contemplate their cosmology and ritual only by extrapolating backward from the Conquest period—presuming a continuity in the people, the languages, and the customs recorded by the Spanish. With the ongoing decipherment of Classic Maya texts and the analysis of the art and symbolism accompanying them, we can access the ancient cosmology directly. We can discover the words and concepts the Maya themselves used to describe their cosmos fifteen hundred years ago. It is then possible to come forward in time from the Conquest period to see what happened to these words and ideas after the Spanish arrived, and what they mean to the Maya of today. The cycle begins anew when we return to the past, with the words of the present, in search of analogous words and ideas.

People have been extending analogies from past to present and vice versa for decades with varying degrees of success. For example, when the stone foundations of small, rectangular buildings raised a thousand years ago resemble the foundations of modern thatch-roofed dwellings, archaeologists are able to identify such little stone structures as ancient houses, unless excavation reveals they were something else.[32] Another example can be found in the word *way* (pronounced like the English "why"). In the current interpretation of the glyphs, the Classic word *way* refers to kings, ritual performers, and gods in their magical alternative forms as animals, stars, and fantastic beasts. In ancient times these *wayob* were powerful, terrifying conduits of supernatural power who could defend those who conjured them up, as well as being able to destroy those who opposed them. For the Yukatek Maya of today, *wayob* are witches who turn themselves into animals to annoy, attack, and steal the souls of their neighbors. Although the practice has diminished, the essence is the same.

While Mayan words have changed in form or meaning since the scribes of the Classic period wrote their texts, the ancient and modern forms of words, like *way*, are fundamentally related. Furthermore, the way they have changed and adapted sheds light on the connections between the dynamic living experience of the modern Maya and the rich symbolic record left behind by their ancestors. The connections between words and concepts of the past and the present open wonderful bridges between the ancient Maya world and the modern one. The patterns found at the core of the Maya worldview have lasted, tenacious in their grip on the mind and tongue.

The pathways connecting ancient words, concepts, images, historical analogs, and their modern counterparts are particularly evident in the striking resemblance between the World Tree and the modern Christian-Maya cross, as we have seen with Don Pablo's altar. The first Europeans who saw the images of the World Tree at Palenque called the buildings housing them the Temple of the Cross and the Temple of the Foliated Cross with good reason. These Maya "crosses" had the same basic shape, and were as elaborately decorated, as those gracing the altars of large European churches. The carvings of these ancient trees are outlined with reflective mirrors, and they wear jade necklaces and loincloths as if they were living beings. Modern Christian-Maya crosses both in Yukatan and Chiapas are decorated with mirrors and dressed in clothing, or flowers and pine boughs (Fig. 1:8). They too are considered to be living beings.[33]

The names of the ancient trees are as important as their appearance. The hieroglyphic name of the bejeweled and bemirrored World Tree (Fig. 1:9) was *wakah-chan*. It was written with the number six prefixed to the phonetic sign *ah* and the glyph for "sky," because the sounds of *wak*, the word for "six," and *ah*, are homophonous with the word *wakah*, meaning "raised up." The name of the tree literally meant "raised-up sky."[34] The Classic texts at Palenque tell us that the central axis of the cosmos was called the "raised-up sky" because First Father had raised it at the beginning of creation in order to separate the sky from the earth. Each World Tree was, therefore, a representation of the axis of creation.

Classic artisans and lords also depicted the World Tree as a luxuriant maize plant heavy with ripe ears of corn, often depicted in personified form as the face of the Maize God. At Palenque and Copan, eighth-century scribes called this maize tree (Fig. 1:9b) the *Na-Te'-K'an*, "First-Tree-Precious (or yellow)." By combining the Maya Conquest-period stories of First Father given in the Popol Vuh with textual evidence and

FIGURE 1:8 Crosses from San Juan Chamula decorated with pine tops and bromeliads

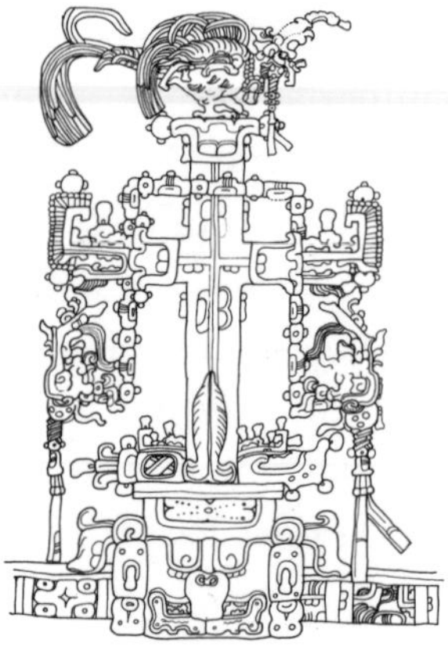

a. *Wakah-Chan* the World Tree

b. *Na-Te'-K'an* the Foliated Cross

FIGURE 1:9

images from the Classic period, we can say that this maize tree symbolizes the original act of creation, sacrifice, and rebirth. First Father was also the Maize God, *Hun-Nal-Ye,* "One-Maize-Revealed," and was depicted both in his human form and as this tree. After First Father's defeat and sacrifice by the Lords of Death in Xibalba, he was reborn as maize, the staple sustenance of humanity and the stuff from which the gods created human beings. Colonial Yukatek Maya referred to both maize and God Almighty as *gracia,* "grace," showing that they understood maize and divinity to be the same substance, a concept deeply rooted in their mythological past.[35]

Don Pablo calls his cross of sticks by the Spanish name, *santo,* "holy thing" or "saint," although Yukatan crosses today are also referred to as *yax che',* "First (or green) Tree." *Yax che'* also happens to be the name of the ceiba tree, and we have naturalistic images from the Classic period of fruit-laden ceibas that represent the World Tree (Fig. 1:10). The working area of Don Pablo's ritual space contains not only the cross but also the arbor-table constructed in front of it (Fig. 1:11).[36] This is the sacred space where spiritual beings arrive[37] to partake of offerings placed on or below the table. This is where Don Pablo places his magical stones and little gourd cups of honey mead, *balche',* to receive the blessed power of the gods. Shamans frequently dip their fingers into these cups to charge them with *flúido,* "spiritual force." This altar is the focus of all prayer and ritual attention. The name of the altar is *ka'an che',*[38] "sky tree" or "elevated wood."[39] Not only is the altar a "sky tree," but the baby maize stalks planted at each of the four corners identify it as a maize-tree place like the Na-Te'-K'an, the foliated cross of Classic-period imagery.

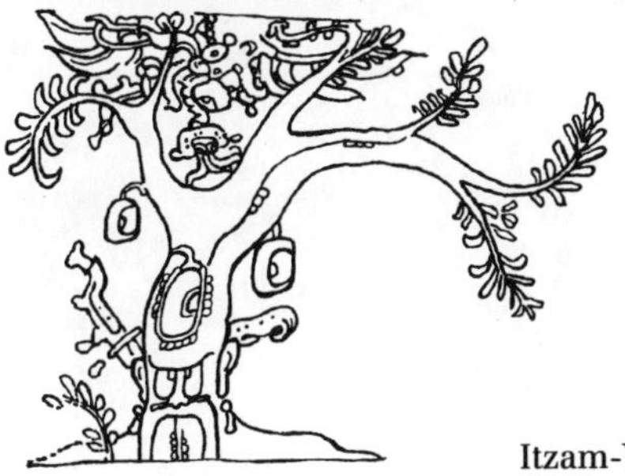

Itzam-Ye landing in the World Tree

Figure 1:10

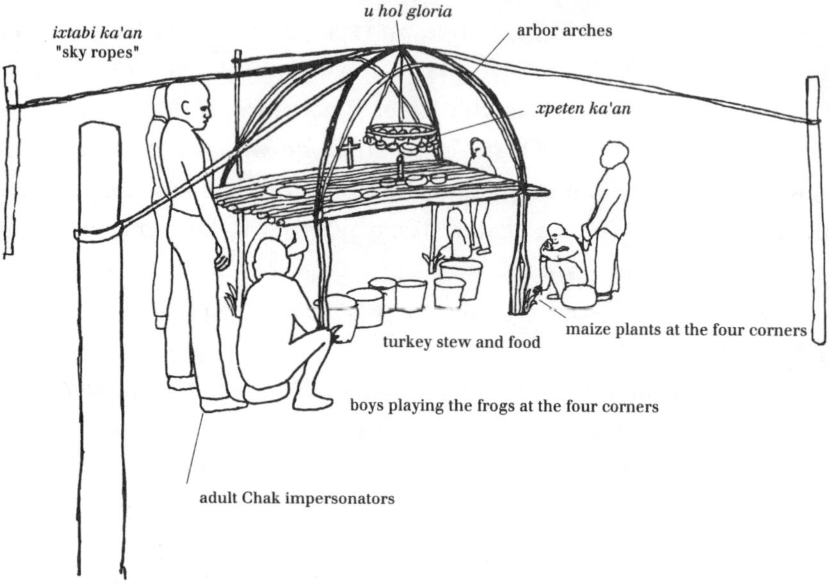

The candle in the center of the table represents the *itz* of heaven with the *itz* of its wax; the hanging platform with thirteen gourds represents the hole *(hol)* in the sky; the cross at the rear of the table is the *santo.*

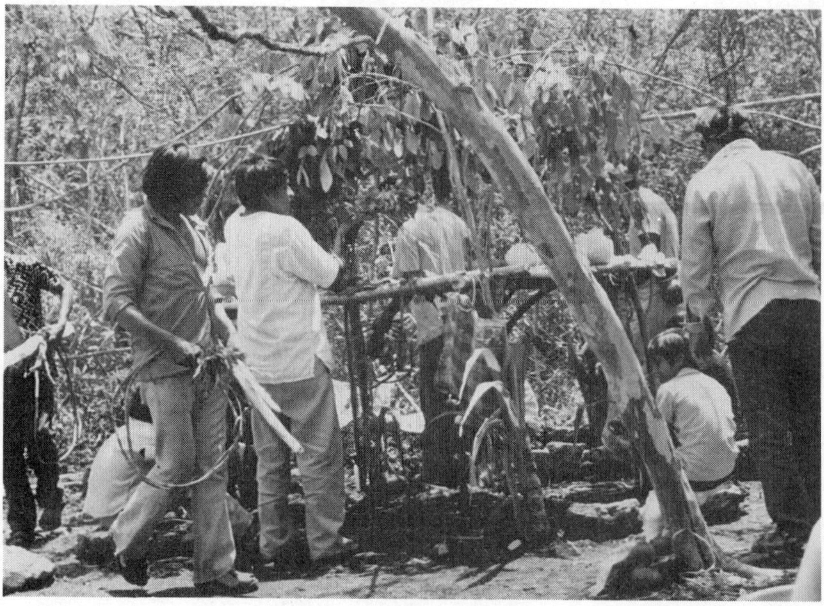

Don Pablo adjusts the *xtab ka'an* as the Chakob and the boys who are playing the frogs move toward their positions.

FIGURE 1:11 The Ch'a-Chak Table at Yaxuna

"Raised-up-Sky," "First-Tree-Precious," "Sky Tree," "Elevated Wood," "Platform of the Sky"—these resonances could be coincidental, but we think they are all names for the same fundamental thing. The names may vary in time and place, but Don Pablo's altar is the present expression of the Maya cosmic center: the *axis mundi*, transferred through practice and apprenticeship from shaman to shaman for nearly a hundred generations.

Sometimes the patterns of modern Maya spiritual belief are embedded in ritual actions and the Classic-period meanings are lost or changed. Sometimes the meanings from Classic times remain but are attached to different symbols. Regardless of the transformational processes at work, we can still discern continuities in the basic structures. And these continuities will often elicit new possibilities of relating the past to the present.

David Freidel has seen this process in action himself, beginning with the first time he participated in a ritual with Don Pablo, back in 1986. For this ritual, Don Pablo built a *ka'an che'* with six legs instead of four. This time the number four was represented by four leafy saplings, instead of the usual six, arching up from the corners of the altar to the peak of the arbor. Four sturdy vines of a kind called *xtab ka'anil*,[40] the "cords of the sky," completed the arches to form the "sky-platform." In Yukatek, the words for "sky," *ka'an*, and for the number "four," *kan*, are near homophones. In Cholan, they are homophonous, and moreover, the glyphs for the number "four," "sky," and "snake" (all *chan* in Cholan or *kan* in Yukatekan) freely substitute for each other in the ancient writing system.

This wordplay between "four" and "sky" seems to have been an important principle in Don Pablo's construction of his altar. He may have reversed the usual order of things by using six platform legs instead of the normal four, and four vines instead of six—but he still retained those numbers as the all-important components of his work.

When Don Pablo reversed the numbers the first time, Freidel took it to be a coincidence. However, in 1989, when Don Pablo made another altar, he built that one with four legs and six arching vines. His curiosity aroused by the possible association with the Classic-period words, David asked Don Pablo about this. Don Pablo confirmed that this four-six construction was exactly what was intended. In the ancient Classic period, *wak* and *kan* ("six" and "four") formed the name of the World Tree—the *Wakah-Kan* (or *Wakah-Chan* in Chol). We don't know if Don Pablo would interpret the play between *wak* and *kan* in exactly the same way that we do, but this incident points out the importance of words and their meanings to the work of epigraphers and archaeologists.[41]

The patterns in words, images, and artifacts are the stuff we study. In order to detect patterns that were meaningful in Maya cosmology throughout their history, many different kinds of knowledge must be brought to bear. When the field was young, masters of the subject could become "experts" on many aspects of the Maya life. Today, we Mayanists no longer have that luxury, because there is simply too much accumulated knowledge to absorb, too many different disciplines to master, and too many opposing methodologies to apply. Instead, each of us must specialize in a few aspects of Maya studies and rely on collaboration with our colleagues to widen our perspective and incorporate as many different viewpoints as we possibly can.

The many experts in the field, both past and present, represent many different perspectives on the Maya way of thinking and behaving. Even in the lifework of a great master such as J. Eric Thompson, many different ways of perceiving the Maya way of being are evident. Over the decades, Thompson managed to collect a wonderful variety of modern creation myths from different Maya societies. In truth, there are many paths to an understanding of the reality of the Maya. These paths all follow a similar direction, but so far, no single road has yet appeared. We sense, however, that the road is there—old, worn, and overgrown, but still able to accommodate the best of the many modern paths into a single, wide thoroughfare. We find evidence of this road in basic Maya ways of looking at the world, like their concept of the fivefold structure (the four directions plus the center) of material and spiritual space; the vitality of their belief in the importance of the ancestors; the Maya one-in-many principle of divinity; the reciprocal nature between sacrifice and nurturing that binds humanity to the gods; and the understanding that people were made from the divine and life-sustaining substance maize. Twenty-five years of study of the ruins, the written history, and the art left on monuments and painted vases, and the more than a century of work by dedicated, meticulous, and visionary professionals, have given us our personal understanding of the ancient Maya cosmos. We believe that the diverse contemporary expressions of these ancient beliefs in the lives of the five million Maya living today bear testimony to the resourcefulness and adaptability of their culture in the face of a difficult and changing world.[42]

CHAPTER TWO

THE HEARTH AND THE TREE: Maya Creation

Here follow the first words, the first eloquence:

There was not yet one person, one animal, bird, fish, crab, tree, rock, hollow, canyon, meadow, forest. Only the sky alone is there; the face of the earth is not clear. Only the sea alone is pooled under all the sky; there is nothing whatever gathered together. It is at rest; not a single thing stirs. It is held back; kept at rest under the sky.

Whatever there is that might be is simply not there: only the pooled water, only the calm sea, only it is pooled.

Whatever might be is simply not there: only murmurs, ripples, in the dark, in the night. Only the Maker, Modeler alone, Sovereign Plumed Serpent, the Bearers, Begetters are in the water, a glittering light. . . .

So there were three of them, as Heart of Sky, who came to the Sovereign Plumed Serpent, when the dawn of life was conceived:

"How should it be sown, how should it dawn? Who is to be the provider, nurturer?"

"Let it be this way, think about it: this water should be removed, emptied out for the formation of the earth's own plate and platform, then comes the sowing, the dawning of the sky-earth. But there will be no high days and no bright praise for our work, our design, until the rise of the human work, the human design," they said.

And then the earth rose because of them; it was simply their word that brought it forth. For the forming of the earth, they said "Earth." It arose suddenly, just like a cloud, like a mist, now forming, unfolding. Then the mountains were separated from the water, all at once the great mountains came forth. By their genius alone, by their cutting edge alone they carried

> out the conception of the mountain-plain,[1] whose face grew instant groves of cypress and pine.
>
> (D. Tedlock 1985:72–73)

So begins the story of Creation in the Popol Vuh, the great genesis myth of the K'iche' Maya of Guatemala. Every culture has its own story of genesis, and by looking at these stories we can learn a great deal about the peoples who make them. We in the West tell Creation stories that show our reverence for science: the Big Bang, the General Relativity Theory, the Theory of Chaos. We have created these cosmic models of reality to explain how things got to be the way they are, and how the basic stuff of the universe works, where we came from, and where we are going. We use science and its tools to try to understand our world, our nature as living beings, and to optimize the chances that the future will work in the way we anticipate will be best for us collectively and individually. The Maya myth of Creation is no different—whether told in its sixteenth-century highland K'iche' form or in the version inscribed on sixth-, seventh-, and eighth-century stone monuments in the ruins of the lowland royal capitals. The myth of Creation, the symbols that expressed it, and the rituals that celebrated it were the tools the Maya used to investigate the same questions.

We have been studying these ancient Maya records of Creation for many years and thought we understood what they had said about the events that began the world. Oh, how wrong we were! In November 1991, a series of events[2] began that pulled back the veil that had hidden a truly stunning and magnificent understanding of the cosmos. For the ancient Maya, Creation was at the heart of everything they represented in their art and architecture. When we took a second look at their temples, ballcourts, statuary, murals, and ceramic art in the light of our new understanding, we were overwhelmed by how these objects mirrored the Maya's unique vision of reality. Letting this remarkable record of their minds and hearts speak to us has been one of the most exhilarating experiences of our lives.

These new insights into Maya Creation mythology came at the eleventh hour in the writing of this book; nevertheless, they grew out of the patterns we had found and the interpretations we had evolved during the earlier studies we had made. We knew, for example, that important Creation texts are found at several Maya sites. We had written about the most important of these texts. We had associated them with images found on vases and sacrificial plates, and had synthesized an account of the Classic-period Creation story to serve as the basis of our chapter on dedication ritual (now our Chapter 5). These data had given us the principal actors of Creation, the dates of the events, and their mythological contexts. For example, the K'iche' Popol Vuh told us that the world had been created, destroyed, and re-created at least three times before the present Creation, the one in which we now live.[3] In both the K'iche' and Classic-period version of the story, important protagonists were male and female Creators born just before the current Creation. They were the instigators of the world in which we are now living. In the Popol Vuh they are called Xpiyakok and Xmukane. They are also:

> Maker, Modeler, named Bearer, Begetter,
> Hunahpu Possum, Hunahpu Coyote,
> Great White Peccary, Tapir,[4]
> Sovereign Plumed Serpent,
> Heart of the Lake, Heart of the Sea,
> Maker of the Blue-Green Plate,
> Maker of the Blue-Green Bowl,
>
> as they are called, also named, also described as
>
> the midwife, matchmaker,
> named Xpiyakok, Xmukane,
> defender, protector,
> twice a midwife, twice a matchmaker.
>
> (D. Tedlock 1985:71)

Since we still do not know how to read the name of the Classic-period mother goddess, we will call her First Mother. She appeared to human beings in the form of her avatar, the moon. Her husband was named Hun-Nal-Ye, "One-Maize-Revealed."[5] He was the Maize God and the being who oversaw the new Creation of the cosmos.

We also realized that the ancients regarded the day of this Creation, the world of human beings, as an extraordinary point in the cycles of time. In three magnificent texts at the site of Koba, scribes recorded it as one of the largest finite numbers we humans have ever written. According to these inscriptions, our world was created on the day 4 Ahaw 8 Kumk'u. On this day all the cycles of the Maya calendar above twenty years were set at thirteen (Fig. 2:1)—that is to say, the cycles of 400 years, 8,000 years, 160,000 years, 32,000,000 years, and so on, all the way up to

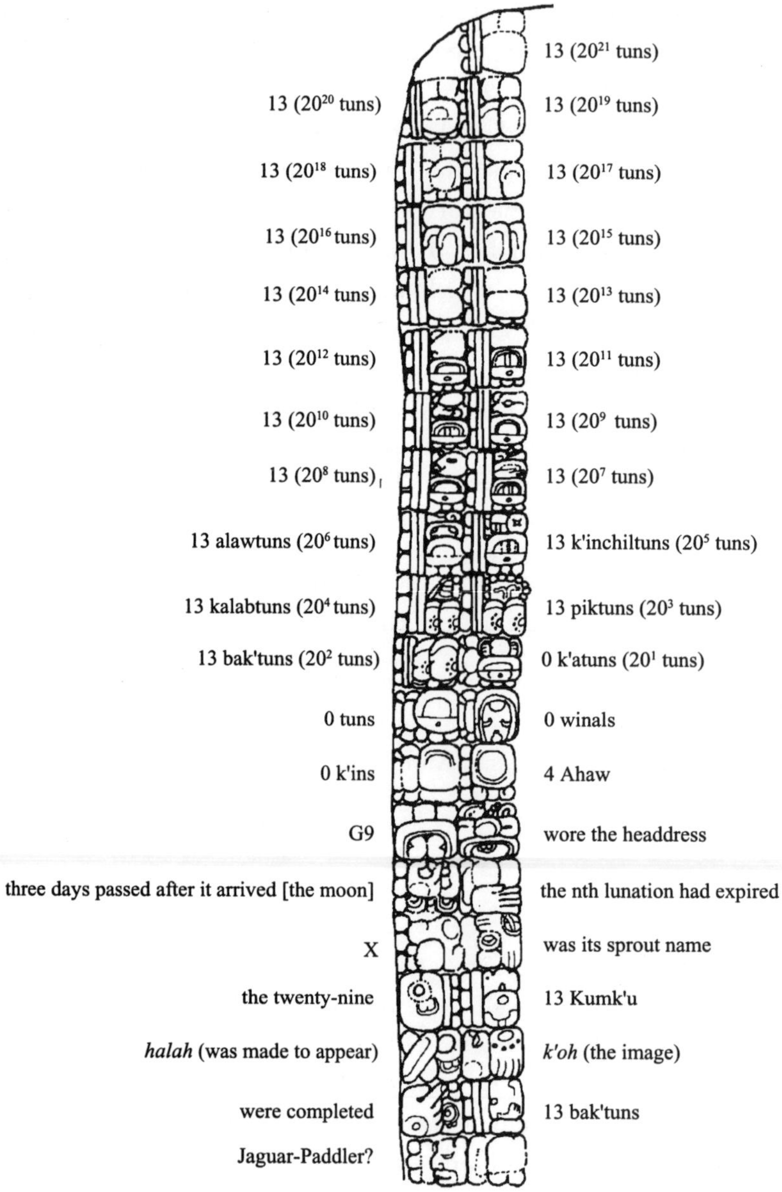

FIGURE 2:1 Koba Stela 1, Creation Day

a cycle number extending to twenty places (20_{21} x 1 360-day year). In our calendar, this day fell on August 13, B.C. 3114 (or September 20, -3113 in the Julian calendar).

To understand what this means, we need a little scale. The thirteens in this huge number act like twelve in our clocks—the next hour after twelve is one. Thirteen changed to one as each of these cycles in the Maya calendar was completed; therefore, we have the following sequence:

13.	13.	13.	0.	0.	0.	1	5	Imix	9	Kumk'u	(Aug. 14, 3114 B.C.)
13.	13.	13.	0.	0.	1.	0	11	Ahaw	3	Pop	(Sept. 2, 3114 B.C.)
13.	13.	13.	0.	1.	0.	0	13	Ahaw	3	Kumk'u	(Aug. 7, 3113 B.C.)
13.	13.	13.	1.	0.	0.	0	2	Ahaw	8	Mak	(May 1, 3094 B.C.)
13.	13.	1.	0.	0.	0.	0	3	Ahaw	13	Ch'en	(Nov. 15, 2720 B.C.)
13.	13.	13.	0.	0.	0.	0	4	Ahaw	3	K'ank'in	(Dec. 23, A.D. 2012)
13.	1.	0.	0.	0.	0.	0	10	Ahaw	13	Yaxk'in	(Oct. 15, A.D. 4772)
1.	0.	0.	0.	0.	0.	0	7	Ahaw	3	Zotz'	(Nov. 22, A.D. 154587)

Each of the years, called a *tun* by the Maya, in these dates is composed of 360 days. If we return to the Creation date with its twenty cycles set at thirteen, we see that it will take 41,943,040,000,000,000,000,000,000,000 tuns for the highest cycle to change from thirteen to one.[6] This huge number functions on several levels. As you've already seen, it allows linear time to unfold in a cyclic structure. As the cycles in the Creation number get larger, we humans can perceive them only as a tangent in an unimaginably huge cycle. To draw an analogy from our own culture, we experience the earth as flat even though we know it is round. Our own Big Bang theory of Creation follows this same pattern. Our own cosmologists contemplate that this universe is only one in a series of many and that the matter in this one will eventually collapse back into a monoblock and explode once again. Comparisons between cosmologies are not really the point here, however. What *is* important is that the Classic-period Maya conceived time on so grand a cyclic scale. To the Maya, time only *appears* to move in a straight line. The Creation date is a point on ever larger circles within circles within circles of time.

The symmetry between the Maya date of Creation and the structure of the Maya calendar is also important. On the first day of the present Creation, all twenty of the progressively increasing cycles were set at thirteen. The most ancient way of reckoning time in Mesoamerica, the way shared by all the peoples of the region, consisted of thirteen numbers combined with twenty day names to give a cycle of 260 days. Creation day was this sacred calendar writ large upon the face of the cosmos—twenty cycles set at thirteen.

We have no Classic-period book equivalent to the Popol Vuh, which was written by the K'iche' people after the Spanish Conquest. But we do have the Creation story as Maya kings inscribed it into their royal monuments. At Quirigua, a small but important town on the Motagua river in Guatemala, a powerful lord named Kawak-Sky raised a series of enormous stelae, stone trees to display himself arrayed in the imagery of the cosmos. On one of these, prosaically dubbed Stela C, Kawak-Sky's scribes wrote precisely what the Classic-period Maya believed happened in the first moments of the current era.

We're going to take you step by step through these opening words of creation on Stela C to share with you what we knew when our adventure with Creation began. People sometimes get the impression that decipherment is the work of one extraordinary individual who comes along and decodes the entire story. Nothing could be further from the truth. The translation of even a single text almost always involves the combined efforts and insights of many different scholars working together or building on each other's work. The same is true here.

The text on Stela C (Fig. 2:2) starts with a shorthand notation of the

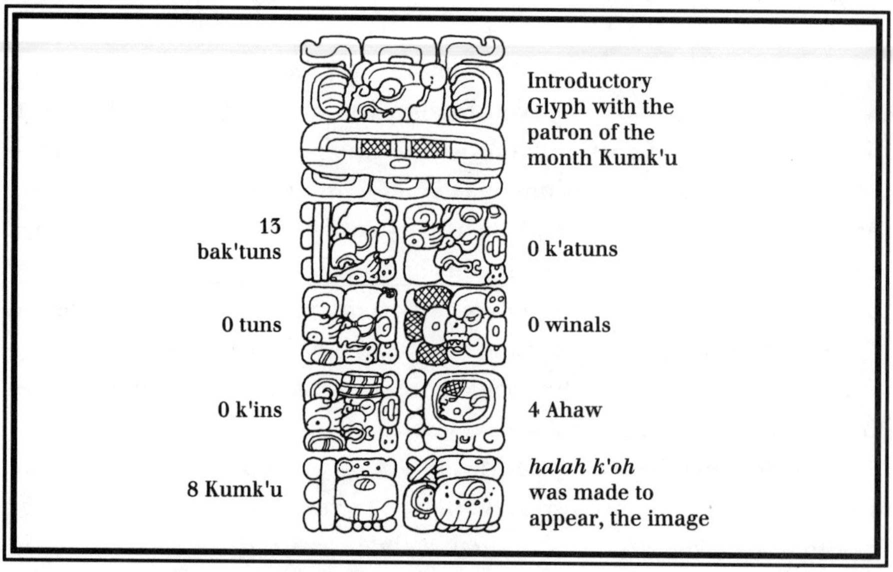

FIGURE 2:2 Opening Phrase of Quirigua Stela C

day—short compared to the immense date recorded at Koba, that is. Kawak-Sky's scribes recorded the birthday of the contemporary cosmos as 13.0.0.0.0 4 Ahaw 8 Kumk'u. The phrase that follows this date is found in almost all Classic-period Creation texts. Our friend Barbara MacLeod[7] deciphered the first part of this phrase as *hal*, which means "to say" and "to make appear." As the opening quote from the Popol Vuh explains, Creation began with the utterance of a word and the appearance of the thing embodied by the word. The ancient Maya apparently thought of the process in the same way.

We learned how to read the second glyph of the Quirigua Creation story through a remarkably serendipitous accident. In February 1992 Linda opened a package containing a book from Austria. It wasn't until later that she realized the parcel had been addressed to her German friend, Nikolai Grube. By then it was too late. Inside the package was Karl Herbert Mayer's[8] new volume of unprovenienced Maya monuments. The book arrived in the middle of a hard day of writing. At first Linda decided to put it aside, but then she began thinking about the odd quirk of fate that had placed this book in her hands. Maybe it had arrived for a reason. She opened it and began to look through the inscriptions it contained. Sure enough, one of the monuments in the book had a Creation text that she had never seen before. That text held the phonetic key that allowed her to decipher the second glyph in the Quirigua Creation story as *k'oh*, "image or mask."[9]

Furthermore, the full reading of the inscription on the stela in Mayer's book (Fig. 2:3) was *ilahi yax k'oh ak chak k'u ahaw*, "was seen, the first turtle image, great god lord." This turtle is a very special one (Fig. 2:4) associated with the Maize God, First Father. In the Popol Vuh, First Father was killed in Xibalba, the Maya Otherworld, by the Lords of Death. They then buried his body in a ballcourt. His twin sons went to Xibalba, defeated his killers, and brought him back to life. Classic-period artists depicted First Father being reborn through the cracked carapace of a turtle shell, often flanked by his two sons. The text on the new stela in Mayer's book told us that the main event of Creation was the appearance of this turtle shell.

On Stela C, however, the Quirigua scribes evoked a much more commonly used Creation image—that of "three stone settings" (Fig. 2:5). They named each stone, and told us who set them and where they were set, as follows:

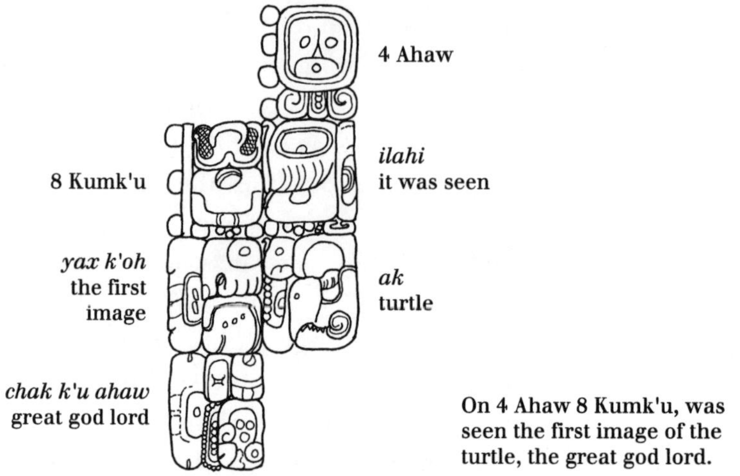
4 Ahaw
8 Kumk'u
ilahi
it was seen
yax k'oh
the first
image
ak
turtle
chak k'u ahaw
great god lord
On 4 Ahaw 8 Kumk'u, was seen the first image of the turtle, the great god lord.

Figure 2:3 A Creation Text

The Jaguar Paddler and the Stingray Paddler seated a stone.
    It happened at Na-Ho-Chan, the Jaguar-throne-stone.
The Black-House-Red-God seated a stone.
    It happened at the Earth Partition, the Snake-throne-stone.
Itzamna set the stone at the Waterlily-throne-stone.

After reading the books written by our colleagues who specialize in the study of the ideas and practices of contemporary Maya peoples, we

The Maize God is reborn from a turtle carapace by his sons, the Hero Twins

Figure 2:4 The Rebirth of the Maize God

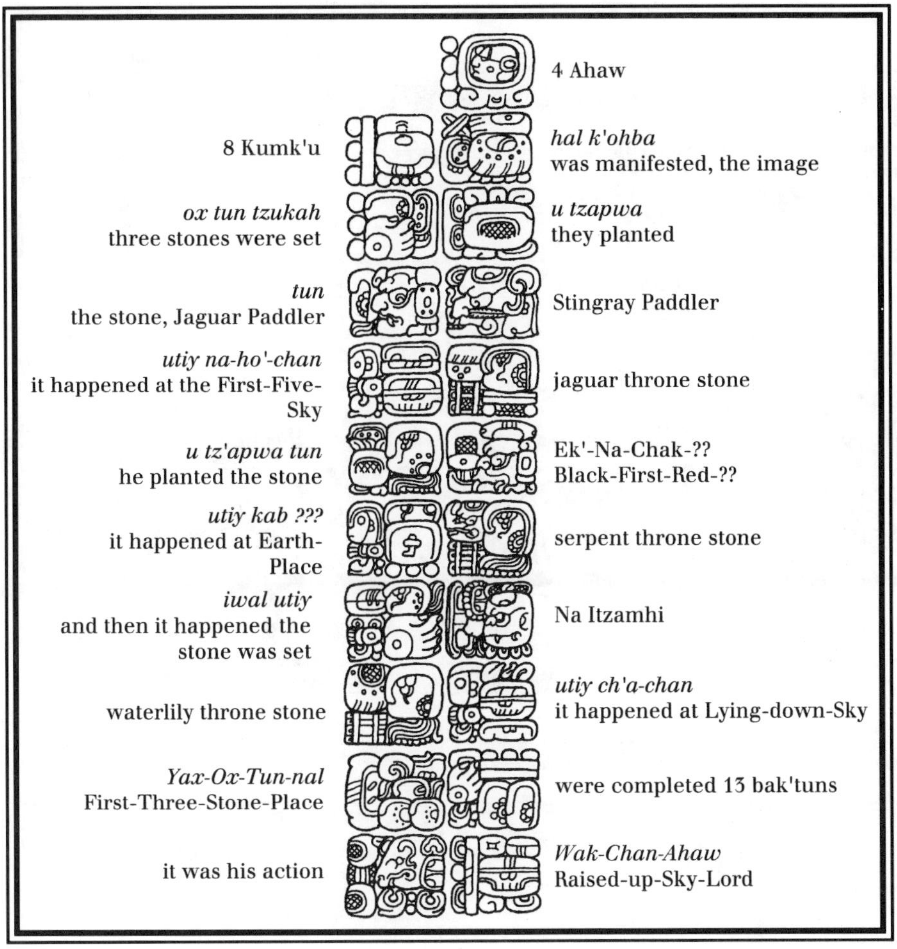

FIGURE 2:5 The Creation Passage from Quirigua Stela C

deduced that these three stones of Creation are symbolic prototypes for the hearthstones used in Maya homes for over three millennia. As the hearthstones surround the cooking fire and establish the center of the home, so the three stone thrones of Creation centered the cosmos and allowed the sky to be lifted from the Primordial Sea. The text on Stela C goes on to tell us many more things—that all these actions happened at a much larger place called "Lying-down-Sky, First-Three-Stone-Place" (*Ch'a-Chan Yax-Ox-Tunal*), that "thirteen cycles ended" on that day, and that these activities were done because of a being called "Six-Sky-Lord" (*Wak-Chan-Ahaw*).

The setting of the first of these three stones of Creation is shown on an extraordinary pot (Fig. 2:6). The scene on this pot depicts six gods

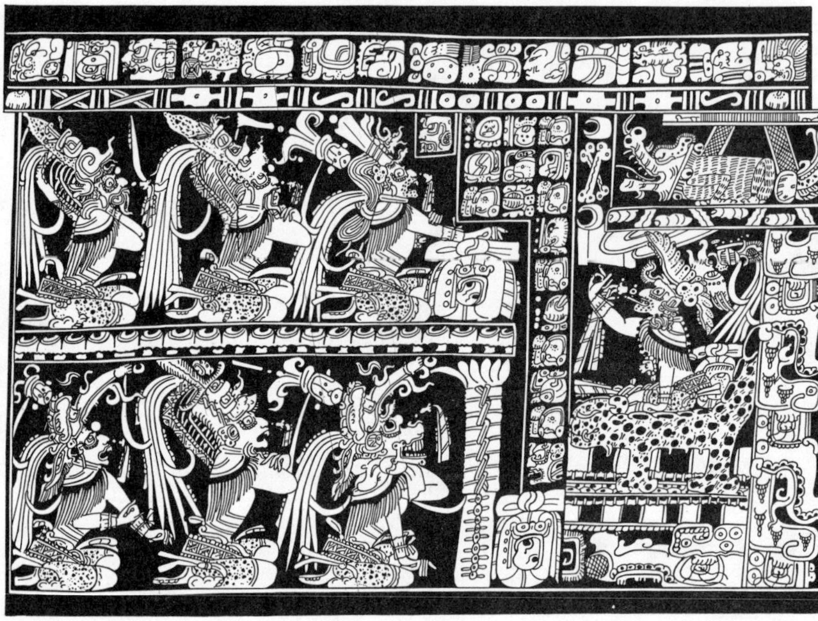
on 4 Ahaw 8 Kumk'u was ordered, Black-Is-Its-Center
the six seated gods are named in the middle set as the beings who put Black-Is-Its-Center in order
jaguar throne
stone

(courtesy of Michael Coe)

FIGURE 2:6 Pot of the Seven Gods

seated in front of a wizened personage whom scholars have dubbed God L—one of the principal denizens of the Otherworld. Large bundles, two marked with a Nine-Star-Over-Earth glyph, rest on the floor in front of them. The aged God L, complete with his cigarette and Muwan Bird headdress, sits inside a house made of mountain-monsters with a crocodile on its roof. Behind him is a bundle called *ikatz,* "burden,"[10] representing the weightiness of his office. He sits on the jaguar-covered throne mentioned in the Quirigua account of Creation.

The text that accompanies the imagery on this remarkable vase begins with the Calendar Round date 4 Ahaw 8 Kumk'u, assuring us that we are, in fact, seeing the Creation. The verb in the first phrase reads *tz'akah,* which means "to bring into existence" and "to put in order."[11] The place "brought into existence" is called *ek' u tan,*[12] "black [is] its center." Remember that in these first moments of the Creation the sky is still "lying down" on the face of the earth, so that there is no light. The black background of this vase is meant to express that darkness. In this case, the things being put in order are the seven gods themselves, all described as *ch'u*, "god" or "holy being." They are organized in the heavens in the following sequence: first God Sky, then God Earth, God Nine-Footsteps, God Three-Born-Together, God Ha-te-chi, and the Jaguar-Paddler, who is named as one of the stone-setters on Stela C at Quirigua. The name of a seventh god[13] didn't fit at the bottom of the list and was put in the corner in front of the upper row of gods. In the Classic texts, as well as in the Popol Vuh, Creation is not the work of a solitary being, but a great effort brought about by many beings who plan, discuss, and act together. This philosophy is still an important part of Maya community life. The special work of the shaman in contemporary Maya ritual is always done within the framework of a supporting and witnessing group.

On the other side of the Maya world from Quirigua, scribes at the royal capital of Palenque carved their own version of Creation on the Tablet of the Cross. This inscription, along with the inscriptions in the Temples of the Foliated Cross and the Sun, are among the most important surviving Classic-period texts. Not only do these texts provide vital details of the Creation story that were not recorded in the public records of other kingdoms, but they also parallel the Popol Vuh in its incorporation of the Creation story into the political charter of a Maya state.

On the Tablet of the Cross, the Palenque account begins with the birth of First Mother six years before the Creation on 12.19.13.4.0 8 Ahaw 18 Tz'ek (December 7, 3121 B.C.) and five hundred and forty days earlier, the birth of First Father, Hun-Nal-Ye, on 12.19.11.13.0 1 Ahaw 8 Muwan (June 16, 3122 B.C.). His birth (Fig. 2:7a) is connected to the Creation day, 4 Ahaw 8 Kumk'u (August 13, 3114 B.C.), when, according to the inscription, thirteen cycles ended. The text mentions an action that First Father accomplished on Creation day, but we haven't been able to decipher what it was.[14]

However, in the next clause, the Palenque scribes repeated Creation again and described it as "it was made visible, the image at Lying-down-Sky, the First-Three-Stone-Place." Then we learned that five hundred and forty-two days later (1.9.2 in the Maya system), Hun-Nal-Ye "entered or became the sky" (*och ta chan*). This "entering" event occurred on February 5, 3112 B.C. (Fig. 2:7b).

The action of "entering the sky" is recorded on another extraordinary painted pot (Fig. 2:7c). This pot depicts one of the Hero Twins (One-

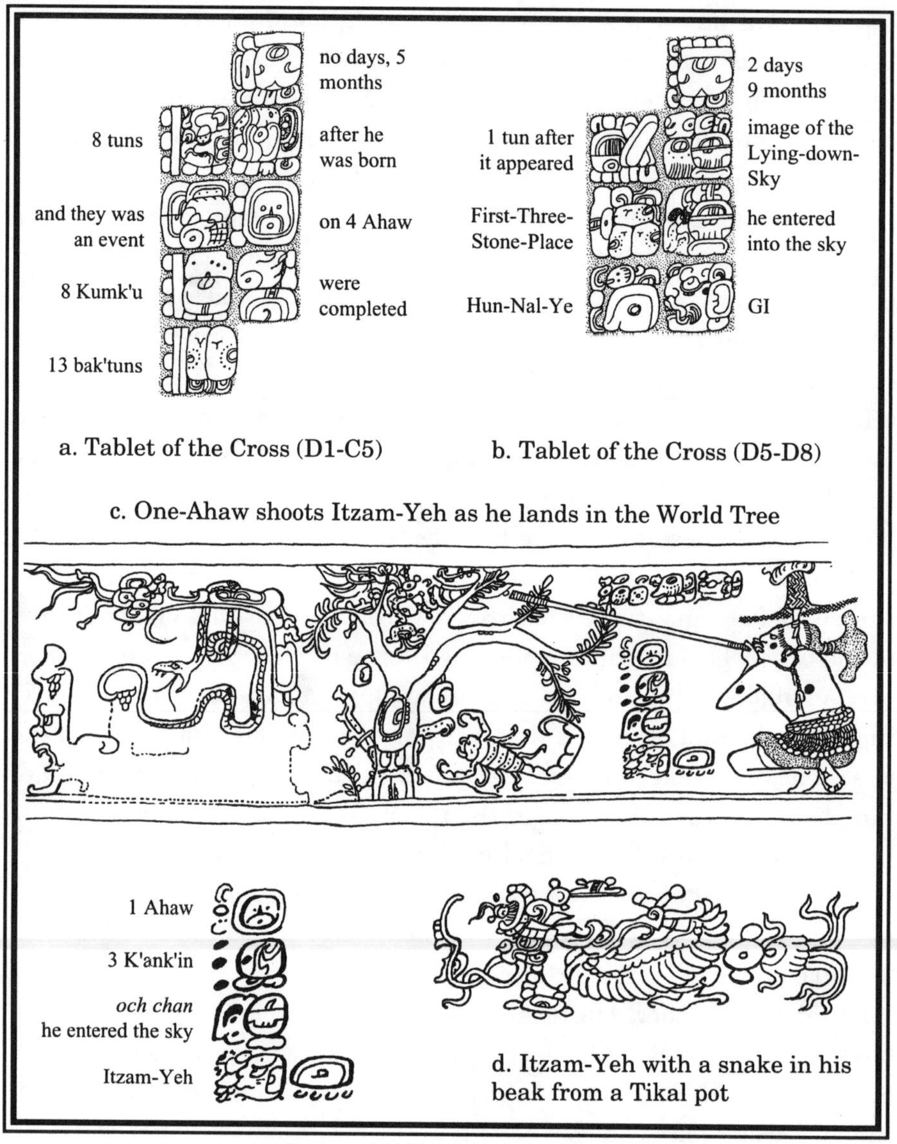

a. Tablet of the Cross (D1-C5)

b. Tablet of the Cross (D5-D8)

c. One-Ahaw shoots Itzam-Yeh as he lands in the World Tree

d. Itzam-Yeh with a snake in his beak from a Tikal pot

FIGURE 2:7 Entering the Sky

Ahaw in the Classic texts and One-Hunahpu in the K'iche' Popol Vuh) and a great bird who is trying to land in a huge ceiba tree[15] heavy with fruit. This mythical bird is Itzam-Yeh, Classic prototype of Wuqub-Kaqix, "Seven-Macaw,"[16] of Popol Vuh fame. In that story, in the time before the sky was lifted up to make room for the light, the vainglorious Seven-Macaw imagined himself to be the sun. Offended by his pride, the Hero Twins humbled him by breaking his beautiful shining tooth with a pellet from their blowgun (Fig. 2:7d). This pot shows One-Ahaw aiming at the bird as he swoops down to land in his tree. As Itzam-Yeh lands on his perch, the text tells us he is "entering or becoming the sky."

This particular "sky-entering" is not the one mentioned in the Palenque text. It is the final event that occurred in the previous creation before the universe was remade. Before the sky could be raised and the real sun revealed in all its splendor, the Hero Twins had to put the false sun, Itzam-Yeh, in his place. If the date on this pot corresponds to that pre-Creation event, as we believe it does, then Itzam-Yeh was defeated on 12.18.4.5.0 1 Ahaw 3 K'ank'in (May 28, 3149 B.C.). After the new universe was finally brought into existence, First Father also entered the sky by landing in the tree, just as Itzam-Yeh did.[17] This is a strange image to associate with a human being, but we will soon see how the Maya imagined such an event.

When the Palenque scribes of the Tablet of the Cross continued the inscription, they used one of the most common and important Mayan literary conventions—a couplet repeating the information from the preceding passage in complementary form. The first passage told us how First Father entered the sky five hundred and forty-two days (1.9.2), after the "image appeared at Lying-down-Sky, First-Three-Stone-Place." The second version of this event told us the formal name of the day—13 Ik', the lying-down of Mol (Fig. 2:8a)—and a second version of what happened. This time around, First Father's entering into the sky was portrayed in the poetic metaphor of a house dedication.[18]

The second passage reads *hoy wakah chanal waxakna-tzuk u ch'ul k'aba yotot xaman*,[19] "[it] was made proper, the Raised-up-Sky-Place, the Eight-House-Partitions, [is] its holy name, the house of the north." In this way we learn that First Father's "entering the sky" also created a house in the north and that it was made of eight partitions. The name of that house, "Raised-up-Sky," is the reciprocal opposite of the "Lying-down-Sky" of the 4 Ahaw 8 Kumk'u place. "Raised-up-Sky" (*Wakah-Chan*[20]) is also the name of the World Tree at the center of the panel in the Temple of the Cross (Fig. 2:8b). Moreover, the image of this primordial tree[21] rises out of the original offering plate of sacrifice. Thus we know that the Maya thought of the entire north direction as a house erected at Creation with the World Tree, the Wakah-Chan, penetrating its central axis. First Father "entered the sky" by raising this tree out of a plate of sacrifice.

We believe that the eight partitions of First Father's cosmic house correspond to the eight divisions in a diagram of the cosmos from the Madrid Codex (Fig. 2:8c) and to the directions recorded on the walls of

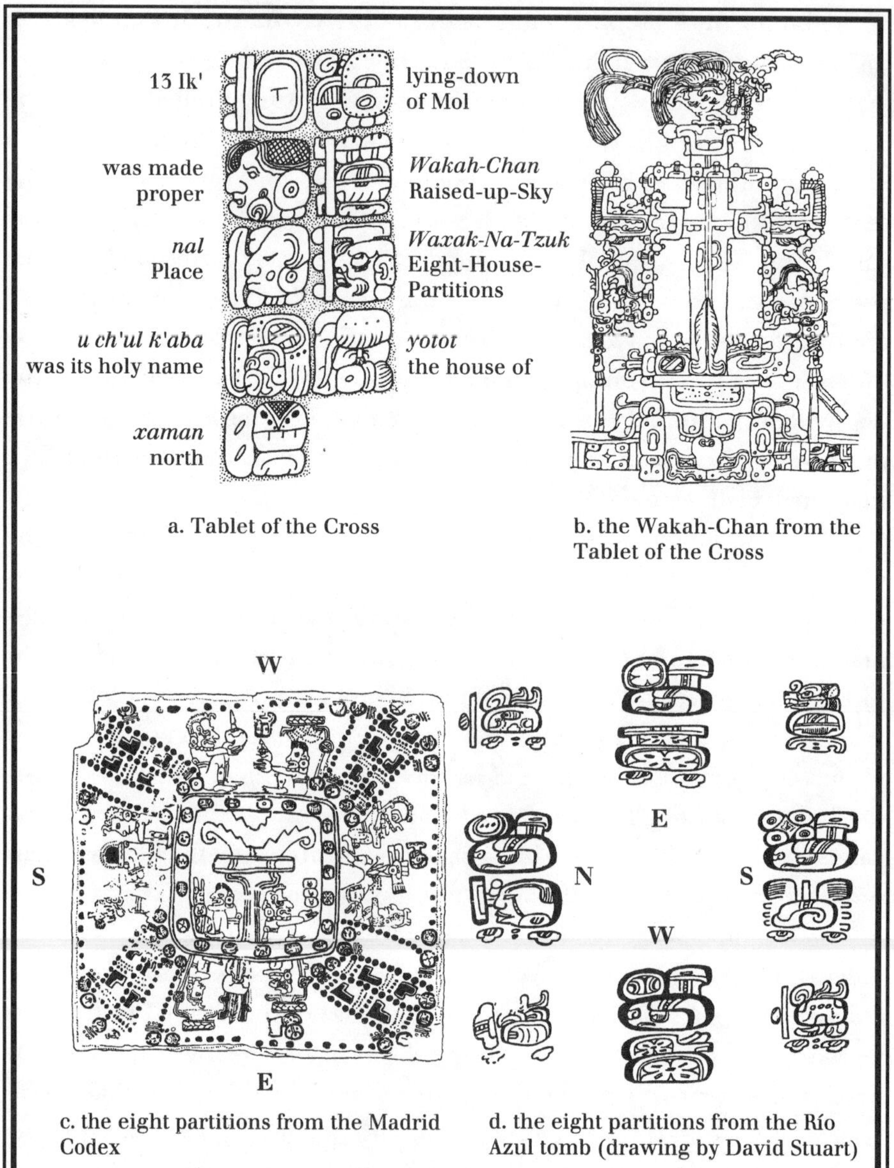

a. Tablet of the Cross

b. the Wakah-Chan from the Tablet of the Cross

c. the eight partitions from the Madrid Codex

d. the eight partitions from the Río Azul tomb (drawing by David Stuart)

FIGURE 2:8 The House of Creation

a tomb at Río Azul, an Early Classic capital in northeastern Peten (Fig. 2:8d). The glyph phrases denoting the cardinal directions of east, north, west, and south are written on the proper directional walls of the tomb. But the four intercardinal direction phrases are written in the room's corners.[22] At least one of these inscriptions refers to the *Wak-Nabnal*, "Raised-up-Ocean-Place," and another reads "Crocodile-Sky."

Taken together, these phrases comprise the eight directional partitions of the world that were established when the World Tree of the Center was erected. The center tree, the Raised-up-Sky tree, itself often has the head form of the glyph *tzuk*, "partition," inscribed on its trunk to mark it as yet another world partition.[23] First Father's house thus orders the entire upper cosmos, the world of humanity, of plants and animals, and of the sky beings, by establishing the center, the periphery, and the partitions of the world. Even today the Maya practice this partitioning and ordering of the world in their rituals.[24]

The Maya conceived of the roof of this house of eight partitions as the dome of heaven, but they also specifically called it *yotot xaman*, "house of the north." For the Classic Maya, the central axis of the cosmos did not run from the zenith of the sky to its nadir—that is, from the point in the sky directly over our heads to the point exactly under our feet.[25] Instead, it penetrated the heavens at the north celestial pole,[26] which today lies near Polaris, the North Star. In the tropics, the North Star and the pivot it marks lie much lower in the sky and closer to the horizon, so that the rotation of the stars across the night sky resembles the interior view of a barrel turning on its long axis.

Furthermore, the name of the Palenque Creation house allows us to unravel the identity of the mysterious *Wak-Chan-Ahaw*, "Six-(or Raised-up)-Sky-Lord," of the Stela C account at Quirigua. Palenque leaves no doubt who he was. Hun-Nal-Ye, First Father, was the being who "raised up the sky" and "Raised-up-Sky-Lord" was the person who according to Stela C, also caused the three stones to be planted when the sky was still "lying down."[27] That he was also the Maize God is confirmed on a

FIGURE 2:9 Tikal MT 140 Offering Vessel from Jaguar-Paw's Tomb

offering vessel from Tikal, MT 140 (Fig. 2:9), that names the Maize God as the *Wak-Chan-Winik*, the "Raised-up-Sky-Person."

On the Tablet of the Cross at Palenque we found yet a third repetition of the events of Creation. In this passage the scribes wrote *pethi Wak-Chan-ki*,[28] "turned, the Raised-up-Sky-Heart" (Fig. 2:10). The *ki* phonetic sign has two references here. One is to an object, that is, a "heart," that

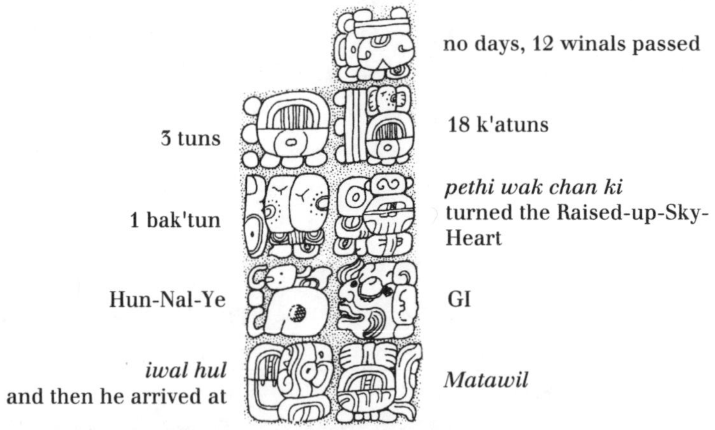

a. Turning the sky passage from the Tablet of the Cross

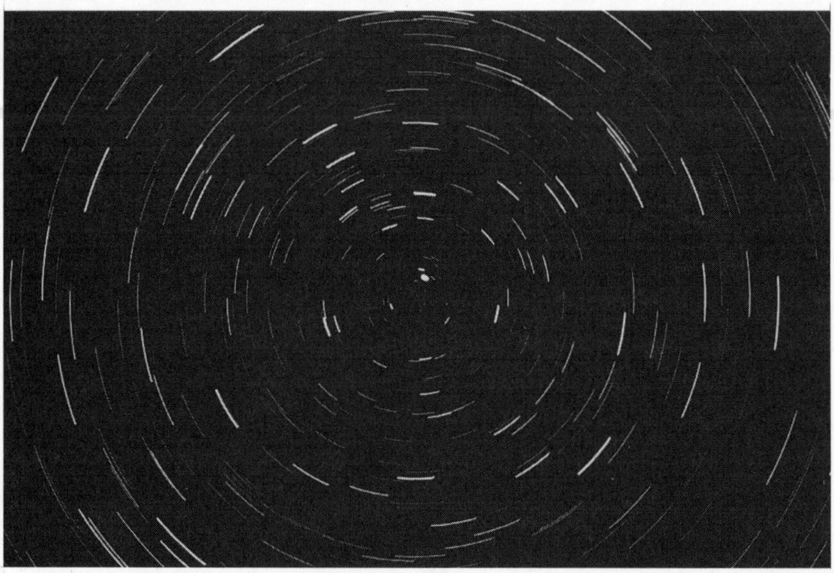

b. The circumpolar star trails photographed in Baja California, December 1, 1989 (© E.C. Krupp, Griffith Observatory)

FIGURE 2:10 The Turning of the Sky

Itzam-Yeh often grips in his beak, and the other is to the heart of heaven. Today, that heart of heaven would be the North Star, Polaris, but in Maya times, the north pivot of the sky fell in a dark area. This black void was the heart of heaven. The "turning" motion described is the movement of the constellations around it. The standing up of the *axis mundi* not only lifted the sky from its lying-down position on the earth but it imparted motion to the star fields. This motion was the beginning of time and space, for it is through the movement of the stars, the Milky Way, and the planets that we human beings calculate the passage of time.

This, then, was the Creation story as far as we knew about it. Under the aegis of First Father, One-Maize-Revealed, three stones were set up at a place called "Lying-down-Sky," forming the image of the sky. First Father had entered the sky and made a house of eight partitions there. He had also raised the Wakah-Chan, the World Tree, so that its crown stood in the north sky. And finally, he had given circular motion to the sky, setting the constellations into their dance through the night. This is where we were when the depths of meaning hidden in these events suddenly began to reveal themselves to us. Since most of it unfolded in Austin, we'll tell the story through Linda's words.

LINDA'S ENCOUNTER WITH CREATION

The key to the unfolding of Creation was given to David Freidel in a hotel lobby in Chicago during the American Anthropological Meetings in November 1991. After the sessions that David had organized were over for the morning, he stood for a while talking to Bruce Love, an epigrapher from the University of California at Riverside. Bruce just happened to mention in passing that a scorpion appeared among the zodiacal constellations represented in the Paris Codex, one of the four Maya books surviving from Precolumbian times.

David immediately remembered that the blowgunner pot shows a scorpion at the base of the Wakah-Chan tree. Maybe, he thought, this scorpion represented the same constellation as the one in the Paris Codex. When he tried out his idea on me, I resisted it, stubbornly and vociferously. During regular phone conversations we conducted in writing this book, we had been arguing about the nature of north and south in Classic Maya thought. David argued for the widely held view that north was conceived of as "up" and associated with the zenith, while south was "down" and linked to the nadir.[29] I argued for their being associated with horizontal directions. If the head of the World Tree pierced the north pivot of the sky, he asked, couldn't the scorpion be opposite it in the south? Yes, I replied, as I remembered the image of Scorpius poised over the mountains south of the Group of the Cross at Palenque.[30]

His idea grew on me through the next month, enough that I presented it at a small conference we attended in San Juan, Puerto Rico,[31] in January 1992. There I revised my former position and said that the ancient Maya conception of the sky was a tunnel stretching from the north polar axis to the Scorpion in the south. When I returned to Austin shortly thereafter, I had to put the finishing touches on a paper on Maya cosmology I had written for my friend Nikolai Grube for an exhibition he was curating in Germany.[32] As I worked, words from another friend who had been at the San Juan conference, Johannes Wilbert, the great ethnographer of the Warao of Venezuela, came back to me. He had admonished me always to look to nature for the source of mythological symbolism. At the last minute, I decided to follow his advice and look at a map of the sky. I wanted to find out what it looked like when Scorpius was opposite the North Star.

I rummaged through my bookshelves and found a book of star maps, Menzel's *Stars and Planets.* Unfortunately, this book printed the south and north views of the sky on two separate pages, so I had to find some tracing paper and place it over the page showing the north sky. I drew it and then shifted the tracing paper to the south page and joined both sides of the sky together. I drew in Scorpius and the North Star and then looked at the result. My heart jumped into my mouth. There it was. The Milky Way (Fig. 2:11) stretched south to north from Scorpius past the North Star. The Wakah-Chan was the Milky Way.

When I presented this new insight at my next seminar, one of my graduate students, Matthew Looper, contributed yet another critical idea to the growing pattern. While I was telling the class about the Wakah-Chan World Tree being the Milky Way, I heard him murmur from the back of the crowded seminar room, "That's why he entered the road." His words exploded like a lightning bolt in my mind. The great image of Pakal's sarcophagus at Palenque (Fig. 2:12) shows him at the moment of his death falling down the World Tree into the Maw of the Earth. The expression the ancient Maya used for this fall was *och bih,*[33] "he entered the road." The road was the Milky Way, which is called both the *Sak Be*

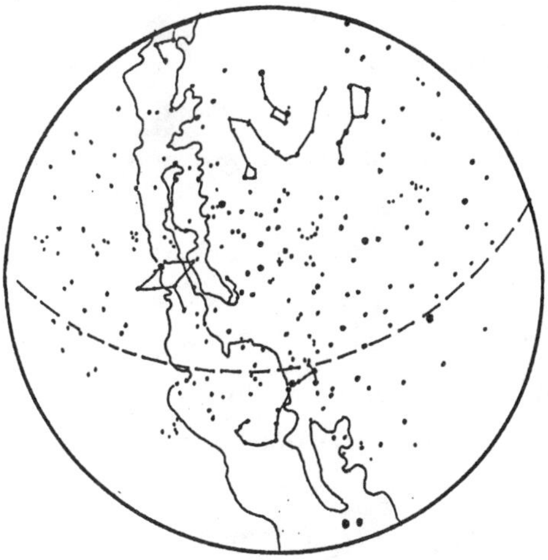

The shape and orientation of the Milky Way at 20° north latitude when Scorpius is opposite the north celestial pole

FIGURE 2:11 The Wakah-Chan

verb for death on the sides of the sarcophagus

och bih
"he entered the road"

As he dies, Pakal falls down the Milky Way into the southern horizon. This falling was called *och be,* "he entered the road." The Maya word for the Milky Way was the *Sak Bih* ("White Road") and *Xibalba Bih* ("Road of Awe").

FIGURE 2:12 Palenque Temple of Inscriptions Sarcophagus

("White Road") and *Xibal Be,* "Road of Awe," by the Maya. Pakal enters this road in death.

I went home that night a little awed by the discoveries and called David to infect him with my excitement. He immediately locked on to the idea, but countered that John Sosa, a brilliant young ethnographer who had been working with Yukatek Maya on their cosmology, had concluded that the ecliptic was symbolized by a double-headed animal. According to Sosa, this was none other than the sun-plate-skull icon we call the Quadripartite Monster joined by a snake body to the same head on the other horizon. I didn't believe that the Quadripartite God was the ecliptic, but the idea of a double-headed snake symbol began to make sense in terms of what I knew about the ecliptic.

The ecliptic is the line of constellations in which the sun rises and sets throughout the year. We divide this band into twelve zones that gives us our zodiacal birth signs. At night, these ecliptic constellations create a path across the sky which marks the track of the sun in its daily and yearly movement. The planets and moon also follow this path, which snakes from north to south and back again as the year proceeds. In the tropics, the ecliptic actually crosses directly overhead and occupies the zenith position of the sky.

I thought about the image of the World Tree and realized that the Double-headed Serpent (Fig. 2:13) that was draped around and through

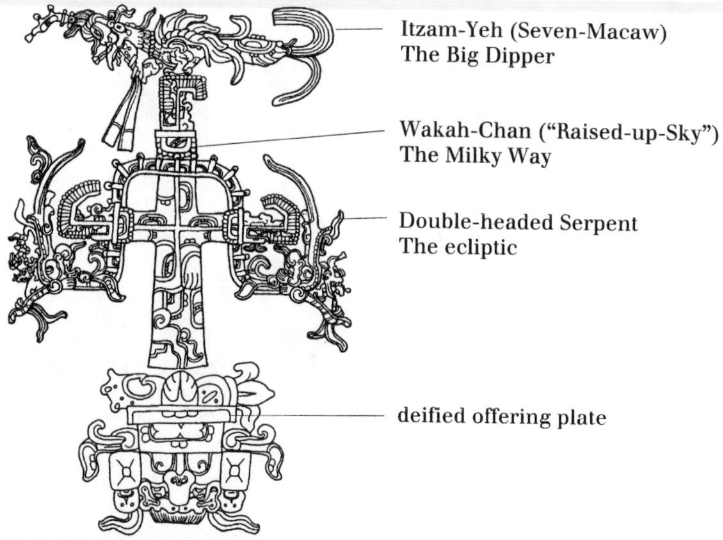

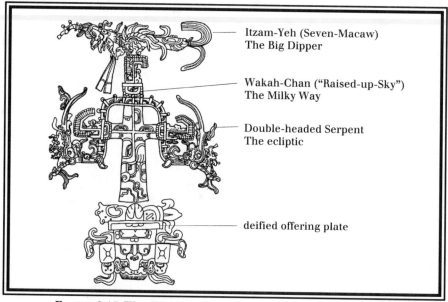

FIGURE 2:13 The Wakah-Chan Tree and the Ecliptic Snake

its branches had to be the ecliptic.[34] Later I realized this new identification also explained the dedication pot from Tikal that had allowed us to identify the Maize God as the main actor in Creation (Fig. 2:9). In the scene, he sits at the place of Creation holding the ecliptic snake in his arms with the Stingray Paddler emerging from one mouth and the Jaguar Paddler from the other. This little pot actually shows the moment of creation when First Father stretches out the path of the sun and the planets across the sky.

The next day Matt came by my house and told me about two more observations he had found in the work of Dennis and Barbara Tedlock.[35] They had discovered that Seven-Macaw of the Popol Vuh was the Big Dipper in the north sky, and that his wife, Chimalmat, was the Chimal Ek' or the Little Dipper.[36] I found that information interesting, but not really mind-blowing. But Matt wasn't finished yet. "Did you know that Dennis Tedlock identified Orion as the three stones of the hearth?" he asked innocently.

"No," I answered. "Where'd he say that?" And we rushed into my library to find Tedlock's Popol Vuh. There it was: "Today [Alnitak, Saiph, and Rigel in Orion] are said to be the three hearthstones of the typical Quiché kitchen fireplace, arranged to form a triangle, and the cloudy area they enclose (Great Nebula M42) is said to be the smoke from the fire" (Tedlock 1985:261).[37]

I was stunned. These had to be the same three stones that were laid at Creation when the sea was still lying down on the face of the earth. The first act of the gods was to create the hearth at the center of the universe where the first fire of Creation could be started (Fig. 2:14). In fact, John Carlson later pointed out to us that the Aztec's name for the belt of Orion was their word for the fire drill that created new fires.[38] I realized that since the sky was still lying down upon the earth on 4 Ahaw 8 Kumk'u, the act of seating the stones in the triangular pattern of the hearth created an image on the face of the earth and in the sky at the same time. The two were, after all, then joined together.

The next bombshell came from another of my students two days later. Khristaan Villela had come over that afternoon to file away some of the stacks of papers that decorate my every table. After he was done, he asked, "What about the Bonampak' murals?" Oh dear, I thought to myself, are they going to fit into this pattern also?

My friends Mary Miller and Floyd Lounsbury[39] had proposed many years earlier that the cartouches over the judgment scene in the middle

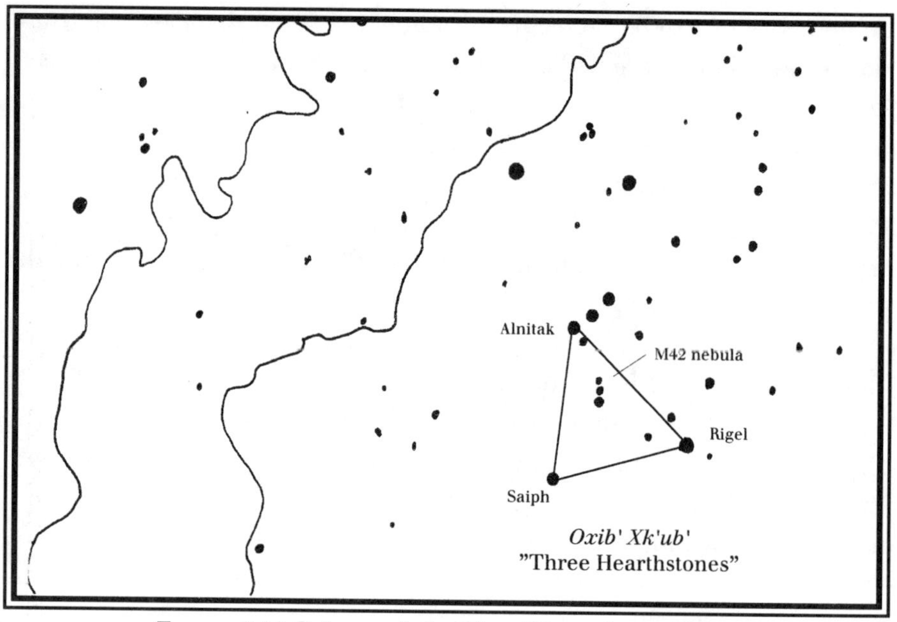

FIGURE 2:14 Orion and the First-Three-Stone-Place

room at Bonampak' represented constellations that were seen at dawn on the day of the battle. At Khris's urging, I checked out the sky on that day, August 6, A.D. 792, with *EZCosmos,* the computer program I use for astronomy. Again I was stunned by the results. In the hours before dawn (Fig. 2:15), the constellations of Gemini and Orion had hovered above the eastern horizon with Saturn and Mars above and between them. The cartouches of Bonampak' show copulating peccaries in the cartouche at the left, a turtle with three stars on the right, and between them two anthropomorphic figures throwing star signs into the scene. Lounsbury had already published evidence that some sources called Gemini the "turtle star," while others identified the turtle with Orion.

As I looked at the computer screen, everything clicked. Gemini had to be the copulating peccaries and Orion the turtle. The three stars on its back in the cartouche at Bonampak' were placed in the exact pattern of Orion's belt. But more important, the Maize God was reborn from the cracked carapace of a turtle, and we had a text that said that the first image of the turtle was seen at Creation (Fig. 2:4). The two anthropomorphic figures between the peccaries and the turtle had to be the personifications of Saturn and Mars. And we already knew many examples of other planets, such as Venus and Jupiter, that were represented by anthropomorphic figures. We seemed to have two more here. Not until

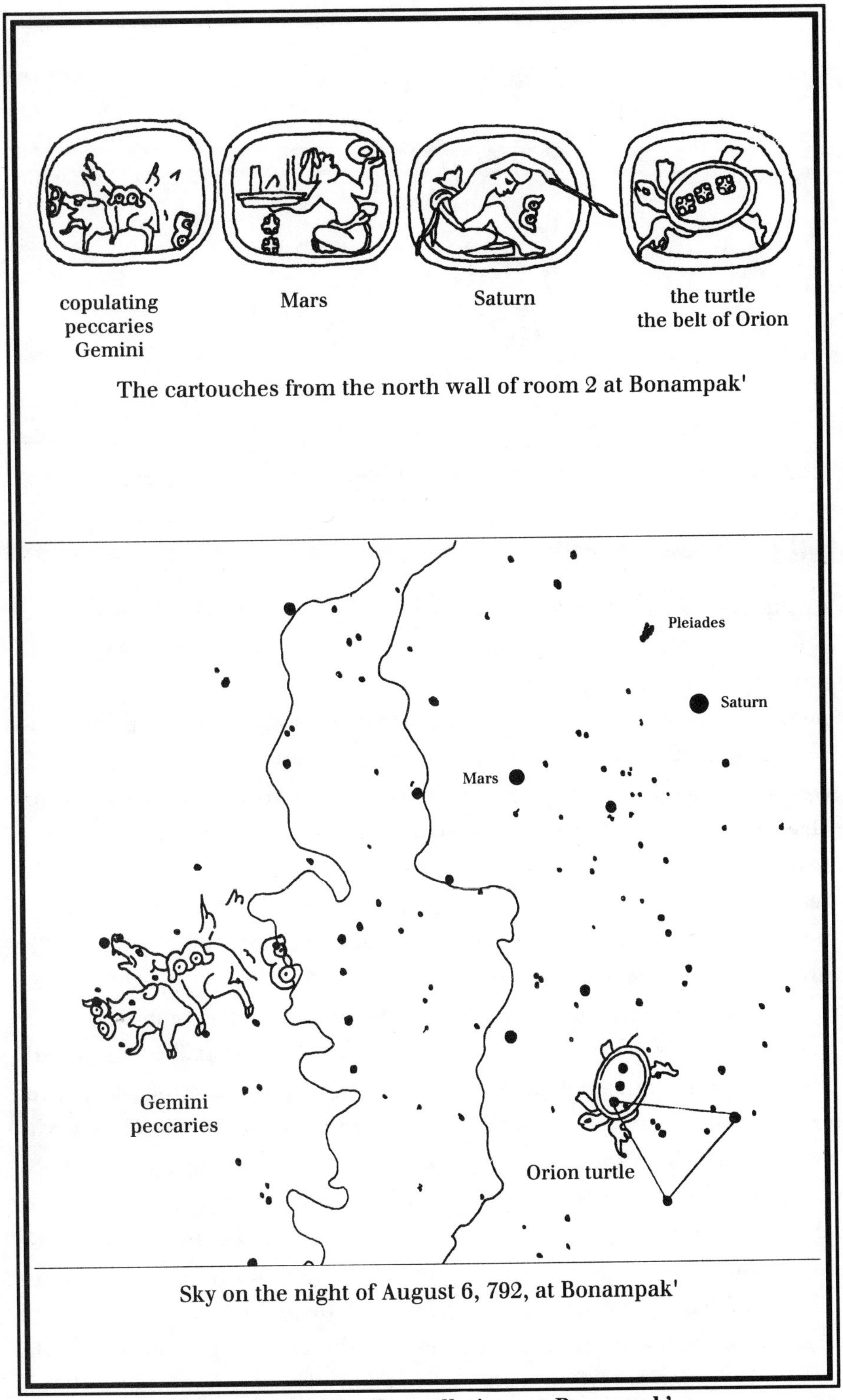

FIGURE 2:15 The Constellations at Bonampak'

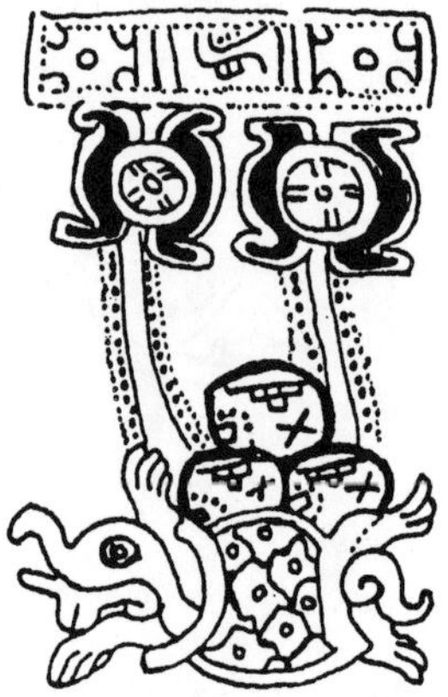

FIGURE 2:16 Detail from the Madrid Codex

much later did it hit me that August 6, the date of the zenith passage at Bonampak', was only seven days before the Creation day of August 13.

Matt Looper provided the confirmation by calling my attention to a picture from the Madrid Codex, another of the four Maya books. It shows a turtle (Fig. 2:16) with a triangle of stones on its back. The turtle in the codex is shown suspended from cords tying it to the skyband because Orion hangs below the ecliptic. Clearly Orion was the turtle from which the Maize God rose in his resurrection (Fig. 2:17). The Milky Way rearing above the turtle had to be the Maize God appearing in his tree form as he does on the Tablet of the Foliated Cross at Palenque. The image of the first turtle really is in the sky. **Plate 34**

The identification of Gemini as copulating peccaries made sense in another curious way. For years we have seen pots that have these odd modeled peccary heads as their feet. All of these pots have the sea painted or drawn on their bottoms and an extraordinary one in the Dallas Museum of Art shows the sun god paddling a canoe across this sea (Fig. 2:18). I realized that this must be the sun riding the ecliptic across the Peccaries of Gemini. I ran into a bit of trouble at this point because there was a lot of counter evidence that identified Gemini as a turtle. The Yukatek Maya of Chan K'om[40] call Gemini *Ak Ek',* "turtle," and they say that the three stars in the middle of his back are the *choch,* "intestines," of the turtle. Fortunately, I found the solution to this paradox close at hand. Floyd Lounsbury had pointed out that *ak* is the word for both "turtle" and "peccary." Thus, *ak ek'* can be either "turtle star" or "peccary star," and you don't know which unless you ask. Furthermore, when I saw the pot again in April, I realized the peccary heads have a trifoil sign in their eyes that specifically identifies them as the *ak* peccary.[41]

Justin Kerr, another friend, gave me some incredible evidence that showed that the Maya thought of the constellations in both ways. I already had pictures showing Orion as a turtle, and Gemini appears as the peccaries at Bonampak', and in the Paris Codex zodiac as a turtle. When I was at Justin's house in late February 1992, he used the enormous database of pottery images he has created to find images of peccaries. One of these images occurred on a shell published twenty years ago in Mike Coe's *The Maya Scribe and His World.* I looked at the photo and felt the magic stir again. Here was a peccary with a Waterlily Monster (Fig.

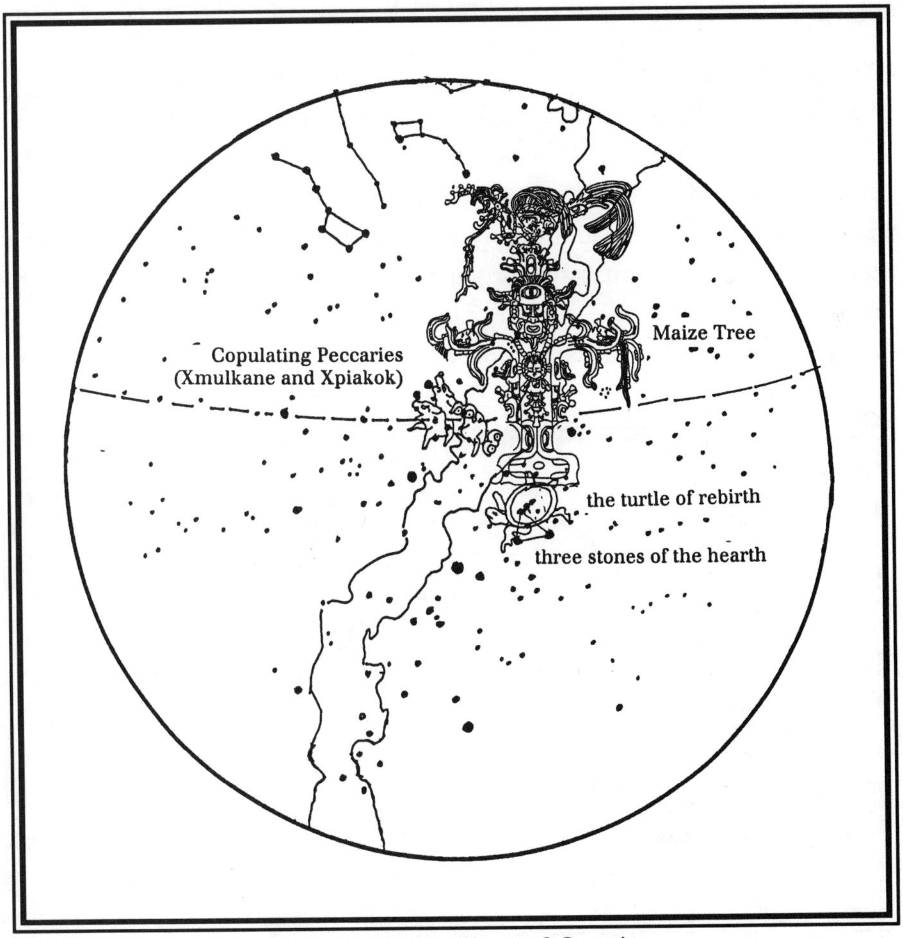

FIGURE 2:17 The Place of Creation

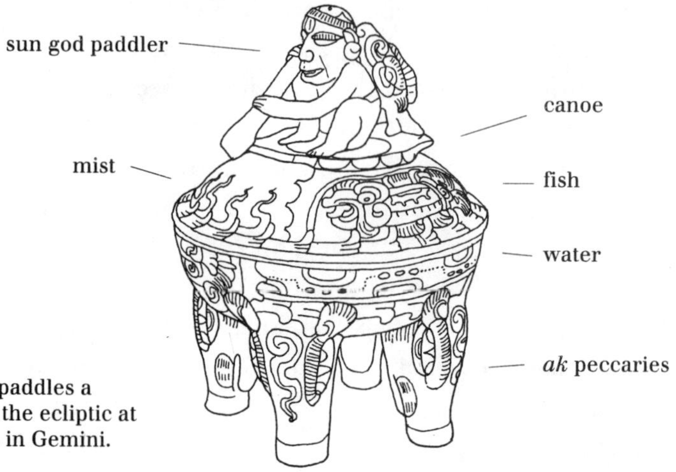

The sun god paddles a canoe across the ecliptic at the peccaries in Gemini.

FIGURE 2:18

2:19a) attached to its haunches just like the one attached to the Maize God's turtle shell (Fig. 2:4). It was split open just like the turtle shell and out of the cleft rose the Maize God carrying a brush and paint pot in his hand so that he could paint the images on the sky. Here was a cleft peccary directly substituting for the cleft turtle. They had to mean the same thing.

Justin then called my attention to yet another pot from *Maya Scribe*

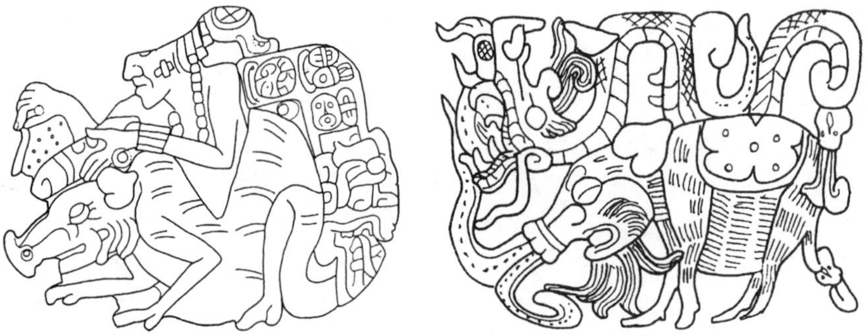

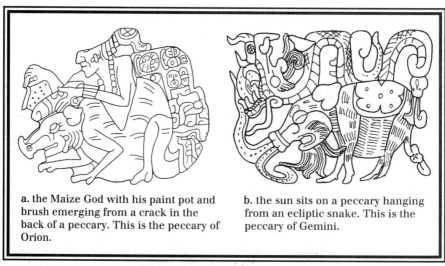

a. the Maize God with his paint pot and brush emerging from a crack in the back of a peccary. This is the peccary of Orion.

b. the sun sits on a peccary hanging from an ecliptic snake. This is the peccary of Gemini.

FIGURE 2:19 Sky Peccaries

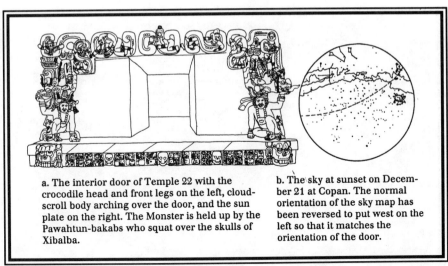

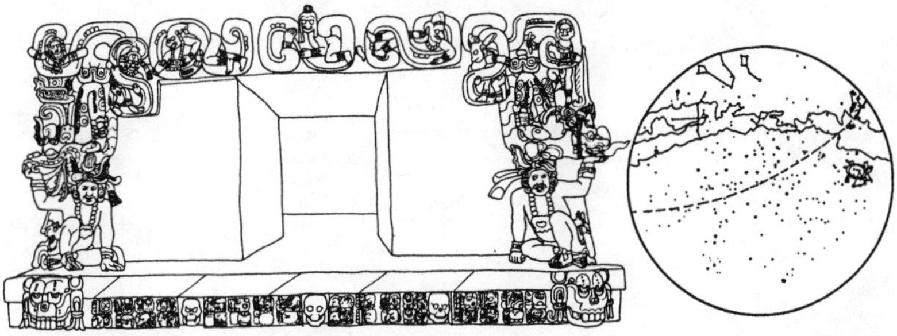

a. The interior door of Temple 22 with the crocodile head and front legs on the left, cloud-scroll body arching over the door, and the sun plate on the right. The Monster is held up by the Pawahtun-bakabs who squat over the skulls of Xibalba.

b. The sky at sunset on December 21 at Copan. The normal orientation of the sky map has been reversed to put west on the left so that it matches the orientation of the door.

FIGURE 2:20 The Milky Way Monster

showing a peccary with the four-petaled flower that represents the sun on its back (Fig. 2:19b). The ecliptic snake undulates along its back, confirming that the peccaries are on the ecliptic rather than below it. Gemini is both the turtle and the peccary. Likewise, the belt of Orion is both the turtle and the peccary, but this constellation also contains the place of the three stones of the Creation hearth.

I was breathless by this time, for the more we found, the more we seemed to discover. It reminded me of a magic time in Copan in 1987 during the last day of my six-month Fullbright. I had gone up to Temple 22 to photograph the great doorway sculpted with the Cosmic Monster (Fig. 2:20a) because on that day, December 12, the sun was on the south side of the building. Since I was usually at Copan in the summer, all my photographs showed the door in shadow. I wanted to get it in the light for the first time. After taking my photographs, I went into the inner sanctum of the temple to look around. When I turned to leave, I noticed that there were almost no shadows cast by the doorjambs (Fig. 2:21). Surprised, I looked at my watch and found it was 11:15 A.M., almost noon only ten days before the winter solstice. "Ah-ha," I thought to myself as I took record photographs. "This is a winter solstice building."

Five years later, I took that insight to the computer to find out what the sky looked like at sunset, midnight, and dawn on the night of the winter solstice. I didn't have to go beyond sunset. The Milky Way in the hour or so after sunset on winter solstice stretches across the sky from east

FIGURE 2:21 The doors into the interior of Temple 22 as photographed at 11:15 A.M. on December 8, 1987

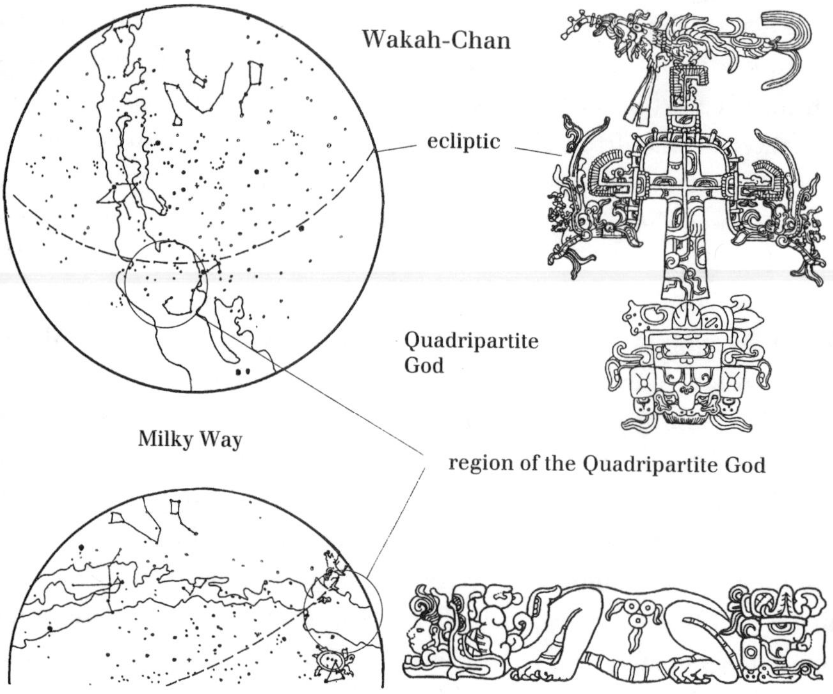

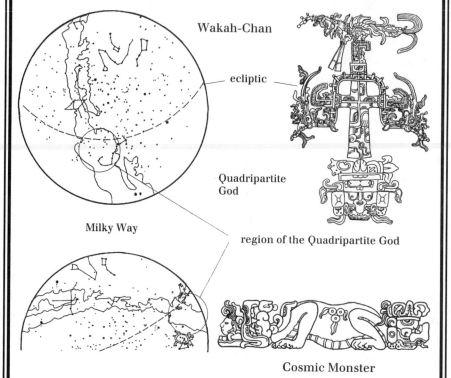

FIGURE 2:22 The Junction of the Ecliptic and the Milky Way

to west with its western side split into the jaws of the crocodile (Fig. 2:20b). The Cosmic Monster is also the Milky Way.

With that discovery, I realized that every major image from Maya cosmic symbolism was probably a map of the sky. The personified, sun-marked offering plate depicted at the base of the Raised-up-Sky World Tree, when the Milky Way is in one of its two north-south positions, also occurs on the eastern rear end of the Cosmic Monster. And both of these points in the sky are where the ecliptic crosses the Milky Way (Fig. 2:22).

I knew I had to find out about all the forms of the Milky Way. I spent two hours on the floor of a local bookstore until I found an astronomy book that showed its position in the nighttime sky every month of the year.[42] I made drawings of each of these star maps, removing the outlines of *our* constellations, keeping only the Maya ones we had identified and the outline of the Milky Way. After I adjusted the maps for 15° north latitude, I had a Milky Way calendar.

With these new maps, I found out some other interesting things. When the Milky Way is lying down flat, rimming the horizon, the area overhead is completely dark (Fig. 2:23a). This is the portal into which Pakal falls on his sarcophagus lid and out of which beings of the Otherworld emerge. The Maya called this dark place the White-Bone-Snake, but it was also called the *Ek'-Way*,[43] the "Black-Transformer" or "Black-Dreamplace." That realization brought to mind another important clue, an incised peccary skull from Copan (Fig. 2:23c). For the first time it made sense that this exquisite carving graced a peccary. The incised scene shows the dedication of a stela, but the point of view represented is from the other side of the portal. It is as if the gods and the ancestors were looking through at humanity from the Otherworld. Some of the animals cavorting around this portal are peccaries. I checked the Ek'-Way position of the Milky Way and found that it has the peccaries of Gemini above its northeastern horizon and the scorpion of Scorpius peeking up out of its southwestern horizon. I compared this sky pattern to the drawing on the skull from Copan and found, to my delight, that peccaries were sitting on its northeastern side just as they do in the sky on this night. The greatest of the Classic Maya portals to the Otherworld is also found in the night sky.

I looked at the sky maps again and realized that I saw the Milky Way forming another pervasive Creation symbol—a tree with a crocodile head at its roots (Fig. 2:24a). As the Milky Way turns from its north-south

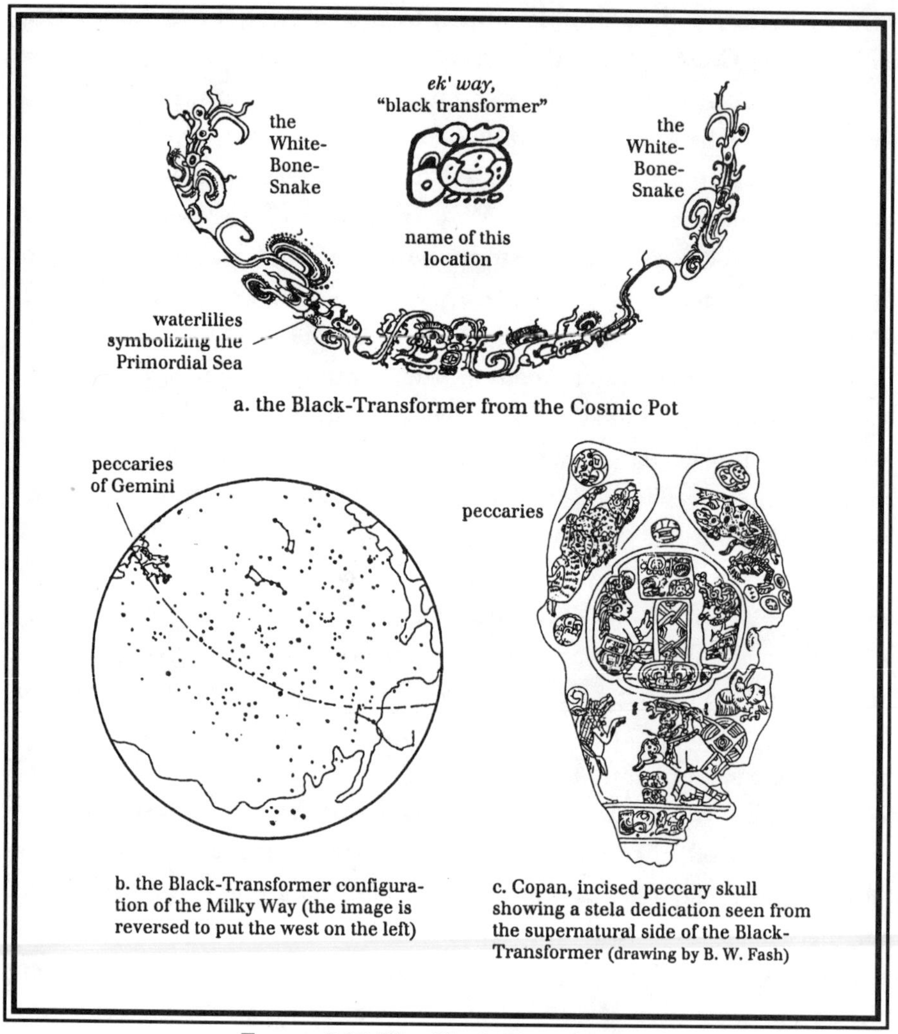
ek' way,
"black transformer"
the
White-
Bone-
Snake
the
White-
Bone-
Snake
name of this
location
waterlilies
symbolizing the
Primordial Sea
a. the Black-Transformer from the Cosmic Pot
peccaries
of Gemini
peccaries
b. the Black-Transformer configuration of the Milky Way (the image is reversed to put the west on the left)
c. Copan, incised peccary skull showing a stela dedication seen from the supernatural side of the Black-Transformer (drawing by B. W. Fash)

FIGURE 2:23 The Black-Transformer

position as the World Tree to its east-west position as the Cosmic Monster, it arches from the southwest to northeast quadrants of the sky. At that time it looks as if the crocodile is riding just above the swollen bulge of its base. This "crocodile tree" is a very ancient image among the Maya and is featured in the Popol Vuh version of Creation. In fact, a monument (Fig. 2:24b) from the very early ceremonial center of Izapa in Chiapas shows this crocodile tree along with a picture of one of the Hero Twins after his arm has been ripped off in a struggle with Itzam-Yeh, the Classic name for Seven-Macaw. With his good arm he is holding a pole on which a Late Preclassic-period version of Itzam-Yeh perches.[44] If Itzam-Yeh was symbolized by the Big Dipper in the Classic period, this picture also happens in the sky. When the Wakah-Chan Milky Way moves from its erect north-south position and becomes the Crocodile Tree, the Big Dipper dives downward until it touches the horizon. It disappears as the Crocodile Tree changes into its east-west Cosmic Monster (Fig. 2:24). I believe this movement may correspond to the defeat of Seven-Macaw by the Hero Twins.

At this point I thought I had better take a look at the famous canoe scenes from Burial 116 at Tikal in Guatemala. These extraordinary carved bones were found in the tomb of Hasaw-Ka'an-K'awil,[45] one of the most important kings in the Late Classic history of Tikal. Four of the bones are incised with canoe scenes. One pair shows a canoe with two paddlers, First Father as the Maize God, an iguana, a monkey, a parrot,

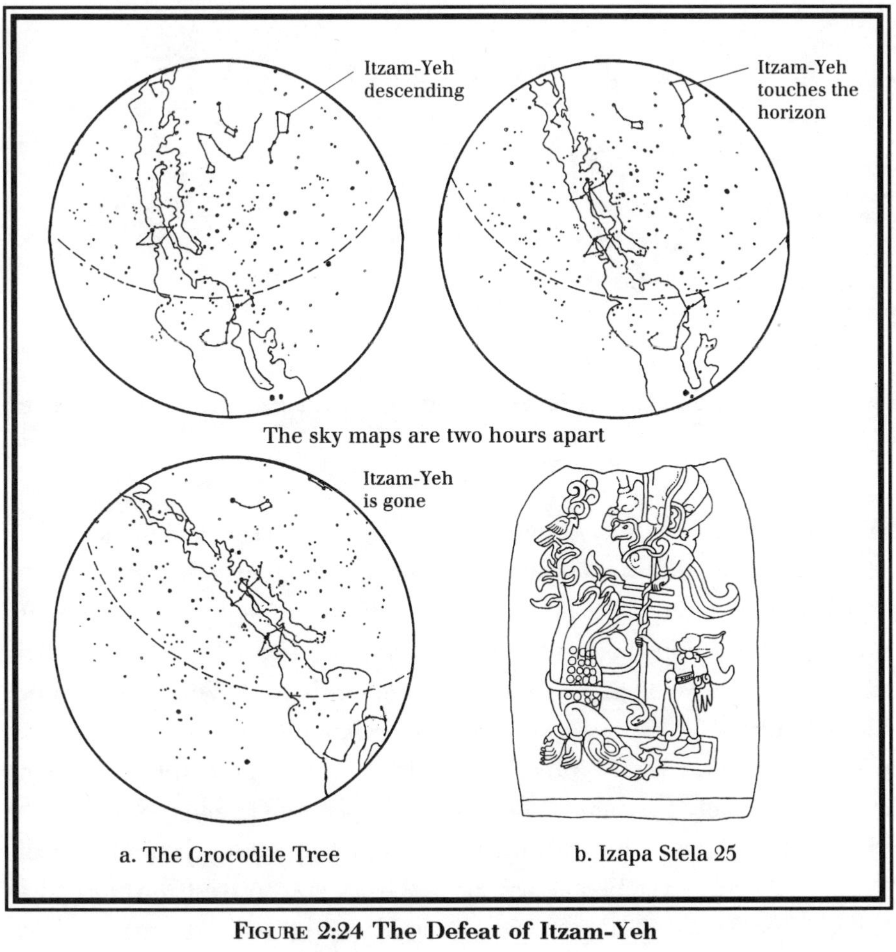

FIGURE 2:24 The Defeat of Itzam-Yeh

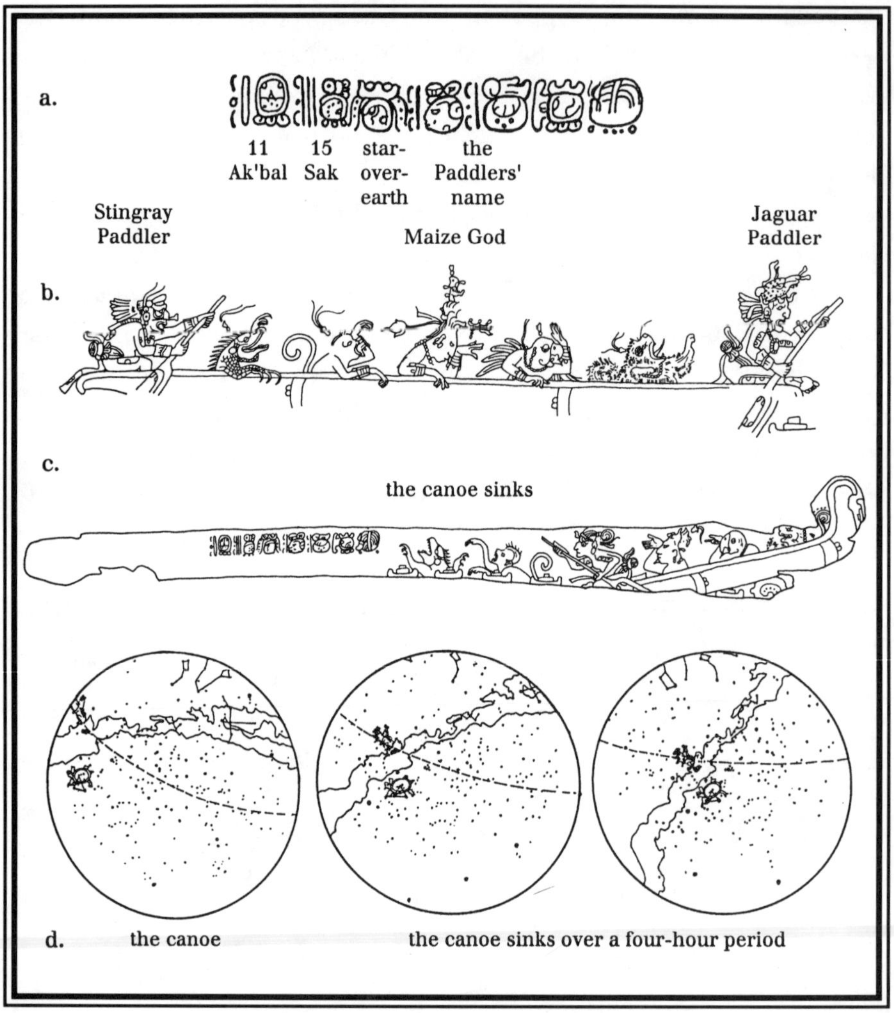

FIGURE 2:25 The Sky Canoe

and a spotted dog. A second pair of bones shows these passengers flailing at the water as the canoe sinks out from under them (Fig. 2:25b–c).

We have known for a long time that these scenes had to have astronomical associations because the accompanying text (Fig. 2:25a) has the star-over-earth verb that appears very frequently with Venus and Jupiter hierophanies. No one, however, had ever been able to associate the date, 6 Ak'bal 16 Sak, with an important station of either planet. I checked the date's most likely position in the Maya Long-count Calendar, 9.14.11.17.3 (September 16, 743), and found something remarkable. At midnight the Milky Way was stretched across the sky from east to west in the form of the Cosmic Monster (Fig. 2:25d). It suddenly occurred to me that this configuration could also be the canoe, and that during the four hours after midnight the Cosmic Monster canoe "sinks" as the Milky Way turns and brings the three hearth stars of Orion to the zenith just before dawn.

There was a lot of evidence to support this idea. I had published a discussion of a fabulous black eccentric flint from the Dallas Museum of Art. This flint showed a crocodile canoe sinking downward with its passengers, just like the Tikal canoes. In fact, another of the Tikal bones shows a canoe with a crocodile face on its bow, replicating the mouth cleft at the western end of the Milky Way in its east-west position in the sky (Fig. 2:26a). When I was drawing *this* bone, I noticed that the paddler in the picture was Itzamna, not the Old Jaguar Paddler or the Stingray Paddler from the other bones. Then the use of two different paddlers in these canoe scenes made me realize something. We have Classic-period pottery images of Itzamna riding a peccary (Fig. 2:26c). When the Milky Way is in the Cosmic Monster–canoe position, east-west across the sky, the peccary stars of Gemini are at the eastern end of the Milky Way (Fig. 2:26b). The peccaries are exactly where Itzamna sits as he wields his

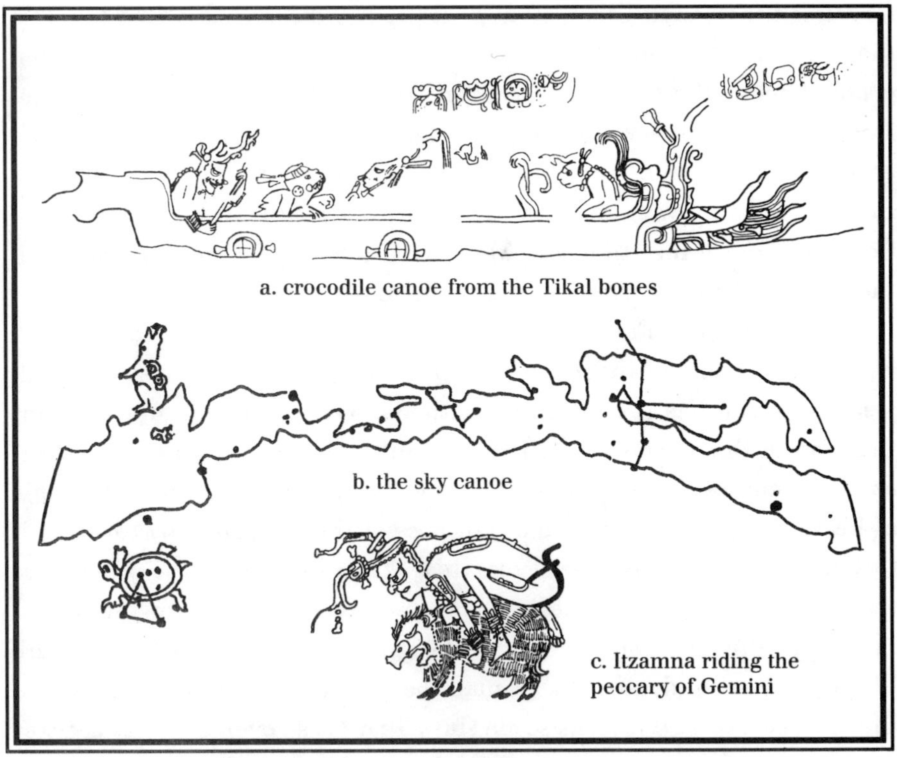

Figure 2:26 The Crocodile Canoe

paddle. Also, Itzamna and the two Paddlers are three of the four stone setters that are named in the Quirigua Creation text on Stela C. They are the makers, along with First Father, of the First-Three-Stone-Place raised up into the night sky on 4 Ahaw 8 Kumk'u.

These Paddlers, then, are central actors in the celestial play of Creation. Like Itzamna, the original shaman, the Paddlers are up in the sky riding the Milky Way to the place of Creation where they will set their stones in the hearth of Orion. They propel the Milky Way canoe with its precious cargo, to the same location. I realized that the Paddlers bring the Maize God to the place of the three stones of Creation and to the turtle carapace, the belt stars of Orion, so that he can be reborn and create the new universe. He is the Wak-Chan-Ahaw who made everything happen.

Two painted pots depicting this same canoe scene confirm its association with the myth of First Father as the Maize God. One (Fig. 2:27a) has a black background showing that the action occurs in the lightless time before First Father raised the sky. The narrative scene includes three episodes from the story. In one, the Jaguar and Stingray Paddlers paddle a canoe carrying the Maize God to the place of Creation. He carries a seed bag on his chest so that he can plant the seeds that are the Pleiades when he raises the Wakah-Chan, or as Enrique Florescano suggested to us, so that he can use them to form the flesh of human beings after Creation is done. Below the canoe, the reclining figure of the Maize God emerges from the mouth of a Vision Serpent in a position that mimics exactly the emergence of a child from his mother's birth canal.[46] The third episode shows two young women helping him dress after his successful resurrection. Here is the rebirth of the Maize God into the Primordial Sea. Both the birth and the dressing scenes occur underwater.

On another pot (Fig. 2:27b), the reborn infant reappears carried in a huge plate by one of his sons. Around the baby rests the jewelry he will wear when he has grown into adulthood. When Nikolai and I taught Creation to thirty Maya in Guatemala in the summer of 1992, a *chuch-kahaw*, "mother-father," named Manuel Pacheco told us that the people of his town (Chiniki, Guatemala) dispose of the afterbirth by carrying it to a burial place in a plate just like this. When they enter the cave where they intend to bury the afterbirth, they carry the plate over their head in exactly this position. The other scene on this pot shows the mature Maize God being dressed by woman, but the fish in the first scene and the text associated with this scene show that these events also took place underwater.

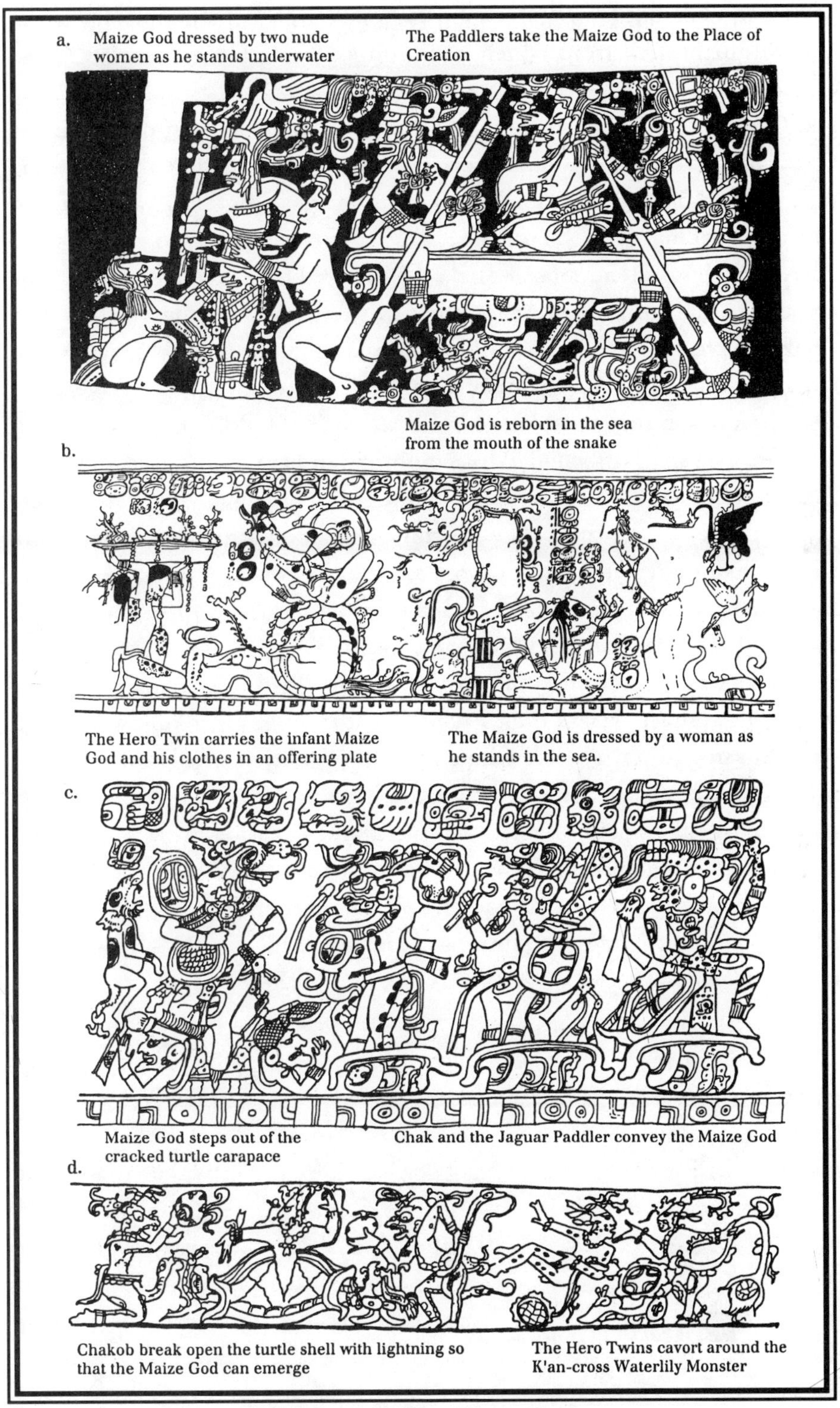
a.
Maize God dressed by two nude women as he stands underwater
The Paddlers take the Maize God to the Place of Creation
Maize God is reborn in the sea from the mouth of the snake
b.
The Hero Twin carries the infant Maize God and his clothes in an offering plate
The Maize God is dressed by a woman as he stands in the sea.
c.
Maize God steps out of the cracked turtle carapace
Chak and the Jaguar Paddler convey the Maize God
d.
Chakob break open the turtle shell with lightning so that the Maize God can emerge
The Hero Twins cavort around the K'an-cross Waterlily Monster

Figure 2:27 The Canoe Pots

Yet another pot (Fig. 2:27c) shows the Maize God stepping from the cracked turtle shell from which he is being reborn. Here, there are three canoes instead of one—each marked with a *tzuk*, "partition," sign. The god Chak stands in the first, holding a stone quatrefoil on his shoulder. Chak, like Itzamna, can be associated with peccaries,[47] but more important, as Karl Taube explained to us, another pot (Fig. 2:27d) shows two Chakob cracking open the turtle shell so that the Maize God can emerge. They are thus attendants of the rebirth. As David pointed out in his account of the rainmaking ceremony in Chapter 1, people personifying Chakob have been the helpers of priests and shamans in Yukatekan Maya ritual at least since the Spanish conquest.[48]

As he rides in the central canoe, the Maize God holds a huge K'an-cross with a turtle shell attached. This combination of symbols corresponds to the K'an-cross Waterlily Monster found at the base of the Foliated Cross at Palenque, and on the pots with the Chakob, on which the sons of the Maize God, the Hero Twins, cavort among the pads and branches of the same plant. Often the turtle carapace from which the Maize God sprouts in rebirth also carries the K'an-cross. David also realized that when the three stones of Creation come to the zenith, this cross sits at the point where the ecliptic crosses the Milky Way at right angles. And we learned from Susan Milbrath[49] that when this configuration comes to zenith at midnight, the sun sits below the earth at nadir. The K'an-cross is a kind of "X marks the spot" symbol of rebirth and Creation.[50]

The beings who paddle the canoe fall into a special category called "sky artist." Following a clue I had given her earlier, my friend Barbara MacLeod brought me inscriptions that described the Paddler Gods as *Na-Ho-Chan Itz'at* (Fig. 2:28). Na-Ho-Chan is the place where they set

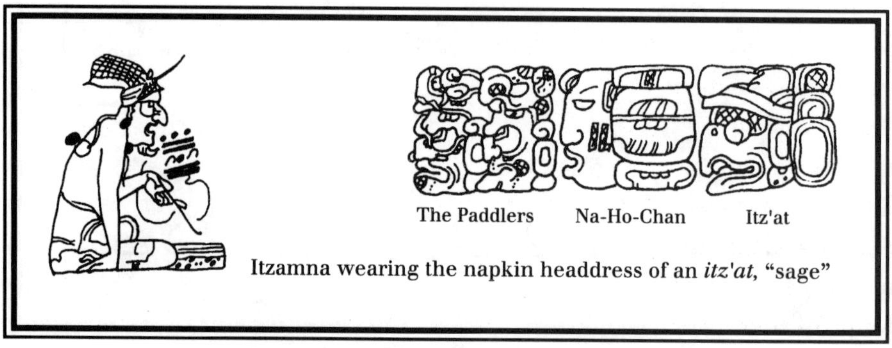

Figure 2:28 The Artists of Creation

up their stone, and *itz'at* is the word for "artist" and "sage." Then Justin Kerr reminded me that Itzamna was also shown wearing the special net headdress of an *itz'at*. All three gods carried the special title of "artist" because they set the stones of Creation, probably by drawing their images in the constellation of Orion. As Barbara pointed out, this *itz'at* title also matches the Popol Vuh's nomenclature for the creator gods: modeler, shaper, sculptor, and mason.[51] **Plate 36**

As all of these things came together with the help of my many friends, I realized that these patterns in the Milky Way and the constellations were directly related to the Maya vision of Creation. Using my computer and my newly made Milky Way clock, I sat down to see what the sky looked like on the two days of Creation, August 13 and February 5. Quite frankly, I was scared. I had found incredible patterns, but if any one of my basic assumptions was wrong, the entire thing would come tumbling down like a house of cards. As we will see, the sky matched the myths perfectly.

I planned to include these discoveries in the 1992 workbook for the Texas meetings and had already written up twenty pages of information. I printed them out and sent them to ten people I trusted to be tough but open-minded critics. I got a lot of feedback from them, and some sage advice from Anthony Aveni, the most respected archaeoastronomer in the field. He urged me to check my ideas out as soon as possible in a planetarium and to make sure that the theory worked in the field. I followed his first suggestion in the company of Justin Kerr at the Hayden Planetarium of the American Museum of Natural History in New York on February 21, 1992, and again at the Hansen Planetarium in Salt Lake, Utah, on May 3, 1992. Everything checked out. I was able to follow his second suggestion in the company of Susanna Ekholm and Duncan Earle at San Juan Chamula, México, on March 3, 1992, and with Nikolai Grube on August 18, 1992. To my relief and exhilaration, the results I had gotten on my computer checked out in two planetariums and in the field with the real stars.

In working these dates, I came to realize that the critical component was the day of the year, regardless of what year it was. The Classic Maya Creation days were August 13 and February 5.[52] However, it seemed fairly clear to me that the Maya hadn't made the connection between these days and the sky constellation in 3114 B.C., their year of Creation. First, it's doubtful there were people we could call Maya around at that time. And if there were, the sky would have been very different in 3114 B.C. because of something the astronomers call precession. Because the axis of the earth wobbles a little, the sun migrates through one constellation on the ecliptic every two thousand years or so. This shift changes the times that a particular constellation would pass the zenith on a particular night or which form of the Milky Way would be visible at a chosen hour. It seemed more likely to me that the Maya, or perhaps their predecessors, the Olmecs, would have wanted the astronomy to work with the myth when they could see it, rather than in a mythological past. I supposed that they had made the connection during historical times—that is, after 1200 B.C.—and then cast it back to the time of Creation. Nevertheless, to be sure, I checked August 13 and February 5 in 3114 B.C., 1000 B.C., 200 B.C., and A.D. 690, the year in which the Group of the Cross was dedicated.

I decided to start with A.D. 690. On August 13 of that year—and all the others, as it turned out—sunset found the Milky Way starting in the position of the Wakah-Chan, but turning into the angled position of the Crocodile Tree in the two hours after dusk when the sky became fully dark (Fig. 2:29). David suggested to me that the bulge at the base of the tree might also have been seen as the severed head of First Father waiting for his sons to come rescue him. But I think the Milky Way, as it became visible in the darkness after sunset, was more likely to have been interpreted as both the Crocodile Tree and the Hero Twins' confrontation with the cosmic bird, Seven-Macaw. This was one of the final actions that prepared the old universe for the creation of the new.

Throughout the night on August 13, the tree turned slowly, transforming itself by half past midnight into the canoe carrying the Maize God to the place of Creation. By 2:30 A.M. the canoe was sinking, and from 4:30 A.M. until just before dawn, the three stones of creation were south and east of the zenith. The Pleiades, which the Maya of today call a handful of maize seeds (also the rattles of a snake), were just to the west of the zenith. Dennis Tedlock also identifies the Pleiades as the four hundred boys[53] of the Popol Vuh who were killed by Seven-Macaw's son.

Thus, the final events of the third Creation were all played out in the sky. At sunset, Seven-Macaw was put in his place; during the hours of the early morning, the Maize God was delivered to the place of Creation. At dawn, he was at the cracked turtle shell from which he was resurrected. From there he rose from the K'an-cross where the ecliptic crosses the Milky Way. There he became the Maize Tree the ancient Maya called the Na-Te'-K'an, First-Tree-Precious. The three stones of Creation were in place close to zenith at dawn. It is significant that in most Mayan

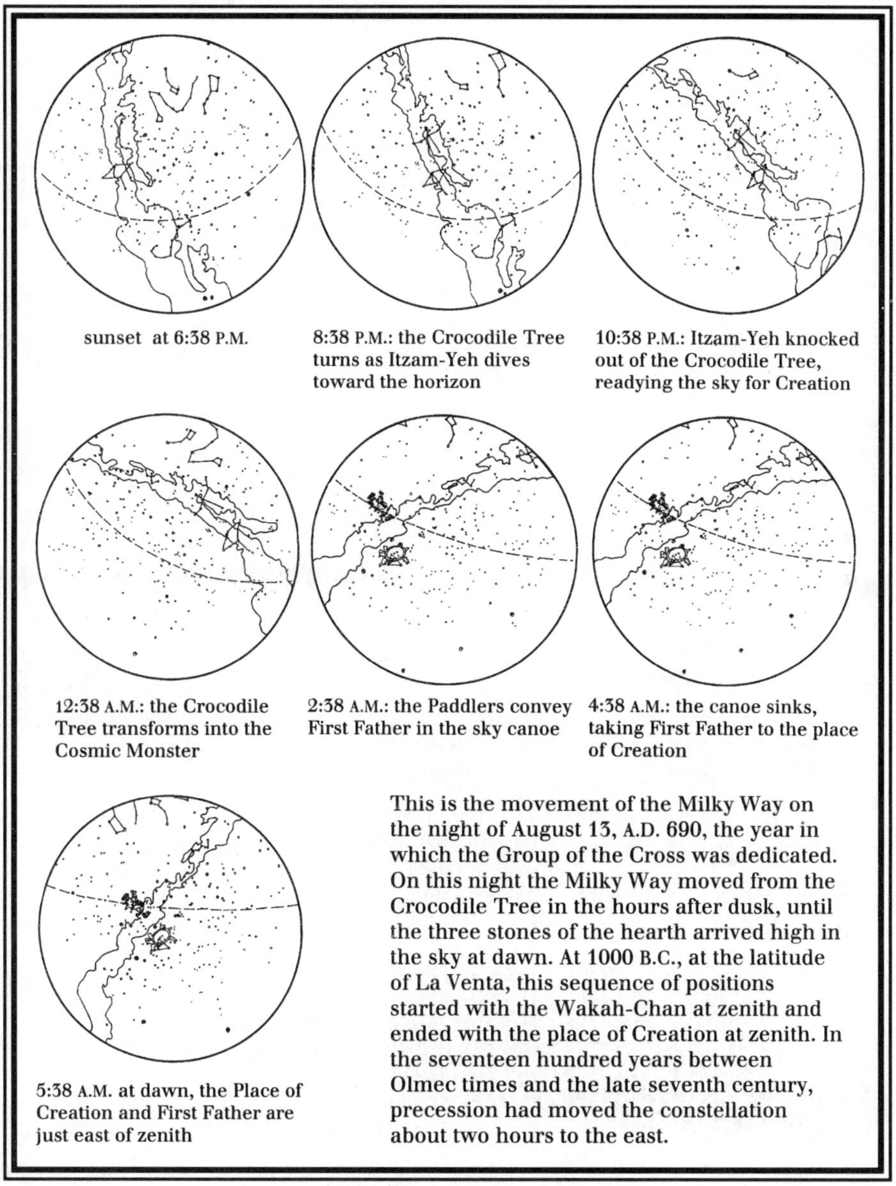
sunset at 6:38 P.M.
8:38 P.M.: the Crocodile Tree turns as Itzam-Yeh dives toward the horizon
10:38 P.M.: Itzam-Yeh knocked out of the Crocodile Tree, readying the sky for Creation
12:38 A.M.: the Crocodile Tree transforms into the Cosmic Monster
2:38 A.M.: the Paddlers convey First Father in the sky canoe
4:38 A.M.: the canoe sinks, taking First Father to the place of Creation
5:38 A.M. at dawn, the Place of Creation and First Father are just east of zenith
This is the movement of the Milky Way on the night of August 13, A.D. 690, the year in which the Group of the Cross was dedicated. On this night the Milky Way moved from the Crocodile Tree in the hours after dusk, until the three stones of the hearth arrived high in the sky at dawn. At 1000 B.C., at the latitude of La Venta, this sequence of positions started with the Wakah-Chan at zenith and ended with the place of Creation at zenith. In the seventeen hundred years between Olmec times and the late seventh century, precession had moved the constellation about two hours to the east.

FIGURE 2:29 The Night of August 13

languages, the word for "to dawn" is also the word for "to create."[54] August 13 was, and is, also the day of zenith passage at Copan and other sites that lie along 15° north latitude.[55] Thus, at noon, the sun, one of the children of the Creator couple, sat in the exact center of the sky.

I then checked the February 5 date in my computer and at the planetarium. I found it was the reciprocal opposite (Fig. 2:30). At sunset the three stones of Creation and the Pleiades lay near, if not exactly in,

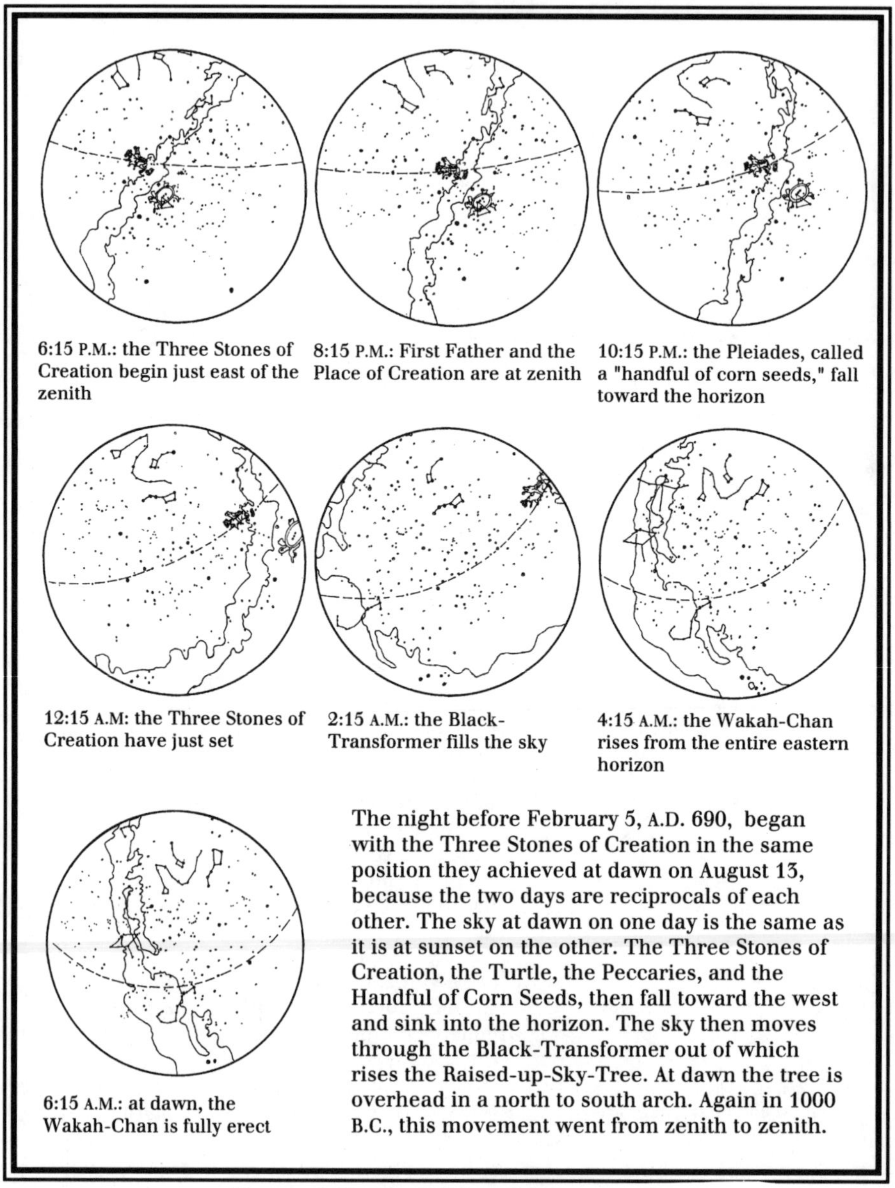

6:15 P.M.: the Three Stones of Creation begin just east of the zenith

8:15 P.M.: First Father and the Place of Creation are at zenith

10:15 P.M.: the Pleiades, called a "handful of corn seeds," fall toward the horizon

12:15 A.M: the Three Stones of Creation have just set

2:15 A.M.: the Black-Transformer fills the sky

4:15 A.M.: the Wakah-Chan rises from the entire eastern horizon

6:15 A.M.: at dawn, the Wakah-Chan is fully erect

The night before February 5, A.D. 690, began with the Three Stones of Creation in the same position they achieved at dawn on August 13, because the two days are reciprocals of each other. The sky at dawn on one day is the same as it is at sunset on the other. The Three Stones of Creation, the Turtle, the Peccaries, and the Handful of Corn Seeds, then fall toward the west and sink into the horizon. The sky then moves through the Black-Transformer out of which rises the Raised-up-Sky-Tree. At dawn the tree is overhead in a north to south arch. Again in 1000 B.C., this movement went from zenith to zenith.

FIGURE 2:30 The Night of February 5

the same positions they occupied at dawn on August 13. By midnight, the handful of maize seeds and the three stones of the hearth had set in the west, and by 2:00 A.M., the sky had changed into the *Ek'-Way* pattern that meant transformation for the ancient Maya, the portal through which things emerged from the Otherworld. By 4:00 A.M., the Milky Way had risen erect from the eastern horizon and arched over the dawn sky. The top of the tree edged past the heart of heaven in the north, while the scorpion sat at the base of the tree in the south. This movement through the night was truly the Raising of the Sky, and you can still see it today.

The last component of this amazing month of discovery came after I had presented all the information above to the 1992 Texas Workshop on Maya Hieroglyphic Writing. A week earlier I had received a cryptic letter from my friend Nikolai Grube suggesting that the famous zodiac pages in the Paris Codex recorded events at Na-Ho-Chan when the universe was made. During the workshop I was too exhausted to think and did not have time to ask him what he meant until the week following.

When we finally spoke, he told me about a black-background vase that has three gorgeous young gods reclining among entwined cords ending in snake heads (Fig. 2:31). The text on this pot recorded a birth that happened at *Na-Ho-Chan-Witz-Xaman*, the "First-Five-Sky-Mountain-North" place. The pages preceding the Paris zodiac have the same black background with entwined cords that link floating beings together (Fig. 2:32). One of these cords is clearly an umbilicus[56] emerging from the belly of the Maize God, so we can be sure that these scenes were associated with death and the rebirth of the Wak-Chan-Ahaw who made Creation happen.

Plate 4

If Nikolai is right, and I think he is, then the pot and the Paris Codex

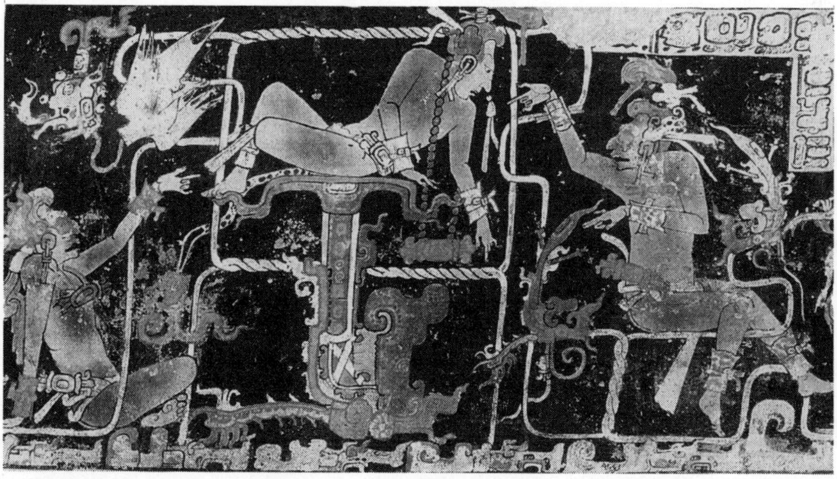

A god is born at a place called *Na-Ho-Chan-Witz-Xaman*, "First-Five-Sky-Mountain-North." The twisted cords are the sky umbilicus.

FIGURE 2:31 The Sky Umbilicus

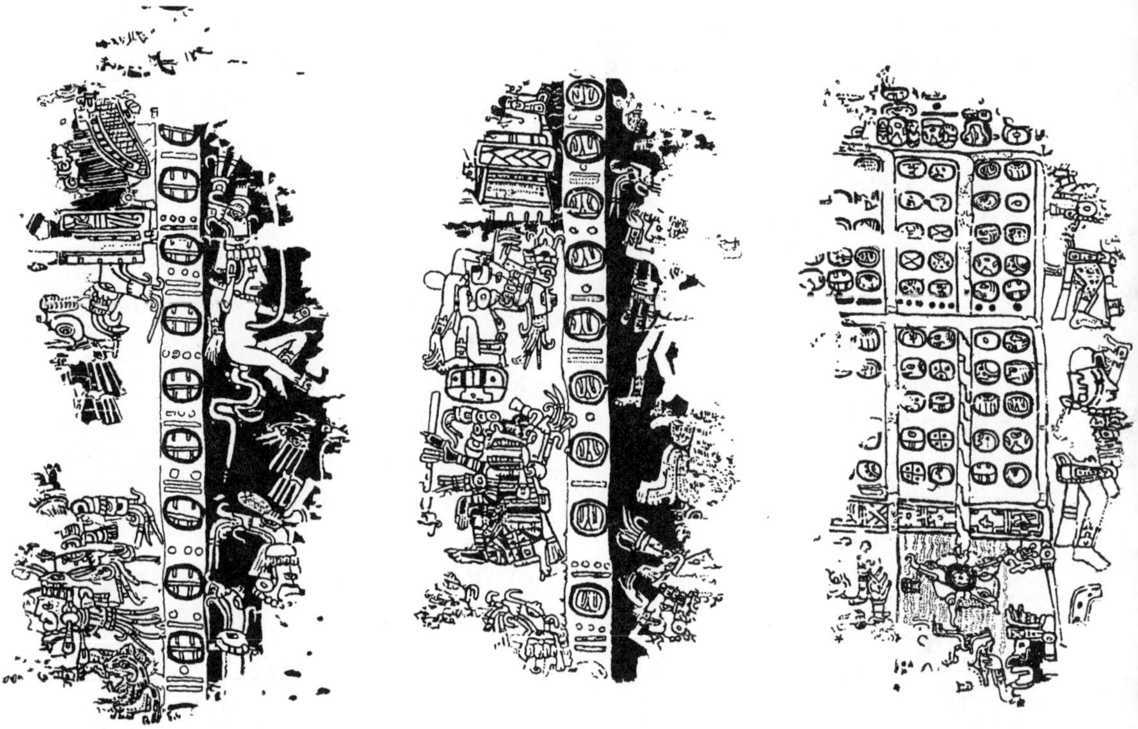

pages 19–20: An umbilicus leaves the Maize God's belly. He carries out various actions with birds and other beasties against a red background on the left and a black background on the right of each page. Nikolai Grube suggested that this place is Na-Ho-Chan. Each page covers 540 days.

page 21: Time is organized inside the umbilicus. The dates included in the grid cover at least forty tzolkins.

FIGURE 2:32 The Paris Zodiac

scenes show both one of the places of Creation and one other critical event—the laying out of the ecliptic. Furthermore, the Paris zodiac may be less an almanac for predicting where eclipses can occur, as David Kelley has suggested, and more a recounting of the laying out of the path of the sun at the time of Creation. Intriguingly, the text above the zodiac includes the phrase "was skyed, three stones," in what I take to be a reference to the three hearthstones set on the first day of Creation. One of the recurrent actors here is Itzamna, who is recorded as a major player at Quirigua, the location of Stela C discussed above.

The story of my own encounter with the Paris Codex zodiac began with yet another suggestion from Germany—this one from the epigrapher Werner Nahm. When Werner read my identification of the Double-headed Serpent as the ecliptic, he faxed support for the idea by informing

page 22: The umbilicus meanders through the sky, while below two Death Gods emerge from the mouths of serpents materializing from the eyes of a monster

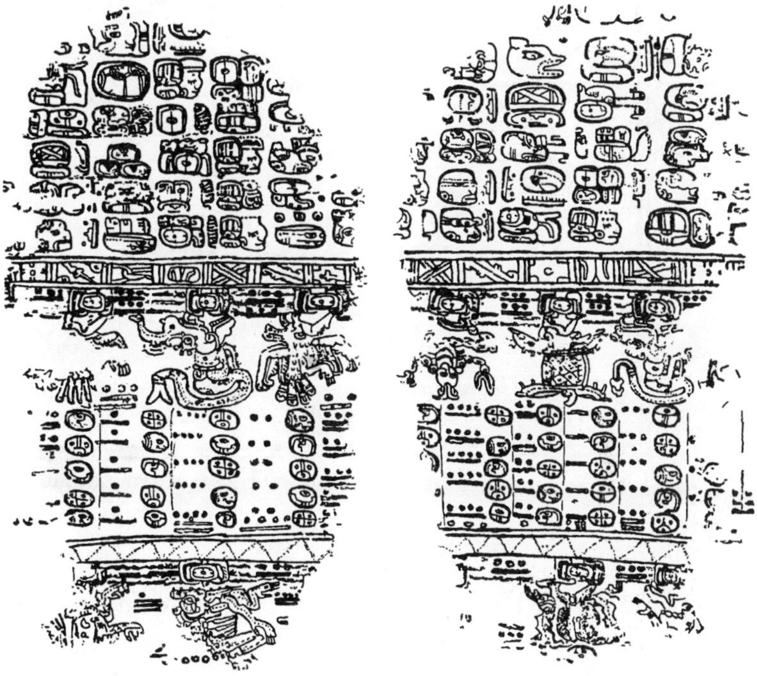

pages 23–24: The constellations are distributed along the ecliptic in 168-day intervals covering a 1,820-day cycle. After three runs, the table was corrected by using the green numbers at the top of columns 2 and 5 and the bottom of all other columns.

me that the little creatures shown climbing a serpent body on the Late Preclassic monument called the Hauberg Stela and on Tikal Stela 1 were constellations. Moreover, he had found that they matched the same order as the zodiacal animals in the Paris Codex. I knew when I read that fax I could no longer avoid the Paris zodiac.

For help in figuring out how the zodiac works, I went to David Kelley's *Deciphering the Maya Script*. He had noticed that there are distance numbers of 168 days between each of the pictures. These intervals served to place the constellations on opposite sides of the sky rather than right next to each other. Since numbers simply do not speak to me, his explanation did not conjure up any pictures in my mind, but I knew how to find these pictures. I went back to *EZCosmos* and plotted out the sky on March 21, 690, at Palenque. Then I added enough days to get the sun to rise in

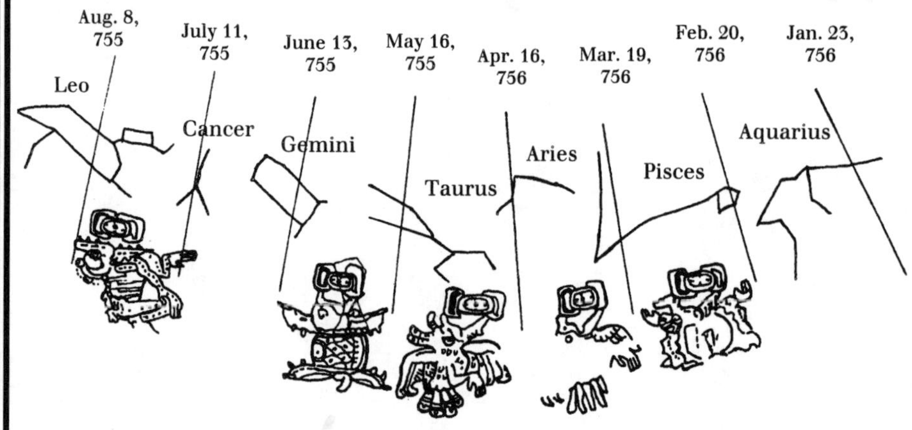

The lines represent the position of the sun at twenty-eight-day intervals beginning with a position on the west edge of Scorpius. The thirteen zones approximate the

FIGURE 2:33 The Maya Zodiac

Scorpius, the only constellation whose identification I was really sure of, and printed out the sky at dawn on that day. I added 168 days and printed out the dawn sky again. I kept this up until I had printed out the full thirteen stations. After tracing each of the constellations in which the sun rose, I matched the drawings I had made to the ones in the codex and had my zodiac.[57] Later, my friends Richard Johnson and Michel Quenon used a slightly different technique and worked out the boundaries of the thirteen Maya constellations (Fig. 2:33).

Then I checked the sky on the dates of the two stelae Werner Nahm had mentioned and found that he was exactly right. At critical times on both nights—midnight and dawn—the Milky Way had been in its north-south position as the Raised-up-Sky Tree, the Wakah-Chan.[58] If, as Werner suggested, we assume that the king corresponds to the Wakah-Chan and that the scorpion is not shown because it was directly on the Milky Way, then the rattlesnake and jaguar constellations are on his right and the fish and the Chak-peccary constellations are on his left (Fig. 2:34). Sam Edgeton, a Renaissance-art historian who has followed up on this work, pointed out to me that the Blowgunner's Pot also follows this pattern. I had forgotten the image was painted on a cylinder. The scorpion is on one side of the tree and a snake is on the other (Fig. 2:7c). These are our constellations of Sagittarius and Scorpius.

David Freidel, with his usual ability to glean the important things I

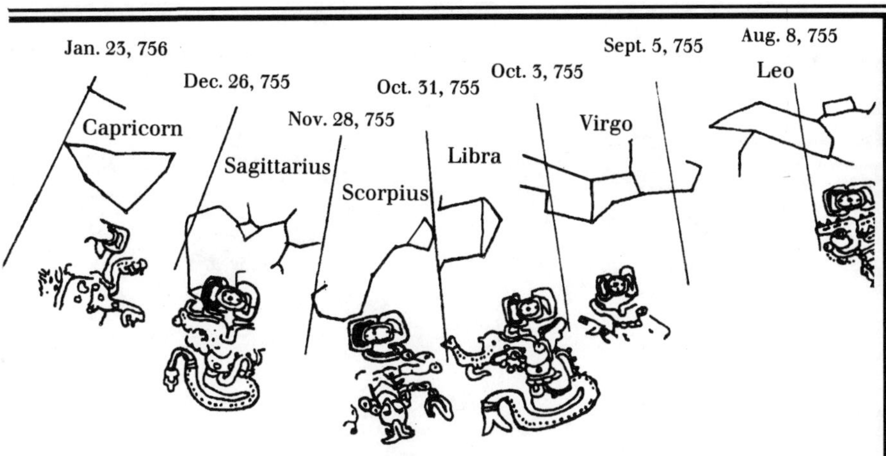

zodiacal sequence used by the Precolumbian Maya (based on a chart calculated by Richard Johnson).

missed, faxed me seven pages of ruminations on the cords of the heavens. As we described in Chapter 1, the modern Yukatek Maya have "sky ropes" attached to their "sky tree" altars which radiate out in four or six directions. He also remembered a riddle in the Book of Chilam Balam of Chumayel in which the heart of heaven is described as cordage, *k'aan*, a near homophone of *k'an*, as in the Na-Te'-K'an. He suggested that the cords in these zodiac scenes are related to heavenly umbilical cords and to the rebirth of First Father as the Maize Lord at the k'an-marked place, where the ecliptic crosses the Milky Way close to the zenith position in the sky. As usual, I resisted his ideas until he struck at my stubbornness by saying, "There has to be something with those cords and they have to be umbilical cords that give birth." As usual he was uncannily accurate in his speculations.

The day we had that discussion, I received a paper from José Fernandez, a young Spaniard teaching archaeoastronomy at Baylor University. José has studied the role of astronomy in the alignment of Utatlan, the capital of the K'iche' at the time of the conquest,[59] finding that all the major temples were oriented to the heliacal setting points of stars in Orion.

He also commented on K'iche' beliefs about an umbilicus connecting the navel of the sky—the North Star, Polaris, to the K'iche'—to the Otherworld. Polaris is the main axis of the sky. Because it never sets, it

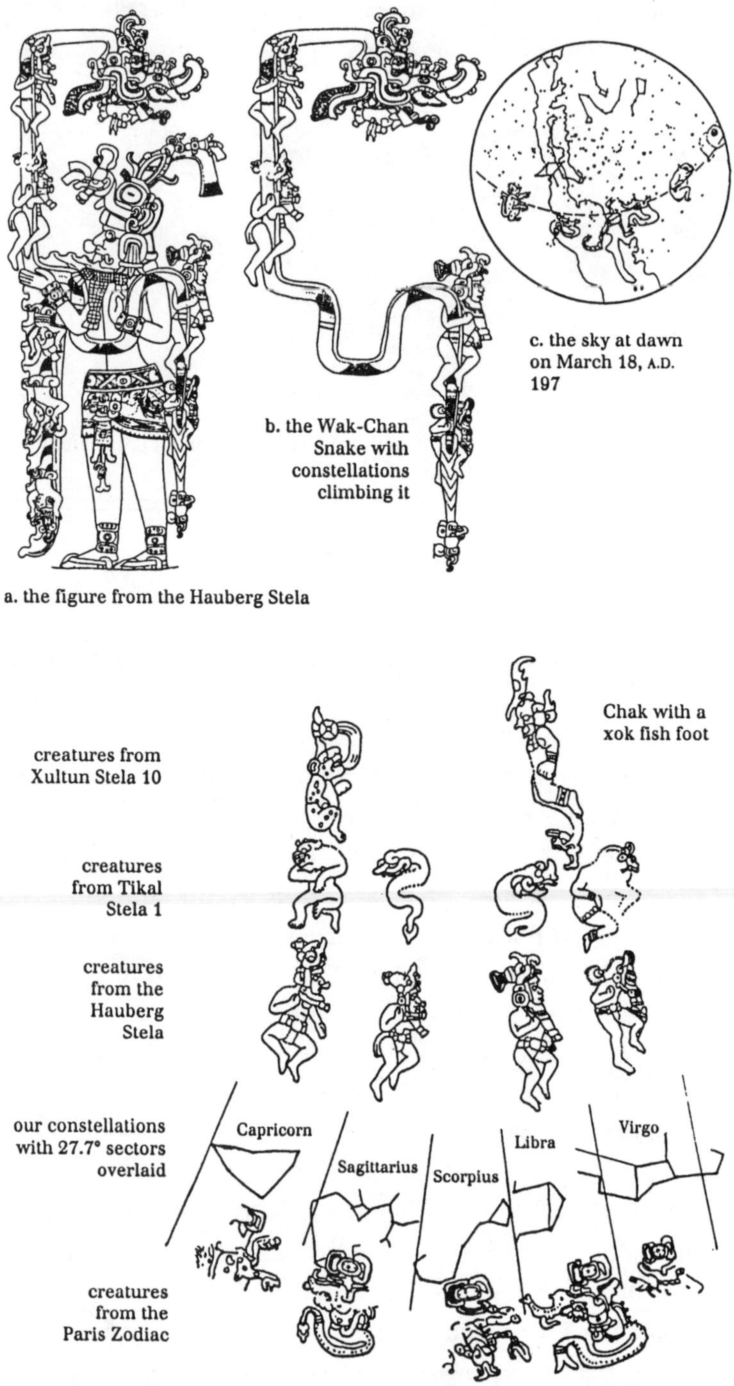

FIGURE 2:34 The Hauberg Stela and the Zodiac

provides the center around which all the constellations rotate. Modern K'iche' also call the Big Dipper "cupped hands." Fernandez believes they conceive of these hands as receiving a newborn sky from the umbilicus stretching from the sky navel. He says that "as the Dippers are rotating endlessly around the star Polaris, they are sustaining the very center of the celestial vault until the end of time." He associates the concept of the umbilicus with the Milky Way, but to us his descriptions of this umbilicus seem very close to the cords of Na-Ho-Chan and the birth scenes of the Paris Codex.

During and before the Classic period, Polaris was not the North Star. Instead, the north polar axis fell in a dark area. This makes little difference to the overall vision that Fernandez described. The Classic Maya apparently called this center darkness the *Wak-Chan-Ki*, "Raised-up-Sky-Heart" or the *ol*, "the heart," but it functioned just like the navel of Polaris. Furthermore, at the 1992 Texas meetings, Barbara Tedlock told us that K'iche' bury the umbilical cord and afterbirth from a female child in the center of the hearth, while a male's umbilical cord is placed in the top of a tree.[60] When we examine these actions against the backdrop of what we know of sky symbolism, we see that the hearth is in Orion, while the top of the tree is in the region of the Wak-Chan-Ki, the raised-up heart of the sky. The cords in the Na-Ho-Chan are the umbilical cords emerging from the Maize God's body at the Orion-Gemini Place of Creation. They also emerge from the Raised-up-Sky-Heart at the north celestial pole. In the Classic Maya mode, they have snake heads at each end. David Freidel suggested to me that these snake-cords could be stretched out to become the Double-headed Serpents of the ecliptic. I think he is right, and suggest that the Paris Codex records the laying out of the constellations along this path.[61] Maybe this is also the action of the Maize God depicted on the Tikal cache vase (Fig. 2:9).

After this I found a set of papers that Peter Dunham, an archaeologist friend of ours, had sent to me just before the Texas meetings in March. As a student, he had compiled information on a sky umbilicus in Yukatekan sources and from modern informants. He quoted an account of Creation recorded by Tozzer (1907:153–154) as follows:

> . . . there was a road suspended in the sky, stretching from Tuloom and Coba to Chich'en Itza and Uxmal. This pathway was called the *kuxan sum* or *sakbe*. It was in the nature of a large rope [*sum*] supposed to be living [*kuxan*] and in the middle flowed blood. It was by this rope that the food

> was sent to the ancient rulers who lived in the structures now in ruins. For some reason this rope vanished forever. This first epoch was separated from the second by a flood called *Halyokokab.*

Karl Taube[62] helped us understand this flood and how it related to the pages in the Paris Codex that Nikolai had associated with the place of Creation. He showed me that we had, in fact, been looking at the New Year's pages of the Paris Codex. 1 Pop, the first day of the Maya 365-day calendar, can fall on only four days—Lamat, Ben, Etz'nab, and Ak'bal—because of the structure of the Maya calendar.[63] These four days are called the Year Bearers. Karl showed me that the columns of days surviving on these pages were two of the four Year Bearers—Ben and Ak'bal. The other two—Et'znab and Lamat—have been destroyed, but can be deduced from the pattern. Each day is repeated thirteen times with numbers arranged in exactly the pattern we expect for a full set of fifty-two years. To read the table you begin in the upper left corner with a now-destroyed 5 Lamat, then read across the top row, which once continued with 6 Ben, 7 Etz'nab, and 8 Ak'bal. Then one reads across the next row below and the next until all fifty-two years in the cycle have passed.

the Cosmic Dragon and Milky Way belching out the waters of the flood

the Old Moon Goddess

God L

page 74 of the Dresden Codex: The flood that ended the last Creation

Figure 2:35

My surprise at Karl's revelation led us into a discussion of New Year's ceremonies in the codices and ethnohistorical sources. He told me they are consistently associated with a destructive flood, just like the one mentioned by Tozzer. In fact, he showed me how the revised pagination of the Dresden Codex places the destructive flood (Fig. 2:35) immediately before the New Year's pages. The Maya—then and now—understood the end of one year and the beginning of the other to be dangerous times when one world was destroyed and another created. It was a replay of the original creation.

This period has been the most exciting and intensive time of my intellectual life. During it I have seen all of the major Maya cosmological symbols woven together into a coherent, eminently logical pattern that is, in essence, a map of the sky. That first discovery of the scorpion sucked everything I knew about the Maya into the "Black-Transformer" and spit it out changed. And most of all, my friends and I affirmed in powerful and dramatic terms that this story of Creation is still meaningful for the modern Maya of today. Creation and re-Creation were and are at the heart of being Maya.

CREATION IN THE POPOL VUH

At the time the K'iche' Maya of the Guatemalan highlands were recording their own version of this Creation myth, the details had changed and been adapted to differences in language, history, and time, but the main story remained the same. Since we recount parts of this great myth in subsequent chapters, we will begin here with a general retelling to flesh out the missing parts of the Classic Creation myth we've told so far.

The Creator Couple in the Popol Vuh are named Xpiyakok and Xmukane. They were responsible for the first and for all subsequent Creations of the universe, or the "sky-earth" as the K'iche' call it. That first Creation consisted of:

> the fourfold siding, fourfold cornering,
> measuring, fourfold staking,
> halving the cord, stretching the cord
> in the sky, on the earth,
> the four sides, the four corners
> as it is said,

by the Maker, Modeler,
mother-father of life, of humankind,
giver of breath, giver of heart,
bearer, upbringer in the light that lasts
of those born in the light, begotten in the light;
worrier, knower of everything, whatever there is:
sky-earth, lake-sea

(D. Tedlock 1985:72)

Next they created the land, sea, and sky as we recounted at the beginning of this chapter, and filled them with animals and birds, who squawked, chattered, and howled but were unable to speak the names of their makers. It so dismayed the Creator Couple that the animals were unable to pray or to keep the days that they limited their dwellings to the wild places of the forest and canyon and told them that their service would be to provide food for the beings who would come later—the beings who would be able to praise and nurture the gods.

Xpiyakok and Xmukane consulted together to create this provider and nurturer, this being who would invoke and remember them. First they shaped people from mud, but the new beings were too soft. They crumbled, became lopsided, and dissolved. Dissatisfied, the Creators made a divination and tried again. This time they made manikins from wood. The wooden people could not remember their Creators and were stiff. They had no feelings and were unable to care for anyone, not even the animals that served them, so the Creators sent a flood to wash them away. They became monkeys, and today are a sign of the previous Creation.

The third Creation is the world of the Hero Twins, Hunahpu and Xbalanke (whose Classical names are One-Ahaw and Yax-Balam). This part of the story begins with the tale of their fathers, One-Hunahpu and Seven-Hunahpu, who were great ballplayers. One-Hunahpu had already fathered one set of twins named One-Artisan and One-Monkey, who were great artists and scribes. One day One-Hunahpu and Seven-Hunahpu were playing ball. The noise of their ball bouncing in the ballcourt disturbed the Lords of Death, who lived just below the playing floor. They summoned the noisy culprits to Xibalba, the Maya Otherworld, to answer for their misbehavior. After defeating the Twins with a series of traps and trials, the Xibalbans killed them. Afterward they decapitated One-Hunahpu and hung his severed head in a gourd tree as a warning to other people. Then the Lords buried their bodies under the floor of the ballcourt in Xibalba and went about their business.

Eventually the daughter of one of the Xibalban Lords got curious and went to see the skull in the tree. When she spoke to the skull, it asked her to hold out her hand. The skull spit into her palm and impregnated her. When her father found out she was pregnant, he was furious. He ordered the owls of Xibalba to kill her, but they took pity on her and helped her escape into the middle world where people lived. After finding shelter with her babies' grandmother, the girl, whose name was Blood, gave birth to twins she named Hunahpu and Xbalanke.

As the boys grew into maturity, they went through a series of adventures that explain how the world got to be the way it is. Eventually, they caught a mischievous rat who had been stealing seed from their milpa, as Maya cornfields are called. The rat purchased his freedom by telling them where their father and uncle had hidden their ballgame equipment.

The Twins were delighted and immediately began to play ball. But they disturbed the Xibalbans below, just as their ancestors had before them. They too were summoned, but before they left for Xibalba, each planted a maize ear in the center of his house, telling their grandmother that when the maize dried up they would be dead.

Hunahpu and Xbalanke took their blowguns and left for the Otherworld, where the Lords of Death were preparing a number of traps for them. But these Twins were more clever than their father and uncle had been. As we relate in Chapter 8, each time the Xibalbans tried to play a trick on them, they evaded the trap. Each day they confronted the Xibalbans in a ballgame, and played them to a scoreless tie. Each night they were put in a different house to face a trial designed to defeat and kill them. But each night they outwitted the Xibalbans and survived to play another game.

Eventually, the Hero Twins made the Xibalban Lords so angry that they knew they would be killed. Realizing what was coming, they worked with Xulu and Pakam, helpful diviners, to plan a way of dying that would let them come back to life. The next morning, the Xibalbans summoned them to an oven filled with a raging fire. Knowing this was the death that they themselves had arranged, the Twins jumped gleefully into the fire and died. The Xibalbans crushed their bones and threw the powder into the nearby river, as the diviners had told them to do. Five days later, the Twins emerged as fish men.

Next they dressed themselves as vagabonds and began dancing and making miracles so fabulous that everyone was amazed. They burned houses that were not consumed and sacrificed animals and people and brought them back to life. Finally, they were commanded to perform before the Lords of Xibalba. Through their miraculous dances, they tempted the Lords into allowing themselves to be sacrificed, and when the Lords were dead, the Twins did not revive them. Thus did they defeat death and banish the Xibalbans from the world of humans.

After chastising the other Lords of Death and limiting their domain to Xibalba, the Twins went to the Ballcourt to resurrect their father and uncle, who represent different forms of the Maize God. Unfortunately, **Plate 42**
when One-Hunahpu's head and body were reunited and animated once more, he could not say the names of things properly. The Twins decided to leave him there in the Ballcourt, where he would be forever worshiped by future generations of humanity.[64]

According to Dennis and Barbara Tedlock, the final event in the third Creation was the Hero Twins' confrontation with Seven-Macaw, a bird who had beautiful blue teeth and metallic eyes. As the Classic texts tell us, Seven-Macaw was so beautiful and vain that he declared himself to be the sun and demanded that everyone worship him. Vanity and pride ran in the family, because his sons Sipakna and Earthquake claimed that *they* had made the earth and could bring down the sky. Offended by this hubris, the Twins decided to teach Seven-Macaw and his sons a lesson. They knew the bird always ate at a nance tree, so they lay in wait for him and shot him with their blowgun. The pellet struck one of the bird's beautiful teeth, which abscessed and caused him great pain. In desperation, Seven-Macaw asked an old grandfather for help. The grandfather said his teeth and eyes would have to be removed. Overcome by the pain, the old bird agreed, and when his teeth were pulled and the metal trimmed from his eyes, everyone saw him for what he was and his greatness left him.

His sons were no better. Sipakna lured four hundred boys into his house and then collapsed it, killing them all. Determined to stop this terrible person from harming others, the Twins built a huge false crab from bromeliads, placed it in a tight canyon, and lured Sipakna down to it. Hungry, Sipakna struggled to get inside the shell to get at the tasty meat, but became trapped as the Twins had planned all along. The mountain settled down on his chest and he turned to stone. Sipakna's brother, Earthquake, was also defeated by the twins.

After their adventures with the Lords of Death and the resurrection of their father and uncle, the Twins rose from Xibalba. It is said that the sun belongs to one and the moon to the other. The four hundred boys killed by Sipakna rose with them and became the Pleiades.

The scene was now set for the fourth Creation—the one in which we are living today. Once again the Creators—the Bearer, Begetter, Makers, and Modelers spoke:

"The dawn has approached, preparations have been made, and morning has come for the provider, nurturer, born in the light, begotten in the light. Morning has come for humankind, for the people of the face of the earth."

As the sun, moon, and stars were about to appear over their heads, Xpiyakok and Xmukane stood thinking about what they would need to create humanity. They knew that true people would need to be made from yellow and white corn, but where to find it? Fortunately, a fox, a coyote, a parrot, and a crow came to help them, bringing news of a **Plate 16** mountain called Split-Place (the Classic Maya called this cleft mountain *Yax-Hal-Witz*, "First-True-Mountain").

The mountain was full of fine foods of all sorts—maize, cacao, sapotes, ananas, jocotes, nances, matasanos, all fruits still eaten in the area today. Xmukane, the First Mother, ground the yellow and white corn nine times. The ground corn became human flesh and the grease from the water in which she washed her hands became human fat.

She molded the first four human beings, who were called Jaguar-Kitze, Jaguar-Night, Mahukutah, and True-Jaguar. They were the first mother-fathers (ancestors) of this Creation: "They were simply made and modeled, it is said: they had no mother and no father."

They were perfect beings. "Perfectly they saw, perfectly they knew everything under the sky, whenever they looked." Like the gods, their knowledge was limitless and they could see through everything, "through trees, through lakes, through seas, through mountains, through plains."

The first human beings turned to their Creators and thanked them. The Creators were pleased that these humans could speak and express gratitude, but at the same time they were a little frightened. These humans could see as well as the gods themselves—all the way out to the four sides of creation, to the four corners of the sky and on the earth. Modeler and Maker did not like this. Human beings were, after all, artifacts created by the gods. Yet the Creators did not want to destroy these perfect creatures they had wrought. Instead, they decided to change them, just a little.

"And when they changed the nature of their works, their designs, it was enough that the eyes be marred by the Heart of the Sky. They were blinded as the face of a mirror is breathed upon. Their eyes were weakened. Now it was only when they looked nearby that things were clear." Thus was humanity given permanent myopia.

CREATION AND THE SKY

When we put together the Creation images and texts of the Classic period with the myth of the Popol Vuh, the miracle of the entire story touched us to our deepest core. The great cosmic symbols of the ancient Maya are a map of the sky, but the sky itself is a great pageant that replays Creation in the pattern of its yearly movements. Sunset on August 13 darkens the sky to reveal Itzam-Yeh (Seven-Macaw) falling from the tree after he has been shot by the firstborn twin, One-Ahaw (Hunahpu). Then the Crocodile Tree transforms into a canoe. This canoe runs east to west, propelled by the Jaguar Paddler and the Stingray Paddler, or by Itzamna, the original shaman. As the canoe sinks under the water, it brings the Maize God to the Place of Creation, the space between Gemini and Orion. The Paddler Gods and other Itz'at of the Sky center the sky by setting the three stones of Creation. This is also the hearth where the first fire is drilled. They draw the picture of the turtle and the peccaries on the earth and the sky simultaneously, at First-Three-Stone-Place at Lying-down-Sky. There also, Hun-Nal-Ye, the Maize God, rises reborn from the cracked turtle shell. The copulating peccaries of Gemini (remember that the Creator Couple in the Popol Vuh were called Great White Peccary, Great White Coati) lie nearby. Perhaps the Maya also once shared the Aztec definition of Gemini as a star ballcourt—the one where the Maize God stayed after he was reborn so that humanity could worship him. As the Maize God is reborn from the Turtle-Peccary of Orion, his umbilicus stretches out to become the ecliptic on which his sons, Venus and the Sun, and his wife, the Moon, will travel through the new Creation. When the day dawns, the first acts of Creation come to fruition.

On February 5, sunset finds First-Three-Stone-Place still at the center of the sky. It sinks toward the west, taking the handful of seeds (the Pleiades) to be planted in the earth. This is also an image of the four hundred boys falling to their death. After midnight, the Black-Transformer appears and from this great sky-wide portal the Wakah-Chan rises, along the entire eastern horizon until it is arching from north to south across the heavens. First Father has raised the sky in the form of the Milky Way, a great tree with its buttress roots in the south. This Wakah-Chan arches north to touch the heart of heaven, the black void that was celestial north in ancient times. When he raised the sky, First Father created a house in the north made of eight partitions. On earth these unfold as the *kan tzuk, kan xuk,* "four partitions, four corners." When all was finished, First Father started the constellations moving in the circular motion "that sustains the very vault of heaven until the end of time"—until the next Creation. The gods wrote all of these actions in the sky so that every human, commoner and king alike, could read them and affirm the truth of the myth.

OTHER CREATIONS IN TIMES MODERN

The Maya of today still affirm the myths of Creation, although they sometimes do not remember how old the roots of that sky tree are. There is diversity in the various forms of Creation celebrated by the modern Maya. The old traditions and myths have fused with the new. David Freidel, for example, found this description of an agricultural ritual practiced by the Mopan-speaking Maya of San Antonio Corozal, Belize. J. Eric Thompson, the famous Mayanist, stayed there and wrote an ethnography of the people in 1927. In the ancient myth, the Wakah-Chan was raised on February 5. The ritual Thompson described took place on February 8, three days later.

> The agricultural year of the Maya of San Antonio may be said to commence in February. Until recent years it was customary for the whole population of San Antonio to repair to the church on February 8 each year for an all-night vigil. This service appears to have no connection with any ceremony of the Catholic church; in fact the Catholic priests were quite unaware that it took place. Within the last few years there has been an attempt to shift this service, which is an intercession for a successful agricultural year, to the first Saturday in February. The instigator of this change was an ex-alcalde, Janaro Chun, who had been educated in Belize, and was opposed to all the old customs that savored of paganism. There was considerable opposition to the change, and since Chun lost his post of alcalde, there has been a return to the old date of February 8. Until recent

> years attendance at the vigil was compulsory, and non-attendants were liable to a fine. At nightfall the whole village proceeded to the church, where fifty or sixty candles were lit. Prayers were offered to the Almighty and Huitz-Hok [*Ah Witz-Hok'*], the Mountain-valley god. The ceremonies were enlivened by the drinking of considerable quantities of rum. At midnight the more devout members of the congregation filed out of the church and, kneeling or standing, offered more candles and copal, as they prayed anew. The prayers have now, unfortunately, fallen into disuse except among a few of the older generation, and some of the people in the remoter alquilos. I was unable to obtain any texts, merely the statement that God and Huitz-Hok were invoked. One informant added that the morning star and the moon were appealed to as well to send good crops. All-night vigils are a part of many Maya ceremonies of aboriginal origin, and except for the fact that this intercession takes place in a church, and that the God of the Christians is added to the old list of pagan deities invoked, there is no evidence of the ceremony being of post-Columbian origin. This is the only ceremony of supposedly Maya origin that is fixed in the calendar. According to the Goodman-Thompson correlation of the European and Maya calendars, the days clustering around the position occupied by February 8 in the tropical year were more emphasized than any other date in the year during the [Classic period]. . . . There seems then a strong possibility . . . [that] this festival still maintained at San Antonio is the same as that observed over 1,200 years ago by the Mayas of the so-called Old Empire in this same general area. Bishop Landa states that either in Chen or Yax the temple of the Chacs was renovated, and their idols, braziers, and pottery were repainted or renewed. At the time of the conquest these two months corresponded to the period January 2 to February 10 (Landa, chap. 40). The Chaks were agricultural gods, and therefore there might be some connection between the modern and the ancient ceremonies. However, in Bishop Landa's time it would appear to have been a movable feast. Among the Lacandones the ceremony of renewing the incense vessels commences in the middle of February and continues until nearly the end of March, but the time depends to a large extent on the ripening of the products of the milpa.
>
> (Thompson 1930:41–42)

The Chortis of the circum-Copan area also mark the beginning of the agricultural year. Alfonso Morales, one of Linda's students, compiled the following descriptions and translations from the work of Rafael Girard.[65]

On February 8 the Chorti Ah K'in, "diviners," begin the agricultural year. Both the 260-day cycle and the solar year are used in setting dates for religious and agricultural ceremonies, especially when those rituals fall at the same time in both calendars. The ceremony begins when the diviners go to a sacred spring where they choose five stones with the proper shape and color. These stones will mark the five positions of the sacred cosmogram created by the ritual.

When the stones are brought back to the ceremonial house, two diviners start the ritual by placing the stones on a table in a careful pattern that reproduces the schematic of the universe. At the same time, helpers under the table replace last year's diagram with the new one. They believe that by placing the cosmic diagram under the base of God at the center of the world they demonstrate that God dominates the universe.

The priests place the stones in a very particular order. First the stone that corresponds to the sun in the eastern, sunrise position of summer solstice is set down; then the stone corresponding to the western, sunset position of the same solstice. This is followed by stones representing the western, sunset position of the winter solstice, then its eastern, sunrise position. Together these four stones form a square. They sit at the four corners of the square just as we saw in the Creation story from the Classic period and in the Popol Vuh. Finally, the center stone is placed to form the ancient five-point sign modern researchers called the quincunx.

Girard emphasized that this action not only starts the new calendar, but also reenacts the Creation of the World: "The Chorti said that this action was done a long time ago and that it is being reenacted by the same gods, who now have different names. These gods do it by 'measuring the space and squaring their measurements.' They explained to me that this was the way it was done at the beginning of the world. The priest in charge 'makes the world' as it was made by the gods the first time and they continue doing it now" (Girard 1966:33).

Later on in this series of rituals, the Chorti go through a ceremony they call raising the sky. This ritual takes place at midnight on the twenty-fifth of April and continues each night until the rains arrive. In this ceremony two diviners and their wives sit on benches so that they occupy the corner positions of the cosmic square. They take their seats in the same order as the stones were placed, with the men on the eastern side and the women on the west. The ritual actions of sitting down and lifting upward are done with great precision and care, because they are directly related to the actions done by the gods at Creation. The people represent the gods of the four corners and the clouds that cover the earth. As they rise from their seats, they metaphorically lift the sky. If their lifting motion is uneven, the rains will be irregular and harmful.

The Chortis, like the K'iche', pay great attention to the Milky Way. Their diviners call it the "Road of Santiago," and they identify Santiago as the god of storms who sends the thunder and commands the rains. He also controls the Milky Way and the rainbow, which are described as giant white and multicolored snakes that move through the sky.

A lot of attention is paid to the positional changes of the Milky Way, especially in relationship to the sun. On the saint day of Santiago (July 25), the sun runs from east to west and the Milky Way from approximately north to south. On that day, the ecliptic and the Milky Way make a giant cross in the sky. This is the Wakah-Chan of the ancient Maya.

This pattern also signals the beginning of the *canícula*, a short period in the middle of the rainy season when the rains cease and the temperature rises. At this time, the Chortis say the rain gods "rest" and make a little dry season. When the Milky Way tilts into the position the ancients saw as the crocodile tree, it signals the renewal of the rainy season. This position coincides with the second of the zenith passages of the sun—August 12 or 13 in the Chorti region. For the modern Chorti, this signals the beginning of the second planting season. For the ancient Maya, this was the day when the sky artists placed the three stones of Creation in the hearth of Orion. August 13 is the day of Creation.

At the end of February 1992, just as the dizzy intensity of discovery was giving way to the more normal pace of life, Linda went to Chiapas, México, to observe the Festival of Games held in the *ch'ay k'in*, the five "lost days," that mark the end of the old year and the beginning of the new year. During that time the people of San Juan Chamala destroy the old world and create a new one. They re-create Creation. This adventure had its own component of serendipity, because at the 1991 Texas workshop, Duncan Earle, an ethnographer who had just that year begun to teach at the rival institution of Texas A & M, mentioned that he knew the outgoing Pasión of the San Sebastián barrio. Duncan had been invited to attend the pageant, and he threw out a half-serious invitation for Linda to come with him, which she half seriously accepted. By the middle of February, as Creation unfolded before her stunned eyes, Linda's intentions became very serious indeed and she bought plane tickets for both of them. Here's what happened.

THE MONKEYS OF THE BEFORE TIME

(as told by Linda Schele)

Duncan and I had arrived in San Cristóbal de las Casas very late on Thursday night, after a hard afternoon in the México City airport waiting for a connecting flight. We found the old colonial house our friend Alfred Bush had offered, got a night of sleep, and puttered around the city until noon finding friends and getting ourselves oriented. After lunch, we took a *colectivo* taxi (a VW van stuffed with Chamulas and two oversized gringos) out to Chamula to find the house of the outgoing Pasión so that I could be introduced to his family and the other officials who were part of his entourage.

During the next three days, we spent most of our time among the Chamulas. Duncan speaks Tzotzil and wears Chamula clothes with the knowledge of one who has worked among them. His blue eyes and light hair always gave him away and the Chamulas kept asking him if he was Lol, the name they gave to Thor Anderson, who had once been allowed to make a film about their Festival of Games.

I could not sink so well into the background, for I am tall even for a woman from our world. Among the Chamula I am a giant who acts funny and does not speak *batzil k'op*, "true human speech." I had another skill, however, that fascinated them. I could draw. They were suspicious of my drawing book at first, but as they gathered around me, they were able to see what I was drawing and began talking among themselves. To them the image "came out" of the page. Through Duncan, I asked many questions, and with those Chamulas who spoke Spanish, I could even communicate. Soon word spread of my strange antics, and the moment I pulled out my notebook and pen, dozens of Chamulas closed in on me until I was packed in like a hot dog in a bun. Before long, I got my Chamula name—*Mukta Antz Chingon* or Large Woman F****r. Believe it or not, the name is a compliment, for they called me that when they admired my drawings.

For three days, Duncan and I watched the Pasiones and the Flowers[66] run the flowered banners that are the Sun-Christ around the square of Chamula. We attended the public banquets and tasted the food the Pasiones provided to all the people of the community. We watched as the banners were fed with clouds of smoke from hand-held braziers. We watched the civil officials feed their *bastones* and join the authority of their offices to the Creation of the world. Most of all, we watched men and boys dressed in costumes resembling those of nineteenth-century French grenadiers cavort through the crowds, playing musical instruments and singing the same song—Bolon Chon—over and over again.

These oddly dressed apparitions are the Max, the Monkeys, who are creatures from a previous Creation. In the Popol Vuh myth, the wooden men of the second Creation, who were dispersed by a flood, turned into these animals. Some of the Chamula Monkeys range free, playing tricks on the crowd, misbehaving in many ways, making off-color jokes, and making sure no foreigner takes an unauthorized photograph.

Other Monkeys are in the service of the Pasiones. We watched them dance the Sun-Christ banners around two clay drums in endless circles for five long days. They call this movement *hoy*, the same word used for the act of dedicating the Wakah-Chan on the Tablet of the Cross. I drew the Monkeys time and again and everyone recognized my images of their high-pointed, beribboned hats. I had read about the Monkeys, but they cannot really be described in words to one who has not experienced them. While I watched them, Duncan's powerful and vivid words brought home to me what they are and what they do. They are pre-cultural beings who tear down the order of the world in order to prepare for its re-creation when the Sun-Christ is brought out of the Otherworld. Duncan explained that I too am classified as a precultural being from another Creation. All of us strange people from beyond the Chamula world are from that past Creation. The wonder of it struck me, for there I was, a being left over from a before-time, watching the Chamulas create the world just as their ancestors had for thousands of years. More than that, I realized I had been thinking of them in exactly the same way. I study the ancient Maya world almost as if it were a past creation of another time. Until that moment, I realized, I had thought of the Chamulas and all of their Maya cousins as beings who had inherited wisdom from that past Creation. We are not so different after all.

One of the most important ceremonies in the five-day sequence of the Festival of Games took place the night before the Fire Run on Tuesday, March 3. Starting just before midnight, the Pasiones and Flowers, who are the principal officials in these rituals, take their people to nearby hills that bear great thirty-foot-high crosses on their summits. They are called Calvarios, and each barrio has one of these sacred summits. I wanted very much to see this part of the ritual and so did our hostess, Susanna Ekholm, a friend and archaeologist who has lived in San Cristóbal for over twenty years. Carol Karasik, a friend of hers who had edited a set of Tzotzil stories and dreams, joined us at the last minute.

Duncan and I spent Monday in easy activities, resting when we could for we knew we would spend the entire night at Calvario. As we ate supper at Susanna's house, we would step out periodically into her courtyard to follow the progress of the hearth stones in Orion across the sky. There was too much pollution from San Cristóbal's lights to see the Milky Way, but Orion and Gemini were clearly and prominently visible.

At midnight, we piled into Susanna's car and drove to Chamula. We were wrapped in multiple layers of clothes to keep warm, but were resigned to being miserably cold and tired. We arrived to find our Pasión and his people running their banners around the square. They were soon joined by the Pasión of San Pedro, who added their ranks to the running crowd. Being a middle-aged gringa, I thought it prudent to leave for Calvario with the musicians, independent monkeys, and fireworks specialists. It was a steep hill. Susanna and I had to stop several times to catch our breath, but we made the summit and had regained our strength by the time the Pasiones and their entourages ran the banners to the summit of San Pedro's Calvario.

We spent the next two hours watching the banners circle the drums, walking among the people, observing the Monkeys feeding the huge crosses. I drew what I saw in the near darkness. Rockets arched high into the black sky, thundering their booms off the hillsides, punctuating the murmur of the crowd and the drone of the Bolon-Chon song. Duncan and I circled through the crowds twice, returning each time to where Susanna and Carol waited. On the third circuit they joined us in our walk. When we saw the fireworks handlers and musicians gathering their things, we joined their exodus to the Calvario of San Sebastián.

Susanna and I wound our slow way up the concrete stairs of San Sebastián Calvario to find a place on the rail overlooking the town and the other Calvario. Chamula threw up as much light pollution as San Cristóbal had, but it didn't make any difference because clouds had covered the stars and hidden the Milky Way from us. Rockets arched sparkling trails toward the clouds while we watched the flashlights of the Pasiones and their entourages leaving San Pedro Calvario to cross the valley past the ruins of the old church and the cemetery and mount the hill where we stood. I imagined how it would have looked when they used torches to light their way. This time two musicians announced the arrival of the banners with the long doleful cry of conch-shell trumpets. The trumpets were not very well made, for the trumpeters had to struggle mightily to make some very uncertain sounds. The drummers arrived with their net-bound water drums, and soon the banners continued their endless, circling path through the Dance of the Warriors.

As we watched the ceremonies and tried to see what was occurring near the crosses, we were surprised at how passive the Pasiones had become. They took no part in the dancing nor did they participate in feeding the crosses. It seemed to us that the Monkeys had taken over and were in charge, but neither of us knew why. Later, Gary Gossen, an ethnographer who has spent his life studying the Chamulas, told me that we were seeing the triumph of chaos over order. The Monkeys had won the battle and banished the Sun-Christ from this world.

We watched for hours. I filled pages in my book with spidery lines hastily drawn in near-complete darkness. Several Chamulas helped me by holding their flashlights on my page. Finally, about 4:30 in the morning, I gave in to exhaustion and sank to the ground next to Susanna, who was sitting on the stool built into my pack.

The cloudy night had disappointed us all, and I had given up, resigning myself to the probability that I would not be able to fulfill Tony Aveni's admonition to check the astronomy in the field. Well, I thought to myself, I'll be back and there will be other times. Suddenly, as we sat there in our exhaustion, eyes heavy and minds filled with cotton, the wind blew away the clouds. The sky opened up before us in all the glory it could steal from the electric streetlights of San Juan Chamula. The light pollution didn't matter. I looked up and saw Scorpius poised exactly behind the huge cross of San Sebastián Calvario (Fig. 2:36). The Wakah-Chan was arching over my head, and while I could not see its shape because of the amount of light, I knew exactly where it was because of the scorpion. As I drew it in my book, one of my kibitzers reached across my shoulder, tapped the image of Scorpius with his finger, and said "Tz'ek." Duncan told me it is the Chamula word for "scorpion." Thus the Chamulas today see the same scorpion in the sky as did their ancestors a thousand years ago.

We drove back to San Cristóbal about an hour later and stopped on the side of the road where we hoped the light pollution would be less intense. We had no luck with that, but I saw something else that I think I was supposed to see. As the sun stained the rim of the night with a huge red arch, filling much of the eastern horizon, we watched Venus hovering in

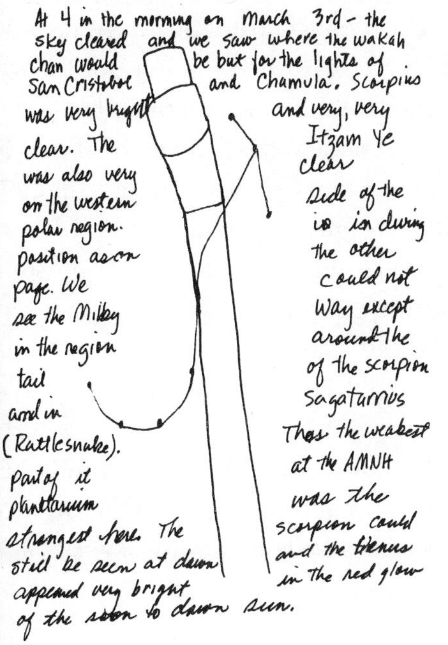

Two old crosses that have been taken down from the Calvario of San Sebastian.

FIGURE 2:36 Calvario and the Wakah-Chan from Linda's journal

the southeastern sky. It was inside the red stain, glittering with a brightness and size that rivaled the scorpion hanging to its right. I turned to the northern sky and saw the Big Dipper, the great bird Itzam-Yeh, diving toward the horizon. Both of these constellations, as well as Orion on the opposite side of night, are visible until the sun peeks its face above the surface of the earth.They are the brightest constellations in the sky.

At first I was disappointed that I had not been in Chamula or some other place in the Maya region on February 5 to see Creation as the ancients had seen it. Later it occurred to me that the mother-fathers had conspired to get me there exactly when I was supposed to be. You see, since the eighth century, precession has moved the constellations just enough to make the night sky on the date I was there correspond to what the ancients saw.

I had to wait six more months to see the tree and the bird in all their glory. It happened with my friend Nikolai at the sacred place of Tixkakal Guardia in southern Quintana Roo, a place we will discuss in the next chapter. Nikolai was living in nearby Señor, the town established by the Cruzob Maya early in this century. On the second night of my visit, we waited until the full darkness of night had banished the light of the sun, and drove out of Señor halfway to Tixkakal in order to escape the electric lights of the town. There among the vegetable fields planted by the Maya as a cash crop, we saw the sky in its full glory. It was August 18, 1992, just five days after the Creation day. We got out of the car and let our eyes adjust to the dark. There it was, arching across the sky from south to north—the great tree in a crystalline glory more powerful than anything I had visualized in my mind. The scorpion blazed away in the south next to the rattlesnake in Sagittarius, and below them the Milky Way spread outward in the form of the buttress roots of the tree. To the north, I saw the great Bird diving toward the horizon as it had at Chamula half a year earlier, but this time it was surrounded by the sparkling light of a billion stars. In the center of the sky above our heads, we saw the dark cleft that the Maya read as the head of the crocodile. The tree, the scorpion, the crocodile, the bird—all of the images I had seen on computer screens, on the pages of astronomy books, and in the pictures created by the ancient Maya, were there. And they were so much bigger than I had imagined for I had grown used to seeing them at the scale of a page I held in my hand. It truly was the Path of Awe painted with all the brilliance the Itz'at of the Sky had originally given it.

CHAPTER THREE

CENTERING THE WORLD

THE NAVEL OF THE WORLD

(as told by David Freidel)

Evon Vogt—"Vogtie," as his graduate students affectionately called him—stopped our car at the summit of the pass and got out. The dust swirled around us as we stretched our cramped legs and followed him to the edge of the bumpy dirt road. I squinted until the air cleared, looked around, and wondered why we had stopped here. Below us lay the valley in a broad swathe of checkerboard green that rose westward until it was lost in the forested mountains of highland Chiapas. At the far end of that vista, I could just make out the red houses and white churches of the town.

Smiling with delight, Vogtie motioned us over to a roadside cross made of rough wood. Our teacher looked relaxed and natural in the black wool mantle of a Zinacanteco.[1] In his hand he carried a bottle of clear, home-made liquor, *posh*, the local lubricant of ritual life. Handing each of us a small jigger of liquor, he instructed us to first spit out a little for the Earth and then swallow the rest in one gulp. Then he turned and gazed at his beloved *htek-lum*, "town." His students called it "Z center."

"Here at this cross," he said, indicating the shrine in front of us, "we enter the world of Zinacantan. This is the foot of the sacred mountain, *muxul vitz*, and up there at its summit"—he pointed upward toward a promontory above the road, where the pine trees met the powder-blue sky—"is another cross shrine. Over there," he said pointing to the east and south, "is *bankilal muk'ta vitz,* Senior Large Mountain."[2]

As I stared at the breathtaking volcano that towered over nine thousand feet above sea level, Vogtie explained the Zinacantecos' belief that it housed the corrals where the Father-Mothers, the ancestral spirits guarding the village, keep the more than eleven thousand wild-animal companion spirits of the people. Swinging his arm southward, he indicated the other mountains and shrines ringing the valley: *sisil vitz,* the steep-sided pyramid-shaped mountain; *kalvaryo,* the great shrine that lies along the road to the outlying settlement of Nakih; and so on, all around the valley. "Each mountain has its ancestral gods and its doorway cross shrines where the 'seers,'[3] the shamans, speak for the people." Plate 1

Then he pointed down into the valley below our feet. "There," he said, "is *mixik' balamil,* the navel of the world."

"Where?" I said.

"There, that little hill," he replied.

Puzzled, because I really couldn't be sure where he was pointing, I finally decided he must mean the cross shrine near the edge of the village. We got back into the car and proceeded over the bumpy road into the small town, the center of the larger *municipio* and home to a cadre of professional ritualists and a larger number of temporary religious and civic officeholders, or cargo bearers, as they are called. We toured the many crosses of the town; admired the Church of San Lorenzo, the Church of San Sebastián, and the Chapel of Esquipulas, the Christ who bartered for the souls of the Zinacantecos. We also visited the great *bolom ton,* the jaguar rock where a Chamula boy feigns sacrificial death every Fiesta of San Sebastián.

But I could not understand why, with all the elaborate forms that the sacred geography of Zinacantan could take, that the navel of the universe was a little nondescript bump outside the town (Fig. 3:1), truly an earthen belly button.

This was my first encounter with the sacred world in which the Maya live. Clearly the small size of the *mixik' balamil,* the center of the world, was not a problem to Evon Vogt or his friends, who knew a lot more about such things. I put the matter away in the sock drawer of my mind, the place where I keep a disorganized collection of handy thoughts that might occasionally generate an insight or two. In the years afterward, the

FIGURE 3:1 The Navel of the World at Zinacantan

"belly button of the world" surfaced from time to time, but it was not until my first large project at the ancient Maya city of Cerros in Belize that I met the mystery once again, face-to-face.

The smallest and least impressive pyramid at the Preclassic center of Cerros, a building dwarfed by a cluster of massive acropolises, turned out to be the pivotal building that anchored everything else in this coastal community. As the kings of Cerros expanded and elaborated upon the sacred precinct of their town, they shifted the active, accessible, and presumably public ceremonial space outward to the south and east of this little founding pyramid and plaza group. I discovered the nature of this original conceptual center (Structure 5C) by accident. Because the pyramid was small, I decided to document it in its entirety, to clear a large portion of the earth and rock covering its surface. My students and I subsequently stumbled into a fantastic discovery: the carefully buried stucco facade of an early royal temple. That small mound turned out to be the heart of the town, dating from its most ancient beginnings as a royal capital.[4] Like the *mixik' balamil* of Zinacantan, it was a small, unimpressive bump off to one side of the sacred landscape.

When I returned to the Maya area to undertake my second major research project, at Yaxuna in Yukatan, I had a similar experience. One of my first jobs was to direct the layout of a survey control grid on a cruciform axis encompassing the ruined city. Standing at the approximate center of the group of pyramids, on the east-west axis, I searched for a line of sight that could become the north-south axis. A large acropolis, hard to climb and difficult to survey, stood in the way of the sight line some distance to the north. I didn't want my crews struggling to chop a straight line up and down the steep sides of that acropolis, so I edged to the west a little to miss it with this main survey trail. When at last I found the clear line of sight I needed to stake the north-south trail, I shoved the long metal survey pin into the ground. And as I looked down, I saw a large black spider in an enormous web right next to my hand. The spider had bright orange markings on its back and eight legs clustered into pairs radiating out toward the four directions.

I called to Don Sefarino, who was chopping brush on the east-west trail nearby, and asked him what the Maya called this spider.

"It's an *am*," he said, "quite poisonous."

In spite of that, I took the spider as a good sign. *Am* is also an ancient name for Maya divining stones of the sort used in casting lots and foretelling the future. The spider stones of antiquity also helped shamans map out the four perimeter points to open their portals.[5] Modern shamans like my friend Don Pablo still use a variety of magical stones. Yukatekans use their clear "stones of light" to discern the true location of the four corner points in an area that must be cleansed of evil or made ready for ceremony. My survey eventually revealed that the true center point of ancient Yaxuna was a small stepped platform dwarfed by a huge nearby pyramid and that the survey benchmark and the black spider were just a few feet north of this little pyramid.

OTHER KINDS OF CENTERS

Since the Spanish Conquest, most Maya towns have been stamped with the grid of the European worldview—straight streets, and churches and government buildings arranged around a square. But Vogtie showed his students, Freidel among them, how the patterns intertwine with the

European overlay; how the metaphysical dimensions of the Maya world, the boundaries of the four directions and the center, exist in relation to one another. How the wild world of the forests, mountains, and ancestral abodes and the tame world of homes, churches, and community are woven together in the pilgrimages of the shamans, or the *h'iloletik*, as the Zinacantecos call them.

A few years after the experience Freidel describes above, Vogtie committed his perceptions of the Zinacanteco cosmos to paper in one of the clearest analyses of Maya reality ever written, *Tortillas for the Gods*. There he explained how the center relates to the four directions:

> Houses and fields are small-scale models of the quincuncial cosmogony. The universe was created by the VAXAK-MEN,[6] gods who support it at its corners and who designated its center, the "navel of the world," in Zinacantan Center. Houses have corresponding corner posts; fields emphasize the same critical places, with cross shrines at their corners and centers. These points are of primary ritual importance.
>
> (Vogt 1976:58)

We now know that the first act of Creation was to center the world by placing the stones of the cosmic hearth. The second was to raise the sky, establish the sides and the corners of the cosmic house that is the sky. The Maya at places like Cerros, Yaxuna, and Zinacantan have been centering the world and creating the four sides ever since. The center could be grand both in scale and execution, or like the navel of a human being, it could be a faint, vestigial marker of the remains of the umbilicus that was once connected to an original source of creation and sustenance. It could be created by ritual wherever the Maya needed one. Each household shrine in the outlying hamlets of Zinacantan is central to the family that worships there. Each water hole shared by families living together for generations is central to their lives. Each of the great mountain homes of the Father-Mothers is central when its crosses are adorned with pine tips and carnations, the offerings are arranged, the portals are open, and devout descendants kneel before them in prayer while the ancestors partake of the offering meal. In fact, Vogtie told us, the three peaks of Senior Large Mountain, the most important of these mountain shrines, are called the three stones of the hearth.

Within the sacred mountains live the ancestors, accessible at their cross-shrine portals. Below is the "land warmed by bones"—the deep earth that rests within the dark embrace of the Earth Lord, *yahval balamil*, who is powerful, dangerous, and potentially evil. The Classic

Maya thought of these underworld beings as manifesting in disease and affliction. The Earth Lord of the Zinacantecos, not too surprisingly, takes the material shape of a fat, rich Ladino, the colonial oppressor incarnate. The weak-willed sell their souls to him for temporal fortune, only to serve, after death, on his netherworld cash-crop farms until their iron sandals wear out. High above the land of the living, the saints preside in their heavenly chapel homes, allies and protectors of the devout.

In the highest levels of heaven dwells Almighty God the Sun who traverses his flowery path across the sky once a day. The rising of the sun is the daily affirmation of the dynamic and participatory presence of beneficent spiritual forces in the lives of the people.[7] This general concept is universal among the Maya. The sun is so central in the mythology of the Tzotzil Maya that they believe north and south, the "sides of heaven," were first defined when the sun made its original journey across the cosmos. To these contemporary Maya all the directions have sacred properties. West is the entryway into the earth where the Sun-Christ had to go before he could rise in triumph. South is nadir, the darkness where the Sun-Christ first traveled before arising from death in the east. Ascending to the zenith in the north, Christ slew his mythological enemies with his curative heat (Fig. 3:2).[8]

For the Zinacanteco Maya, the central point where the horizontal and vertical axes intersect, the *mixik' balamil*, is the navel of the world, the belly button whose very name evokes the image of a life-sustaining cord traversing the layers of the cosmos, connecting humanity to the gods, the source of life, and the gods to the human sustenance they require—processions, prayers, and offerings that flow to them when the Otherworldly portals are opened. The Precolumbian Maya represented this conduit between the supernatural and human worlds as a snake-headed cord that emerged from the belly of the Maize God and the sacred place they called Na-Ho-Chan. Classic Kings carried it in their arms in the form of the Double-headed Serpent Bar. The descendants of the Maya who fought the Caste War of Yukatan call it the *kuxan sum*,[9] and they believe that it was cut by the Spanish invaders. Old men who spoke of this umbilicus with our friend Nikolai Grube said that it lies dormant under the Ballcourt of Chich'en Itza. One day, they believe, a Maya king will reign again, and when that happens, the cord will emerge from the great cenote and join the Maya once again to the original source of sustenance.

Just as the gods marked the periphery by placing the four sides and corners[10] around the center, the Maya shaman creates a five-part image

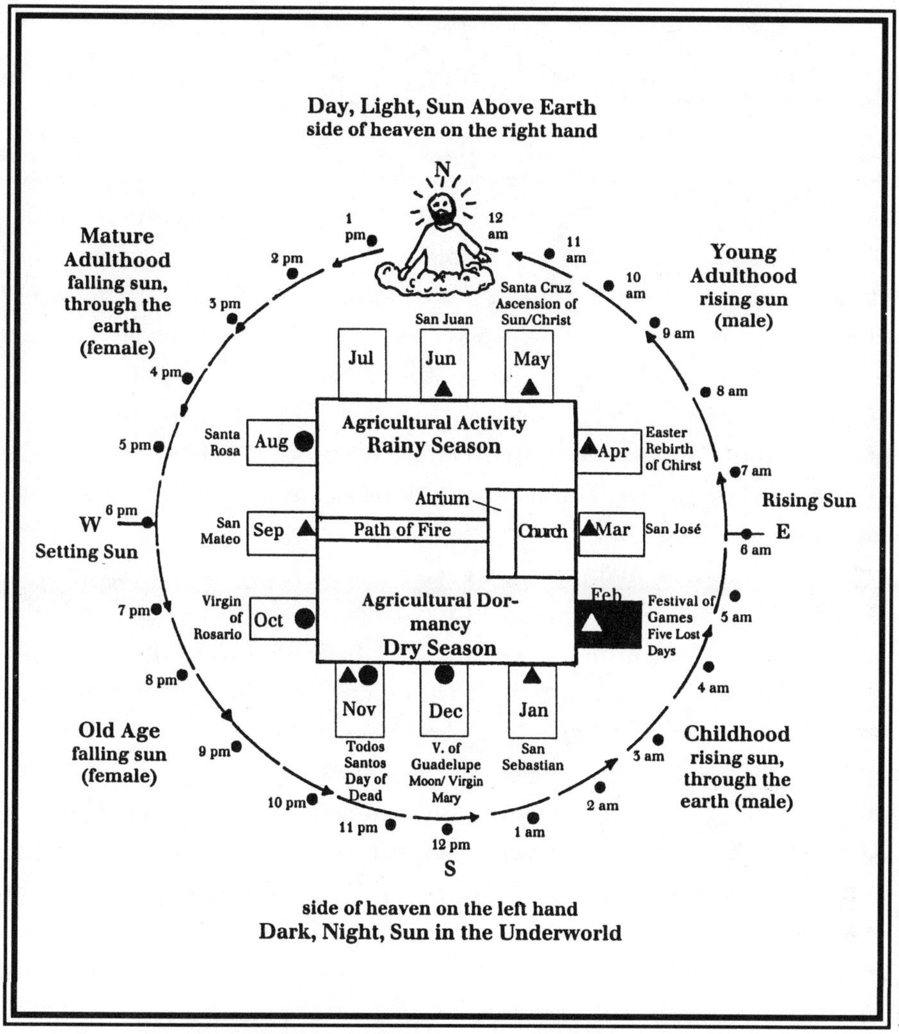

(after Gossen 1986)

FIGURE 3:2 The Chamula Model of the Cosmos

to sanctify space and open a portal to the Otherworld. Mayanists have adopted the Latin word *quincunx* for this five-point-plan concept, although the Maya have many ways of expressing it in their own languages.[11] The discerning of the four sides or the four corners and the establishing of their position relative to the center point is what we mean by "centering." The Yukatek farmers today "center" their fields ritually even before they begin to cut them out of the fallow brushland.[12] They mark off their fields and the units within them with small piles of stones, just as villages mark off their lands from those of neighboring communities with large piles of stones.

The very act of preparing a plot of land for growing food—the clearing and measuring out of rectilinear spaces—echoes Creation mythology thousands of years old. Before cutting down the trees and brush, a devout Yukatek farmer will make offerings at the center of his field. His field has four corners and four sides like the original order established at Creation. The farmer centers the field by piling up the stones to mark the center—properly a layer of three followed by a fourth and then a fifth one stacked on top. This centering transforms the land from wild forest to cultivated land. Like his wife who starts the day by lighting a fire in the three-stone hearth of the house, the farmer repeats the acts of Creation first enacted by First Father when he set up the first three stones of Creation to establish the cosmic center. He marks the corners and sides of his field, just as First Father lifted up the sky and created a house with four sides and four corners.

The Maya field and house are analogs of these cosmic structures. William Hanks (1990:349) says, "Altars, yards, cornfields, the earth, the sky, and the highest atmospheres are described in terms of the five-point cardinal frame." According to him, these concepts are built into the very language itself. Thus, the basic work of making the world livable—building houses, planting fields—is the everyday experience of all Maya; and it is the same work that the gods undertook at the beginning of everything. These ideas are woven together in the quincunx pattern so prevalent in Maya imagery and symbolism. The Classic-period glyph that included this quincunx pattern in its center reads *be*, the word for "road" or "path." Hanks's shaman informant says that he "opens the path" when he lays out the cardinal locations on his altar.[13]

In present-day Yukatan, Maya divinities and spirits regularly conform to this five-point pattern of the cosmos. The *balamob* or jaguar-protectors,[14] the *babatunob*[15] or sky-bearers, and the *chakob* or rain gods are all fourfold beings associated with the four directions. The jaguar-protectors are the most intimately involved because they operate at the level of the human landscape. In the Classic system, the Chakob, the K'awilob, the Pawahtunob, and many others were also fourfold gods.

This quincunx view also has a part in healing a Maya house of afflictions. Hanks describes how Yukatek shamans use their crystals, their "stones of light," to discern where evil is located in the domestic space. Called a "solar," this space includes all the buildings and grounds inside the family's stone-walled enclosure. First, the shaman "fixes earth," *hetz luum*, because it is the earth that is in need of treatment, irrespective of the location of the afflicting spirit within the yard.[16] He causes his crystals to "dawn" or become illuminated so that he can see something he calls *butz'* or "smoke" inside them. This "smoke" identifies the afflicting spirit and "corners" it in one of the four corners of the house-lot space. To contain this evil, and then force it out of the family space, he raises guardian spirits in the four cardinal directions[17] by pointing out the boundary stones to each guardian, except for the one who resides in the place where the evil spirit is "cornered." When the shaman is ready, he gets the guardians to "drop" the evil spirit and to cast it out into the wilderness where it can be locked into an abandoned underground place called a *chultun.*[18] As we shall see, the conjuring of spirits is a very ancient Maya practice indeed.

Centering the world is thus a way of re-creating a spatial order that focuses the spiritual forces of the supernatural within the material forms of the human world, rendering these forces accessible to human need. Because centering the world requires movement to, from, and around the designated center point, the processional route humans use to define the center is as important as the center itself. The traditional label "ceremonial center," whether it refers to modern places like the town of Zinacantan or to the pyramids and plazas of the ancient cities, accurately reflects the function of these places. These locations are not so much centers *for* ceremony as they are centers *because* of ceremonies performed in them by ritualists who center the world each time they create sacred space and open the portals to the Otherworld.

This work of creating centers, of marking off their corners, of encircling them in order to "bind" them up,[19] of moving in and out of them, has an effect on the shape of time as well as space. The ancient Maya were experts in discerning complex and intricate patterns of repetition and symmetry in both human time and cosmological time—the movements of the planets across the house of heaven. They codified these patterns into dozens of calendrical cycles. The days when these cycles overlapped formed a matrix of complex ritual in which the rhythms of village life, elite politics, intercommunity warfare, trade, and interactions with Otherworld beings played themselves out. The famous Maya fascination with time is no more than a preoccupation with discerning and codifying the patterns that give time and space meaning.

THE OLMEC AT LA VENTA

The Maya not only centered their world through ritual in their homes and in their fields; they also replicated the sacred landscape of Creation in their cities. They did not, however, invent the idea of centering or the components of that sacred landscape. Instead, they inherited these ideas from an even older civilization called the Olmecs. Our guide through their world has been Kent Reilly,[20] a colleague who has penetrated many of their mysteries. We draw upon his work for much of our discussion of Olmec cosmology.

For the Olmec, the place of Creation was a huge volcano today called San Martín Pajapán. Located in the Tuxtla mountains on the Gulf Coast of Veracruz, this volcano towers above its neighbors and dominates the land around the sacred lake of Catemaco. The Olmec who lived in the region knew these volcanic mountains intimately; they quarried the dark, fire-born stone to make their sculptures. No doubt they had witnessed the frightening miracle of the earth bleeding stone in slow, molten sheets from the craters on top of the great cones. Volcanoes were, in the Olmec experience, the clearest example of the world being born out of the Otherworld below. No people who have seen the sky turn black in billowing clouds of eruption and then rain stony fire and desolation onto the fertile, surrounding countryside could doubt that mountains contain spiritual forces capable of dispensing prosperity or disaster in human lives. Perhaps because of this, volcano and cleft mountains are a prominent feature in Olmec art. They carry sprouting vegetation and represent openings between the earthly plane and the world below it, and between this world and the Otherworld. This, we think, is the Olmec prototype for the Maya clefted First-True-Mountain that we mentioned in Chapter 2 as the place where humanity was first created out of maize.[21]

In 1897, an engineer surveying the region[22] found a life-sized statue sitting in the saddle between the two highest peaks of San Martín's crater. The statue (Fig. 3:3) depicts a kneeling Olmec ruler wearing a headdress decorated with a cleft-headed god and a maize plant. Kent realized that this god is grasping the trunk of the World Tree, ready to lift it into an upright position. Like the Maya First Father, the Creator of the Olmec cosmological imagery raised the sky away from the earth by setting the World Tree upright.[23] For the Olmec, the ordering of the earth and sky

FIGURE 3:3 The figure from the crater atop San Martín Pajapán

apparently took place atop the great volcano that was the source of creative force within their world.

The Olmec who lived in a town today called La Venta built a replica of this creation mountain and surrounded it with equally potent cosmic structures. Built between 1000 and 600 B.C., La Venta sits on a small island in the Tonalá river in the steamy, hot swamplands of coastal Tabasco.[24] Although the rulers of La Venta created their city in a swampy, stone-poor environment, their people met the challenge of their environment with ingenuity and determination. Their master builders worked with what they had, raising platforms and excavating sunken courts by shaping the natural sandy soil around them. They capped floors and walls with brightly colored sands and clays. They transported hundreds of tons of basaltic stones, cut from the volcanoes of mountains. Floating huge stones down rivers and through swamps, they dragged them to their final resting place in La Venta, where they were carved into images of gods and kings. The largest of these stones weighed a hundred tons, and the moving of it was a marvelous engineering feat. From even farther away in the mountains of southern México, these lords imported vast quantities of precious greenstone and serpentine, which they arranged into massive, thick floors composed of many layers of brick-sized stones. These huge mosaics created power patterns, focusing cosmic forces within their architectural designs.

The Olmec of La Venta built an effigy volcano at the southern end of

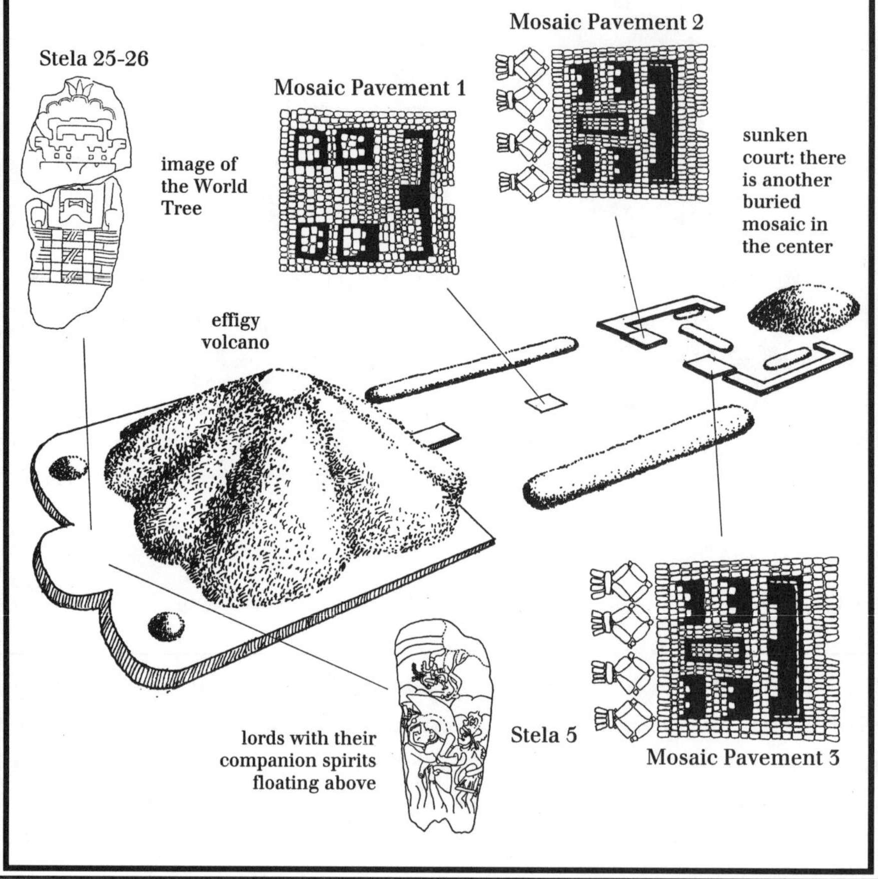

FIGURE 3:4 La Venta Complex A

the set of buildings known today as Complex A (Fig. 3:4). They used local clay to create a huge earthen pyramid and fluted its sides[25] to make it resemble a volcano. But the La Venta pyramid was not just a replica of a volcano; it was intended to echo the mountains made by gods and to show that a similar thing could be made by human hands. It was a fiery portal to the Otherworld, harnessed to the will of living kings. The Maya who came after also built their pyramids to resemble sacred mountains and called them by that name.

Extending the symbolism further, the lords of La Venta embedded huge stone stelae into the base of their ceremonial volcano-pyramid. One portrayed lords conjuring their companion spirits, while another depicted the World Tree. Like the Maya who came after them, the Olmec portray their kings contacting the Otherworld and they gave physical substance to the central axis of the world. Their stelae were meant to represent beings who stood at the corners and the center of Creation. The stelae often represented their king as the embodiment of the World Tree. We don't know what the Olmec called their stelae, but the imagery they carved on at least one of the stone slabs they erected at the base of this human-made volcano at La Venta showed they also represented cosmic trees.

At the north end of Complex A, opposite the volcano (Fig. 3:4), the builders shaped a court enclosed by clay walls. These walls symbolically cornered the space and made it into a sunken plaza—a place "beneath the waters" of the Otherworld.[26] Throughout its construction history, the court had three small platforms arranged in the triangular form of the three stones of Creation.[27] At the center of this arrangement was an enormous pavement made of brick-sized serpentine blocks. In the final phase of this plaza's construction, the Olmec builders added two large adobe platforms on either side of the entrance into the court and placed inside them massive deposits of serpentine. Built by the citizens of the city as an act of devotion,[28] this greenstone gateway at La Venta was sacred power incarnate—its modern analog would be an entrance into the heart of a nuclear reactor.

Teams of men laid down the heavy serpentine blocks in layer upon layer, forming deposits that were meters deep. Finally, they capped each of these platforms with a final mosaic symbol that directed its massed energy down into the Otherworld below to fulfill its specific function—the conjuring up of supernatural beings. Kent realized that these images were the Olmec version of the Maya quincunx.[29] The mosaics, placed as they were on either side of the entrance, established a physical gate, a threshold joining the sunken court to the plaza path. This path, in turn, spanned the space between the sunken court and the volcano. Attached to the southern edges of the fivefold mosaic patterns capping the serpentine offerings are fringed, diamond-shaped tassels representing aquatic vegetation and flowers floating on the surface of water.[30] People who passed through this serpentine-empowered gate into the sunken court entered onto the surface of the Primordial Sea of the supernatural Otherworld.

Two long, narrow mounds formed the path, an alley between the volcano-shaped pyramid at the south end of Complex A and the sunken court at the north end (Fig. 3:4). In the center of this alley, the La Venta

a. God of the center with a tree in his crown, a scepter in his hands, and the sprouts of the four corners around him

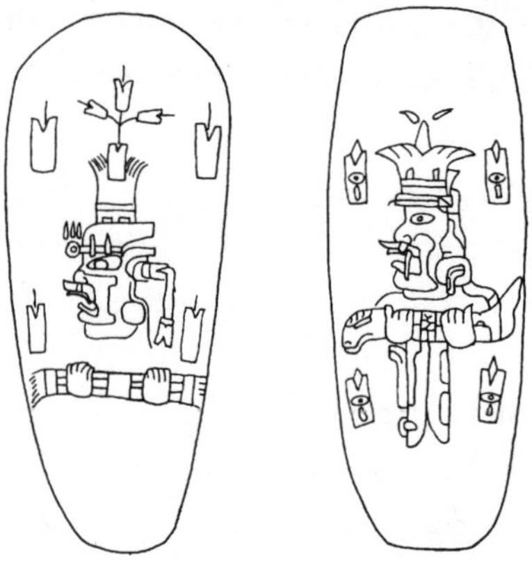

b. God of the center with a tree in his crown, a scepter in his arms, the sprouts of the four corners, and the crocdile head as his legs

FIGURE 3:5 The Olmec World Tree and Sprouts of the Four Corners

crocodile tree from Izapa Stela 26

crocodile tree from a Maya pot

FIGURE 3:6

shadows cast from the figures to show their relative position

FIGURE 3:7

Olmec constructed another immense serpentine pavement, incorporating the quincunx designs flanking the gateway to the sunken plaza, but without the tasseled flowers. An earthly portal rather than a watery one, this pavement showed the abrasion of long-term use. The other serpentine pavements we described had been constructed and immediately buried, but this one had felt the feet of many people running or walking across its surface.[31] Kent suspects that this may have been the center marker of an Olmec ballcourt.[32] Whatever its true identity, it created a path, a threshold, and an edge, a liminal space between the Primordial Sea of the sunken plaza to the north and the Creation mountain to the south.

In the midst of this sacred landscape, the kings of La Venta, like their Maya successors, placed themselves at the pivot of this cosmic structure and the quincunx[33] that centered their universe. This concept is beautifully illustrated by two small stone celts from Arroyo Pesquero (Fig. 3:5). On both, an Olmec god is depicted cradling a scepter in his arms and wearing the headband Olmec rulers wore to mark their status. Both figures may, in fact, be rulers wearing the mask and accoutrements of the god. A budding tree sprouts from their heads to identify them as the World Tree of the Center.

Reilly discerned two things about these images that demonstrated their connection to the worldview of the Maya. First, the feet of the full-figure version of the man on the celt are depicted as the head of a crocodile diving downward. Reilly knew that in later Izapan and Maya imagery (Fig. 3:6), the roots of the World Tree are often transformed into a crocodile head. Second, he guessed that the other symbols were not floating above his head and below his feet, but were surrounding him on a flat surface. As an experiment, Kent moved the five images apart and cast shadows from them. The transformation was amazing. Suddenly, the center tree and the four corners were visible (Fig. 3:7). The Olmec artists surrounded their king with sprouting seeds set at the four corners, defining the periphery of the human world and containing the sacred space of the center. The Olmec king was the embodiment of the World Tree of the Center. The Maya after them inherited this concept of kingship and placed it at the heart of their worldview. Just as our culture sees the Greeks as the source of much of our fundamental culture and philosophy, so the Classic Maya saw their source as the Olmec.

THE MAYA INHERITORS

The Maya inherited more from the Olmecs than the idea of the king as the World Tree. When they built their cities, they also replicated the sacred landscape of the Creation. In the Maya story, the gods planted the three[34] stones of the heavenly hearth and raised the sky, and then formed the landscape of mountains, lakes, and forest that became the world in which their new creatures would live. Then they created the first humans at the sacred mountain that was called Split-Mountain in the Popol Vuh. Palenque's artist made an image of this mountain as a living being with its name glyphs sitting inside its eyes. It was the *Yax-Hal-Witz*, the "First-True-Mountain" (Fig. 3:8) of this Creation. At Bonampak', it has

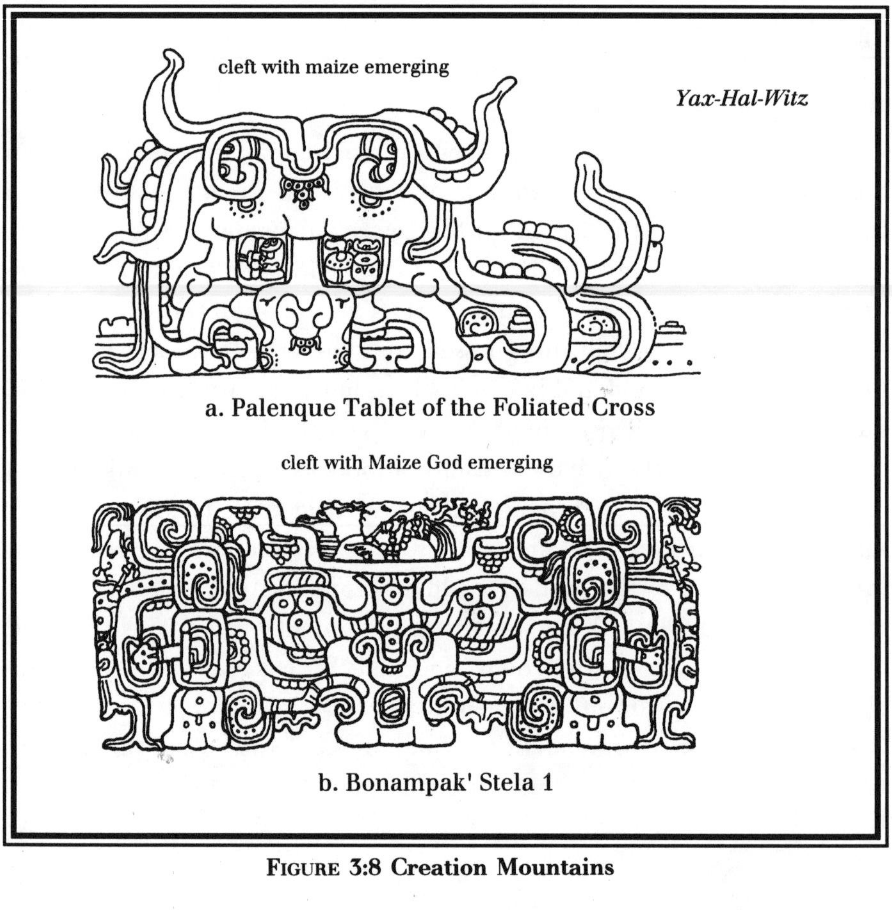

Figure 3:8 Creation Mountains

a cleft opening on its summit from which Hun-Nal-Ye, the First Father, emerges in rebirth, while at Palenque the king in the guise of the Maize God stands on it. Here was the place where First Father and First Mother fashioned the flesh of human beings from maize dough.[35] The precious maize was located in a pool of still water at the heart of a cleft mountain.

The representation of a sunken court as the surface of a pool of water and a portal to the Otherworld also has its analog in Maya architectural symbolism. In Classic Maya texts, *nab*[36] was the word both for "plaza" and for large bodies of waters, including lakes, rivers, and the ocean. The Maya conception of a plaza as a "watery" place through which spiritual communion with the Otherworld could occur was anchored in the precedent of the Olmec. Both saw plazas as a place where people could "swim"[37] through the incense in the ecstasy of dance. The ballcourt as a portal to the Otherworld and as a piece of pre-Creation space and time is also found at most Maya sites.

The lowland Maya began raising sacred mountain-pyramids[38] of their own by at least 600 B.C. They soon became masters at sculpting in plaster, the abundantly available architectural material of the region.[39] By around 300 B.C.[40] Maya master builders were decorating their human-made mountains with extraordinary cosmic images. These artisans, like their Olmec counterparts, did not yet put lengthy texts on their buildings. Nevertheless, we can identify the meaning of their designs because the Classic-period Maya who followed them retained the strategic symbolism of the early Maya builders quite faithfully. And these later descendants put texts on much of their art—ranging from single glyphs to entire histories. We can extrapolate the meaning of these symbols and images *backward* into the Preclassic Maya world with considerable confidence.[41]

WAXAKTUN

In Group H at the city of Waxaktun in Guatemala, Guatemalan archaeologist Juan Antonio Valdés uncovered stunning and beautifully preserved buildings from the earliest centuries of Maya kingship. One of these buildings depicts a *witz*[42] that we think is the First-True-Mountain of the world rising directly out of the primordial waters of Creation (Fig. 3:9)—or more literally, out of the "watery surface" of the plaza in front of it.[43] Fish are shown swimming among the water scrolls and liquid dots below the fangs of the personified mountain sculpted as a mask on the lower terrace of the pyramid, while vegetation grows from the monster's sides and cleft. A second monster mask looming above the first has a Vision Serpent undulating through its mouth bearing a head on its tail. In Classic-period imagery, the Vision Serpent was invoked during the ritual of communion between this world and the Otherworld. It was the embodiment of the path to and from the Otherworld, and ancestral figures were often shown leaning out of its open jaws to communicate with their descendants.

The open maw of the upper monster grips a small face, a little god with a mirror in its mouth. Recently Nikolai Grube and Linda Schele[44] realized that the mirror and its head variant read *tzuk* (Fig. 3:10), the word for "partition." This partitioning was a primordial act of Creation and here refers to divisions of the earth, the sky, the sea, and the Waxaktun kingdom. Most important, this word refers to the idea of the marking off of the four corners or directions in relation to the center.

This Waxaktun monster mask contains one of the earliest known examples of this partition symbol in the lowlands, but this symbol soon appeared on every image that had to do with the center, the corners, the directions, and the partitions of the world. This symbol marks the belly[45] of the great magical bird, Itzam-Yeh (Fig. 3:11a), represented on monuments at Kaminaljuyu, because this bird of Creation sits in partition trees. It was also engraved on the trunk of the great World Tree, Wakah-Chan because this tree marked the partition of the center as well (Fig. 3:11c). It sat on the noses of Mountain Monsters (Fig. 3:11b) because they marked the center and the corners of sacred space. At Waxaktun, the personified form of this symbol sits in the mouth of the mountain because that mountain is at the center of the world.

Creation shows up in this group at Waxaktun (Fig. 3:12a) in another way—the triadic arrangement of the first three stones of the cosmic hearth.[46] Here we see triads upon triads—each associating its components with the beginning of time and space. Three mini-acropolises (Fig. 3:12b) are arranged in a triangle atop a huge foundation platform, while yet other building triads sit upon their summits. The largest of these mini-acropolises lies on the eastern side of the main platform. On its summit is yet another group of buildings surrounding a large court. Like a miniature of the acropolis it sits upon, this upper group also has its largest building on its eastern side. Looking out through the trees that grace its summit, you can see the temples of Tikal twenty kilometers to the south.

A tiny building stood on the western edge of this acropolis opposite the

monster with a Vision Serpent emerging from its cheek and a *tzuk* head in its mouth

Mountain Monster with vegetation growing from its edges

Primordial Sea with swimming fish

stucco masks of the mountain of Creation

FIGURE 3:9 Waxaktun, Structure H-X-Sub-3

introductory glyphs to the Primary Standard Sequence on pottery

sky-earth partitions (Tikal plate)

tzuk-stones carried by Chak

tzuk-te' as the name for a stela (Tonina M26)

the World Tree written as *tzuk-te'* (Yaxchilan)

FIGURE 3:10 Substitution Sets for the *Tzuk* Glyph

a. Itzam-Yeh Bird with a *tzuk* head on his belly

b. Mountain Monster with a *tzuk* head on his nose

c. World Tree with a *tzuk* head on its trunk

FIGURE 3:11

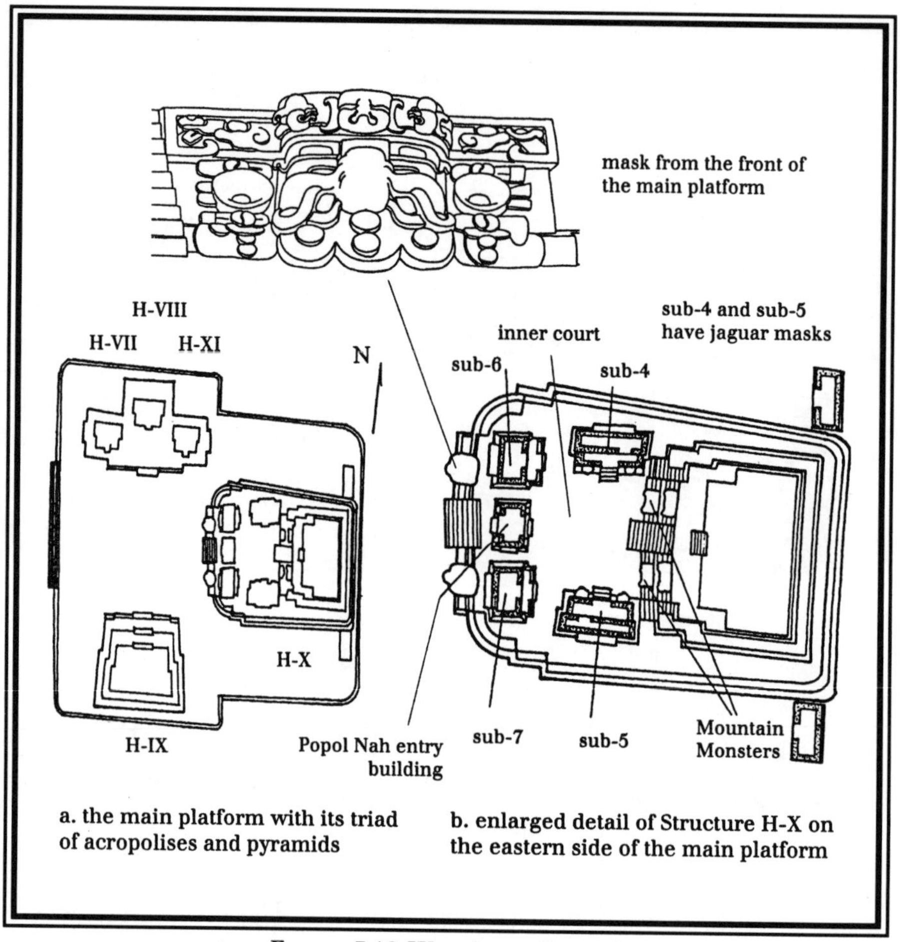

FIGURE 3:12 **Waxaktun Group H**

big pyramid. Doorways in both its eastern and western walls provided a formal passageway between the outer court and the interior of the acropolis complex. The walls of this building are marked with great mats modeled in plaster. On the corners of the building and on the jambs of the doors stand lords wrapped in clouds of *ch'ulel*[47]—soul stuff—conjuring holiness from the Otherworld. Like the Pawahtuns, the age-old burden-bearers who stand at the four sides of the world and hold the heavens up above the earth, these kings stand at the corners of the world (Fig. 3:13).

This little building is more than just a threshold house. The Mayan word for mat is *pop*, and the word for council is *popol*, so that the mat signs on its walls also identify it as a Popol Nah, a community council house,[48] a place where the king interacted with his people—especially through the performance and teaching of sacred dance. The courts of this acropolis, like many others in the Maya world, were the settings for great dance pageants that enabled the lords and their people to travel to the Otherworld to greet the supernatural beings who gave power and legitimacy to the human community. Directly below the Popol Nah and flanking the grand stairway that rises up from the outer court sit two huge masks of ancestral gods (Fig. 3:12).[49] These gods represent the founders of the community and are sacred manifestations of the Otherworld.

So we find that the Olmec and the early Maya defined sacred space in fundamentally similar ways: plazas shimmered with the hidden currents of the Primordial Sea, stairways descending from the summits of Creation mountains shaped paths between worlds. Threshold buildings and ballcourt alleyways[50] marked out the liminal space for dance, ritual sport and—as we shall see—for sacrifice. In each case, we find evidence for centering the world and anchoring it in the original moments of its birth. With the Olmec we must resort to educated guesses when interpreting what they meant to express in their imagery, because there are no texts and no sure descendants left to ask. With the Maya, however, we have written records of their ancient history and mythology, and their descendants are here today to tell us how they understand their world.

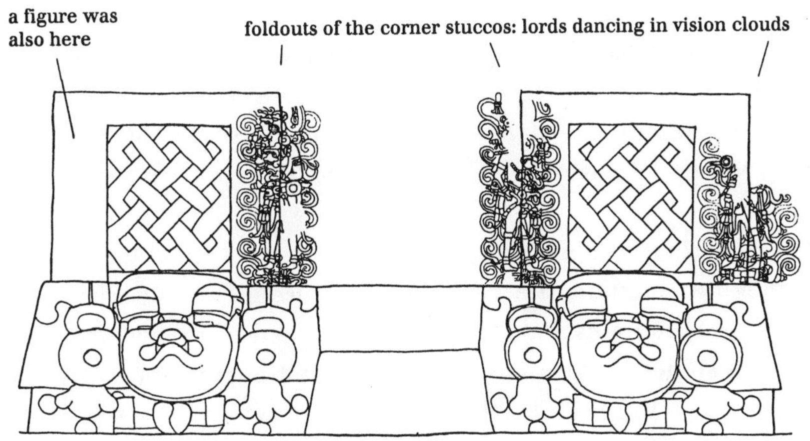

The Popol Nah entry structure on the west side of Structure H-X
(hypothetical reconstruction by Schele based on Valdés 1988)

Figure 3:13 Waxaktun Group H

PALENQUE
Mountains on the Big-Sea

Much of the story of Creation we told in Chapter 2 is recorded in the Group of the Cross at Palenque. As at La Venta and Waxaktun, the architects of the city echoed Creation in its temples and plazas. The city was built on a flat shelf halfway up one of the slopes of a majestic ridge of mountains. From this vantage, the city overlooked a tidal wave of green foliage rolling over the broad plain that edges the Gulf of México. This shelf formed the foundation for a great plaza that runs from one edge of the sacred precinct of the city to the other. In their texts, the Maya of Palenque called the shelf and its plaza *Lakam-Nab*[51] or "Big-Sea" (or *Lakam Ha*, "Big-Water") perhaps in honor of the sea that lay to the north. Plate 9

On this Big-Sea, the kings of Palenque built their homes and their human-made mountain-pyramids surrounded by the forested mountains raised by the gods when they brought forth the land from the sea (Fig. 3:14). Explorers and archaeologists have dubbed the royal acropolis where the kings lived the "Palace." Just to the north of the Palace lies a small ballcourt whose form was first revealed by Mexican archaeologists in 1992. With the low side benches characteristic of the ballcourts at Yaxchilan and Tonina, this playing field contained by this ballcourt lies just off the eastern side of the great processional stairway that mounted the north terraces of the Palace. Like the navel of Zinacantan, it is small, unimpressive, and unassuming compared to the elaborate buildings around it, but terribly important in its time-worn ties to the source of creation and sacred power. It was a portal that led to the time and space of the Otherworld.

Pakal, perhaps the most important king to have ruled the city, raised his funerary mountain, the Temple of the Inscriptions, against the flank of the huge mountains that loom behind it. In the warm evenings, its pale silhouette shines in harmony with the sunset edging the southwestern sky above the Big-Sea. To the east, nestled against a steep mountain the city folk called *Yemal-K'uk'-Lakam-Witz*[52] or "Descending-Quetzal-Big-Mountain," his son Chan-Bahlam[53] recorded the history of Creation inside the three temples his architects arranged in the pattern of the cosmic hearthstones (Fig. 3:15). We believe that this mountain, like the

The Group of the Cross is in the upper left below the Descending-Quetzal-Big-Mountain; the Palace and the Temple of Inscriptions are in the center on the *te'-nab*; the ballcourt is below and just to the left of the Palace.

FIGURE 3:14 Aerial View of Palenque

Senior Large Mountain near Zinacantan center, housed the ancestors of the Palenque people.[54]

These three temples not only record the events that culminated in the Creation of the world, but they replicate these events in their arrangement, their imagery, and their ritual function. Inside two of these temples are exquisitely carved visions of the World Tree that centered the cosmos at the moment of Creation (Fig. 3:11). The Temple of the Cross, at the northern apex of the triangle, recorded the events of Creation and the history of the dynasty, surrounding the image of Raised-up-Sky Tree. When the Tree is erect over the Group of the Cross, the scorpion hovers just above the mountains to the south of the group, as the *Xibal Be* enters into the Otherworld in its Milky Way form. The Temple of the Foliated Cross, on the eastern sunrise side, depicts images of the First-Tree-Precious maize plant, the First-True-Mountain, which contained the maize used to mold human flesh in the last Creation, and the shell that

The Group of the Cross with the Descending-Quetzal-Big-Mountain behind it

FIGURE 3:15

opens into the Otherworld and the sacred space of First Mother and First Father. The theme is the rebirth of the Maize God and the generation of humanity from maize and water. Most of all, the text and imagery of the eastern temple identify the cleft mountain we have seen at La Venta and Waxaktun as the First-True-Mountain of the Creation myth.

COPAN
Centering the Valley

The rulers of Copan created an urban center that used the same elements as described in the cities above, but arranged into their own particular and eloquent pattern. To the south,[55] the earliest rulers built a great mass of pyramids. This human-made sacred mountain range had Witz Monsters modeled into the walls of its temples (Fig. 3:16). Today we call these sacred mountains, comprised of four hundred years of layered, superimposed construction,[56] the Acropolis. Just to the north of this Acropolis, the Copanecs built their own Ballcourt portal to the Otherworld. This Ballcourt was rebuilt at least six times during the history of the city. The three-dimensional macaw heads that served as the bench markers remained throughout every incarnation.[57] The final phase that we see today was dedicated by Waxaklahun-Ubah-K'awil, nicknamed 18-Rabbit, on January 10, A.D. 738. We will return to this Ballcourt in some detail in Chapter 8.

A ruler of great architectural vision, 18-Rabbit shaped his city's sacred center into its greatest and most elegant incarnation. Over a twenty-five-year period, he remade the Great Plaza north of the Ballcourt (Fig. 3:17), erecting one stela after another, each portraying him in the ritual guise of a different god and more often than not as the embodiment of the World Tree. He created a forest of images manifesting the axis of communication and himself in the guises of the actors of the Creation myth. This big-stone forest marks out a ritual pattern timed by the events of ancestral history. It also reflects a cosmic pattern, timed by the movements of Venus and other planets, all interacting with the movements of the Milky Way and the constellations of the ecliptic.[58] Last, these stelae record the unfolding of the k'atuns (a period of twenty years made up of 360 days each).

During the years these stelae were being raised, 18-Rabbit also reshaped the great mountain range of the Acropolis.[59] His most memorable work during this period was Temple 22, a building that has long been admired as one of the finest of all Maya architectural expressions. We know now that by raising Temple 22, 18-Rabbit was re-creating a type

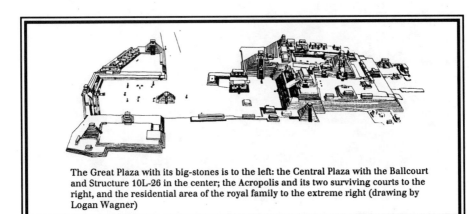

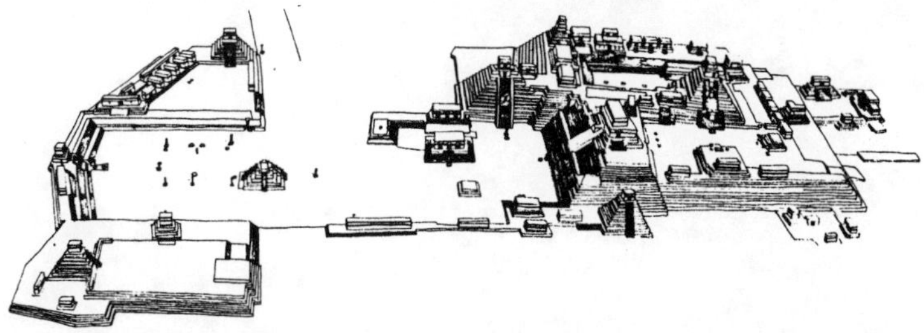

The Great Plaza with its big-stones is to the left: the Central Plaza with the Ballcourt and Structure 10L-26 in the center; the Acropolis and its two surviving courts to the right, and the residential area of the royal family to the extreme right (drawing by Logan Wagner)

FIGURE 3:16 Copan

of structure that had been built many times before by his ancestors,[60] expressing their own special vision of Creation. This temple's last and most ambitious manifestation was constructed with mud mortar, a kind of construction that required constant and careful maintenance to ensure that its plaster seal did not leak and weaken the walls. Once the Copanecs no longer maintained it, it deteriorated and collapsed. Today we have only fragments of the beautiful sculptures that once decorated it—pieces found lying in the grass by the temple's feet—but they are enough to help us contemplate the building's lost beauty and significance.

The Great Plaza with its tree-stones is above and the Ballcourt is below (photograph by William Ferguson and John Royce)

FIGURE 3:17 Copan

Corner masks of the Mountain Monster from Temple 22 with the fangs from the central door monster visible on the right

FIGURE 3:18

A stack of three great stone-mosaic masks representing the Witz, the personified mountain[61] with cleft forehead, adorned each corner of Temple 22 (Fig. 3:18). Maize sprouted from these heads above each earflare. Atop each stack stood a bird[62] we believe to be the great Celestial Bird, Itzam-Yeh. As we saw in Chapter 2, this Celestial Bird is a central player in the Creation myth that united Classic Maya civilization. At Copan, Itzam-Yeh is shown descending from the heavens and landing atop the holy temple-mountain. Other sculptures that decorated Temple 22 include stunning three-dimensional portraits of Hun-Nal-Ye, First Father, in his manifestation as the Maize God (Fig. 3:19), intermeshed with skeletal Venus gods; portraits of 18-Rabbit himself; smaller Witz Monsters; and dozens of other symbols and images that graced the upper half of the walls between the corner masks with the Celestial Birds atop them. Temple 22 was Copan's version of the First-True-Mountain of Creation.

18-Rabbit's master masons shaped the central door of his temple to represent the mouth and gullet of the great Witz Monster (Fig. 3:20).[63]

FIGURE 3:19

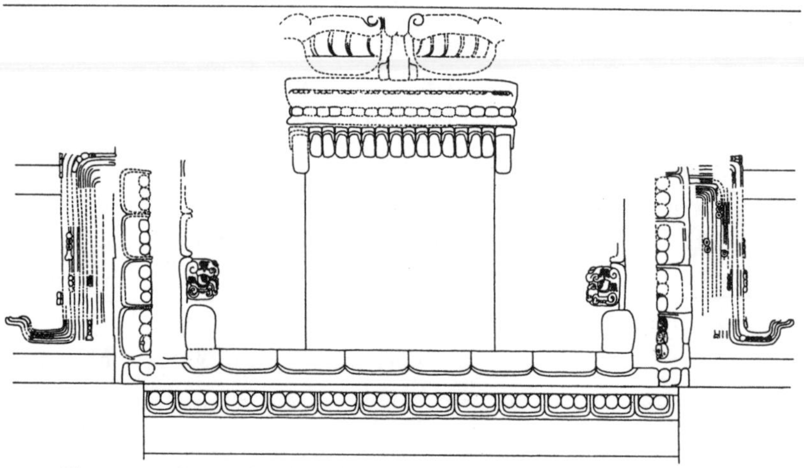

The monster door in Temple 22 (hypothetical reconstruction based on 1986 fieldwork at Copan)

FIGURE 3:20

FIGURE 3:21 Inner Door of Temple 22

This was meant to indicate that the interior of the temple symbolized a living cave that opened into the heart of the mountain. To the Classic Maya, all natural openings into the earth, whether caves or cenotes (sunken waterholes), were portals to the Otherworld. Their architecture echoes this belief. Deep behind this cave door stood the sanctum where 18-Rabbit and his successor conjured up their ancestors and the gods.

The frame of the doorway that led into this inner chamber was one of the most extraordinary architectural compositions ever conceived by the Classic Maya (Fig. 3:21). A Cosmic Monster, representing the arching body of the Milky Way in its east-west configuration, frames the door in such deep relief carving that it seems to be writhing out of the wall. The front end of the monster takes the shape of a crocodilian head and is decorated with the symbols of Venus. Pointing west, the counterpart in the sky of this great crocodile parallels the ecliptic so that planets often travel along its belly following the path of the sun. At the opposite end, the sun-marked plate of sacrifice that sits on the junction of the Milky Way and ecliptic rides the tail of the monster. A huge stingray spine, used to pierce royal flesh and allow the flow of soul and blood to nurture the gods, juts from its head as it opens a portal into the Otherworld.[64]

In the great lazy-S scrolls composing its arching, serpentlike body cavort the beings who have been conjured up by the bloodletting rituals inside the sanctum. These particular scrolls are clouds. For the Maya, clouds of light in the night sky constitute one perception of the Milky

Way. Clouds as metaphors for the heavens still prevail among some modern Yukatek shamans.[65] Clouds, rain-laden, celestial, or in the form of sweet incense smoke, harbor *ch'ulel*, the soul stuff of the living universe. According to John Sosa's shaman teacher in Yalcoba, the pink-tinged clouds of dawn and dusk house the souls of deserving deceased relatives.[66] Here at Copan the beings conjured up in the clouds are spirits called *way* or *nawal* and the serpent-footed god, K'awil—all beings that the king called upon in the exercise of his power.

The cloud-conjured body of the Cosmic Monster is held up by the elegantly modeled figures of two of the Pawahtunob, also called *bakabob*, the world-bearers who hold up the four corners of the sky. Their buttocks transform into the *tzuk*, "partition," glyph[67] to mark them as beings of the four quarters. Here are the bearers of the east and the west, the path of the sun.[68] Below their feet rest skulls, referring to the place of death and to the gaping skeletal maw of the Vision Serpent that opens into the Otherworld.[69] The inscriptions along the tread between the skulls record the completion of 18-Rabbit's first k'atun of reign[70] and also the dedication date of this temple. The inner sanctum of Temple 22 thus recalls the original acts of establishing sacred space at the time of the Creation. It lies within a mountain sprouting maize in a replication of the vegetation-bedecked Creation mountain we saw at Waxaktun.

On the west side of 18-Rabbit's temple at Copan stood the Popol Nah, or "council house," an identification first made by Barbara Fash[71] (Fig. 3:22). As with the Popol Nah we saw at Waxaktun, great mats carved in mosaic stone decorate the upper surfaces of the building. At Copan, the upper wall also displays portraits of lords seated in majesty upon the glyphs that might represent localities in the domain. Like the council house at Waxaktun, the working space of this meeting place was not the close interior of the shrine but a space outside, under the sun and stars. A wide porch in front of the house provided space for the council to meet. Barbara Fash[72] also discovered that the location had a second great function—the teaching and performance of sacred dance. A second low platform, much longer and wider than the porch, lay in front of the Popol Nah, all along the western side of the East Court. It is littered with the remains of stone censers. From this platform, dancers could easily move out onto the terraces that run along the south side of Temple 11, and along the north side of the Acropolis. The ceremonial view of the audiences standing within the East and West Courts may have been restricted, but the pageants that began at the Popol Nah could expand into the far larger and more visible plaza spaces north of the Acropolis.

The East Court itself addressed this dance platform in a very special way. Across from it on the eastern side stood a now-lost building that Barbara Fash has identified as an embodiment of the Bat House in the Popol Vuh.[73] This Bat House completes the court as a representation of the great myths that guided the lives of the Maya, for that was where one of the Hero Twins lost his head during their sojourn in Xibalba. On the northern side of the court sat the First-True-Mountain, which declared the apotheosis of First Father as Maize and held the maize grains that became the flesh of humanity. Within this mountain was the place where the king conjured the gods and the ancestors into this world through

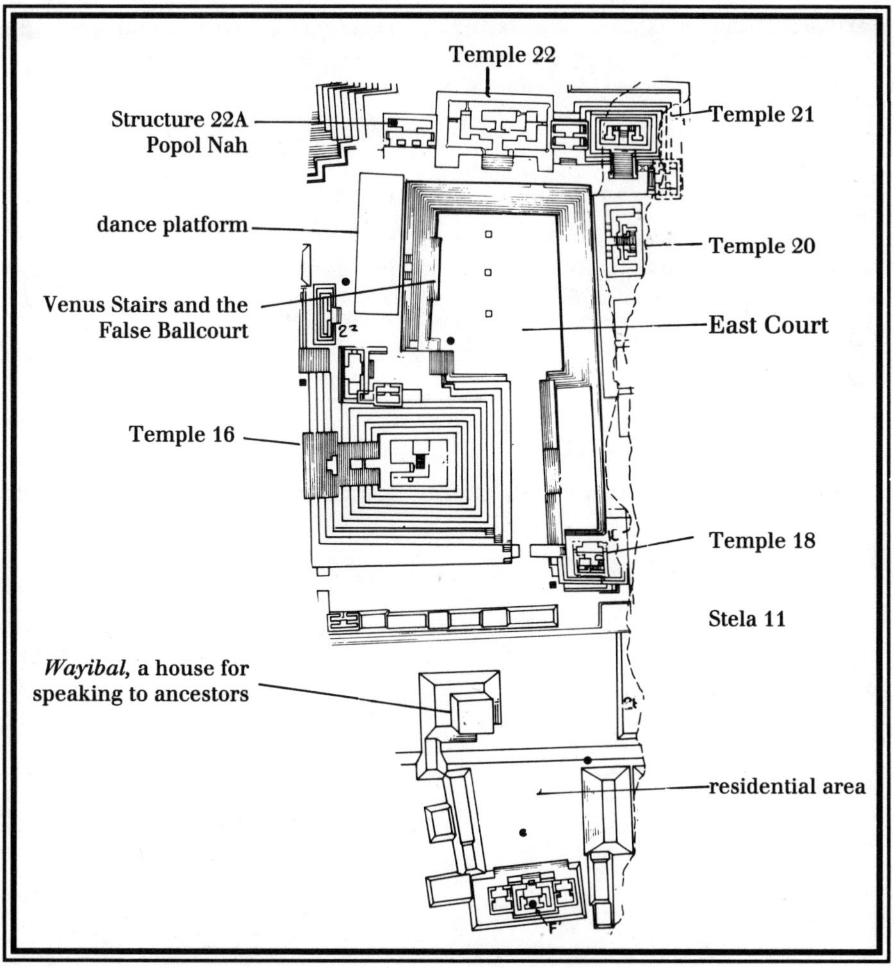

Figure 3:22 Copan, the East Court of the Acropolis

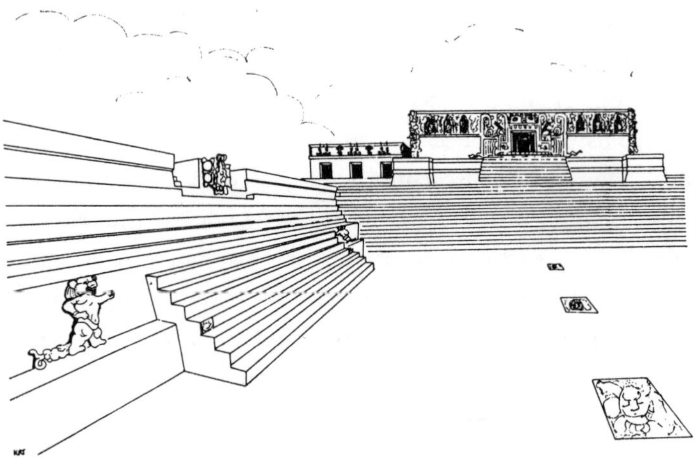

Reconstruction drawing of the Popol Nah and Temple 22 above, with the Venus Stairs on the left and the ballcourt markers on the right (drawing by Kathryn Reese-Taylor)

FIGURE 3:23 East Court of the Acropolis

sacrifice. On the western side, directly under the dance platform, was a false ballcourt (Fig. 3:23) marked by a great Venus God[74] emerging from the jaws of an ecliptic snake. Flanking the stairs, which constituted the "playing area" of this false ballcourt,[75] are rampant jaguars, the spots on their pelts originally rendered with inset obsidian disks. They dance, just as the Hero Twins danced in victory over the Lords of Death. The three

Marker from the false ballcourt in the East Court with the head of Hun-Nal-Ye hanging in the vines of the gourd tree

FIGURE 3:24

markers that lie in the center of the East Court[76] depict the severed head of First Father (Fig. 3:24) tangled within the twisted vines that the Lords of Death used to hang it in the gourd tree at the edge of the Place of Ballgame Sacrifice.

The East Court was a place of communion where the lords of Copan and their people reenacted the stories of the Creation, of First Father and the Hero Twins. These lovely buildings, like their predecessors fashioned by the Olmec kings and earlier Maya lords in the lowlands, provided the interior and the exterior sacred, magical space in which pageant and ritual could unfold, joining together the worlds of human and divine experience. Seen symbolically, the heart of Copan contained the Primordial Sea in its plazas, the portal that pierced through to the Otherworld in its Ballcourt, the forest in its big-stones, and the mountains in the temple range of the Acropolis. To the south of the Acropolis, royal lords and their families dwelt in their palaces close to the homes of their gods and ancestors, near the navel of their world.

CHICH'EN ITZA
Mountain, Path, and Pool

After over a thousand years of success, most of the kingdoms of the southern lowlands collapsed in the ninth century. In the wake of this upheaval, the Maya of the northern lowlands tried a different type of government. They centered their world around a single capital at Chich'en Itza.[77] Not quite ruler of an empire, Chich'en Itza became, for a time, first among the many allied cities of the north and the pivot of the lowland Maya world. Yet it also differed from the royal cities before it, for it had a council of many lords rather than one ruler.[78] Even with this form of government, the traditional components of sacred place remained the same. The Pawahtunob-Bakabob are still there, raising the sky just as they do in the sanctum inside the Creation mountain at Copan. They squat on their carved jambs, straining against massive wooden lintels in the doorways of the temple on top of the four-sided temple-mountain called, by today's modern pilgrims, the Castillo (Fig. 3:25). This mountain was a Creation mountain complete with Vision Serpents, just like the one we saw at Waxaktun.

Instead of a single stairway that rose to its summit, this pyramid has four, one running up each of its four sides. Here we see yet another

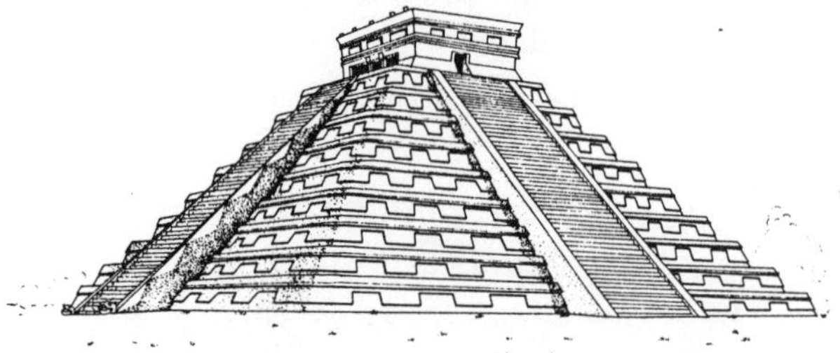

The Castillo at Chich'en Itza with the spring equinox shadow (drawing by Kathryn Reese-Taylor)

FIGURE 3:25

symbol of the fourfold partitioning of the world at Creation. Unlike the temple-mountains of Copan or Palenque, however, this temple sat at the center of time as well as space. The axes that run through the northwest and southwest corners of this pyramid are oriented toward the rising point of the sun at the summer solstice and its setting point at the winter solstice. This declaration of the four corners as solstitial points makes the pyramid a massive sun dial for the solar year.[79]

The balustrades along the four stairways descend in the shape of enormous Feathered Serpents, their mouths gaping at the bottom of the stairs. Cutting across the dominant four-part pattern, the main doorway of the outer temple on the summit is on the northern side, and the sanctum sanctorum of the temple on the summit opens only to the north. The northern stairway is clearly the principal sacred path: on the evenings of the vernal and autumnal equinox, the stepped terraces of this pyramid cast triangular shadows across the northeast balustrade wall, manifesting a serpentlike diamond pattern of the rattlesnake this sculpture represents.[80]

A broad plaza surrounds the Castillo on four sides. We can tell by the northern orientation of the innermost chamber inside the temple on the summit of the Castillo that the plaza on the north side of the Castillo was a major location of ritual activity (Fig. 3:26). This, like the plazas in the other Maya cities we have discussed, is Chich'en Itza's Primordial Sea of Creation. In the center of this north plaza stands a low platform with four stairways oriented to the solstitial directions, symbolizing the four partitions set by the gods at the beginning of the world.[81] The walls of the platform carry elaborate reliefs celebrating the Venus war god, a supernatural we will meet again in Chapter 7. This plaza connects the Castillo,

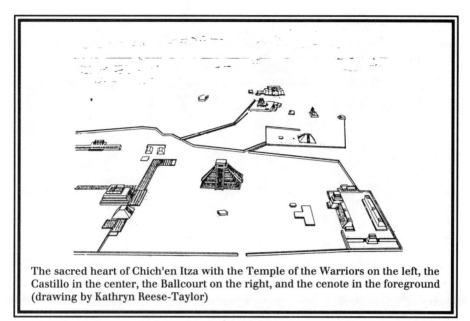

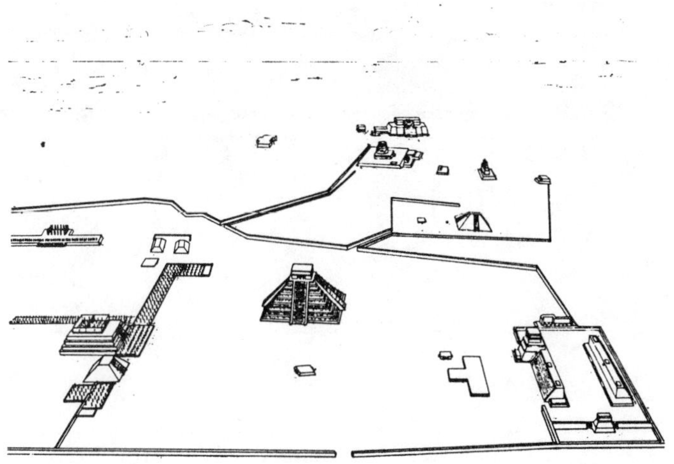

The sacred heart of Chich'en Itza with the Temple of the Warriors on the left, the Castillo in the center, the Ballcourt on the right, and the cenote in the foreground (drawing by Kathryn Reese-Taylor)

FIGURE 3:26

as Creation mountain, with two other portals, one natural and one human-made. The first, an enormous cenote known as the "great well of the Itza," is reached by a causeway leading northward from the Venus platform. Here the earth is pierced by a natural portal, revealing the watery depths that lie beneath the surface of the earth—certainly a dramatic symbol of the Maya view of the world.

At the west end of the plaza lies the second portal, the largest ballcourt ever built in Mesoamerica. Here the people of Chich'en Itza celebrated the myth of the Hero Twins through sacrificial pageant. Here they also painted murals documenting the founding wars of the Itza that gave the kingdom its sacred charter.[82] Their descendants today believe that the cosmic umbilicus lies under the floor of the ballcourt and will emerge again one day through the great cenote.

On the eastern side of the plaza stands the Chich'en Itza equivalent of a Popol Nah, the Temple of the Warriors, its porch held up by the Northwest Colonnade. Although this meeting house (Fig. 3:27) is not marked with the huge mat signs we saw at Waxaktun and Copan, it bears its own equivalent of the lords represented on the roof of the Copan shrine. Converging on the central staircase of the Chich'en Temple of the Warriors are portraits of the council members, resplendent in the regalia

FIGURE 3:27 Temple of the Warriors

of their offices, carved on the stone columns of the outer colonnade. To enter the council chamber above, the living council members had first to pass the stone lords in the shadows of the lower chamber, then ascend to the upper chamber by means of a steep stairway. For witnesses on the plaza below, a procession up that stairway would take on a magical appearance as it passed through the huge Feathered Serpents whose rattle tails held up the lintel of the doorway of the upper sanctuary. Such a vision might call to mind the images of their ancestors and the spirits of their prophets and leaders, thought to hover above the roof of the Popol Nah during times of ritual.

Feathered Serpent columns, with jaws open and menacing,[83] were the sentinels of the upper chamber (Fig. 3:28). They symbolized both the War Serpents of the Itza and the ancient Maya concept of the path of communication to the Otherworld. Pawahtunob, holding up the stone doorjambs, flanked them on either side. Ornate long-beaked masks of the Itzam-Yeh bird marked off the corners of the pyramid, establishing the fourfold space of the sacred mountain's summit. On the outer walls of the upper chamber, we find the time-honored theme of the bird of creation descending to rest on the mountain. Several mosaic sculptures of creation birds are represented as screaming, sharp-taloned battle beasts flanking the doorway and guarding the four sides of the temple. The heads of warrior ancestors emerge from their beaks. The stone councillors frozen in their colonnades stride up the stairs beneath them. Plate 44

As described in ethnohistorical sources, the meeting places of Maya councils were also the sites of much ceremonial dancing. We think the dancing place of Chich'en Itza was right next door to the Temple of the Warriors, a building found within the structure called the Group of the Thousand Columns. Images of dance processions and of other ceremonies conducted there are carved on the benches lining the colonnade. It is much more difficult at Chich'en Itza than at other sites to identify which buildings were the palace homes of the councillors. But a building called the Mercado was probably residential. It forms, along with the Temple of the Warriors and the Group of the Thousand Columns, the southern side of a vast quadrangle. The design of this building, an open interior patio surrounded by roofed space, is replicated in smaller residential structures throughout the city.[84]

If Chich'en Itza did without a dynastic monarchy, seat of Maya worldly power for a thousand years, it still clung fervently to the transcendent cosmology that had buttressed those lineages of holy lords. The watery surface of the Otherworld revealed itself naturally and magnificently in the great cenote at the northern apex of the ceremonial center. The white stone causeway traveled south from this portal to the Venus platform, then farther south, onward up the stairway of the Castillo, the Creation

The outer walls have Itzam-Yeh belching up a *way* from the Otherworld, while the piers in the door are feathered Vision Serpents. The Pawahtun on the right occurs on the doorjambs.

FIGURE 3:28 Temple of the Warriors

mountain, into the sanctum at the summit. The mountain, the path, and the pool are connected here by the same spatial axis as at the ancient Olmec site of La Venta. The north-south axis of Chich'en Itza's center is not limited to the city's ceremonial center. This road continues southward from the sacred precincts through the dense residential areas of the city. Just as significantly, the Temple of the Warriors faces westward across the plaza to the Lower Temple of the Jaguars on the Great Ballcourt. Its doorways held up by aged Pawahtunob and skull-headed, bare-breasted women, the Lower Temple of the Jaguars declares the presence of the Father-Mothers, the ancestors, in this portal as it faces the dawning sun rising in the east. Just as the four partitions were set at Creation, so the four partitions and the center had to exist for the life of humanity to exist in balance with the cosmos.

Mesoamericans still care deeply for such spatial patterns and orientations in modern communities. When the La Venta kings and the rulers of Chich'en Itza laid out their mountain, path, and pool on the north-south axis, they forged these symbols into a coherent cosmogram endowed with the power and integrity of the sky axis itself. But neither the Olmec nor the Maya after them allowed this pattern to become frozen and inflexible when they created their ceremonial centers. The ceremonial centers of the ancient world display not a single rigid pattern but the unique, personal expression of the Maya cosmic vision as created through the imaginations of each reigning king.

As we shall see in later chapters, each king also had his own political agenda when he raised his sacred buildings, his own problems to solve. These sacred spaces were arenas not just for religious pageant but also for the political activities of the kingdom—for the Maya made no separation between these domains of activity. Pageant always had both functions, so that architecture that replicated the time and space of Creation sanctified all the activity that took place within it. A king who controlled Creation and the power of the Otherworld was by definition a successful king. There was also an aesthetic element involved in the crafting of these sacred spaces. Both king and community sought to enhance and amplify the natural potential of their landscape. As the landscapes varied, from mountainsides to valleys to plains and coastal swamps, so did the centers. The miracle of Chich'en Itza is that even as they declared the death of the holy kingship, its lords managed to reach deeply into their heritage for the vision needed to forge a new, revolutionary and hopeful future in the shadow of the social collapse occurring in the south.

COZUMEL AND THE PATTERNS OF A RURAL AREA

No matter how ingenious their vision, the revolutionaries of Chich'en Itza would have failed without the cooperation and understanding of the commoners. This broad class of people comprised the real Primordial Sea from which Maya government would rise again, even after the fall of that city. As David Freidel discovered when he was a surveyor for the Harvard-Arizona Cozumel Project in the early 1970s, the humble as well as the mighty designed their communities upon the principles of sacred space—the four directions and the center. When David began his surveying of the northern interior area of the island, location of Cozumel's ancient central community, it was buried under the kind of dense secondary bush that fills milpa fields after they are abandoned. In simple terms, you couldn't see your feet in front of you unless you cut a trail with a machete. Cutting trails was a treacherous affair, for the forest holds an extravagant variety of thorny bushes and trees that seem to lunge and snap at anyone trying to make a way through them. Because of this problem, it took months to make sense of the patterns of ruins that sprawled across the land of the modern ranch of San Gervasio. Only gradually did the details begin to emerge as maps were made and the data from grid after grid was added to the whole.

When the ancient community at San Gervasio finally came into focus, the map revealed a Late Postclassic center that was probably still the capital of the island when Cortés landed on the shores of Cozumel. Its four noble families had constructed their households at four points around a common public quadrangle crowded with temples and colonnades. This pattern conformed to the age-old concept of quincunx-ordered space. The layout of this center intentionally integrated residential space with public space in such a fashion that the residential compounds of the four families stand at the locations of the four burden-bearers—the protective, prosperity-bringing Chakob honored by the Yukatekan Maya of the contact period.

The Cozumel leaders designed their houses with an open patio at the front and a private one at the back, facilitating the combining of both public and private ritual. This house style was originally developed at the Late Postclassic community of Mayapan, capital of the last major confed-

residential compounds and the road system at San Gervasio

a small shrine building

FIGURE 3:29 Cozumel

eracy of Maya states in northern Yukatan. Most of these kinds of households had shrines both inside and out. The outer shrines were always at the edge of the front patio, facing the residence. Sakbes—ceremonial roadways or paths—led from the main temples and colonnades to the household shrines, connecting each household to the center of the city.

Other sakbes connected San Gervasio with other towns all over the island. One of the larger of the roads was clearly important in the ritual life of the region. As it led out from the city, it passed under a fine masonry arch, then ran past a small ceremonial temple on a raised platform. Other small stone temples studded the length of this road until it arrived at the northern towns on the island (Fig. 3:29). All along the coast of Cozumel, temples and shrines perched above the dunes and crags of the hills, forming a perimeter of portals to the Otherworld that integrated the sacred life of the whole island. This intercommunity sakbe and the others of San Gervasio served as sacred paths for processions that traveled from temple-shrine to temple-shrine. These little temples were much like the cross shrines of modern Zinacantan and those that mark the perimeter paths and entryways into Yukatekan towns on the nearby mainland. In both ancient and modern times, the connection of home and shrine by means of paths indicates how the spaces of public ceremonies embraced and ordered the places of daily living; and how, in turn, the places of daily living nurtured the spaces of public life. The pillars that upheld the four corners of the cosmos on Cozumel were private homes. This principle extended from the homes of the four ruling families down through all the strata of society, to the homes of their constituents in the towns of the island.

We used to think there was an abyss separating the way the Maya in the time of the great Classic kingdoms thought about the cosmos and the religious beliefs of their Postclassic descendants. Postclassic homes *did* have shrines inside them or next to them on their patios. But J. Eric Thompson argued that the high religion of the Classic period had been vulgarized in the Postclassic. It had fallen from the grace of the glorious urban centers to the dwellings of individual families. Now, however, from studying the archaeology of Copan, Tikal, and Palenque we know that Classic homes also had shrines. The household devotion of the Maya did not dilute the central ceremonial practice; rather, it integrated the interests of the people with those of the king, and integrated the community itself through cosmic spatial patterns.

Postclassic Maya decorated their offering vessels and incensarios with three-dimensional pottery images of supernatural beings, the notorious "idols" destroyed in the autos-da-fé of the Spanish priests. Thompson thought that these pottery images bespoke a move from public to private ceremonies.[85] Now we know that these images were also fashioned and used by Preclassic and Classic-period Maya in their homes, and that they represent the same gods and ancestral heroes found decorating the great pyramid mountains.[86]

AFTER THE CONQUEST

The Maya continued to replicate the space of Creation in the mode of their ancestors until the Europeans arrived in their land, surrounding forever the outward patterns of their lives with Christian doctrine and modern development. Learning the language and ways of the Maya was integral to this conversion process, and what resulted was an interesting hybrid of the two religions. In his *Relación*, Diego de Landa described the centering of sacred space in indigenous rituals of baptism[87] conducted for young children. To cleanse the space in which the ritual was to take place, the priest chose four old and honorable men to assist him by fulfilling the role of the Chakob. First the house or courtyard where the ceremony was to take place was swept clean and spread with fresh leaves. To drive out any hostile or evil spirits that contaminated the space, the four Chakob sat on stools placed in the four corners of a plaza and stretched a cord between them, enclosing the periphery of the space.

When all the people who had been purified by fastings and other cleansing rituals had stepped across the cord into this secured space, a fifth stool was placed in the center. There the ritualist himself sat down before a brazier. First, he burned ground maize and copal incense. Then he called each of the children to be baptized to come forward and drop their contribution of maize and incense onto the coals. When he deemed the space cleansed, a man carried away the cord that had bound the space, the brazier with its coals and ash, and a cup of what Landa called "wine" into the wild, untamed forest beyond the periphery of the human community. The individual who performed this ritual was bound to refrain from drinking the wine or looking behind him as he returned, lest the evil attach itself to him and thus recontaminate the community.

In sixteenth-century Yukatek rituals, especially the New Year's celebrations, Landa describes how processions moved from the center of the community to its periphery, and then from the periphery back to the center again, along each of the cardinal directions. Remember that New Year's ceremonies included the concept of the destruction of the old world and the creation of the new. To accomplish this, gods were then carried from the outer boundaries—the sides—of the community into the center. During the Muluk years, for example, a god called Bakab-Kan-Siknal was carried to the center where it was placed in a temple along with a bird god called Yax-Kokah-Mut, "Green-Firefly-Bird."[88] In the court in front of this temple, the Maya made a figure of stone upon which they burned copal incense and rubber while they sang and prayed. They also offered this idol squirrels and plain cloth woven by old women, whose duty it was to dance in the temple before the image of Yax-Kokah-Mut. The day continued with a dance on stilts and more offerings of turkeys, tamales, and drinks of maize gruel. The old women danced again, this time with small clay dogs, after which a small dog with a black back was sacrificed. The most devout drew blood from their bodies to anoint the gods.[89]

The god Yax-Kokah-Mut was still around over a century and a half later when Avendaño visited Lord Kan-Ek' in his capital on Lake Peten-Itza.[90] In the center of Kan-Ek's palace stood a round stone pedestal surmounted by a column called the *Yax-Cheel-Kab,* the "First Tree of the World." The western side the pedestal base was sculpted with the image of Ah-Kokah-Mut, a variant name for Yax-Kokah-Mut, the bird god the Yukatek Maya brought to the center during the New Year's ceremony for the installation of the Muluk years.

ENCOUNTERS WITH THE MODERN MAYA WORLD

The components of sacred space that gave form to the Olmec and Maya worlds three thousand years ago still govern the shape of Maya communities today. Nikolai Grube eloquently brought this reality home to Linda Schele during the summer of 1991. Linda and Nikolai were together in Antigua, Guatemala, working on the inscriptions of several Classic kingdoms in between visits to highland sites like Mixco Viejo and Iximche', the capital of the Kaqchikels at the time of the Conquest.[91] One afternoon, Nikolai had arrived late after meetings in Guatemala City to find a contemplative Linda brooding over the structure of this very chapter. Their conversation led Linda to explain the pattern she had discerned in the organization of the sacred space of Copan and other Maya cities. As she talked, Nikolai's face lighted up with excitement and he began telling her about Tixkakal Guardia, the modern Cruzob village where he had been working for the past five years. He agreed to contribute the following description of Tixkakal Guardia to our discussion of sacred place.

THE SPATIAL ORGANIZATION OF TIXKAKAL GUARDIA

Tixkakal Guardia is one of four Shrine Villages of the Cruzob-Maya of East Central Quintana Roo in México. The Cruzob Maya, who today live in several small villages dispersed throughout the forest, are the direct descendants of the Maya of Chan Santa Cruz, who in 1847 staged the largest Indian uprising against white domination in the Americas. This war, during which the Maya almost reconquered the whole peninsula of Yukatan,[92] is known as the Caste War of Yukatan. The rebellious Maya had almost conquered the capital of Yukatan, Mérida, when the rainy season began and the Maya soldiers were forced to return to their milpas to plant their crops for the coming year. According to local oral tradition, the return of the Maya to their fields allowed the Mexicans to drive them out of the more developed and densely populated northwest part of the peninsula. They had to flee into the jungles in the southwest, which were virtually free of permanent Spanish or Mexican settlements. In that remote area, the Maya remained independent for the next fifty years. These independent Maya named their capital *Chan Santa Cruz,* "Small Holy Cross," after a wooden cross that became the object of veneration central to their religion because it made prophecies and gave religious advice. For the Maya of Chan Santa Cruz, the Talking Cross became the Indian counterpart of the Christian Messiah.

Their independence did not last long. In 1901 Mexican troops conquered the capital of the Chan Santa Cruz Maya and, once again, the Maya fled into the deep jungle and settled in small villages. Although the victorious Mexicans burned the Talking Cross of Chan Santa Cruz, the Maya had already split into several small groups. Each of these groups had its own capital, complete with a shrine to house one or more "new" Talking Crosses. Today, the descendants of the Chan Santa Cruz Maya call themselves Cruzob after the most important material component of their religion. They are organized into four groups and live in the state of Quintana Roo on the eastern side of the Yukatan peninsula.

A view toward the Iglesia and the Popol Nah in Tixkakal Guardia

FIGURE 3:30

The shrine village of Tixkakal Guardia is the religious and political center for a group of Cruzob living in about ten villages. The small permanent population of Tixkakal Guardia lives around the sacred precinct in traditional houses, called *xaanil nahoob,* "thatch houses," in their Mayan language. The center of Tixkakal Guardia is the shrine itself, a large oval masonry building with a roof thatched like all of the Maya houses in Yukatan. The shrine is called Iglesia (Fig. 3:30), although the religious ceremonies taking place inside it only superficially resemble Catholic rites. All important ceremonies, from weddings to *okotbatam* **Plate 29** rain ceremonies and *matan* offerings, are conducted inside the shrine.

Three portals lead from the plaza into the shrine, one large front portal and two smaller portals on either side of the building. No one is allowed to enter the shrine wearing shoes, dirty clothes, or when obviously drunk. Armed guardians carefully watch that all entering persons behave according to the rules. The rear of the shrine contains the actual sanctuary where the crosses and some of the holy books are kept. These books are still being used, written, and read publicly by the scribes of the Cruzob.[93] Another portal leads from the publicly accessible part of the Iglesia into the rear sanctuary, which can be entered only by persons holding a religious office.

Next to the Iglesia is a building called the Popol Nah or Community House. While the Popol Nah is the same size as the Iglesia, it is made completely of wood and palm leaves. The function of the Popol Nah is to house all instruments used in the ceremonies of the Cruzob, but it is also a place where the leaders of the communities meet and where political activities, like the election of new officeholders, take place. Like the Iglesia, the Popol Nah has three entrances protected by guardians.

In front of the Popol Nah and the Iglesia is a large platform. Many ritual activities, especially during the week-long fiesta, take place here. One of the main functions of this platform is to serve as a stage where dances can take place. The most important of these begins with the dance of the *wakeras*—old, honored women who are joined, one by one, by all the men of the community. The platform is also the place where individuals are named who are to enact the main roles in the various activities of the next fiesta. This ceremony is called *eleksyones* and lasts an entire night, until all of the participants get drunk in the early morning hours. During the *eleksyones* hundreds of spectators watch the ceremonies, carefully noting the names of the few who have been elected for this honor.

In front of the dance platform is the corral, or bullring, which has become the modern analog of the ancient ballcourt. The corral is roughly circular and is enclosed by a fence, making it also into a kind of sunken court. Since there are no seats, people climb the fence to watch the spectacle from a safe place.

For several days, all members of the subtribe come together in Tixkakal for the fiesta of the Patron Cross, which is the most important of the Cruzob ceremonies. The Fiesta begins with the young men going into the forest to search for a young ceiba tree between twenty and thirty feet high. The tree, which is called *Yax-Che',* "First Tree," is cut, brought to the bullring, and erected in the center as the modern analog of the Wakah-Chan—the World Axis—from ancient imagery. When the tree is firmly planted, a young man called *chik* or coatimundi[94] climbs the tree and throws candies down to the crowd as the ritual proceeds.[95] With the erection of the *Yax-Che',* the week-long fiesta is formally inaugurated. Every day bullfights take place in the bullring. In the time when the Cruzob were poor and had no contact with the outside world, they could not afford to buy and to kill living bulls. Instead, one of the men was placed inside a framework of vines, wood, and sacking in the shape of a bull and simply played the part of the bull. Today, the Cruzob can afford real bullfights. Often, after the bull has been killed, it is divided among the various lineages, who then prepare it as a food offering.

Finally, the four corners of the sacred precinct containing the Iglesia, the Popol Nah, the dance platform, and the bullring are marked by four altars bearing crosses. Around the periphery and outside the boundaries of these corner crosses sit the houses where members of all the lineages from the various villages reside during the fiesta. In the time between the two fiestas, every male member of Cruzob society has to spend at least a week in the shrine center performing religious service and protecting the Cross. During this time, people reside in the house of their lineage.

As Nikolai told his story and drew a diagram of Tixkakal Guardia (Fig. 3:31) in Linda's notebook, they both realized that it reproduced in detail the pattern of the East Court at Copan. There Temple 22 and Structure 22A, the *Yax-Hal-Witz* and the Popol Nah, correspond to the Iglesia at Tixkakal with its sanctuary holding the Talking Cross and the sacred books, and the Tixkakal Council House. In front of the buildings at Copan sat a huge dance platform and a sunken court marked as a false ballcourt where sacrifice took place. A large dance platform also sits in front of the

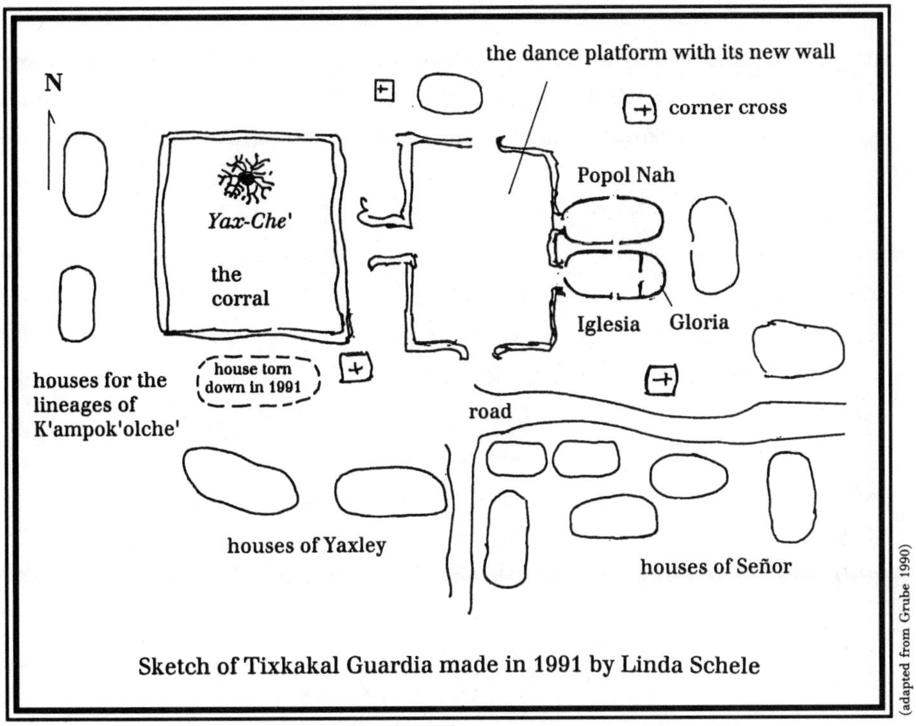

FIGURE 3:31 Layout of Tixkakal Guardia

Iglesia and the Popol Nah at Tixkakal. The Ballcourt at Copan has its counterpart in the bullring, for although the game has changed, the purpose and the end result—sacrifice for the sanctification of ritual—is the same.

This chapter's journey through the sacred landscape of the Maya began with David Freidel's first encounter with the navel of the world. Here at the end, in good Palenque fashion,[96] we close this chapter with a similar experience Linda had in the summer of 1991, twenty years after the incident David described. At the time, Linda was involved in research for this book. She was studying Barbara Tedlock's work at Momostenango, a town in the Guatemalan highlands, where priest-shamans, called *chuchkahawib,* "mother-fathers," undergo long training and initiation rites.

Barbara Tedlock herself became a chuchkahaw and has described the process of training and initiation in her 1982 book *Time and the Highland Maya.* Among the series of lessons, ceremonies, pilgrimages, and transformation rituals she underwent in the process of becoming a chuchkahaw, there is one ceremony that involves the idea of being at the center of everything. During the 260-day period after initiation, the new "mother-father" must learn about a place called *waqibal* or "Six-Place." *Waqibal* is located on the summit of the hill in the center of Momostenango. Called Paklom, this hill "is the 'heart' (*k'ux*) or center of the Momostekan world, spiritually connected to four corner hills of the four directions or corners, each located within a radius of about three kilometers" (B. Tedlock 1982:71). Within this landscape—given shape by these sacred places—the mother-fathers visit shrines that are sacred to individuals, lineages, and the town itself, timing their movements and rituals according to the rhythms of the sacred calendar.

In the summer of 1991, Kaqchikel and K'iche' friends invited Linda to attend rites at Momostenango in celebration of Waxaqib Batz' (8 Monkey), the day when the Momostekan calendar begins and the mother-fathers are initiated. Many highland Maya have adopted this day as their own and celebrate it as a day of renewal of their cultural identity with both traditional and modern rituals.

THE SIX-PLACE

(as told by Linda Schele)

I drove to Momostenango in a rented car, picking up one friend in Tekpan and another at a dusty little crossroads called Los Encuentros. The three of us carried a little food and baskets of offerings for the long night ahead. We followed the paved highway and then turned off onto the washboard dirt road that led to Momos. Along the way, we saw Duncan Earle, the friend and ethnographer who would accompany me to Chamula six months later. Duncan had driven past us on the road, recognized me through the steady drizzle that had fallen throughout the day, stopped his van, and come running back to stop me. The apparition of his purple-coated body loping down the road waving his arms to get my attention brought me to a skidding stop. He told us that there would be a ritual that night atop a mountain at a location called *nima sabal* or "Nine-Place," and that he would meet us there. We continued on and entered the crowded center of Momostenango about an hour before dark. Parking the car at the edge of the town plaza filled with the makeshift shelters of the Sunday market, we made our way through the crowd. Pakal and Saqahix, my Maya friends, led the way.

I told Pakal that I wanted to see Paklom and explained that I knew that place was important from my reading of Barbara Tedlock's work and from studying Maya Creation for this book. He quickly asked around and found that Paklom was located under the local radio tower. Somehow I had visualized the center of Momostenango a little to the side of a valley just like Zinacantan's *mixik' balamil*, the "navel of the world," but I was wrong. Pakal, with all the energy of the young, promptly led us toward the steepest hill around—mounted by a tile-covered road. I followed my companions upward, accompanied by two other North Americans we had met along the way—all of us gringos huffing at the steepness of the hill and the altitude. Climbing past stores and homes, we finally breasted the hill and walked out onto the cleared area at the base of the local radio tower. Screaming children kicked an aged soccer ball back and forth across the small field while we looked around for *waqibal*, the center of Momostenango's sacred landscape.

It was there, singularly unimpressive for so sacred a place, located across from the makeshift soccer field at the edge of the hill (Fig. 3:32). On one side was a man-high rise scarred by shallow pits, blackened by

(photograph © Robert Carmack)

Figure 3:32

centuries of burning copal and covered by the broken remains of hundreds of clay pots. A mother-father and his assistants knelt before a pillar of smoke rising from the copal incense they burned, chanting prayers against the shrill background of the playing children. Piles of copal pellets bound into corn-husk tubes sat beside him ready for his all-night vigil. Across a shallow depression on the other side of the hill was another small rise marked by a small concrete pedestal and three soot-blackened crosses. A family knelt in prayer before the crosses, burning their own incense to blacken the hallowed altar of that place.

We stood there, caught halfway between the most sacred piece of landscape in the valley and the thoughtless enthusiasm of the playing children, between the beliefs and rituals of the traditional world of the Momostekans and the modern world we had left behind only hours before. For reasons I still do not understand, a powerful presence embraced me in that moment, evoking an expanding emotion that eased out of my conscious control and brought tears to my eyes. I was at the place I had read about. I knew what it meant from the outside, from the detached view of the anthropology created by our worldview—that thing we call science. Yet, at that moment, I was neither detached nor objective.

I knew where I stood in time and space. For the Momostekans, I was at the center of everything—their version of the *mixik' balamil*, the navel of the world. They call it *waqibal*, "Six-Place," and I knew that the Classic Maya called the tree at the center of their world the Wakah-Chan, "Six-Sky" or "Raised-up-Sky." I truly did stand at the place of the Creation, not only in terms of space, as the Momostekans understand it, but in terms of time. The axis of *waqibal* pierced straight back through three thousand years to the kings on their stone images at La Venta. *Waqibal* truly was and is hallowed ground and in my heart I centered the world.

CHAPTER FOUR

MAYA SOULS

IDOLS AND SAINTS

(as told by David Freidel)

Yaxkaba is a sleepy little town off the main road between Chich'en Itza and Mérida, the capital of Yukatan. It is the county seat for Yaxuna, a few kilometers to the east. Driving into the dusty center, one is immediately struck by the vision of a massive, beige-limestone eighteenth-century church. Still graced by its original three spires, it rises majestically from a plaza adorned with red-flowered trees. It is a grand and desolate monument to the Christian god and his saints. I found a piece of Yukatek history under Spanish rule hidden in this quiet place.

In the summer of 1992, I took a small group of American tourists to Yaxkaba to show them the church. The young priest is a local Maya, gentle, caring, and learned. In one auxiliary building next to the church, he showed us the place where, during the Caste War, the militia of the nation fortified graceful arches to protect themselves against the Chan Santa Cruz rebels lurking in the forests to the east and south. Pillboxes still edge the churchyard, and gun slits are visible in the front facade of the church. Here for a time in the nineteenth century was a border between the free Maya and their enemies. The church, the best lookout tower around, became a symbol of that struggle. The priest who was our guide had thought about having the defensive walls removed so the arches could stand clear again, but these walls are part of history and must stay.

We went into the cool interior of the church and slowly walked its cathedrallike length. Under its broad barrel vault, huge wooden altars edge the high white walls. One altar holds an image of the Virgin in mourning black lace, hovering above a Christ in white winding sheets, his image laid out in a glass coffin. Across the way is a large cross painted bright green with a bloody red heart at its center. The cross is decorated with the multicolored symbols of the religious brotherhoods of the community, the *cofradías*. At the time of our visit, this cross was dressed with a beautifully embroidered banner draped across its arms. The words on the banner thanked the cross for answering the prayer of a supplicant and gave the date of the offering.

At the front of the church, we found ourselves at a cluster of benches huddled at the foot of the high altar. This altar is a glorious and elaborately carved wooden original, rising to the ceiling with masterfully sculptured portraits of saints who still bear flecks of their original gold and painted colors. At the center of the altar stands the more freshly painted image of San Francisco, the patron, inside a niche framed with glass doors. The priest took us around to the back of the church where there is a stairway leading up to a small door. This door leads into the niche holding San Francisco so that he can be turned around to face inward on Good Friday before the midnight resurrection of Christ, or removed for redressing and processions.

We walked out into the morning glare and stopped briefly to look at another building along the side of the church. I was moving on when several of the tourists stopped me and pointed to the crosses carved above the archways: they looked as if they had ears of maize on the ends of their arms. Yes, said the priest, they were maize adornments. The main arch carried the date of the building, 1789, and we wondered at finding yet another bridge between the Classic Maya maize tree and the cross of Christianity.

I next took our group over to the library at the edge of the plaza. There we looked at a carved bas-relief depicting a seated lord of the Terminal Classic period (A.D. 800–1000) wearing a god mask. The glyphic text, I explained, showed that he was acceding to the title of *sah* or Fearful One, a rank second only to Ahaw, "king," in parts of the southern lowlands. The style of the god mask was particular to the lords of this local area in the time of Chich'en Itza's great power, I said, so this was probably one of the lords of Yaxkaba in that period. An unusual feature of the relief portrait was a circular hole, about two centimeters deep and ten centimeters in diameter, carved in the lap of the seated lord. This hole is about the size of one of the mosaic mirrors people used during that time and I speculated that this relief carried such a mirror. These mirrors, found at nearby Chich'en Itza, served as portals for ancestors and gods.[1]

I trundled everybody out into the van and we drove over a bumpy dirt road to the abandoned church on the edge of town. Sitting in an overgrown farm field at a place called Mopila, this building was probably built in the seventeenth, or even the late-sixteenth, century. The roof is mostly gone over the nave, but the gray stone walls are still standing and some great black wooden beams, from the time the roof collapsed, are still scattered about. Over the apse, the roof is still intact, and protects a weathered and dilapidated wooden altar, a smaller version of the one in the main church. The niche is empty and the glass broken out of its door, but the saints are still here (Fig. 4:1). At the base of the altar stands an old wooden image. Somebody once decapitated this saint; then later people took the headless body and replaced its missing head with a plain stone. It makes for a faceless head but is sufficient for the purpose. Next to the saint is a large and simple wooden cross. The area around the altar is swept clean. Neatly arranged masonry blocks from the ruins, arranged in a semicircle before the altar, function as seats. Traces of burned incense and candles, corn husks, and gourd bowls, show that people from the vicinity still come here to pray before these images under the open sky.

After a few quiet moments, I took the group around to the outside of the church and pointed to a hole high up in the wall directly behind the niche of the wooden altar. That's where the image of the seated lord had been before the townspeople took it to the library for safekeeping. This older, now abandoned church had been built by the Maya under the direction of their new masters shortly after the Conquest. At first appearance its basic structure looks a lot like the design of the newer church, but the stone of its walls came from old Maya structures—buildings that had been sacred places for generations before the Spanish had come. Maya masons had mortared one of their own cherished heirloom images, a lord from a time of past glory, into the wall of the new foreign place of worship. In the church the Maya use today, a doorway sits behind the image of San Francisco. In the original church, the saint sat in front of a proud Maya ancestor, perhaps cradling his mirror portal to the Otherworld. Who indeed received the prayers of those Maya, and who receives them now?

* * *

The altar in an abandoned church at Mopila, Yukatan. A wooden cross sits next to a saint whose head has been replaced by a stone from nearby ruins.

FIGURE 4:1

From the moment the first of the Spanish invaders stepped ashore on the east coast of Yukatan, they began using words that would automatically classify Maya religion as pagan and worthless. Maya day-keepers, doers, seers, and mother-fathers were *brujos*, "witches," while Spanish priests were *padres*, "fathers." The statues in a Maya temple were *ídolos*, "idols," while statues in a Spanish church were *santos*. Maya gods were demons, although the closest word in their languages to the European concept is *kisin*. *Kisin* ("farter") does not connote evil in the Western sense, but is closer in spirit to the ideas of the Greeks, who regarded their daemons as supernatural beings who mediated between the gods and humankind. The Yukateks also called their statues *k'ulche',* "divine tree" or "holy thing of wood," a term that is still used these days among the

Maya for saints and crucifix crosses. Finally, an idol was simply a *k'ul*, a "holy spirit." The contrast between the material object, and the spirit manifested through it, was and is real to the Maya, but so is the union of the two.

The Catholic Spanish had no difficulty with such distinctions, for they worshiped their crosses and saintly images in much the same manner as the Maya did their own images of the gods. The remarkable congruency between the worship of saints and crosses, on the one hand, and the worship of idols and standing stones or posts on the other, fostered the eventual substitution of the former for the latter in Maya thought.[2] Indeed, the ancient Maya word *k'ul*, "holy spirit" or "god" (which makes up part of *k'ul ahaw*, the title of rulers during the Classic period), has been replaced in some contemporary Yukatek communities by the word *le santoho*, "that saint."[3] Modern Maya crosses are called *santo*[4] or *santoh de che'*. In this context "saint," used as an adjective, means "derived from God," the sacred source of all lifegiving things. Farmers can gratefully observe, "Here comes the saint rain." What began in Catholic teaching as an image portraying a supernatural being now denotes the sacred quality of all important things. The ancient Maya used the word *k'ul* in their texts in much the same way.

Contemporary Maya regard images, crosses, or saints not as alive or sentient but as places where sentient spirits can abide. John Sosa describes how the tree-crosses,[5] *santoh de che'*, at the perimeter paths of Yalkoba in Yukatan have slabstone altars at their bases where *yum balam*, "father guardian," rests and receives offerings of little pebbles touched to the bottoms of the sandals of passersby. "Father guardian" is neither the cross nor the altar, but he can be present there on his *k'an che'*, his "bench."

The Maya relationships between spirit and object, between soul and person, are dependent upon the relationship between human agents and supernaturals. The actual operation is very similar to what we would call, in our culture, "channeling." For every talking cross, stone, or object there is a human agent who acts as the mouthpiece through which the supernatural spirit can speak. The Talking Crosses of Quintana Roo could also deliver messages in written form. We suspect the old statues also spoke through the written word. The old word for the human agent was *chilan*, "translator."

On much rarer occasions, the object itself speaks without any outside help. Sometimes communication is unexpected. Several times since the Conquest, supernatural beings have manifested directly in objects that "talked," and most of them exhorted their followers to throw off the yoke of the foreigners.[6] The tradition of talking statues certainly goes back to Precolumbian times. The Oracle of Ix Chel on Cozumel, a life-sized clay statue of an old woman, was such a talking object. The Spaniards who saw it[7] said there was a small door so that a Maya priest could enter it and speak for the goddess unseen by the supplicants. This oracle was no doubt only one of several on the eastern coast of Yukatan in the Postclassic period. The most famous Maya talking image is the Talking Cross of the Chan Santa Cruz Maya. Although the first "translator" was killed by Mexican soldiers, another took his place and the Cross continued to communicate both verbally and in written letters. For more than ten years, it was the effective ruler of the rebellious Maya of the Chan Santa Cruz. Today that Talking Cross sits in Tixkakal Guardia. It still speaks to the descendants of those rebels through the descendants of the original translator.

In 1867, a Chamulan girl in highland Chiapas saw three stones drop out of the sky while she was tending her sheep. She took them home and gave them to a local Maya official, Pedro Diaz Cuscat who, some weeks later, created a cult around them. He claimed that the stones talked and that the girl, Augustina Gómez Checheb, had given birth to them; therefore, she was the Mother of God and the stones were saints. So began another Maya revolt. Just as the original Talking Cross of the Chan Santa Cruz Maya embodies the symbolism of the World Tree, so the three stones represent the three hearthstones of Classic Maya Creation cosmology.

Stones continue to talk among the modern Maya, although they do not currently threaten the civil peace. Paul Sullivan, in his sensitive and penetrating book on the Chan Santa Cruz Maya,[8] relates the tale of a boy who in 1985 was addressed by a stone figure who wished to adopt him. The stone also spoke to the boy's father, a shaman, who transformed their home into a chapel, which was soon a center for devotees from all around the surrounding countryside. David Freidel has confronted this special feeling about stones in his work as an archaeologist. He describes it this way.

THE TALKING STONES

(as told by David Freidel)

When I read Paul Sullivan's book, it helped me understand something I had witnessed among the village people of Yaxuna who worked with me on the nearby ancient city. When excavation first began, the villagers were deeply concerned that we might try to remove stones, especially carved stones, from the ruins. I had difficulty understanding their anxiety. I explained to them that sometimes artifacts had to be removed for analysis, but that they would be returned faithfully when safe storage could be built for them. The matter was of such importance to the villagers that finally Don Pablo, the local shaman, took it personally upon himself to ensure that no carved stones be removed from the site. There were some strained moments when the archaeologists of the Mexican government insisted that carved stones be taken to safekeeping and the Yaxuna people insisted that they stay; but the tensions were finally resolved. The stones of Yaxuna are still there, under the watchful eyes of the villagers, and I now know why the matter loomed so large: such stones are likely *k'an che',* seats of supernaturals.

I had one other encounter with Don Pablo and talking stones. One day in the summer of 1989, after he had done some work on the camp kitchen, I found a clear glass marble in the area. Thinking it belonged to Don Pablo and was one of his *saso'ob,* the "lights" he used when focusing spiritual forces, I took it next door to him that evening. He took the marble and inspected it carefully.

"Yes," he said finally, "this is a stone of light."

Then he smiled, "However, it won't speak until it has been soaked in maize gruel, *sak-a',* and then it will only speak Maya."

WORKING PRAYERS

Spirits speak, whatever their material appearance, because words are a fundamental medium for Maya communion between this world and the Otherworld. They are not merely preamble to some magical action or a way of describing things that are manifested through tangible supernatural and natural forces like rain, lightning, and thunder. They are, rather, an essential conduit of those forces.

The Creator gods of the Popol Vuh made humans so that they might have intelligent, nurturing beings to communicate with. Even so, it is important to stress that the means through which the gods and ancestors convey their messages and guidance are still human agents, the Maya shamans and priests who relate these insights to their people. This same rule held in antiquity, when kings and other exalted individuals could become, under special circumstances, the talking idols of their people. Because their bodies were the vessels of life's sacredness, they acted as the conduits of divine revelation. They recorded their visions in words of stone that today open the world of their ancestors and gods to their descendants and to those of us who seek to understand their world.

John Sosa's Yukatek h-men teacher refers to the rituals he performs as "sacred work"—and it is indeed work to bring gods and saints among living people. Stones and crosses may speak first to innocents, but the miracle of Maya visionary communication is rarely a matter of spontaneous expression. The circumstances in which an individual can enter into vision, see apparitions, or communicate with the divine requires diligent work and attention on the part of many people. In the deep past, individuals—whether kings, shamans, or patriarchs—who could enter and leave the world of the spirit at will, manipulating its forces and bringing back its wisdom to the rest of the community, stood at the center of social and political power. In a world of this kind, soul and the supernatural were dynamic phenomena that fulfilled a role in the material world that it is difficult for the Western mind to fathom.

When Sosa's friend prays, he regularly includes the phrase *sak-a' che' t'an,* "clear water tree speech." *Sak* is "white, clear, human-made"; *a'* is *ha',* "water"; and *sak-a'* is a special gruel of maize and water used in ritual offerings. When interpreted in the context of Classic-period symbolism, this combination of words brings together the idea of maize gruel, the stuff and sustenance of human life, and the World Tree, the portal of communion with the divine. Just as maize is the basic food of life and the dough that humans are made from, maize gruel is the primeval sacred liquid. Several of the elaborately painted vessels used in royal banqueting and ritual in the Classic period have been identified as vessels made to hold *sak-a'.*[9]

In the Popol Vuh, the Hero Twins jumped into a great fire pit where they were immolated. This is a direct analog to the creation of a special maize gruel by the modern Maya. Today in the community of Santiago Atitlan, this maize gruel, called *maatz*, is made by roasting the maize kernels. Just as the bones of the Hero Twins were ground and cast into the river, so the kernels of burned maize are ground and mixed with water. The sacrificial acts of the Hero Twins set the stage for their return to life and eventual triumph over death. Later, human beings of the new Creation were made of maize and water drawn from the pool at the center of the Cleft Mountain. In that episode, the water in the gruel is explicitly declared to be human blood.[10]

To the Maya of Guatemala and Chiapas, the gruel is like semen and is the source of rebirth. In the Atitlan ritual, the association is explicit. Linda Schele experienced this sacred substance directly when served maize gruel at San Juan Chamula in March 1992. The gruel at Chamula is made by the Pasión himself, who must stand in a hot, steaming shed stirring a huge vat of the stuff for hours while it thickens over a fire. The resulting mixture is thick and almost gelatinous in consistency, translucent and milky-white in color, and sweet in flavor. Like Jell-O, it flows only slowly. The resemblance to semen is immediate and inescapable.

Just as in the ancient past, so today the words of the translators, the doers, and the mother-fathers who speak to the ancestors and to the spiritual beings who inhabit the crosses and statues of the Maya world are associated with Creation and to the substance that gave form to human flesh.

ON SOULS

The Maya conceptions of soul are complicated and multifarious. Still, they sometimes find partial parallels in Western experience. For example, John Sosa's informants in the Yukatan culture conceive of two *pixom*, "souls"—one good and the other bad.[11] The contrast of good and evil is familiar to us. While the concept of two souls is not, this idea points to a fundamental and distinctive duality of spirit inherent to Maya reality. Zinacanteco Maya also have two souls,[12] as do many other Tzotzil-speaking peoples of highland Chiapas. For them, one of these souls is invisible, indestructible, and divided into thirteen parts. A scary accident, like a fall on a mountain path, can cause a person to lose a piece of this soul and become ill. Evil witches can also steal pieces of soul and sell them to the Earth Lord. Shaman seers can find the missing piece with proper preparation, the cooperation of the subject, and supernatural help and restore the human soul. Zinacantecos call this kind of soul *ch'ulel,*[13] which has the same root as *ch'ul* or *k'ul,* a word used by ancient Maya scribes to describe "holiness" and "divinity."[14]

Although subject to temporary damage, the *ch'ulel* is eternal. When a person dies, this soul hangs around the grave for some time, then becomes part of a pool of such souls in the care of the Father-Mothers, the ancestral gods, who may decide to plant it again in the body of a newborn baby. Parents have to be especially careful with these babies: "A small child is especially susceptible to 'soul loss,' as the soul is not yet accustomed to its new receptacle" (Vogt 1976:18). Ch'ulelob can move around outside the body, in dreams during sleep, and visit the ancestors. The real power of the experience of soul, however, is revealed in its universality:

> Virtually everything important and valuable to Zinacantecos also possesses a CH'ULEL: domestic animals and plants, salt, houses and household fires, crosses, the saints, musical instruments, maize, and all the other deities in the pantheon. The most important interaction in the universe is not between persons and objects, but among the innate souls of persons and material objects.
>
> (Vogt 1976:18–19)

It is hardly surprising that the Classic Maya called their kings *ch'ul ahaw,* "lords of the life-force."

The second type of soul conceptualized by the Tzotzil Maya is called *chanul,* derived from the word for animal. This concept may also be related to the word *kanul,* which means, in some Yukatek Maya communities of Quintana Roo, "supernatural guardian" or "protector."[15] This is a supernatural companion, which usually takes the guise of a wild animal and shares *ch'ulel* with a person from birth. The fates of the baby and the animal spirit are intertwined, so that what befalls the one affects the other for good or ill. Usually the ancestral Father-Mothers take care of the *chanul* and keep it in one of their magical corrals inside the mountains. But if they get angry, they can punish someone by neglecting their *chanul.* People learn the nature of their *chanul* in dreams or through the insights of the shaman who cures them of an illness or guides them on a spiritual journey. Like the human body, the *chanul* is subject to mortal danger.

The Classic Maya also used *ch'ul* as one of their words for soul, but like

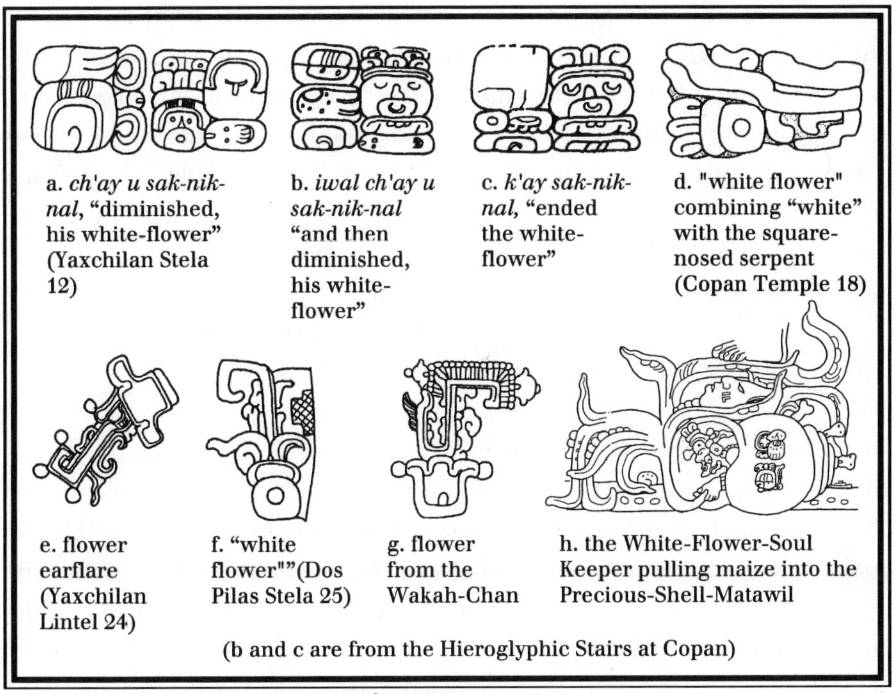

FIGURE 4:2 "White Flower" as the Term for Soul

the Tzotzil and the Yukatek, they apparently believed in more than one kind of soul. Another word they used for soul combined the signs for "white" with the signs ahaw, *ik'* and *li.* Proskouriakoff[16] first recognized this phrase in the death statement she identified in the inscriptions of Yaxchilan, but her idea that it was related to soul came from the appearance of the syllable *ik'* which means "breath" and "spirit." It turns out that her idea was correct, but her reading wrong. Nikolai Grube and David Stuart have independently found clues to the correct reading: *ch'ay sak-nik-nal,* "expired, the white-flower-thing" (Fig. 4:2).

For us, the most important context for this new reading is found on depictions of the World Tree. The bell-shaped objects on the ends of its branches (with square-nosed serpents emerging from them) are the iconic representation of these white flowers.[17] When all of this is put together, it tells us that the souls of human beings were created when First Father raised the great tree. We are the blossoms of that tree. All that remained was to shape human flesh from maize and create the host for these flower souls. Very probably this maize came from the First-Tree-Precious depicted in the Temple of the Foliated Cross at Palenque, the tree that rises from the three stones of Creation in Orion.

Moreover, the Temple of the Foliated Cross tells us what happens to this white-flower soul when its owner dies. On the main panel, the personification of the flower soul appears in the mouth of a huge conch shell where it is shown pulling in a maize plant. This great conch carries glyphs calling it *K'an-Hub-Matawil* (Fig. 4:2h), "Precious-Shell-Matawil." Matawil[18] is the place where First Mother and her children—the Palenque Triad gods—reside. Thus, the keeper of the white-flower souls takes them to dwell with First Mother in the Otherworld.

These three ways of defining the life-force resonate in the contemporary communities of other Mayan-speaking peoples, as well as in the world of their ancestors. The Tzotzil of Chamula speak of a sky tree of many breasts that suckles the souls of unweaned babies, while the people of Zinacantan say that the souls of dead babies change into flowers tied to the celestial cross.[19] The K'iche'-speaking peoples of highland Guatemala conceive of *nawals*, and in this way differ from both the Tzotzil and the Yukatek Maya. Scholars generally use the term "nawalism" to describe the notion that an animal or a spirit companion is linked with a human being from birth.[20] For the K'iche' the nawal is the "spirit of the day" on which a child is born.[21] These spirits are also associated with the powerful deities of the four quarters, the great Mundos—"worlds"—the Earth Lords who ascend as nawals into the midnight rafters above shaman mediums to communicate in noises and voices to the frightened audience seated on the floor.[22] For the K'iche' the word also applies directly to the souls of their deceased ancestors, thus blurring the distinc-

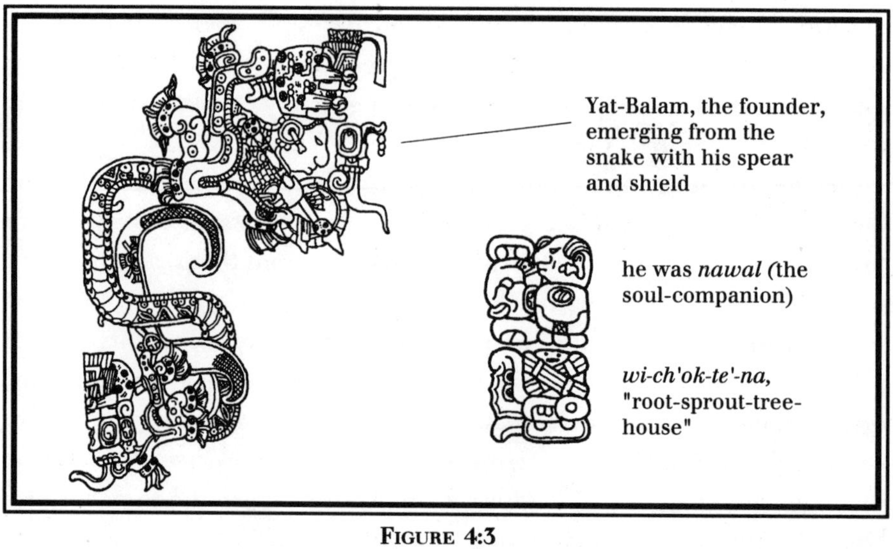

FIGURE 4:3

tion between human souls and animal-spirit companions made by the Zinacantecos. Calling an ancestral soul a nawal also has precedent in the practices of the ancient Maya. The image of Yat-Balam, the founding ancestor of the dynasty of Yaxchilan, emerging from the mouth of an enormous Vision Serpent is called "nawal of the founder"[23] (Fig. 4:3).

In the summer of 1992, Linda Schele got the chance to visit the most sacred of these ancestral shrines—the cave under the ruins of the K'iche' capital. Today tourists call it Utatlan, its Aztec name, but the K'iche' call it Qumaar Kah. Here is her story:

THE CAVE AT UTATLAN

(as told by Linda Schele)

I had been in Antigua for only a short afternoon when Duncan Earle arrived at my door to remind me of the trip he had planned for us. It was time for him to return to his chuchkahaw mentor, Don Lucas, for the next stage of prayer and sanctification in his own journey as a Maya-trained shaman, and he wanted me to go with him. The next morning he arrived in the huge Texas A & M van he had brought down for his field school. It was loaded with his wife, herself an experienced ethnographer, their two children, and a graduate student who had agreed to help baby-sit.

Off we went very early on a Saturday morning, out of Antigua, along the winding Panamerican Highway to the fork that took us to Chichicastenango and beyond to Santa Cruz K'iche' and the little town of Chinik'.[24] There, just outside of town, we came to a long series of houses built along the edge of the road. These were the houses of Don Lucas and all the members of his family. Duncan saw Manuel Pacheco, the oldest son, and invited him to lunch with us in the village. Afterward we returned to the house compound, and Don Lucas made a small ceremony at a shrine in the arroyo behind their house.

Then we piled into the van along with Don Lucas, Manuel, and Manuel's teenaged daughter to go to Utatlan. For the ritual Don Lucas brought his huge divining crystal and Duncan brought moonshine from México. We drove back to Santa Cruz K'iche, stopped to buy the candles and copal we needed, and then headed for the grass-shrouded ruins of Qumaar Kah. Unlike the Kaqchikel capital of Iximche', the K'iche capital has never been cleared archaeologically and consolidated. Its mounds lie under a pine grove, their forms and functions hard to discern beneath five hundred years of deterioration, neglect, and the use of its stones for making the streets of colonial towns.

After parking, we walked past the low mounds along a dirt path that leads along the side of the central plaza, ringed by high temples, until we dropped over the edge of the plateau upon which the city was built and climbed down to the cave. The entrance was blackened Plate 43 by centuries of burning copal and surrounded by piles of corn husks that had once held disks of copal. In the cave ritual, the people take the pellets out of the husks before throwing them into a fire. At Momostenango, I had seen the chuchkahawib throwing the entire packets—husks and all—into the fire. While we were waiting for a group of Ladino schoolchildren to leave the cave and give us some peace, three college students from the United States arrived at the entrance. They were clearly curious about us and respectful. The questions they asked led Don Lucas to invite them in with us.

The cave was narrow, deep, and very dark.[25] I distributed the four or five small flashlights I regularly keep in my pack for the tunnels in Copan to people in our group. We entered, feeling our way along the uneven floor until we arrived eventually at a tight little chamber at the end of the cave. There was the altar in the form of a large stone covered with thick layers of candle wax and surrounded by soot-blackened walls. In fact, I got more than one warning about leaning up against the walls and getting the soot on my jacket. On the altar lay the head and feet of a chicken that had been sacrificed that morning.

Don Lucas squatted in front of the altar and prepared to begin the ritual of the next stage of Duncan's preparation as a shaman. He carefully avoided disturbing the remnants of the morning ritual. He made sure Duncan, who was squatting next to him, had the videocamera turned on, and then he began his prayers. Don Lucas wanted the rituals and prayers recorded so that they wouldn't be lost. No one but Duncan and Manuel and his daughter understood the words Don Lucas chanted as he lighted candle after candle. He would soften the base of one candle over the flame of a previous one and stick it upright on the altar. Before long, a row of burning candles illuminated the prayer-filled darkness.

When Don Lucas finished the ceremony some twenty minutes later, he led Duncan into a side corridor that angled off to another chamber even smaller than the first. The rest of us followed and quickly realized there was no room. We retreated one by one to the main altar where we waited for them to finish. In the candlelit darkness, we all fell into our own thoughts. Some of us spoke, but the conversations were whispered, as if we would break the magic of the moment if we made too much noise. Duncan's wife, Erica, was trying to explain a little to the students so that they would understand what Don Lucas was doing. I tried to help too.

One of us, I don't remember which, switched into Spanish to ask Manuel a question about the altar. Manuel has his own bundle and is a chuchkahaw, a shaman, in his own right. In the flickering light, he stepped forward and answered our question. Then he began speaking in his soft voice, volunteering to share his learning with people who clearly cared and wanted to know.

We translated for the students as he began to explain that the chicken had been sacrificed in the morning by people from the nearby region. They had come to ask the Mundos, the Earth Lords, who lived in the cave for permission to kill animals so that their families could eat. The Mundos also had to give permission for the farmers to disturb the earth when they planted maize. The Mundos required a gift in exchange for the bounty they supplied the people. The chicken, he said, had been the gift the farmer had given to the Earth Lords.

I asked him about the Mundos. Who are they? He said there was Tekun Uman, the most important and powerful one. I knew that Tekun Uman was the name of the last K'iche' king who had ruled there, but for Manuel and his people Tekun had become an Earth Lord, a spiritual being who protected the land, its bounty, and the human community. It was as if Manuel's concept of history—his Creation—began with the death of the last independent ruler of his people. He explained more about how the Mundos worked, about the special power that the chuchkahawob drew from the cave. This was the most sacred place in all the K'iche' world. Later in the week that followed, when he came to the workshop on Creation that Nikolai Grube, Federico Fahsen, and I gave in Antigua, Manuel told us that the people of his town prefer to bury the afterbirth of their children in the cave at Utatlan because of the power of the ancestors there. The souls of many, many chuchkahawob abide in and near the cave ready to help their successors in their work. For the K'iche', the cave is alive with the most powerful energies of the Otherworld.

* * *

In death, ordinary K'iche' souls linger in prisonlike places, released only on special days to commune with their descendants at grave sites or in churches. They receive prayers and offerings from those who remember them. Great souls of shamans cluster at the shrines they used in life, comprising the *mai,*[26] the generations, recalled in the prayers of the living. In this way, the shrines constantly increase in power and efficacy with the lives and deaths of the shamans who use them. Among the K'iche', many of the shamans are also calendar priests and leaders of their lineages. It's not surprising that many of the important shrines are dedicated to lineage ancestors and that shamans visit them according to carefully calculated calendrical schedules. In principle, each K'iche' lineage in Momostenango has nine of these special shrines. The Zinacanteco notion of reincarnation may be echoed in the presentation of pregnant K'iche' women before lineage shrines, so that the current lineage priest-shaman can "sow or plant" the soul of the new family member into her womb from the pool of ancestral souls.[27]

The K'iche' lineage shrines reveal something of the integration of sacred and natural spaces in the highland Maya world. In fact, many are placed in low spots close to natural springs or other water sources, while others are placed high on mountain tops.[28] Here as elsewhere among the Maya, the shrines represent points in a grand pattern of procession and visitation, timely ritual action, and prayer. They are holy places along the path of words.

The K'iche' shrines also serve as important markers on our own journey into the Maya past. Their word for lineage or *cofradía* shrine, where the dead souls are propitiated, is *warabal ja,*[29] literally "sleeping house." This concept has echoes in the Maya Classic period. The logic of calling the shrine home of a dead soul a "sleeping house" is given by one of Garrett Cook's informants in Momostenango:

> "This then is what I believe. It is as we say when we dream of someone who has died. We have known him and we have a dream. Then we say that the spirit comes down visibly to our spirits in this way when we sleep, and so the two spirits converse, the living and the dead."
>
> (Cook 1986:146)

When we examine Classic Maya glyphic texts, we can see that the modern K'iche' term for sleeping house is derived from the glyph *waybil. Waybil* refers to a number of important classic structures, including the Temple of the Inscriptions at Tikal.[30] *U waybil ch'ul,* literally "the sleeping place of god," is the name of a class of little stone houses found near residential groups at the Classic-period city of Copan (Fig. 4:4a). Nikolai Grube and Linda Schele proposed that these stone house effigies were gifts from the king of Copan to close associates and relatives. The *warabal ja*, the lineage shrines of the K'iche', are typically small, box-shaped altars made of slabstones. In Momostenango, these little altars were reputedly gifts to the lineages from the founder of the new town in the seventeenth century.[31]

The identity of these Copan house models as lineage altars was confirmed when Will Andrews and his team from Tulane University excavated the residential area south of the Acropolis. In 1991, he excavated a small L-shaped building located on a terrace above the residential compound and below the Acropolis. The little house models had been

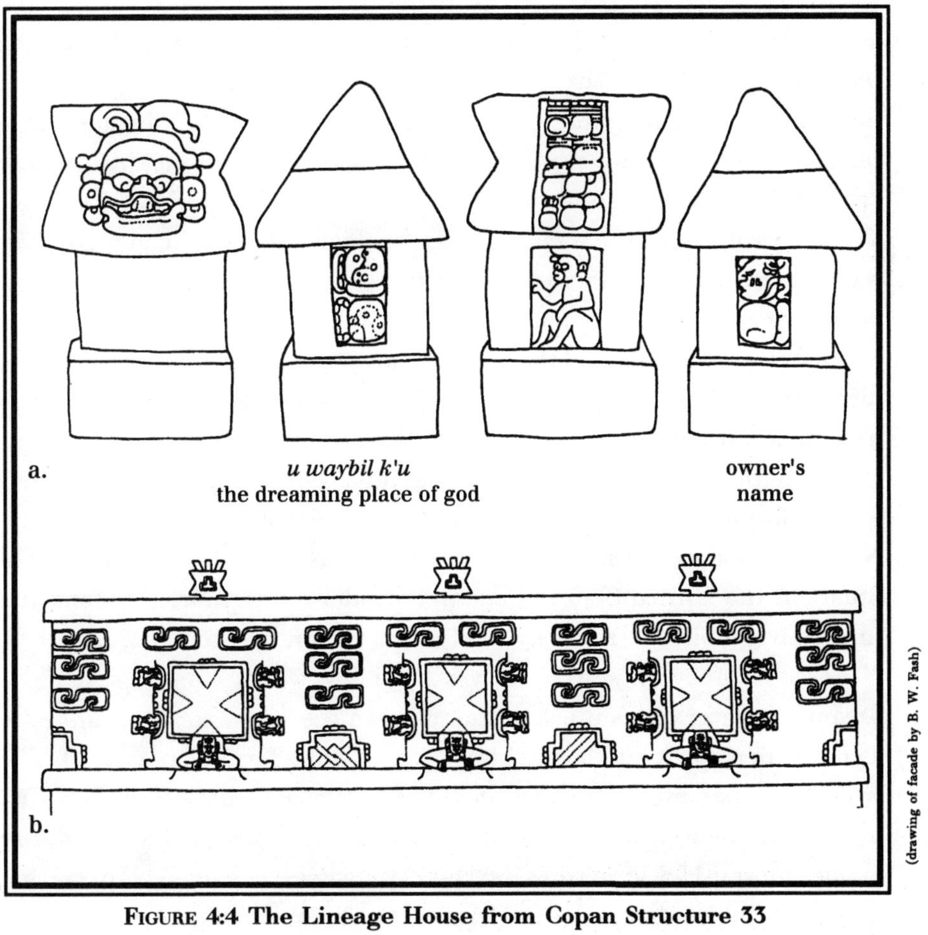

FIGURE 4:4 The Lineage House from Copan Structure 33

found on the east and south sides of this building. By a miracle of time and luck the whole rear wall of the building had fallen intact. With exemplary technique, the Tulane team excavated the fallen wall stone by stone, carefully marking its exact position in the fall. Then the inspired eyes of Barbara Fash and Rudi Larios managed to put most of it back together again.[32] When they were done, they had a wall (Fig. 4:4b) with ancestor cartouches surrounded by S-shaped cloud scrolls. In Yukatek, *tzak*, the word for "conjure," also means "to conjure clouds." These may be clouds in the sky or clouds of incense smoke, but they are the medium in which the vision comes. Here they float all over the *waybil* in which the ancestors of Copan's royal house were conjured.

Although the *waybil* structures and altars of the Classic Maya may be the prototypes of the modern K'iche' lineage shrines, it is likely that the ancient altars also housed gods intimately related to the lineages as well as lineage ancestors. A death and resurrection memorial of a king of Palenque, K'an-Hok'-Chitam (Fig. 4:5), refers to the "housing" of the deceased in the *waybil* of a god named Ox-Bolon-Chak under the auspices of Pakal, the reigning king of that city. The text tells us that someone "stepped on the base of the mountain"—an expression referring to dancing—in a Plate 37
ceremony with an actor named *u ch'ul hil u tz'akhi* K'an-Hok'-Chitam—"the soul of the deceased one, the replacement of K'an-Hok'-Chitam." When we connected the concept of the *waybil* lineage shrine to this dance, we realized that this ritual signified the process by which a descendant became the replacement of his ancestor. In other words, Pakal's second son became the replacement of the ancestral king K'an-Hok'-Chitam, who had died exactly ninety-four years earlier. The rites necessary to this transformation occurred in front of a *waybil* shrine.

The notion of powerful humans having special soul-bonds with gods is now documented in Classic Maya thought by more than just imagery. Nikolai Grube in Germany and the team of Stephen Houston and David Stuart in the United States[33] have independently deciphered a centrally important glyph that reads *way* (Fig. 4:6) or "animal companion spirit." *Way* as "companion spirit" derives from the words "to sleep" and "to dream." In later times it also came to mean the act of transforming into a companion spirit and bewitching (in the Maya form of wizardry). The words for the lineage shrine—*waybil* in the ancient inscriptions and *warabal* in modern K'iche'—derive from the same word. The word *way* applies to Kan-Hok'-Chitam's dance in two ways—his lineage shrine was a *waybil* and he danced as the *way* of Ox-Bolon-Chak.

och ?? tu waybil Ox-Bolon-Chak
entered into his dreaming-place, Three-Nine-Chak

tek'ah yok tu witzil u ch'ul ?? u tz'akabih K'an-Hok'-Chitam
he stepped on the foot of his mountain, his god, and then he replaced K'an-Hok'-Chitam.

FIGURE 4:5 The Dumbarton Oaks Tablet

The *wayob* of Classic Maya imagery appeared in many guises, including humanlike forms, animals of all sorts, and grotesque combinations of human and animal bodies. *Wayob* are depicted dancing like human beings, although they can also float in the air above the action. It is interesting that pottery scenes from most of the major kingdoms depict creatures who are the *way* of their ruling lords (Fig. 4:7); but with the exception of the rulers of Palenque, individual kings never recorded the names of their *way* in the texts on their monuments. From this we deduce

Three *wayob,* two of the skeletal type and one a water jaguar who is the *way* of a Seibal lord

FIGURE 4:6

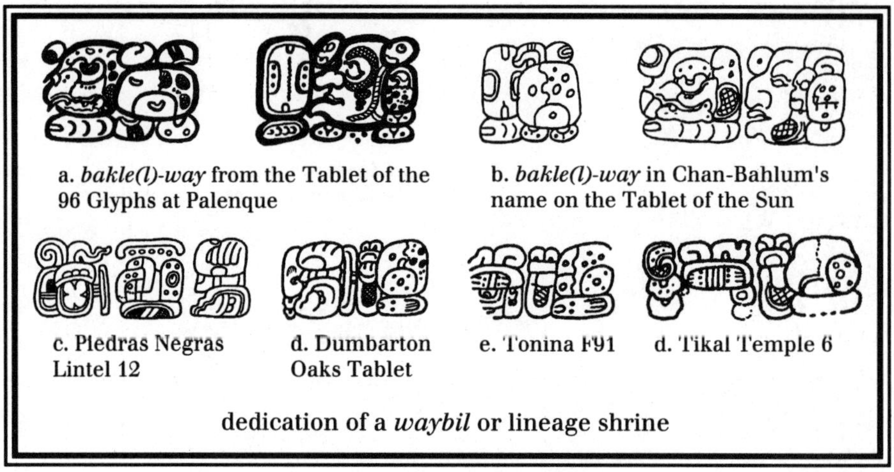

FIGURE 4:7 The *Way* Glyph

that particular companion spirits were associated with particular lineages and kingdoms, and that their names were generally known to the artist who painted the pots. Every rule, however, has its exception. Palenque ruler Chan-Bahlam, K'an-Hok'-Chitam (his younger brother), and a later king, Bahlam-K'uk', recorded their nawal, *bakle(l)-way,* "Boney-Thing" (Fig. 4:7a–b), as part of their names. Nikolai Grube suggested that this *way* might be one of the great skeletal death gods who dance on Maya pottery and that members of the same lineage may even have inherited their *way* from their parents exactly as the Lakandon do today.[34]

The ancient Maya also transformed into their *wayob* when they fought their wars, and they very likely saw the planets and constellations as *wayob* of the gods and their ancestors.[35] Today stories abound throughout the Maya region of people being followed in the night by *wayob* that look like animals. Among the Yukatek-speaking Maya of Quintana Roo, the ancient concept of *way*—the spirit companion of gods, ancestors, kings, and queens—connotes an evil, transforming witch, a person to be feared rather than admired. Nikolai Grube told us of a visit he and his mentor, Ortwin Smailus, made to the most respected and feared shaman in the Cruzob area. When they left on the long walk back to their house, a pig followed them. The people of Señor told them it was the shaman following them in the guise of his *way.*

The modern designation of such beings as evil may have at least partly resulted from the suppression of indigenous practices and beliefs during the era of Christian domination. Nevertheless, we must acknowledge that there is something inherently dangerous and potentially evil in all types of powerful beings, objects, and encounters. Modern shamans are always at least potentially sorcerers and witches because the same forces that can generate prosperity, health, and wealth can also generate poverty, famine, and disease. Shamans have to work at their craft to avoid arousing, even unintentionally, evil in the cosmos.[36]

ON STATUES AND SUSTENANCE

During eighteen years of his reign, Chan-Balam of Palenque created one of the great literary and artistic legacies of the New World. One of his first acts after his father Pakal's death and his own accession to the throne was to decorate the piers of his father's funerary mountain, the Temple of the Inscriptions. These painted stucco reliefs display portraits of his parents and other adults cradling him when he was six years old. As signs of his power and special status as a divine being, a smoking ax pierces his forehead and his leg is transformed into a snake (Fig. 4:8). These divine attributes identify the young boy as the god K'awil,[37] a sacred being who symbolizes the embodiment of spiritual force in material objects.

Adults could also display the smoking ax of K'awil pierced through

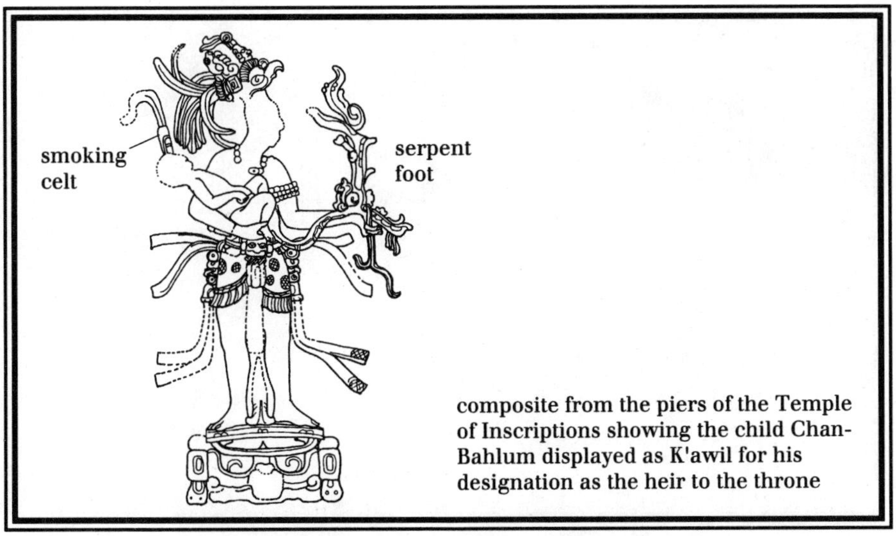

FIGURE 4:8 Temple of Inscriptions Pier

their foreheads, but only after they were dead. Pakal wears it as he falls down the Milky Way into the Otherworld (Fig. 4:9a). At Copan (Fig. 4:9b), the last scion of a dynasty that had ruled for four hundred years wears it as he dances in the portal to the Otherworld at the moment of his own death. The inscription on the rear of the panel tell us that the dynasty has ended with this particular dance. Each appearance of this smoking ax represents a moment of transition—from child to heir or from life to death. As we shall see, to wear the ax through the forehead signaled that the person was in a state of transformation embodied by the power of lightning.

The Temple of the Foliated Cross, a temple commissioned by the adult Chan-Bahlam, records the birth of K'awil, who was the third-born child of First Mother and First Father. Here we learn this god's full name, *Tzuk Yax Ch'at K'awilnal Winik Ox Ahal Ch'u Ch'ok K'awilna(l).*[38] This translates as "Partition[39] First Dwarf,[40] K'awil-born Person, the Third-manifested God,[41] Sprout, K'awilnal" (Fig. 4:10). This is quite an impressive title, but what then does *k'awil* mean?

In Yukatek, *k'awil* means "sustenance" or "alms"—any precious substance, usually some type of plant or body fluid like blood or sap, given freely as thanks for the sustenance provided by the divine.[42] The Popol Vuh records that the gods wanted to create beings who could reciprocate their love and care by returning nourishment to their creators. The gods

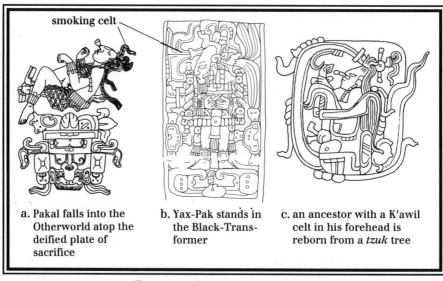

a. Pakal falls into the Otherworld atop the deified plate of sacrifice

b. Yax-Pak stands in the Black-Transformer

c. an ancestor with a K'awil celt in his forehead is reborn from a *tzuk* tree

FIGURE 4:9 Images of the Dead

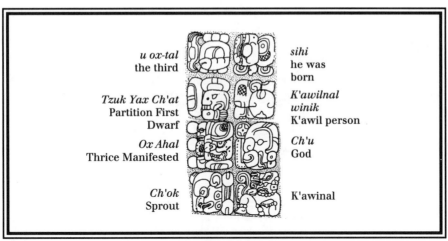

FIGURE 4:10 The Birth of K'awil

failed several times before finally hitting upon the right formula—human beings made from a mixture of maize dough and water. And that final try was successful beyond their wildest hopes, resulting in beings who were capable of prayer and worship, providing sustenance for their gods by "suckling" them, through bloodletting and sacrifice.

So, in the Popol Vuh maize dough is equated with human flesh, and the original waters of Creation with human blood. K'awil, in this sense, represents the contractual obligation bonding people and gods, for the gods receive from people that which they provided in the first place—maize and water transformed into flesh and blood. The gods continue to provide water in the form of rain, springs, rivers, and cenotes, and to engender maize from the ground. Thus the cycle of generations is perpetuated through time. K'awil as "substance" conveys the idea of the magical transformative cycle that changes food (maize) into the flesh of gods and humans, and then back again into food (blood or its equivalent). The central role of food offerings, meals, and ceremonial banquets in Maya religious life hinges on this equation.

The serpent foot we see on the baby Chan-Bahlam on the Temple of the Inscriptions represents another of K'awil's important aspects, his association with Vision Serpents. In one depiction of K'awil, known as the Manikin Scepter (Fig. 4:11a–b), one of his legs has transformed into a serpent, which serves as the handle to be grasped by the ruler who holds the image. Other depictions show the long looping body of the Vision Serpent transforming into the torso of K'awil (Fig. 4:11c), and in still

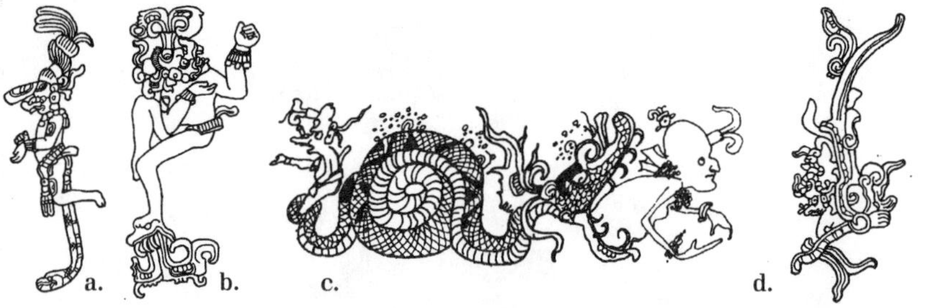

K'awil as a serpent-footed scepter — K'awil merged with a Vision Serpent — K'awil emerging from Serpent Bar

FIGURE 4:11 K'awil and the Snake

others, K'awil emerges from the mouth of the Double-headed Serpent Bar (Fig. 4:11d), which symbolized both the ecliptic and the sky umbilicus.[43] For the ancient Maya, these serpents were symbols of the path along which supernaturals traveled on their way to being manifested in this world. The Maize God was reborn from the serpent's mouth, and according to the Palace Tablet, human souls find the bodies of their newborn owners by traveling along the serpent's gullet. Serpents were also the path of the sun and the planets as they moved through their heavenly cycles. By the time of Chan-Bahlam, the K'awil serpent foot was an old, familiar image, with roots going back as far as the pictures left by the Olmec kings who depicted their lower legs as crocodile heads, to symbolize their status as the World Tree, the *axis mundi* of creation.

Stephen Houston and David Stuart discovered the key to this relationship between K'awil and serpents when they realized that a text at Copan (Fig. 4:12a) explains that K'awil's serpent foot is his *way*, his animal spirit companion. Other texts at Yaxchilan and Palenque confirm this idea. At Palenque (Fig. 4:12b–c), a serpent called White-Bone, *Sak-Bak*, is the *way* of K'awil. At Yaxchilan (Fig. 4:13), yet another serpent known by a different name writhes upward in full-bodied splendor as he spits out the supernatural who has made the journey from the Otherworld. Thus, K'awil could be associated with several different Vision Serpents. He embodied the conduit from the Otherworld no matter the form that conduit took on a particular occasion.

These conjuring actions involve another meaning of *k'awil*, one that survives most clearly in the highland Mayan languages. The terms used to describe ritual practitioners, as recorded in sixteenth-century dictionaries of Poqom Mayan,[44] include *ih cam cavil,* "one who carries the figures of the gods," and *helen cavil,* "cave god." In these entries, *cavil* is the word for the physical statues of the gods, which the Spanish called "idols." With these entries alone, we would not be able to ascertain if this word has the same meaning as the *k'awil* of the inscriptions; however, by comparing Poqom to other highland languages, we have found that this word corresponds to the K'iche'an word *q'abwil.*[45] Even today *q'abwil* is the word for the statue of a saint and its spirit; and in K'iche', Kaqchikel, and their sister languages, it is the word for god. In the Popol Vuh, *q'abawil* are wood and stone images of the titular deities of the K'iche' lineages.

When we found that *k'awil* meant "statue," we realized why the Vision Serpents are the *way* of K'awil. *K'awil* refers to an object made of wood, stone, or some other material, while the *way* is the spiritual being who resides in it. Classic monuments and pots actually show many of these small wooden images being manipulated by lords and shamans (Fig. 4:14). And by an archaeological miracle,[46] we have four *k'awilob* from Burial 195 at Tikal (Fig. 4:15a). Originally made of wood and mounted on staffs,[47] these four statues accompanied the occupant of the tomb into the Otherworld. The actual carved figures had decayed, but our friend Rudi Larios saved their impression by injecting plaster into the void where they had been.

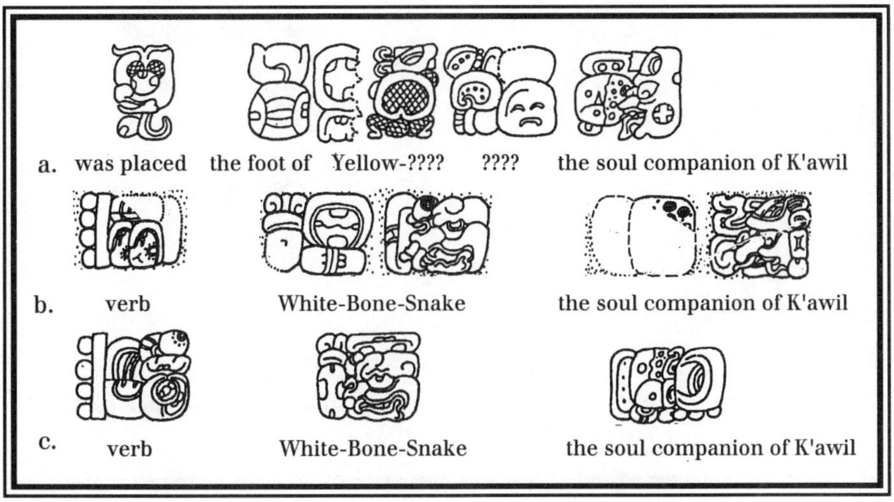

FIGURE 4:12 Vision Serpents as the *Way* of K'awil

tzakah	*Yax-???*	*Na-*	*u way*	K'awil
was	First-???	*chan*	the soul	
con-		snake	compan-	
jured			ion of	

Yaxchilan Lintel 15

FIGURE 4:13 The Rearing Vision Serpent

(photograph © Justin Kerr)

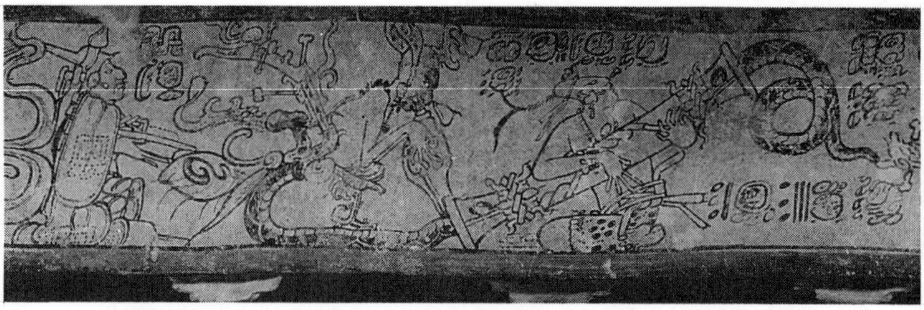

Och-Kan, the *way* of K'awil, emerges from a ceremonial bar
FIGURE 4:14

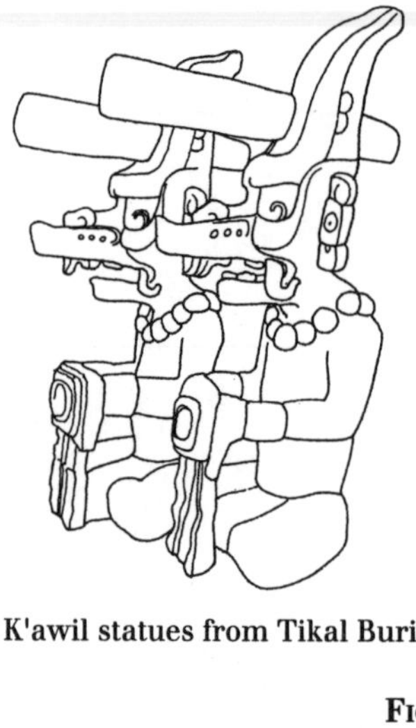

a. K'awil statues from Tikal Burial 195

b. an eccentric flint with human profiles pierced by smoking celts

FIGURE 4:15 K'awil

K'awil also regularly appears on the extraordinary eccentric flints (Fig. 4:15b) placed in cached offerings at cities like Copan. Significantly, most of the K'awil heads on these flints have celts piercing their foreheads, just like the image of the falling Pakal on his sarcophagus at Palenque. The overlapping relationship of these eccentric flint *k'awilob* with the K'awil gods who emerge from both ends of the Double-headed Serpent Bars is reinforced at Copan. There almost every image of K'awil that emerges from either end of a Serpent Bar is carved in the style of an eccentric flint (Fig. 4:16a).[48] The very glyph for the Double-headed Serpent Bar has flint blades in its mouths (Fig. 4:16b). All of this leads us to conclude that the most reliable definition of the term *k'awil* is a statue made of wood and stone into which a ritualist called a supernatural being. The Double-headed Serpent Bar, made of wood and matting, was another such host object.

Yet the word *k'awil* has many subtleties. When we understand *k'awil* as "idol," we have just begun to understand the process by which spirit is manifested within material objects. Dennis Tedlock[49] in his analysis of the Popol Vuh talks about three *q'abawil*—wooden and stone deities called *Cacula Huracan*, "Lightning One-Leg"; *Chipa Cacula*, "Youngest or Smallest Lightning"; and *Raxa Cacula*, "Sudden or Violent Lightning."[50] One-legged gods are a rare phenomenon in Mesoamerica. The only major example besides K'awil is Tezcatlipoca, the Aztec god called Smoking Mirror.[51] Tedlock sees Huracan as the K'iche' analog of K'awil.[52]

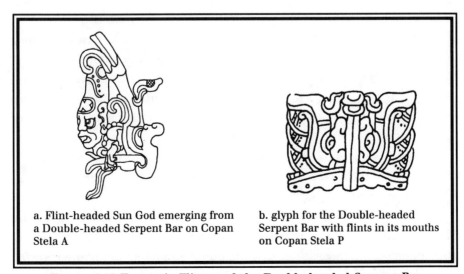

a. Flint-headed Sun God emerging from a Double-headed Serpent Bar on Copan Stela A

b. glyph for the Double-headed Serpent Bar with flints in its mouths on Copan Stela P

FIGURE 4:16 Eccentric Flints and the Double-headed Serpent Bar

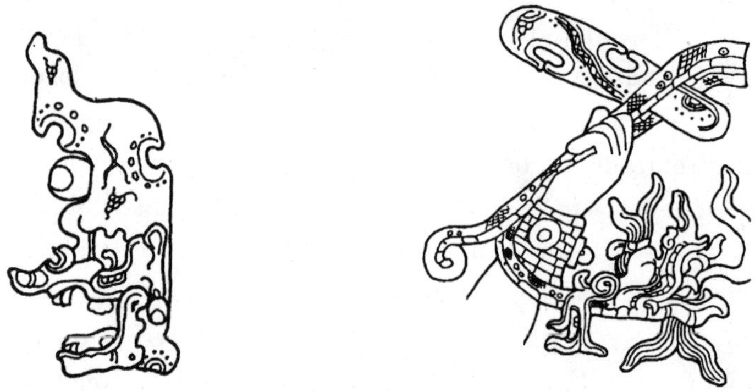

a. personified eccentric flint b. ax with lighting marks on the blade

FIGURE 4:17 Flint

We surmise that the One-legged Lightning of Huracan links to the god K'awil in other ways. From his earliest appearances in the archaeological record, K'awil has a smoking ax or celt through his forehead. Axes in other contexts are marked with "kawak" signs to identify them as products of lightning (Fig. 4:17).[53] Even today stone axheads, flint, and obsidian are thought to be created when lightning strikes the earth. In the Maya calendar Kawak was the day of lightning. The eccentric flints with the silhouettes of human *k'awilob* on them are the materializations, frozen in stone, of lightning itself.

Our friend Barbara Tedlock,[54] who is an initiated shaman in the K'iche' tradition, regards lightning as a root metaphor in the ritual language and beliefs of those highland Maya. As she describes it, Momostenango seers feel lightning within the blood and muscles of their bodies, causing their blood to speak. The ability to divine and reveal the intentions of the ancestors rests on the adept's capacity to sense the lightning in his blood through his own pulse. This lightning in the blood is specifically likened to sheet lightning that flashes over the four sides of the sacred lakes of K'iche' country.

Garrett Cook told us about a character called the Tzitzimit, a shamanistic trickster in the modern K'iche' ritual called the Dance of the Conquest. This character manifests in three different forms, all brothers of Tekun Uman, the chief Mundo or Earth-Lord (and also the last king of the K'iche'). All three dress in red clothing and carry a red hatchet. Two of them, one played by an adult and the other by a child, are named *Ah Itz,* "He, the Wizard," and *Kaqik'oxol,* "Red K'oxol." In myths and stories, **Plate 7**

Saqik'oxol,[55] "White K'oxol," is the guardian of both the treasure and the magical corral beneath the ancient city of Utatlan. The older brother and adult, Ah Itz, carries a small doll, who is called the *Ch'ip*, or "youngest brother." Ah Itz divined the defeat of the K'iche' and the death of Tekum Uman in their battles against the Spanish. **Plate 24** **Plate 25**

Linda Schele found one of these little dolls for sale in Antigua in 1992. The shop owner thought the little wooden statue represented Santiago, but the red color of the face and clothes was unmistakably that of the Kaqik'oxol. The final confirmation of his identity was the little red ax in his hand. Curiously, only six weeks earlier, John Fox had shown her a drawing of a petroglyph at Utatlan showing what may be a scene from the Popol Vuh. In explaining the imagery to her, he pointed out the special crown worn by the K'iche' kings. It had an odd shape that rose to a peak over the forehead of its wearer. The little doll Linda found in Antigua wore exactly the same crown. Even today, with statues made for tourists to buy, the K'iche' hold on to the past with tenacious commitment.

In Momostenango thought, it is K'oxol who awakens lighting in the blood of novice shamans by striking them silently all over with his little red ax.[56] We believe it is no coincidence that one of K'awil's birth names was *Tzuk Yax Ch'at K'awinal Winik*, "Partition[57] First Dwarf, K'awil-born Person."

The grasping of the serpent-footed, axheaded K'awil is one of the most common ritual acts depicted in Classic royal portraiture. Now that we know that the snake foot represents the companion spirit called into the wooden statue during ritual, we can also surmise that the king holding K'awil is grasping the path to the Otherworld, just as he does when he grasps a Double-headed Serpent Bar or a Vision Serpent. Just as K'awil, as lightning, awakens the revelatory potential in the blood of holy people, so the king, who holds him, grasps both the path to the Otherworld and the means by which it is opened.

The Zinacanteco Maya also believe that the blood talks; in this context, however, it is the blood of the patient. A skilled shaman diagnoses soul sickness by taking the patient's pulse at the wrist and elbow. Sometimes patients have lost a piece of their soul (*ch'ulel*), or their animal spirit companion is wandering lost outside the ancestral corral underneath the mountain.[58] Even Evon Vogt, the great ethnographer of the Zinacanteco Maya, is not sure exactly what is speaking to the shaman through the blood, but he has observed that *ch'ulel*, the "inner soul or spirit" of an

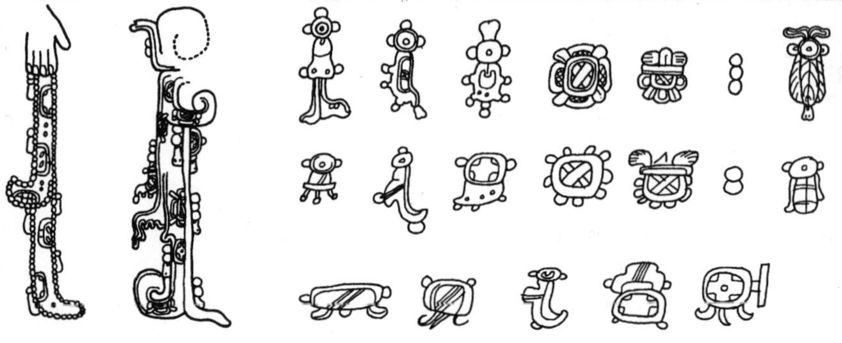

scrolls of *ch'ulel* flowing from the body as blood and some of the signs of *ch'ulel* that can occur in Maya imagery

FIGURE 4:18

individual, abides in that blood, and perhaps it is that soul that speaks.

For the ancient Maya,[59] human beings released *ch'ulel* from their bodies when they let their blood (Fig. 4:18). Through bloodletting, they "conjured" (*tzak*) the *way* and the *ch'u*, the "companion spirits and the gods." They conjured them into the clouds of smoke rising from censers, and into the swirls of mist and clouds hanging in the mouths of caves and drifting over the slopes of ancestral mountains.

The axheaded K'awil not only embodied the spiritual force of blood, but was the instrument by which it was released from bodies to feed the gods. The ax wielded by K'awil and the many Chak gods (Fig. 4:19) was the principal instrument of decapitation sacrifice. One of the words for sacrifice[60] in Yukatek, *p'a chi'*, means "to open the mouth," because people literally smeared blood upon the mouths of the wood and stone images to feed the gods within them. *P'a chi'* also means "to sacrifice and dedicate something to god" and "to discover, to manifest, and to declare."[61] Giving the god sustenance (*k'awil*) by feeding blood to its image (which was also called *k'awil*) allowed the lightning to flow (also a form of *k'awil*) and establish the path of communication, manifested in the image of the serpent-footed god K'awil.

This linking of means and ends in sacrifice was fundamental to Maya thought. Dennis Tedlock says that in the act of creating the world, the gods in the Popol Vuh required three things—their words, their *nawals*, and their *pus*. He defines *nawal* and *pus* this way:

> In K'iche', *nawal* refers to the spiritual essence or character of a person, plant, animal, stone, or geographical place. When it is used as a metonym

for shamanic power, as it is here, it refers to the ability to make these essences visible or audible by means of ritual. *Pus* . . . refers literally to the cutting of flesh with a knife, and it is the primary term for sacrifice. . . . It means that the creation was accomplished (in part) through sacrifice."

(D. Tedlock 1986:79; our orthography)

We think that K'awil in his form as the Manikin Scepter embodied concepts ancestral to both the *nawal* and *pus* of eighteenth-century K'iche' belief.

K'awil was a nexus for powerful phenomena in the Classic Maya world that conjoined spirit to body and sustenance to sacrifice. Evon Vogt sheds some light on the complicated relationship between sacrificial blood, flesh, and spirit in contemporary Maya culture. He describes how a black chicken is sacrificed and its blood and flesh exchanged for a piece of lost soul. The black chicken is the *k'exol* or "substitute" for the patient. The reasoning here seems to be that when people get sick, the gods have taken away part of their soul force, by either stealing a piece of their *ch'ulel* or molesting their *chanul* ("animal companion spirit"). The shaman offers the ancestral gods a substitute for the patient's body in return for the lost soul. This is the same idea that Manuel Pacheco explained to Linda and the other gringos in the cave of Utatlan. The chicken we found sacrificed there had been given to the Mundos who lived within the cave as a substitute for the sustenance people took from the earth and from animals.

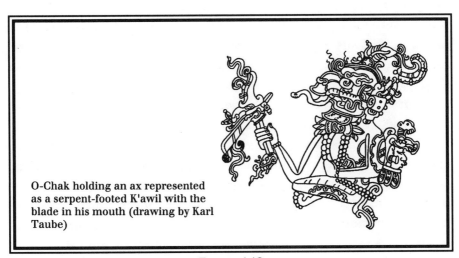

O-Chak holding an ax represented as a serpent-footed K'awil with the blade in his mouth (drawing by Karl Taube)

Figure 4:19

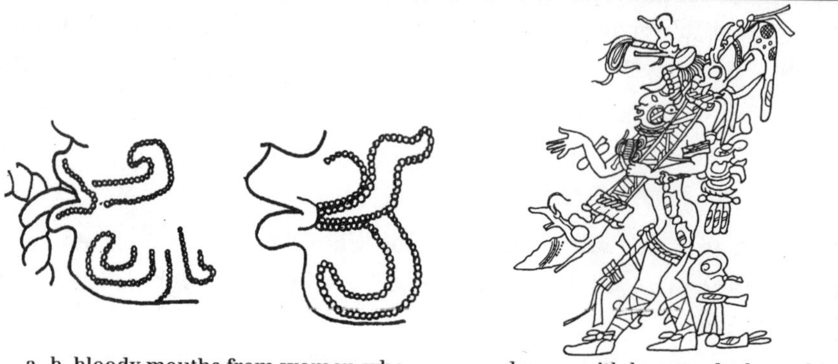

a.-b. bloody mouths from women who have let blood at Yaxchilan

c. dancer with bone awls through his penis

FIGURE 4:20

For contemporary Maya like the Zinacantecos and the K'iche', sacrifice is a two-way process of exchange between people and gods, involving the consumption or absorption of soul and flesh by both humans and supernaturals. One of the significant features of Zinacanteco sacrifice is that blood does not convey simply sustenance but also soul, *ch'ulel*. This connection between blood and *ch'ulel* was also fundamental to the worldview of the Classic Maya and explains their devotion to pious bloodletting. In Classic Maya images, *ch'ulel* appears as stylized droplets of blood that fall from the hands of people letting blood and often stain their mouths, faces, and ears (Fig. 4:20).[62] Other precious substances, such as copal incense, maize dough, rubber, and jade also contained soul and were burned in huge braziers where they were converted to smoke, the form of sustenance the gods could consume. These same soulful materials were buried under floors to endow places with portals to the Otherworld. First and foremost, however, ranks the blood of people, the primary conduit of *ch'ulel*.

When the ancient Maya let blood, they were feeding the gods their *ch'ulel* and giving of their souls. The modern Maya do not practice bloodletting to the same degree but use mainly intermediary offerings—chickens, deer, earth-oven bread, candles, and incense.[63] They view these direct exchanges with the supernaturals as dark and dangerous adventures: they may commune with the Mundo or the K'oxol in the cave under the mountain fortress of Utatlan. The search for supernatural power can be rewarded with earthly treasure, but it can also result in the loss of soul and body and accusations of witchcraft. Surely the stakes were just as high in antiquity, but we know that the ancients did not let fear stop them. Bloodletting was a common practice among adult males of the Conquest period and elite males *and* females were central to bloodletting rituals in the Classic period.[64] Direct communion with the supernaturals and the nurturing of the gods with *ch'ulel*, the "soul-stuff" of human beings, are clearly portrayed in the ancient monuments and recorded in their texts. The *ch'ul ahaw* ("holy lord" or "king"), like the black chicken of modern rituals, was the *k'exol*," substitute," and the *helol*, "replacement," sacrifice for the community.

The complex of images and words that surrounded the bloodletting ritual in Classic-period art shows its profound importance. One of the favored instruments of bloodletting was a small obsidian blade fragment or worked flake. We have archaeological examples of these from Tikal, Waxaktun, Piedras Negras, and many other Classic Maya cities. These pieces of obsidian were chipped into little scorpions and other shapes or incised with images of Ch'ul, K'awil, Sak-Hunal (the Jester God), Hun-Nal-Ye (First Father as Maize), the Moon Goddess, Am (the divination stone[65]), and K'in ("day," "sun") (Fig. 4:21a). We have pictures of these bloodletters in bowls, along with blood-splattered paper and bloody ropes that kings drew through their penises and kings and queens drew through their tongues (Fig. 4:21b).

When this same bloodletting image is used as a glyph in sacrificial

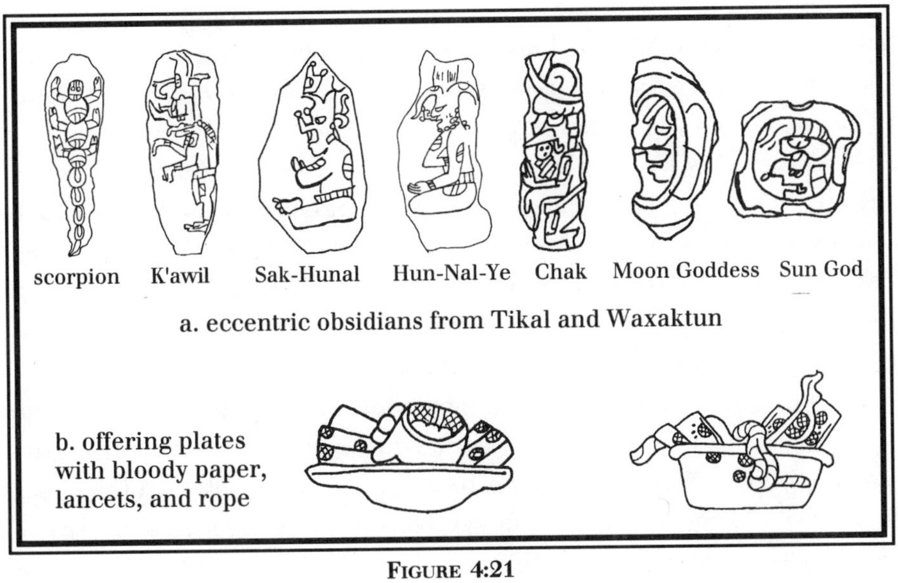

Figure 4:21

phrases, it means *ch'am*,[66] "to harvest." In English the word "harvest" brings up images of wheat being gathered, but to the Maya it means plucking the ears of ripe maize from their stalks, an action with sacred connotations. Karl Taube informed us that the act of husking maize is a perfect analog of penis perforation. In fact, the bone awls the Maya use today to husk maize are exactly like the ancient awls that were inserted into the opened wounds along the top of the penis. Karl[67] says that "during harvest, the ear of corn is dehusked by holding it about waist level and inserting the pointed instrument near the tip of the cob." The harvesting of corn is more than the mundane action of a farmer. It is the analog of the first act of sacrifice in mythological history, the decapitation of the Maize God. It is also the action that puts food on the tables of human beings. And since maize is a plant that cannot seed itself, cutting off the head of the maize is necessary if the seeds are to be removed and planted for the next crop, an action that symbolizes the perforation of a penis. Maize is a marvelous metaphor: sustenance, sacrifice, and rebirth all rolled into one.

To the ancient Maya the blood and spirit given in sacrifice were constantly recycled between the world of humans and the world of the gods. Lords and shamans burned their blood after sprinkling it on paper and mixing it with other materials, like incense. We have pictures of *ch'ulel* pouring off their hands into buckets during rituals. We have other pictures of great scrolling S-shapes rising from bowls of bloody paper, representing the clouds of smoke and incense described from the time of the Conquest and witnessed today all over the Maya world, swirling above the praying adepts. This association is confirmed epigraphically: the same S-shaped scroll is used as a glyph for the word *muyal*,[68] "cloud," in Classic texts. (Fig 4:22)

We learned why clouds and smoke are so synonymous in Maya thought when we began observing and participating in Maya ritual. During the festival of Carnival at San Juan Chamula, Linda saw the banners and the men who carried them all but disappear in clouds of incense rising from two clay censers in their midst. Every day at Chichicastenango, supplicants wave impromptu censers made of tin cans filled with burning incense. The clouds of smoke rise high above the church front, carrying prayers upward. During most of their festivals, the Tzotzil fire rockets into the air so that the fire and smoke will be closer to heaven. In the Classic-period symbolism, these clouds convey *ch'ul*, "soul," just as blood does, for they are marked with the same glyphs and

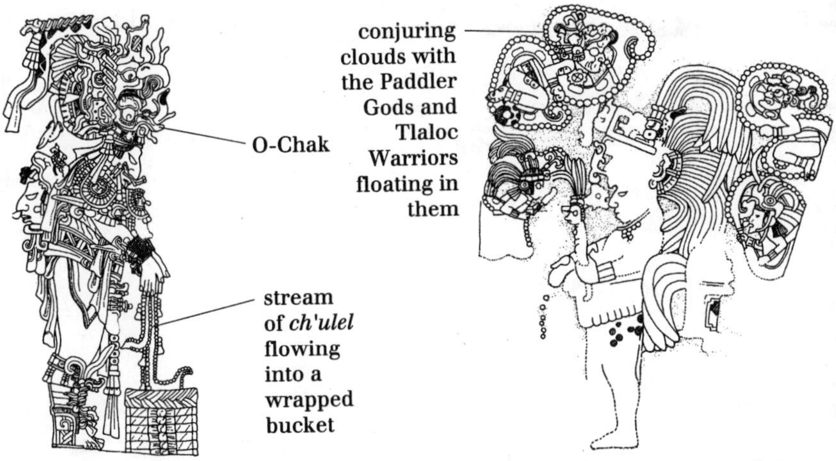

a. Bird-Jaguar scatters on La Pasadita Lintel 1

b. Ruler scatters on Ixlu Stela 2

FIGURE 4:22

elements that connote preciousness and completeness.

Maize consumed by people is again transformed into blood and endowed with soul. We can't prove that the ancient Maya believed that the rains, like blood and the clouds of sacrifice, contained *ch'ul*, "holy spirit," but their descendants in Yukatan do. "Here comes the holy waters, the saint rain," declare the farmers toiling in their thirsty fields, as they watch the black clouds cover the sun. Blood sacrifice for the ancient Maya was necessary to the survival of both gods and people. It was the mindful expression of power that was directly symmetrical to the expressions of divine power in the natural world around them.

VISION SERPENTS AND TRANCING

If the Precolumbian Maya practiced bloodletting and sacrifice to sustain the cycle of the soul, they also did it to enter trance and commune with the gods. Recall that the Vision Serpents conjured up by the ancients in trance rituals have names. Some, like the great War Serpent, the *Waxaklahun-Ubah-Kan*,[69] have special roles in Maya cosmology that we can identify. When the Vision Serpents open their jaws, they convey the gods and the ancestors into the land of the living. One of the most

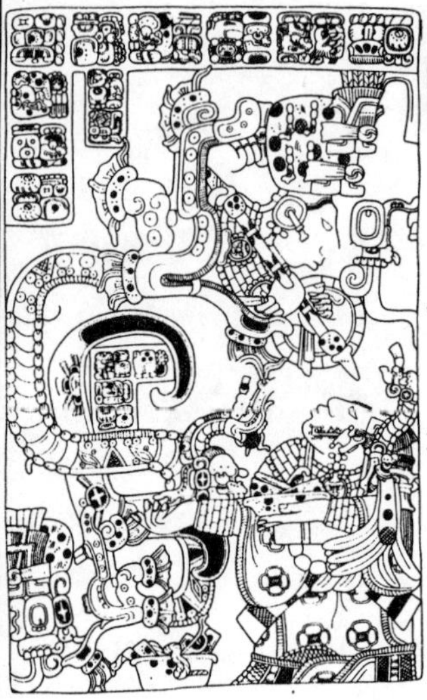

u tzakwa u k'awil he conjured it, the k'awil of	*u tok'-pakal* the flint-shield of	*Ah K'ak' O-Chak* He of Fire, O-Chak	*u ch'ul hul tzak* [is] his holy lancet conjurer

Lady K'abal-Xok conjures up a War Serpent with the *way* of the founder of Yaxchilan's dynasty (drawing of lintel by Ian Graham)

Figure 4:23

masterful representations of this rite to have survived from Classic times is the beautiful Lintel 25 (Fig. 4:23), commissioned by Shield-Jaguar of Yaxchilan. The scene depicts the principal wife of this king conjuring up the founder[70] of her husband's lineage during his accession rites. Wearing the costume of a Tlaloc warrior, this ancestor emerges from the jaws of a frightening, double-headed beastie with a half-flayed body decorated with feather fans. He is Waxaklahun-Ubah-Kan, the Maya War Serpent.

The Vision Serpents of the Preclassic[71] and Classic period Maya survived into the final centuries before the Spanish Conquest. The Feathered Serpent, a Postclassic expression of this idea, was a major god of the Maya in Yukatan and among the K'iche' in the Guatemalan highlands. Today, however, this image is lost. Along with the practice of bloodletting, the vision rites, and ecstatic trance, this dynamic and palpable manifestation of communication with the Otherworld has faded from Maya practice.[72] There is one surviving modern example of the Vision Serpents. A remarkable account of this survival was recorded and published by J. Eric Thompson in his ethnographic research among the Q'eqchi Maya of Belize more than sixty years ago. In this account Thompson described the final initiation of a Q'eqchi' shaman in the village of San Antonio:

> The instructor and the initiate retire to a hut in the bush for a month or so that there may be no eavesdropping. During this period the initiate is

> taught by his master all the different prayers, and practices used in causing and curing sickness. At the end of the period the initiate is sent to meet Kisin. Kisin takes the form of a large snake called Ochcan (*och-kan*), which is described as being very big, not poisonous and having a large shiny eye. When the initiate and the ochcan meet face to face, the latter rears up on his tail and, approaching the initiate till their faces are almost touching, puts his tongue in the initiate's mouth. In this manner, he communicates the final mysteries of sorcery.
>
> (Thompson 1930: 68–69)

Och-Kan is also the name of one of the Classic-period Vision Serpents (Figs. 4:14 and 24).[73] The essential action of communion characterizing this Vision Serpent of antiquity is clearly shared with its modern namesake. Both are conduits of communication and both facilitate the work of the shaman. We find it interesting that the tongue is involved in Thompson's story, because the ancient Maya drew ropes and bark paper through their tongues to conjure up Vision Serpents. Thompson recounted another story of the learning of would-be shamans in the Q'eqchi village of Sokotz, Belize, this one involving a snake and an ant's nest:

> Each ant's nest is presided over by a master, who is inevitably an expert in brujeriá [shamanism]. The grandfather of an informant made this trip in the company of a *Hmen*, who was teaching him. The master knocked three times on the nest, and a serpent issued forth. The master had previously removed all his clothes and was standing nude. The snake came up to him and after licking him all over, proceeded to swallow him whole. A few moments later he passed him out of his body with excrement. The master didn't appear to be much the worse for his adventure. Very similar initiation ceremonies in Chiapas are described by Nuñez de la Vega [1700]."
>
> (Thompson 1930:109–110; brackets ours)

Och-Kan, the *way* of a lord of Kalak'mul

Figure 4:24 A Vision Serpent Named *Och-Kan*

Perhaps this story[74] is related to the emergence of visions from the serpent's mouth. It's possible that there are more examples of this kind of encounter to be found among contemporary Maya, but the subject of magical snakes is a dangerous one that the Maya are not eager to talk about. David Freidel once casually asked Venancio Novello, first foreman of the Cerros Project in Belize, if he had heard of such snakes. Venancio was a generally genial man with a great sense of fun, so when he turned deadly serious at the mention of magical serpents it brought Freidel up short. Venancio said that there were such things, but that even talking about them was to invite trouble.[75] Here, a central public symbol of ancient Maya thought has become marginal, dangerous, and principally associated with evil.[76]

The contemporary Yukatek Maya of Yaxuna, for example, have a number of ideas about the magical properties of snakes. Chip Morris, who has been working with Yukatek craftspeople told us about one of the most interesting. Yukatek Maya women say that they stroke a snake when they learn how to embroider. Moreover, all of the ancient embroidery designs of Tixkakal Guardia are of snakes, such as a boat snake. Thus, even today snakes represent the idea of special knowledge.

ITZ, THE COSMIC SAP

As we discussed in Chapter 1, communication with the Otherworld also involves the powerful concept of *itz.* In the Maya world of today, *itz* refers to excretions from the human body like sweat, tears, milk, and semen. But it can also refer to morning dew; flower nectar; the secretions of trees, like sap, rubber, and gum; and melting wax on candles. In Yukatan the *itz* of melting votive candles is directly analogous to the *itz* (the blessed rain) of heaven that God sends through the portal opened during shamanic rituals.

When celebrating the ends of important time cycles, ancient kings and lords also scattered different types of *itz,* along with their *ch'ul*-laden blood, into large braziers where the *itz* was converted into smoke, the form of divine sustenance. Scholars have long argued about what this scattered stuff shown pouring off the hands of kings in Maya art might be[77] (Fig. 4:22), but we know that it sometimes represents *ch'ul,* the "soul-stuff" residing in human blood. At other times the holy stuff that is being scattered could be a tree-gum incense called *ch'ah*,[78] "copal"; rubber (yet another tree excretion); or maize.

People took blood from their bodies as offerings of sustenance to the gods, but they also regarded gum excretions, *itz*, as suitable additions to—or even as substitutes for—the flesh and blood of sacrificial victims. Blood Maiden, the mother of the Ancestral Hero Twins, escaped the death ordered by her father in the Otherworld when her assigned sacrificers substituted a heart made out of tree sap for her own heart and fooled the Lords of Death. *Itz* and *ch'ul* are fundamentally related substances—magical and holy stuff.

Set in contrast to this positive image of *itz* is its opposite: the term is sometimes translated as "witch" by Maya of the Guatemala highlands. Among the modern K'iche', Ah Itz is the shaman-trickster portrayed in the Dance of the Conquest. In antiquity *Itzam* generally meant "shaman," a person who worked with *itz*, the cosmic sap of the World Tree. Itzamna is the First Shaman and one of the gods who drew the images of the constellations on the sky at Creation.

In Classic-period imagery *itz* has two personified forms—the aged Itzamna and the great Cosmic Bird whose name was Itzam-Yeh.[79] The **Plates 12a, b** bird, whose name means "Itzam Revealed," may actually be the *way* of Itzamna.[80] Itzam-Yeh's analog in the Popol Vuh was Seven-Macaw, the bird who thought he was the sun.[81] The story of Seven-Macaw comes as a curious interlude in the Popol Vuh, following the great flood and the destruction of the earlier experiments at making humanity. It is the last event to take place in the Creation sky (at sunset) before the resurrection of First Father, which takes place at midnight. Seven-Macaw's behavior and defeat is a commentary on essential questions of material and spiritual power. Seven-Macaw was a gorgeous bird, who brightened a dark world with his beauty. He was wealth incarnate—his eyes were bright with silver and jade, his teeth were blue with beautiful stones, and his nose glistened like a brilliant mirror. But he was boastful and prideful. Lost in his arrogance, he proclaimed himself to be the sun and moon. Knowing that Seven-Macaw's claim of wealth and power would inspire vanity and envy in all the creatures of the world, the Hero Twins decided to teach him his true place. They shot him with their blowguns as he **Plates 11a, b** perched in his nance fruit tree, knocked him down, and broke his jaw. Angered, Seven-Macaw grabbed Hunahpu's arm and tore it out of his shoulder (Fig. 4:25a).[82] After their escape the Twins asked help from their

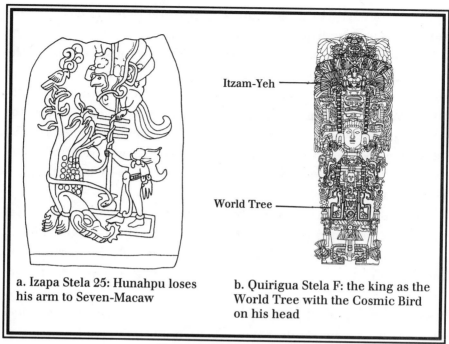

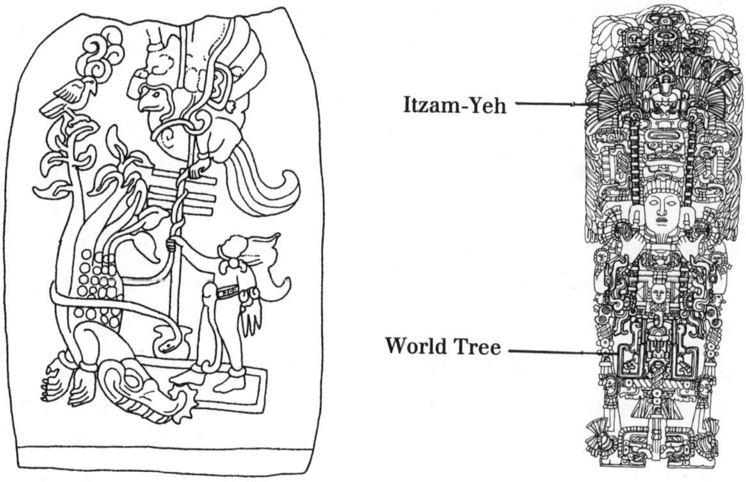

a. Izapa Stela 25: Hunahpu loses his arm to Seven-Macaw

b. Quirigua Stela F: the king as the World Tree with the Cosmic Bird on his head

FIGURE 4:25 Itzam-Yeh

grandmother, Great White Peccary, and their grandfather, Great White Tapir, and eventually defeated the vain bird. With their new guides, the Twins set out to pretend to cure the wounded Seven-Macaw. When they came to his home, they heard him complaining loudly about the pain in his teeth. Pretending to be shaman healers, the twins fooled Seven-Macaw into allowing them to remove his bejeweled, broken teeth and replace them with ground white maize. They also persuaded him to allow them to remove the shining metal from his eyes. Blinded and helpless, Seven-Macaw died. Hunahpu retrieved his arm and the grandparents magically restored it to its place.

As we saw in our description of *sak-a'*, ground white maize is an enduring and important ingredient in the sacred gruels used in offerings described on Classic-period pots and in modern rituals. Seven-Macaw tried to declare himself a god by means of material wealth and physical force. The Twins transformed him through attack and trickery into something comparable to a sacrificial offering. Seven-Macaw personified what the Maya farmer feared to find in his own lord and king: a false god decked out in the trappings of power. When the Hero Twins defeated Seven-Macaw, they opened the way for the eventual creation of humanity. When a Maya king stood up as the World Tree, he wore in his headdress the fabulous plumage of birds, more often than not those of Itzam-Yeh himself (Fig. 4:25b).[83] He, like the Hero Twins, was responsible for keeping the cosmos in its proper order. But in all his glory, the plumage of Itzam-Yeh hovered over his head like the sword of Damocles, reminding him and his people that the coin of power has two sides. The job of the king was to intercede with the gods and ancestors in order to sustain the balance of the world. Properly used, this power preserved cosmos and country. Improperly used, like Itzam-Yeh's arrogance, it became empty strength and a danger to all around.

THE HEART OF SACRIFICE

When people walk into a museum to see Maya art, what they mostly see are beautifully adorned clay pots and plates. We see those things too, but we also see in our minds the simple gourd bowls, plastic plates, tubs, and steel buckets that hold the food offerings, the hopes, and the hearts of Maya farmers today (Fig. 4:26a). Debra Walker, one of David Freidel's students, even noticed that during the Ch'a-Chak ceremony in 1986, the villagers of Yaxuna borrowed black felt pens from our archaeology project to write their names carefully on the small brown gourd bowls holding their maize gruel. In this way, they made the offering a personal covenant just as the Classic Maya lords did with their lovely glyphic inscriptions around the rims of their vessels.

Plates and pots are good to pray with precisely because everyone spends a lot of time looking at plates and cups. Meals are one of the really universal metaphors for sharing life between people. Nowadays, when a Yukatek Maya woman is mad at her husband, she may well throw his stool, his *k'an che'*, out of the kitchen where everyone eats. Maya meals are a deep, commonsense illustration of the sharing of life between people and gods. The vessels Maya nobility used in antiquity for offerings of food and drink were just fancy versions of the same things everyone used for the family meals. But just as the people of Yaxuna, through decorating them and placing them on the altar, transform their pots and plates into precious offerings suitable for blessing, so the ancient Maya regarded their pots and plates as much more than pretty crockery.

The plate that received the offerings of sustenance was a magical instrument. Three primary kinds of plates were used throughout Classic

FIGURE 4:26 Offering Plates, Ancient and Modern

Maya history—a large plate with angled sides called a *lak*[84] or *sak lak*,[85] a plate with three feet called a *hawate*,[86] and a bucket made of clay or stone known as *sak lak* or *sak lak tun*[87] (Fig. 4:26b). We have had hints for a long time that these particular kinds of plates were more than just material things made of clay, wood, woven reed, and stone. Today the Lakandon Maya think of their offering pots as living beings who transfer sustenance to the gods. They are called *läk k'uh*, "plate god," a term that is remarkably close to the old Yukatek month name *kum k'u*, "pot god."[88] We will see how they bring those pots to life in Chapter 5.

It took epigraphers a long time to figure out what the ancient Maya called their most important category of living vessel. Antonio Gaspar Chi, Bishop Landa's informant, however, gave us a vital clue we just recently discovered when he wrote the month Kumk'u in the sixteenth century, he tried to help the confused Spaniard by spelling out the way the

Yukatek Maya pronounced the name of this month. He had to do this because the traditional form of the glyph spelled *Ol*[89] (Fig. 4:27a), which was the sixteenth-century Chol and Poqom name for the same month. That is, the Yukatek speakers shared the glyph for this month with their Chol cousins, but not the word for it. *Ol* was the word for "in the center of" and "heart," not in the sense of an organ but as in "heart of something."

Ol was the name of the Classic-period god pot, but more important, it was the name of one of the most important portals used in the vision rites. In fact, an altar from El Peru tells us that the crack in the turtle's back from which First Father sprang in his rebirth was *yol ak*, "the heart of the turtle" (Fig. 4:27b). Seen from above (Fig. 27c-d), the turtle's crack is the quadrifoil door that has symbolized the portal to the Otherworld from the beginning of civilized life in Mesoamerica.

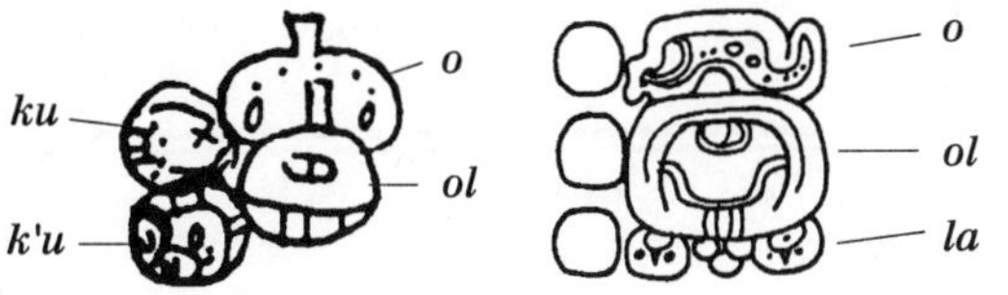

a. Landa's glyph for Kumk'u and the Classic-period version

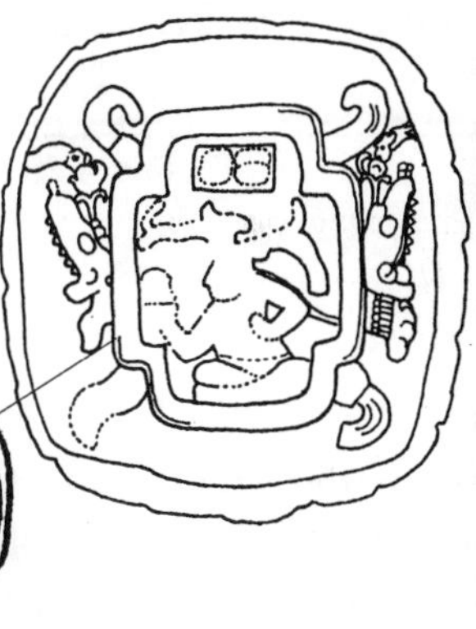

b. *u yol ak,* "the heart of the turtle" is the glyphic name of the cartouche in the center of El Peru Altar

c. the Maize God rise from the crack in the carapace. This is how the crack appears in side view

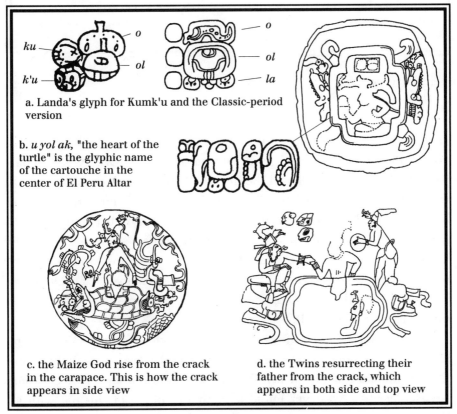

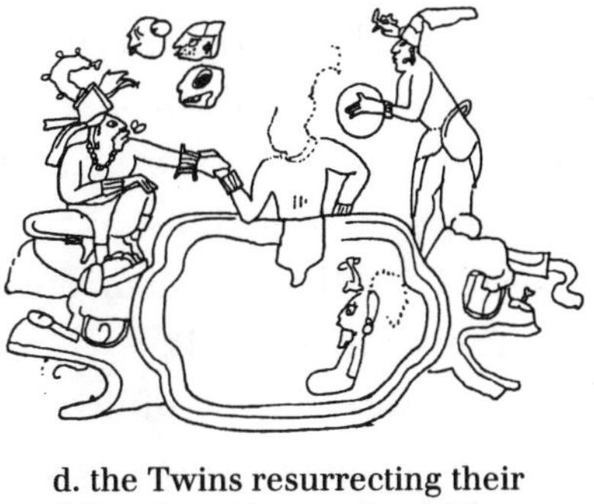

d. the Twins resurrecting their father from the crack, which appears in both side and top view

Figure 4:27 The *Ol* Portal

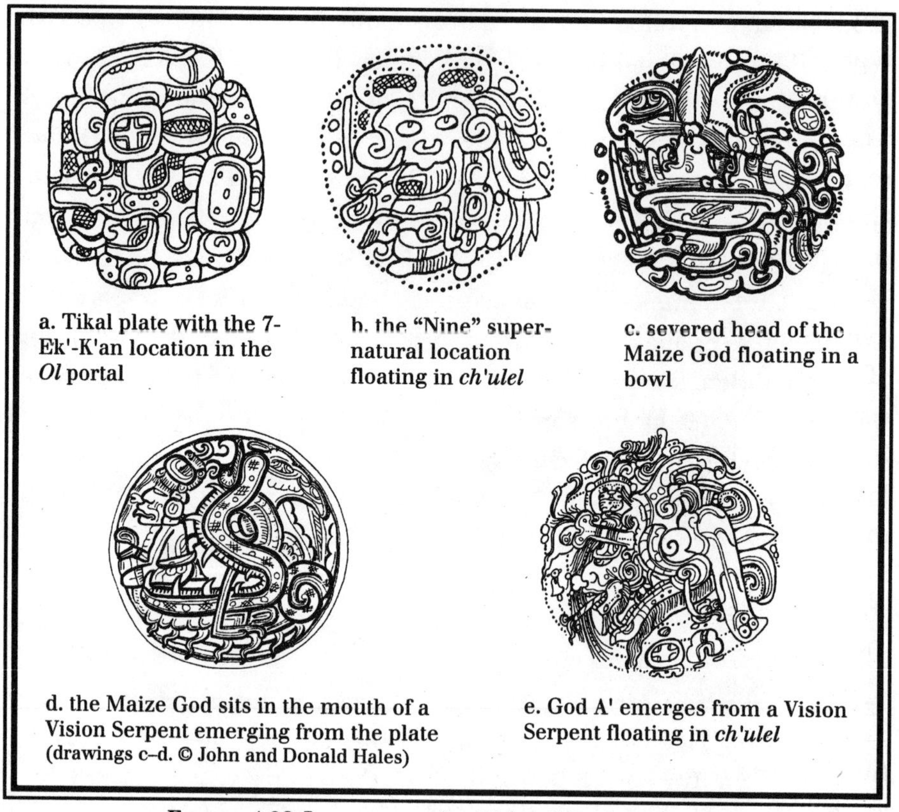

a. Tikal plate with the 7-Ek'-K'an location in the *Ol* portal

b. the "Nine" supernatural location floating in *ch'ulel*

c. severed head of the Maize God floating in a bowl

d. the Maize God sits in the mouth of a Vision Serpent emerging from the plate (drawings c–d. © John and Donald Hales)

e. God A' emerges from a Vision Serpent floating in *ch'ulel*

FIGURE 4:28 Images on the Lids of Cache Vessels

So we think that the offering plates the kings and nobility used in vision rites were both "god pots" and portals to the Otherworld. Their painted plates often displayed the images of locations in the supernatural world (Fig. 4:28a-b). Artists also painted fabulous Vision Serpents rising up out of the plates and carrying emerging gods in their gaping mouths: food in, god out of this primary orifice. Still others showed the severed head of First Father as the Maize God lying in the same kind of plate in the Otherworld (Fig. 4:28c-e).[90] All of these plates opened an *Ol* portal, all reached for the covenantal heart of the matter: sharing life-giving sustenance with the gods.

A special form of the offering plate, called the Quadripartite God by modern researchers,[91] rose sparkling in the night sky at the place where the ecliptic crossed the Milky Way in the cosmic imagery of Creation. Archaeologists have never found an actual example of this pot with all that it contained, which suggests to us that, like the World Tree, its physical forms were metaphors—real pots with objects in them that represented the things symbolized in the images. The artists showed the power in the plate personified by a skeletal head attached to its bottom.[92] This plate's special contents, as artists showed them in painted and carved representations, signaled its function as the holy of holies, the Maya Grail. These contents included three basic objects: a stingray spine or shark's tooth; a bound and tasseled device with crossed-bands or a "kimi"—death—sign worn by gods and spirits as a pectoral; and a spondylus shell. The stingray spine was for drawing blood, and we have plenty of these from actual offering vessels. The pectoral device was worn consistently by *wayob,* the soul companions from the Otherworld and also by the god Chak when he danced as the axwielding executioner. We have lots of pectorals in archaeological contexts, but this one might have been made out of knotted cloth. And the shell was an almost universal offering in burials and a caches, perhaps because it represented *ch'ulel,* the "soul-stuff" of the universe that sacrifice brought forth. We have in this special plate, then, the stingray spines representing the objects that made the sacrificial cut, the blood-red spondylus shell representing the *ch'ulel* given in sacrifice, and the pectoral adornment representing the spiritual transformation that was the result of dressing as a god, dancing, and ultimately sacrificing.

This plate was the exemplary offering. It rested at the base of the Raised-up-Sky as a big bulge in the Milky Way where the Scorpion constellation, another image of a bloodletter, was inside it. This conjunction of the ecliptic with the Milky Way gradually rose westward in the sky as the Milky Way turned to its east-west position. When the Milky Way was in the east-west Cosmic Monster position, the conjunction with the ecliptic shifted from south, the place of death, to east, the dawning place. The great heart-of-heaven plate then rode the tail of the Milky Way as the Cosmic Monster (which faces westward), and so the Maya artists showed it. When the Milky Way was in this Cosmic Monster position at midnight on Creation night, August 13, the Three Stones of the Hearth appeared next to this heart of heaven in the eastern sky. Here the stones emerged to take their journey to the center of the sky for the miracle of First Father's rebirth. The appearance of the Three Stones in the sky marks the beginning of our Creation, an appearance that conjoins with the shifting of the heart of heaven, the Ol, from the western to the eastern side of the sky, from the side of death to the side of life.

The palace artists fashioned plates decorated with images that actually show the *Ol* portal open and in use. On one such plate (Fig. 4:29a), a Vision Serpent is portrayed as if it were erupting up out of its surface. On another (Fig. 29b), the plate god is shown sitting on the portal as the World Tree springs up from the plate. Others (Fig. 29c-d) show the same tree rising from the plate or erupting from the stingray spine itself. Even more telling are the images (Fig. 29e–f) that show the maize tree, flesh of god and humanity, or the jeweled square-nosed serpent representing the soul flowers of the tree actually replacing the stingray spine of sacrifice. People and gods opened the portal in the plate with sacrificial offerings, and through this heart of heaven flowed the miracle of birth and life for both gods and people.

The souls of infants came through the Ol portal to find the bodies growing for them in the wombs of Maya women, making them the progeny of the plate of sacrifice at the heart of everything. The Palace Tablet of Palenque tells us that K'an-Hok-Chitam, the second son of Pakal, was born from the "heart of the center of the Primordial Sea" (*yol tan K'ak'-nab*) (Fig. 4:30a) and that his white-flower soul (*sak-nik-nal*)

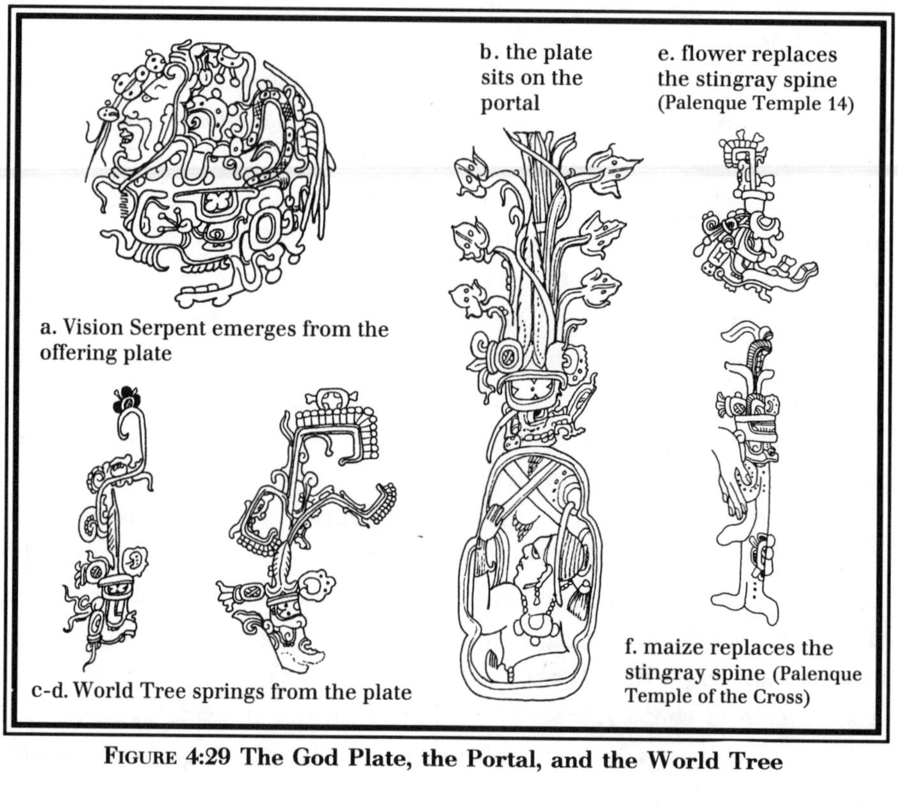
a. Vision Serpent emerges from the offering plate
b. the plate sits on the portal
c-d. World Tree springs from the plate
e. flower replaces the stingray spine (Palenque Temple 14)
f. maize replaces the stingray spine (Palenque Temple of the Cross)

FIGURE 4:29 The God Plate, the Portal, and the World Tree

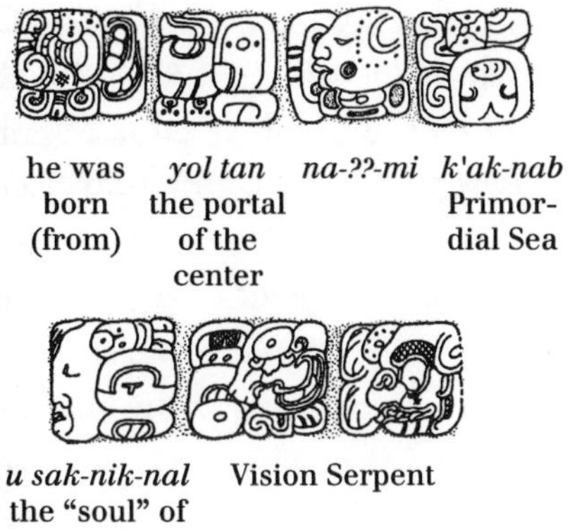

a.-b. references to K'an-Hok'-Chitam's birth on the Palace Tablet

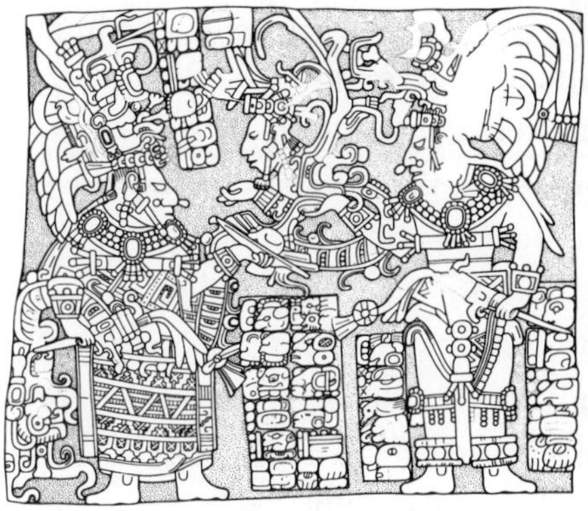

c. Yaxchilan Lintel 13, Bird-Jaguar's son being born from the snake

FIGURE 4:30 Birth Through the Mouth of the Snake

came from a Vision Serpent (Fig. 4:30b). Exactly this kind of birth is shown on Lintel 13 of Yaxchilan where Bird-Jaguar's newly born son emerges from the jaws of a Vision Serpent held in his parents' hands (Fig. 4:30c).[93]

The cosmic path of infant souls can be found embedded in some rituals that Don Pablo still performs in Yaxuna today. Grace Bascoupe, a student of medical anthropology and a trained nurse, spent several months in Yaxuna village during the spring of 1992. What follows is her description of a *Lo K'ex* ceremony. *Lo* refers to the sacred work of shamans; *k'ex* means substitute. As it does among the highland Tzotzil Maya, here it refers to curing of affliction.

DON PABLO'S CURING CEREMONY

(as told by Grace Bascope)

Don Pablo came and got me to see if I wanted to go with him to Santa Maria. There was going to be a *Lo K'ex* for an infant. The baby was a big fat boy, nine months old. The mother said he had been sick for a couple of months. He sounded congested in his chest, but he was happy and playful. He didn't have a fever to the touch.

The mother sat holding the baby in a hammock at one end of the hut facing southward toward the door. Don Pablo and the father went outside and made *sak-a'*, maize gruel. The father had already made little round vine hoops and harnesses in which to hang the gourd bowls that would hold the *sak-a'*. Don Pablo stretched a rope over the baby's hammock, parallel to it, and attached it to the walls on either end. On this rope he suspended the thirteen little gourd bowls. At this point Don Pablo closed the door of the hut.

The *kax* is a small seed pod about the size and shape of a large walnut, although its thick skin is smooth. Don Pablo took several of these *kaxob* and cut off one section of the sphere of each one, making them look like tiny gourd bowls. He placed thirteen of these in the shape of a cross on the floor perpendicular to the baby's hammock, and between it and a little table where he had his large gourd bowl full of his *sastunob*, his stones of light. He also placed on the floor six candles, two at the end of the cross closest to the baby, two at the far end next to his table, and one at the end of each arm of the cross. He then filled each of the tiny *kax* bowls with alcohol (because he didn't have any of the correct cane liquor).

Don Pablo's little table stood about three meters in front of the mother and baby in their hammock. On the table, he had placed his "lights" in a bowl of alcohol, a small switch of leaves to sweep the baby, and a candle, whose light he would use for seeing with his principal crystal. Beside the gourd bowl he put another stone, smooth, oval, and golden brown in color. Don Pablo told me he got it from a cenote and that it is shaped like a fish. He had told David some years ago that he uses it especially for curing babies. He also placed a little tied-up paper of dried herbs on the table. He later gave these to the father with instructions on how to administer them to the baby.

Next Don Pablo took out of his bag a corncob wrapped with thread. He told me later that the thread must be white cotton. He called it the *sakbe*, the "white path." He strung the thread carefully along the rope with its thirteen gourds suspended over the baby's hammock. He passed five times back and forth along the rope so that it was covered with five strands of the white path. Then, from the middle gourd bowl on the rope, he ran the thread down onto the floor and covered each of the little *kax* bowls, passing once over each arm of the cross and then onward to the top of the cross. From there, he passed the thread up onto his little table and into his large gourd bowl filled with his "lights."

All these preparations meticulously checked, rechecked, and completed, Don Pablo seated himself on a low log seat behind the table. He washed his hands, arms, neck, and face with a little alcohol. He sighed and let his face relax. He then waved his hands over the large gourd bowl and prayed. Then he took up his principal light and looked through it at the candle on the table. He looked through three stones in all, frowning with concentration. He smiled and picked up the smooth yellow stone and the little switch of leaves. Holding these in his right hand, he went to the hammock where the mother was holding the baby and rubbed the child with the leaves, around and down the head, from the chest out the arms, down the legs, and from the heels of the feet to the toes. He paid special attention to the feet, rubbing them most carefully with the switch and the stone. At the end of each movement, he shook the switch as if shaking something off onto the cross of tiny *kax* bowls. Next he moved the switch and the stone above the head of the baby in a clockwise motion numerous times, saying prayers as he did.

Don Pablo checked the level of alcohol in the *kax* bowls of the cross to make sure they were still full, then returned to the table to pray and examine his principal *sastun*. Going back to the baby, he prayed over it for some time. He cupped his hands over the baby's head and blew on the top of it. Then he shook his switch at the cross on the floor three times. For a third time he returned to the table, prayed, and consulted his light.

After this he began to take down the white thread, starting with the end in the large gourd bowl that contained his lights and working his way to the end of the little bowls suspended over the baby's hammock. He put the thread into the large gourd bowl in which he and the father had made the *sak-a'*. He was careful not to allow any of the thread to touch the floor or any other object while he was taking it down. He put the thirteen little *kax* bowls into the big gourd bowl, and emptied all the *sak-a'* from the thirteen small bowls suspended on the rope into this same big gourd bowl. Last, he put the switch of *tz'itz'ilche'* leaves into the gourd bowl.

Don Pablo put three of his lights in his pocket, but left the others in the alcohol in the gourd bowl on his table. Leaving the hut and the compound, we walked down the road about a hundred meters, climbed a stone fence, and went up a little hill into the bush. There Don Pablo and the father moved rocks away from a little hole in the side of the hill. They put the contents of the bowl into this hole and re-covered it with rocks. Don Pablo accompanied these actions with prayers.

We then returned to the house. Don Pablo replaced his *sastunob* in the bowl of alcohol and again sat behind the table. Consulting his principal light again, he smiled and pronounced that all was well.

* * *

How much does Don Pablo understand of the ancient numerology and symbolism of this ritual? Grace isn't sure and neither are we, although we know that Don Pablo is a conscientious and well-prepared practitioner. We can see the thirteen suspended bowls as the thirteen constellations suspended on the path of the ecliptic. Perpendicular to this path is the cross, source of baby souls and the symbol of creation and rebirth. Across the suspended rope runs the white path, the umbilicus, folding back five times, just like the "first five sky" of the Creation story, the place where the Maize God floats like a baby attached to his umbilicus (Fig. 2:31). Don Pablo's white path of thread then goes through the cross and into the bowl of magical stones, like the cord that runs around the Wakah-Chan Tree on Izapa Stela 25 (Fig. 2:24). Don Pablo's cross is adorned with six candles—again, the all-important number six that goes with the Six-Sky (Wakah-Chan) Tree. The thread is the path that draws affliction out of the baby and into the substitutes, the maize gruel and alcohol in their little bowls. Rationally, the path that the affliction takes as it leaves the baby should be the same one it took to enter it. This path bears a remarkable resemblance to the one that First Father's soul took back into this world at his rebirth.

OTHER PORTALS

While a sacrificial plate could open a portal at any location, there were permanent portals built into Maya architecture that could be opened, when necessary, by ritual. We have seen that plazas, representing the surface of the Primordial Sea, were portals, and that ballcourts were another kind. In addition, there were fixed portals on top of the temple-mountains. From all of these locations, the great Vision Serpents could emerge into this world. At Palenque, fixed portals were contained in the little sanctuary buildings called *pib na* ("underground house") and *kunil* ("bewitching place"). These little houses (Fig. 4:31) are marked by the great Cosmic Bird Itzam-Yeh hovering above their entrances as places where *itz* was materialized. At Copan (Fig. 4:32), the inner sanctum of Temple 11 was adorned with the imagery of the White-Bone-Snake, the maw to the Otherworld. Also known as the *Ek'-Way* (Black-Transformer), this great portal allowed the human-made *itz* of sacrificial offerings to be exchanged with the cosmic sap, the *itz*, of the Otherworld.

Itzam-Yeh

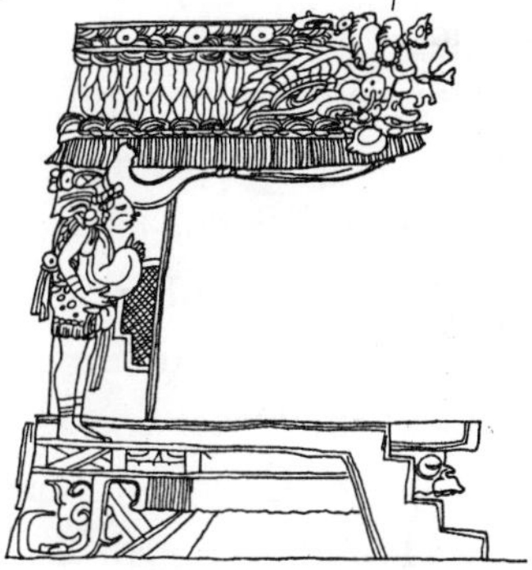

drawing of a *pib na* from a pot

the *pib nail* (*kunil*) inside the Temple of the Cross; the wing of Itzam-Yeh is still visible on the right side of the entablature

FIGURE 4:31 Conjuring Houses

White-Bone-Snake

raised platform under the debris

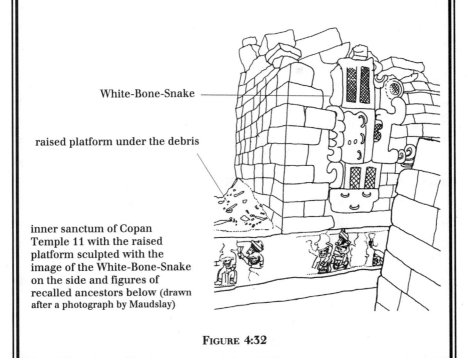

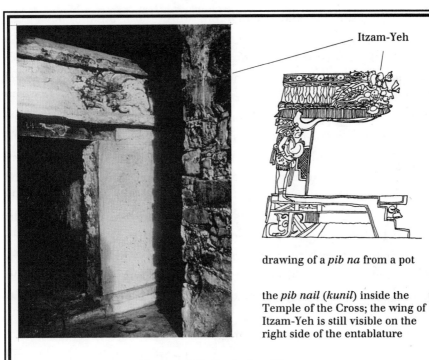

inner sanctum of Copan Temple 11 with the raised platform sculpted with the image of the White-Bone-Snake on the side and figures of recalled ancestors below (drawn after a photograph by Maudslay)

FIGURE 4:32

The cosmic sap was and is dangerous stuff, and shamanic power can be used as a force for evil as well as for good. Surely the ancient Maya common folk understood this contradiction, even when they gazed with envy and admiration upon their nobility decked out in the evidence of prosperity. We understand the Precolumbian Maya conception of soul and its incarnation primarily through the words and images carved by these elites on the objects they owned, the structures they built, and the tombs where they embarked upon the road to the Otherworld. These are the people shown dancing in their *wayob* accompanied by gods and ancestors in the Otherworld. These are the ones shown conjuring K'awil, bringing forth the Vision Serpents and nawals, sprinkling *ch'ul* into the world, and sending *itz* through the portal. We know that they undertook these dangerous transformations on behalf of their people, and with their people's understanding and support. We also know that the descendants of these ancients are still perpetuating important aspects of this complex spiritual and material view of the world.

If the ancient Maya elites received *itz*, magic, from their encounters with the Otherworld, they also accepted the responsibility to wield this power well and not fall into the trap of pride in wealth. The penalty for failure was the withdrawal of the blessings of the gods, famine, disease, defeat in war, and exposure of the all-too-human frailties that afflict leaders, stripping them of charisma and power. Living embodiments of blood and *ch'ul*, the ancient Maya elites grew up understanding these responsibilities. It fell to them to pour out upon their people the divine soul force that flowed freely when they sang the path of words, of memory, and of truth, embracing their ancestors and children in the land of their gods.

When we have been able to participate in the ceremonies of modern Maya communities—David at Yaxuna and Linda at several places in the highlands, we have been given the opportunity to watch these things in practice. Duncan Earle provided Linda with another such opportunity on the day after we went to the cave at Utatlan with Don Lucas and his son.

A DIVINATION

(as told by Linda Schele)

Having spent the night in Chichicastenango, we all got up on Sunday morning, ate breakfast, and prepared ourselves to visit the most famous market in Guatemala. Filling the central plaza of the town, its plastic-covered stalls present an explosion of color and choices to wandering customers—both foreign and local. We spent the morning moving down one tiny people-choked path after another, then visited the western church called Calvario along the way, and stopped at the stalls owned by special acquaintances of Duncan's. I bought more than I should have and less than I wanted, but after a couple of hours I was so loaded down, I couldn't carry more. I thought we were going back to the hotel to drop my loot, but Duncan, who was carrying as many of my purchases as I was, led us through a high wooden fence on the road just down from the church.

Passing through an open gate, we entered a bare courtyard and passed into a narrow ally flanked by double-storied houses. There Duncan left us, feeling uncomfortable under the gaze of the curious inhabitants. Not many of the tourists who frequent the Sunday market at Chichicastenango venture into private spaces like this. Duncan disappeared, and the rest of us settled down to wait politely while we watched the watchers.

After a while Duncan came bounding down a narrow stairway across the alley and motioned for me and his wife, Erica, to come up. Dodging electric wires and potted plants along the way, we climbed up to a narrow porch. At the far end, a small elderly man with a K'iche' woman sitting next to him hunched over a small table. He looked around briefly, smiled to acknowledge our arrival, then returned to an earnest conversation with the woman. Duncan told me that his name is Sebastián Panjoj.[94] He is a highly respected elder of the kind called a *principal* by his people. He was also a very powerful Ah Q'ij, who is also, I learned later, one of the four alcaldes of Chichicastenango. At this moment, he was in the midst of a divination for the woman. Suddenly, I realized that Duncan had arranged to have a divination done for me.

I took off my pack, which has a little seat built into it, and sat down where Duncan directed me—behind Panjoj. Erica sat behind me on one of the small chairs used in many Maya homes. The chuchkahaw's wife and daughters greeted us and then watched us surreptitiously from their kitchen. Duncan's children played with a flea-ridden litter of kittens that scooted from the children's laps to the furniture in a frenzy of play. As I sat there waiting, I opened my journal and began to draw what I saw.

Ah Q'ij Panjoj wore dark maroon handmade woolen clothes. His trousers ended just below his knees and his short jacket had a high collar **Plate 15** that hid some of his graying hair. Beautiful embroidered flowers graced the edges of the jacket, and flowers with sun daggers, an image that goes back at least to Aztec times, decorated the flaps on his trousers. Duncan told me that the more ornate the designs, the higher the rank of the owner. Panjoj's were ornate indeed.

At the corner of his porch, just beyond where his present client sat, was a table about the size and height of an American card table. I could not see everything that lay on it because it was crowded with papers marked with his clients' names and the rituals he had agreed to do for them, with candles of various sizes, and with several jars holding bundles of tall white lilies and freshly cut pine boughs. Tied to the ceiling beams, dried ears of corn hung at the level of the lily blossoms. Pictures, many of them faded with age, hung on the wall. I realized that they showed the Ah Q'ij in many different places and with important people in his community.

Ah Q'ij Panjoj was bending over a smaller table about the size of a TV-dinner serving stand. It was covered with a handwoven cloth with black-lined green stripes against a red background. An odd assortment of rocks, coins, and other objects was piled two or three inches high along the top and left sides of the table. If these coins had been paid by the clients who had come to see him that day, then he had seen many people already.[95] I saw coins from all over North America and Europe and a sprinkling of anomalous objects from our world—a broken watch, a small metal box. His magic stones were of many different shapes and kinds—most of them dark, but some of them crystals. Some had special shapes that could lend themselves to seeing faces or animals. I also recognized fragments of Precolumbian axes and the little adze-shaped stones that must have been used as tools.

This collection of stones and assorted trinkets were the *q'abawilob* he had collected over a lifetime's service as a chuchkahaw. Most of them he had acquired during the 260-day initiation period when he had become an Ah Q'ih. Many of the stones were of the type that the Maya identify with lightning, and they helped him focus the lightning in his blood on the work of divination. These stones were unimpressive at first sight, but I realized that I had seen them before. The white stones and associated green stones, the axheads and celts, the eccentric flints and incised obsidians I had been studying for twenty years had served the same function for the ancient Maya. The k'awil of the Classic period lived on in Don Sebastián.

When the Ah Q'ij finished his work with the woman, Duncan motioned me forward to the little chair next to the table. Putting away my sketchbook and pen, I took my seat and focused my mind. I did not intend to be an objective observer of the divination, but a full participant. For me to do anything else would be disrespectful.

The Ah Q'ij and Duncan spoke for a minute and then Duncan asked for my question and told me to pay the diviner. I put five quetzals (the equivalent of a dollar) on the table, and after a bit of back and forth in English, Spanish, and K'iche', we were all satisfied that he understood the question I wanted answered. I won't tell you what the question was for it was a private thing for me. The Ah Q'ij turned his leathery face toward me and asked my name. He nodded his head, repeated it out loud, and turned to his beans murmuring prayers I could not understand.

He gathered all of the bright red beans together in a single pile and pushed them around by running his fingers through them with a turning motion of his wrist. Then he gathered a group of them with his fingers and pulled them away into a second pile. He did not count them, but seemed to *feel* when he had enough. He arranged the beans and crystals in groups of four until he had two rows lined up on the table. Throughout it all he chanted the prayers and the day names, occasionally pausing as he glanced down at a leg or hand. He was listening to the lightning in his blood.

He gathered the beans together and laid them out three more times in quick succession, asking my name each time but requiring little other information from me. When he was done, he explained the divination to me in Spanish. The answer was utterly unexpected, but eminently satisfactory. Then Erica took my place and he made a divination for her. Here I saw the difference between the two divinations and came to understand that the diviner does not work by rote or formula. The process is different with each of his clients. While my divination had been quick and incisive, hers was far more difficult. Time and again he stopped to ask her questions, and then resumed his work with the beans. Finally he was done. He imparted her divination to her just as he had given me mine earlier.

The Ah Q'ij offered to conduct a ritual for us—a *costumbre* he called it—for the coming twenty-one weeks. This ritual was to reinforce the divination and to counteract the problems he had detected through the beans. Since I knew that a chuchkahaw took on such responsibility at the peril of his soul and health, I readily agreed and we arranged to meet him in the church after lunch.[96] Each of us wrote our name on a piece of paper that he put on the larger table, and then we left for lunch.

We did not find him again until three that afternoon, for he had been seeing people continuously throughout the day. I paid him for the long round of prayers he had agreed to do for us, knowing that a chuchkahaw could not neglect his obligations without bringing illness and disaster on himself and his family. He also had to buy special tallow candles, copal incense gathered from special trees, and aguardiente or white lightning to pour out in prayer. According to traditional values, diviners cannot become rich from their work. The fees they charge the supplicant covers only the cost of their time and the materials used in the prayers.[97] To be a diviner, especially a very powerful one, is a heavy responsibility and one that does not bring wealth to the practitioner, unless he enters into the practice of witchcraft. **Plate 2**

Dodging the flower sellers sitting on the round steps outside the church, Duncan, Erica, and I followed our guide into the church. I was raised a Protestant and have not actively practiced religion for over thirty years so I wasn't sure of what to do. There were simple wooden pews on either side of the central aisle of the church. Six low concrete platforms about five inches high, three feet wide, and five feet long lay on the floor in a line down the center of the church. Huge, very old paintings or cabinets containing the statues of saints lined the walls, a pair facing each other across the nave at each of the six stations. The retablo and altar filled the east end of the church behind the low chancel rail. A small table sat on the nave side of the rail. Thus, combining the front table with the six floor platforms gave the church seven stations for prayer and ritual leading from the altar to the entry portal. **Plate 45**

The Ah Q'ij led us to the platform closest to the chancel, where we knelt on the floor behind the platform and facing the altar. He prayed over Erica for a while, then took three candles, and motioned with them over her head as he chanted another complex series of prayers. These prayers were in K'iche' so I couldn't understand him. When he was done, he touched the top of Erica's head with the candles, then touched each shoulder in turn. He held one side of the candles to her lips so that she could kiss them. Turning them, he then presented the other side to her lips. He bent over the platform for a moment as he lighted them from an already burning candle; he gave two to Erica and one to Duncan.

Next he went to the table by the chancel rail, prayed, lighted a candle, and set it on the table where he left it burning. Moving to the station of Santo Tomás, the patron saint of Chichicastenango, on the southern wall of the church, he lighted another candle and set it before that saint as he prayed. Afterward he crossed to the northern wall and did the same at a station he called *Justicia* or Justice. He came back to us and performed the same series of prayer and ritual for me, only using two candles instead of three. When his prayers were completed, he returned to us and took Erica's two candles and Duncan's one. He heated the bottoms on a burning candle and then stuck them to the platform. He placed my two candles beside them, leaving five candles burning in a row. To complete the ritual at this station, he poured aguardiente, a strong form of locally made moonshine, around the base of the candles and along the edge of the platform. We crossed ourselves and then went to the third platform where the process began again. For some reason, he skipped the second station.

Then we went to the fourth, fifth, and sixth stations in their turn. The ritual at the third platform, he told us, called upon the communal dead, while the one at the fourth contacted all the dead chuchkahawob. A cross stood next to the final station. There he prayed to San Pedro, the patron saint of diviners, and San Ramos, but most important, he left a candle and prayers at the cross that evokes all the days of the sacred calendar.

My knees have badly deteriorated from the years of climbing the high steps of pyramids, so that the kneeling and the getting up and down at each station became more and more painful for me as the ritual proceeded. By the second stop, I could not kneel, but had to sit cross-legged. At the fifth, I felt my legs begin to go to sleep, but I moved to relieve the pressure and keep my blood flowing. At the sixth and final station, I gave in to the experience and held my position as an electric sensation flowed through my legs and expanded to fill my legs from hips to heels. Finally, my legs and feet went numb. In the Western side of my mind, I knew I was feeling nerves protesting when the blood supply was cut off. But in the other side of myself, which I have tried to cultivate for the last decade, the mind that is open to experiences that have no rational explanation, I wondered if this was not the lightning in the blood that the chuchkahawob use to make their divination. It occurred to me that we in the West must feel the same sensations as the chuchkahaw, only we describe them in a different way. Most of all, we don't pay attention to them until they cause us pain—and even then we avoid them if we can.

In the months since it happened, I have thought a lot about my experience. I think I understand why such divination helps the clients of the Ah Q'ij. The questions he asks his clients can be as subtly penetrating as those of a trained and experienced psychiatrist. But most of all, the divination gives clients a chance to focus on their problems, to share them with other people, and to receive advice that often links them back into their community and the greater cosmos. In our world, medicine addresses the body, while divination and healing in the Maya world work with the mind and spirit. I suspect both are necessary for good health, but we have trouble dealing with the second.

I also saw the *q'abawilob* in action and can now picture how the anonymous stones and trinkets in Maya offerings and graves were once used by their owners. I had participated in a process of divination that is millennia old. I know that the details of divination vary today from community to community across the Maya world. And there are surely differences between the methods used by Sebastián Panjoj and the diviners of the ancient world. Nevertheless, the underlying intent; the use of sacred stones that have living power in them; the counting of the days to lock people into the basic structures of the cosmos; the use of sacred places, incense, prayer, flowers, and other sacred things to reinforce the efficacy of the divination—all link back to practices thousands of years old.

Duncan even explained to me the role of sacrificial offerings in the process. The candles that Don Sebastián burned for us were made from the tallow of special bulls sacrificed in Chichicastenango's most important religious ritual. There is even a *cofradía* that has the special responsibility of making the sacred tallow candles. They are the candles everyone in Chichi prefers to use in these sacred rituals. Sacrifice was different in the Classic period, for people drew blood from their own bodies and gave the lives of captives to sanctify their most important rituals. Today the bulls sacrificed at Chichicastenango, San Juan Chamula, and other communities feed the people who participate in the ritual, as well as providing such things as tallow candles. Yet I think the idea is the same. The bull gives his life to nourish and sustain the people in an exchange that is not unlike the ancient story of Creation in which the Maize God died to enable the birth of humanity.

CHAPTER FIVE

ENSOULING THE WORLD AND RAISING THE TREE

A DEDICATION AT IXIMCHE'

(as told by Linda Schele)

In July 1991, my Kaqchikel friend, José Obispo Rodríguez Guajan, now known as Pakal-Balam to his friends, invited me to go to Iximche', the ruins of the Kaqchikel capital destroyed by the Spanish soon after the conquest of the Maya of the Guatemala highlands in 1524. We had already been there together a year before when he and his family had taken me and my friends Nora England and Nikolai Grube on a wonderful visit after a workshop on hieroglyphic writing we had given to a group of Maya. It had been the first time I had seen a Postclassic highland site. I learned a lot about how those very late sites were so alike and yet so different from their Classic-period prototypes. On this trip we weren't going to see the ruins but to attend the inauguration ceremony for a new weaving co-op that had been organized by a group of Tekpan women. Since Doña Juana, Pakal's mother, was one of the women in the group, he invited me and our mutual friends to come to the ceremony. On a bright Sunday morning, I found myself in a rented car driving up the road from the Pan American Highway into Tekpan to find the Rodríguez house.

Don Rodrigo, Pakal's father and a man of more than seventy-four years, stayed at home. The rest of us piled into my little sedan and headed for the washboard road that led from Tekpan to Iximche'. After twenty minutes of bottom-bumping travel over rim-deep ruts, we rolled into the grassy parking lot at Iximche', which was full of cars belonging to both the ceremony participants and the normal cadre of Sunday visitors. I threaded my car into a slot and parked on the grass. Then we made our way to the site of the ceremony just outside the entrance to the ruins and next to the Iximche' museum. The women had lighted fires in a nearby grove of trees to cook their meat and heat the tubs of freshly made tamales that would be washed down by liter bottles of Coke and Pepsi. Paper plates and cups, napkins, and plastic forks lay in ready piles for the hungry hordes who would eat when the ceremony was done.

A band of musicians from Tekpan, sporting two huge marimbas and a set of drums, had arranged their instruments under the porch of the museum. Behind them the brilliantly colored wares of the co-op weavers hung on its whitewashed exterior wall. Rows of metal folding chairs sat in front of the display, soon to be filled with the audience milling around waiting for everyone else to arrive. My friend Federico Fahsen and his wife, Marta Regina, who had just ended her tenure as the minister of culture, came out from the capital to join us for the ceremony. The crowd included people from Tekpan, friends and guests from Antigua and the capital, and tourists who were joining the celebration out of curiosity.

I had never been to a Kaqchikel ceremony of this type, but over the two decades of my work in Central America, I had sat through dozens of other similar ceremonies from México to Honduras. Traditionally, these occasions are times when every politician and petty officeholder in town gives long, flowery, and usually inane speeches to a long-suffering audience waiting impatiently for the food and drink to flow. I had braced myself for exactly the same kind of performance here—but I was wrong. Oh, there were the expected speeches, but they were mercifully short—except for the one given by the head of a local Ladino organization. He managed to offend just about everyone before he ended his speech and left the grassy stage to the Maya women who had organized the co-op.

At that point the ceremony changed into something very different from anything I had anticipated or yet experienced. The marimba band struck up a song—and the women of the co-op appeared dressed in their finest formal garb. They wore their best *cortes*—yards-long, finely woven cloth wrapped around their waists many times and bound tightly in place with woven cinch belts. Their normal huipils—blouses with astoundingly complex patterns woven or embroidered into them—lay hidden under formal capes they wore only on special occasions. Woven from unbleached-cotton threads, the dark brown capes were accented by brilliant patterns of zigzags representing lightning, lines, flowers, and shapes of all sorts blazing in bright whites and hues of undiluted intensity. Delicately woven white cloths, decorated with brightly colored bands, lay folded doubled and redoubled on top of their heads, balanced there only by the women's natural grace and long experience.

They danced out from behind the corner of the museum in a double line, swaying to the complex rhythms of the marimbas, carrying baskets full of wrapped candies to throw into the audience at the end of the ceremony. At first I was embarrassed at seeing adults of such unstudied dignity and personal presence dancing in front of a gawking audience like us. I am very tall for a woman and I have always felt terribly awkward when I have tried to dance. In fact, the only time I've ever enjoyed dancing was when a little alcohol help liberate my self-conscious restraint into uninhibited abandon. Yet there I was, watching this collection of women dance before an audience of friends and strangers. I was a little embarrassed for them.

Gradually I began to realize what was really happening here. These were not untried girls dancing, but women of respected status, economic forces in Tekpan and leaders in their communities and families. The youngest were in their thirties and the oldest in their sixties—and they were people of worth and pride. I have only rarely felt such a concentration of power focused in a single group of people as in these women.

Then, in a moment of insight, I was transported to another time and place, just as I would be at Momostenango three weeks later when I stood on the Six-Place. Doña Juana and her colleagues suddenly transformed **Plate 23**
for me into Lady K'abal-Xok and Lady Sak-K'uk', the wives and mothers of Classic-period kings. Like their predecessors of fifteen hundred years ago, these women were dancing to sanctify a moment of beginning in their world.

I do not know if all modern Maya would dance in the manner of Doña Juana and her partners in the co-op. I have not seen enough modern ceremonies to know even if all highland Maya would do it the same way, but I suspect that these women were not following prescribed rituals dictated by specific local custom of today. I perceived that together they had invented, perhaps recalling ceremonies they had witnessed in childhood, a new ceremony they thought would be appropriate for their purpose. And not only did they dance, but they read from the Popol Vuh. More important, they chose to bless their enterprise not on the grounds of the local church but near the hallowed portals built by their ancestors.

They did what seemed right to them[1]—and what they did linked instantly and directly into a way of understanding the world and making it proper that comes from Precolumbian tradition.

When Doña Juana, her children, and the Maya living throughout the Yukatan Peninsula dedicate new houses, inaugurate new organizations, bless the things they make, and pass on their land to their children, they do these things in ways that were time-worn when the Spanish landed on Cozumel Island. The ancient Maya danced, read from the myth of Creation, and gave their blood and other precious things to vest with soul the objects they made. For them, the beginnings and endings of things were occasions of rich and complicated pageant and ceremony.

Beginnings were important because of the way the Maya thought of the material world. They believed that places and things made by the gods during Creation were imbued with sacred force and an inner soul from the beginning of time. In contrast, places, buildings, and objects made by human beings had to have their inner souls, their *ch'ulel*, put in them during dedication ceremonies. As long as people used these objects, this power was safe, even though it grew through use. But when an object was no longer to be used, this living force could become dangerous. It had to be contained or released in special termination rituals that protected the community. The rituals Maya designed to accomplish these acts of ensouling and terminating objects and places represent a significant portion of the Classic inscriptions and the archaeological record.

In their texts, the Classic Maya described the action of dedication in several ways—"to make proper," "to bless," "to circumambulate (through the four quarters)," "to cense with smoke," "to deposit plates full of offerings," "to set something in the ground."[2] People gave proper names to objects like temples, shrines, houses, stelae, altars, and ritual accoutrements according to the images represented on them. They brought these things to life with pageants in which participants danced, played ball, held rich feasts, took blood from their bodies, and sent messages to the Otherworld on the tongues of the people they sacrificed, all to bring sanctity to the newly made thing.[3] They deposited precious caches that remade the Primordial Sea under the floors of their buildings: jades, spondylus shells from the sea, the red pigments cinnabar and specular hematite, mercury, eccentric flints, obsidian. Sometimes great fires were lighted and jade was thrown in to shatter into hundreds of pieces. And when everything was done, the object had its soul. These deposits were accompanied by ritual and symbolic sequences that reestablished the conditions of the first act of Creation.

So the formulae for dedicating buildings involved far more than simply digging a pit and concealing valuables, however exciting those are for the archaeologists who discover them. The Classic Maya knew that cleansing, preparing, and propitiating were vitally necessary to create ensouled and pivotal places, around which the business of living and governing, of praying and playing, could revolve. The ancestors and the gods required properly made spaces in order to come into the world of people and help those they favored. To conduct life and ritual in a poorly prepared space was unthinkable for the ancient Maya, as it is for their descendants today.

We think dedication rituals pervaded Precolumbian Maya experience because excavation shows us that caches of special sacred objects accompanied most architectural modifications of any significance in the urban centers. We also think that during the Preclassic and Classic phases of Maya civilization, the widely shared cosmology prescribed a widely practiced program of dedication rituals. We envision a grand and elaborate cycle of performances spanning many days, sometimes even longer periods. In the royal capitals, these festivals expressed the universal understanding of the Maya people that nothing important is just made. It also has to be born.

We have only hints of the complex range of the activities involved in these great rituals. The Maya recorded them in their inscriptions and showed them in paintings. But because they also deposited offerings, these rituals yield one of the most important components in the archaeological records of Maya cities and towns. Our view of these pageants is, however, fragmentary, so that we must piece together our picture puzzle by combining the archaeological, inscriptional, and visual record from different cities. By joining them together, we become aware of one of the essential Classic-period ceremonial cycles, the re-creation of Creation. This cycle periodically affirmed that the cosmos lived, just as the dancing Tekpan women at Iximche' affirmed the living truth of their own history and practices.

CHAN-MUWAN'S DEDICATION AT BONAMPAK'

One of the most detailed depictions of Maya house-dedication rituals was painted on the walls of the three rooms in Temple 1 at Bonampak', a small but important capital of the western lowlands on the Mexican side of the Usumacinta River. At King Chan-Muwan's behest, his artists depicted a complex series of events that celebrated his son's formal presentation as the heir and the ensouling of the temple itself. The murals present a dramatically detailed view of Classic Maya ritual. The artists divided the composition of Room 1 into double registers so that they could depict two important historical events at once. In the upper register of the eastern, southern, and western walls, they painted the presentation of the young heir of the king to fourteen high-ranking lords on 9.18.0.3.4 10 K'an 2 K'ayab or December 14, A.D. 790.[4]

The second event, Chan-Muwan's dedication of the building itself, took place two hundred and thirty-six days later, on November 15, A.D. 791. At sunset on that day Venus first appeared as the Eveningstar. The artists depicted this second event, the ensouling *of* the temple, in the entire upper zone of the north wall and on the entire lower register. The narrative scenes[5] concentrate on two moments in the long ceremony—the dressing of three dancers and their public performance.

The dressing scene (Fig. 5:1a) shows a group of people busily accoutring three lords in the elaborate feathered costumes of the dance. Attendants on a lower terrace open bundles and chests and then pass their contents up to their counterparts above. One attendant ties wrist cuffs on the central dancer's arm, while another paints his skin. Various people are straightening out the feathers of a backrack that is about to be inserted into another dancer's belt. A group of bystanders watch as they talk among themselves.

The same three lords reappear in the lower register performing what the accompanying text calls literally a feather dance[6] (Fig. 5:1b). To the right of the dancing rulers, we see thirteen lords—all of them vassals of various ranks (Fig. 5:2)—in an informal processional array, talking among themselves while they move toward the dance area. The first man in line holds a battle standard, which he points straight down toward the floor. The sixth holds a staff over his head, and the ninth and tenth carry

a. Room 1, north wall, upper register showing the adorning of the three dancers

b. Room 1, south wall, lower register showing the three lords dancing

FIGURE 5:1 The Bonampak' Murals

large feathered battle standards attached to long staffs. The battle standards anticipate the war that must follow the first phase of the dedication rites so that suitable sacrificial victims can be taken for later ceremonies.

To the left of the feather dancers, we see a procession of musicians[7] and dancers cast in the roles of gods (Fig. 5:3). The first five wear tall white headdresses and painted leather skirts and they shake large magical gourd rattles.[8] Behind them, a drummer beats on a chest-high drum, accompanied by three other musicians who counterpoint his rhythm on turtle-shell drums beaten with deer antlers. Two other men, hidden by the turtle-shell players, carry more feathered battle standards, echoing the scene on the opposite wall. The masked dancers come next, followed by two musicians sounding large wooden trumpets. At the very end of the procession is a person carrying sacred objects in a bundle.

Room 1, lower register, west side, showing a procession of sahalob and ahawob accompanying the three dancers

FIGURE 5:2 Bonampak' Dedication Dance Procession

Room 1, lower register, east side, showing the procession of the musicians and masked dancers into the main ritual

FIGURE 5:3 Bonampak' Dedication Dance Procession

the masked dancers from Bonampak' and a Yaxchilan ballplayer wearing a similar mask

FIGURE 5:4 The Masked Figures from Bonampak' and Yaxchilan

For our purposes, the masked dancers (Fig. 5:4) are the most important figures in the procession, for they appear often in similar rituals depicted at other sites and on pottery vessels. Dressed in the frightening masks of the monstrous Otherworld beings they have become through trance dancing, they hold ears of corn, bundles, and staffs in their arms, or wave huge crab-claw gloves about. One, wearing the head of a crocodile, sits on the floor next to another person dressed in the costume of the First Father as Hun-Nal-Ye, the Maize Lord[9] who was resurrected from the Otherworld by the dance of the Hero Twins. He and the masked dancers around him wear waterlilies to represent their status as denizens of the watery Otherworld. We'll have more to say about these characters when we detail the rebirth of First Father in Chapter 5. For now, they are important because they bridge to another scene at the nearby city of Yaxchilan on the mighty Xokolha, the Usumacinta River.

In the dedication rituals of a place called the Three-Conquest-Ballcourt at Yaxchilan,[10] the sculptors carved these same masked dancers in a series of scenes on the step of a monumental stairway set on the mountainside above the river. At Yaxchilan the masked performers are each bouncing a big ball from the side bench of the court (Fig. 8:13). Since this Yaxchilan scene is also associated with the dedication of a building—in this case a ballcourt—we surmise that ballplaying was a general part of the cycle of dedication rituals. However, just because ballplaying was part of the complex of dedication rites doesn't mean that the play took place in a formal ballcourt. Although the ballgame depicted at Yaxchilan served as part of the dedication of a ballcourt, the action of the masked players took place not in a ballcourt, but on a hieroglyphic stairway composed of six steps. Called a *Wak-Ebnal*, "Six-Stair Place,"[11] these kinds of structures were "false ballcourts," places where Maya lords delivered their enemies to the Otherworld.

The initial house dedication at Bonampak' shown in Room 1 of this painted temple was followed by a battle on the next zenith passage of the sun, which happened to coincide with an inferior conjunction of Venus as Morningstar. Depicted in Room 2 (Fig. 7:11), the victorious lords of Bonampak' and their allies take captives, who were subjected to torture and humiliation. The miserable captives who survived that event finally died in a sacrificial dance performed along the terraces of a pyramidal platform depicted in Room 3. Sacrifice linked to dance was an integral part of dedication ceremonies for the Classic Maya.

"SOUL-STUFF" IN THE GROUND
The Empty Cache
(as told by David Freidel)

A lot of people worked with me on Structure 5C at Cerros back in the 1970s, but Jim Garber was the main supervisor of this exciting excavation. Jim loved the architecture; he lived the clearing of this masterpiece for months on end and spent days of his life meticulously writing notes and recording plans, profiles, and photographs of it. But for all the wonders of the giant masks, Jim wanted to find a cache to help us date the building more accurately. We joked a lot about how he wanted to tear up every suspicious floor surface, and in the course of investigating, we found a lot of interesting holes. Some were postholes for enormous poles. Another was the dump from a termination ritual. It held shattered stucco torn from the facades of the last building and buried in the corner next to the stairway. The cache we wanted eluded us. Then early one day when I cruised out to the temple to confer with Jim about excavation strategy, I found him sitting at the top of the broad plastered stairs, staring like a gargoyle at the bottom of them. He motioned me up to the top excitedly, turned me around and pointed. It had rained the previous night, and there at the bottom of the staircase, on the smooth plaster surface of the plaza, was a shallow round pool of water. No doubt about it. Jim had found his plaster patch. This had to be the dedication cache. I made him dig it really slowly, recording every detail of the context. When we got down to the two big red plates placed lip to lip under the plaza floor, it got hard even for me to maintain detachment. Finally, Jim got to lift off the upper plate with trembling, sweaty hands: the lower plate was completely empty. "Damn," he said, "it's doin' the nothin'." It was a hard lesson: the point of the cache was to cache, not to hide precious things.

Over the years I've learned what lip-to-lip vessels in the ground mean to living Maya in Yukatan. They are the earth ovens they call *pib* in which people cook special stews in just about every ritual—from the town festivals to the remembrance of a parent on his death day. There in the ground they transform raw stuff into cooked food. This food is blessed with the *itz* that passes into the oven from the altar when the *pib* is built for the Ch'a-Chak ceremonies. So the empty cache at Cerros, devoid of even perishable offerings in the form of carbon or dirt residue, declared its function in a way that it could not have done if it were full of stingray spines and shells and jade. It was a *pib*, pure and simple. When the lords passed over that threshold and danced up the stairway to the temple above, they passed over the *pib* and, perhaps, into it. For *pib na*, underground house, is the other name for the conjuring houses Chan-Bahlam dedicated in the Group of the Cross at Palenque. And there in the *pib* the Palenque king was magically transformed into sustenance fit for the gods.

We have found the remains of these dedication rituals in cached offering deposits at dozens of archaeological sites—at Kolha in Belize, at Waxaktun, Tikal, and other central Peten sites. Lords and shamans placed magical offerings inside these large ceremonial plates and used other plates as lids to cover them. These are *sak-lakob*, "pure or human-made plates,"[12] and the *Ol* "portals" we talked about in Chapter 4. Lip-to-lip caches are among the most ancient expressions of Maya ceremonialism. One such cache vessel, dating to the Middle Preclassic period, was discovered at the stone-tool-manufacturing community of Kolha. It held an extraordinary set of materials that our colleague Dan Potter, who discovered the cache, explained to us.

As Dan reconstructs it, the people participating in this ritual struck a large leaf-shaped sacrificial blade off a core of beautiful honey-brown chert. This first blade broke as it came off the core so the shamans threw it into the bottom of the pit and struck off another.[13] Dan told us of his excitement when he first fitted the broken blade back onto the core and discovered that the second blade fit next to it. After the shamans had successfully struck off the second blade, they threw the remainder of the core into the pit, which they then filled up with dirt and offerings of shattered jade. Then they set the bottom plate into the pit. Someone used the chert blade to draw blood, probably from his penis. The blade was then placed in the plate along with jade beads, shark's teeth, large cut and carved spondylus shells, and other magical odds and ends. Finally, the shamans covered the bottom vessel with its twin and sealed the cache.[14] There it lay under the earth, awaiting the archaeologists who found it again after two thousand years.

Similar lip-to-lip caches at Waxaktun held the severed heads of victims who were sacrificed in the kinds of building-ensouling rituals we saw at Copan and other centers.[15] At Tikal, these rituals involved lip-to-lip caches and buckets placed in pits dug below the floors of buildings and plaza areas.[16] The offerings (Fig. 5:5) included sea creatures brought live from the coasts, spondylus shells, jade, eccentric obsidians with incised drawings of various gods, and bloodletters of various types including stingray spines, obsidian blades, and thorns (today called "cuerno de vaca"). Generally, participants used these sharp objects to draw blood, enabling them to enter the visionary trances necessary to the successful spiritual activation of these places.

At Copan, caches ranged from the simple to the elaborate—objects placed in simple pits dug into the floors and then plastered over, or put into badly made buckets hastily decorated with modeled cacao pods. These humble buckets belie their contents, for the lords of Copan filled them with spectacular treasures and instruments of supernatural power,[17] including pounds of raw or roughly sawn jade, "popcorn" jade shattered in fire pits, earflares, jades and greenstones shaped like charcoal briskets and drilled for stringing together into necklaces, and greenstone images of hunched-over shamans. The latter items were all probably carved in central Honduras. Craftspeople sliced jade pebbles into slabs of various thicknesses and carved them into small images of lords called "Yahaw-Te'," little figures with their hands hooked in front of their chests (Fig. 5:5c), accompanied by images of the great World Tree.[18] Other items the Maya placed into their caches were spondylus shells lined with red pigment, often closed around a jade bead to form a natural jewel box (Fig. 5:5b). These objects were often accompanied by bloodletters of various sorts—stingray spines and flint blades—and sets of extraordinarily complex and beautiful items called "eccentric flints"—branching pieces of flint knapped to create multiple human profiles. Some of these flints were manifestations of K'awil, the god of spiritual embodiment (Fig. 5:5g).

At the Late Preclassic site of Cerros at the summit of a huge acropolis, David Freidel and his colleagues uncovered a cache bucket that was not just "doin' the nothin'." It contained many of the same objects described above (Fig. 5:5a). The time frame for the interring of this cache falls roughly halfway between the period of the Kolha cache and the Copan cache.[19] The Cerros cache contained five beautifully carved jades that had once served as the headband and chest pectoral of a king.[20] These stones of prophecy and symbols of the kingship were arranged in a quincunx pattern at the bottom of the bucket. The king who made this temple covered the heirlooms with layers of mosaic mirrors made of blue hematite crystals, originally glued onto mother-of-pearl backings. The remain-

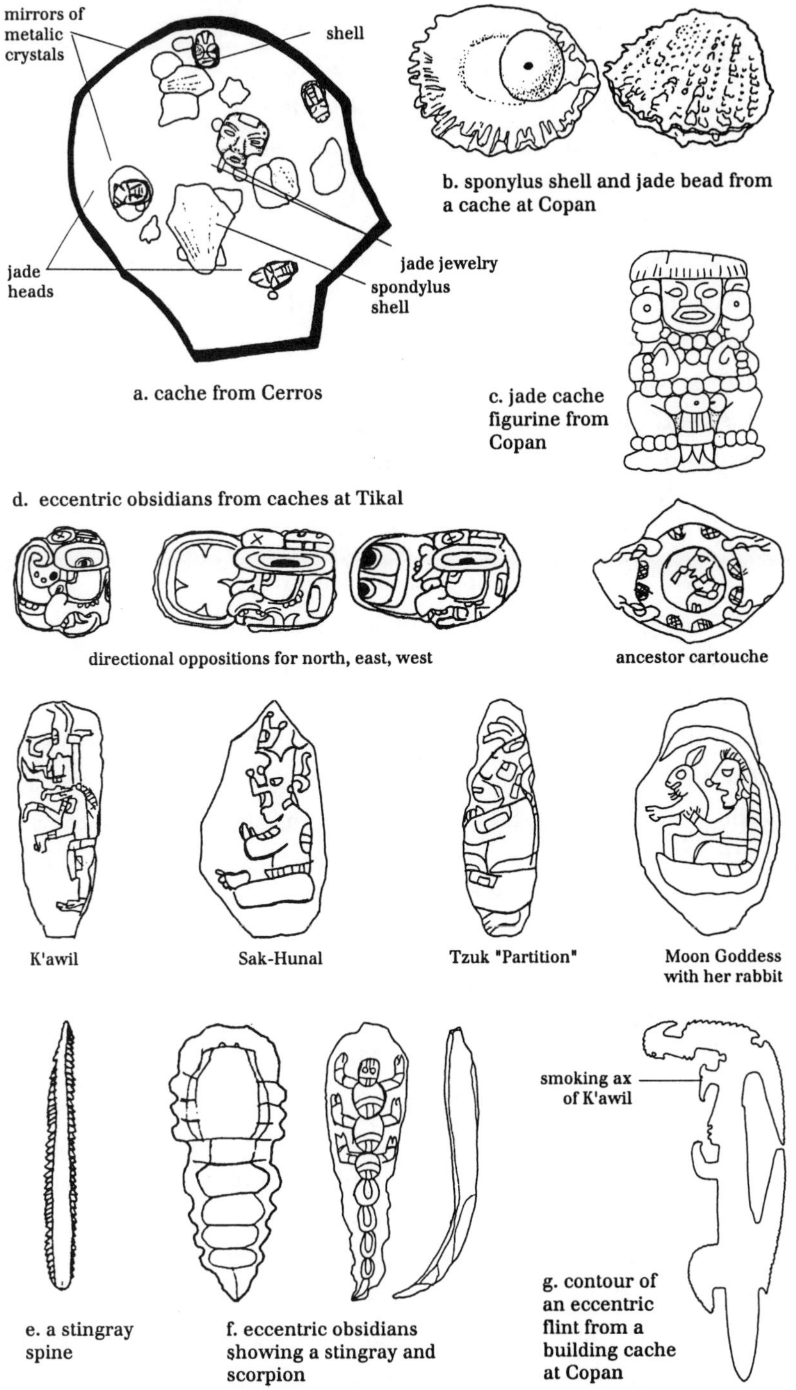
mirrors of metalic crystals
shell
jade heads
jade jewelry
spondylus shell
a. cache from Cerros
b. sponylus shell and jade bead from a cache at Copan
c. jade cache figurine from Copan
d. eccentric obsidians from caches at Tikal
directional oppositions for north, east, west
ancestor cartouche
K'awil
Sak-Hunal
Tzuk "Partition"
Moon Goddess with her rabbit
smoking ax of K'awil
e. a stingray spine
f. eccentric obsidians showing a stingray and scorpion
g. contour of an eccentric flint from a building cache at Copan

FIGURE 5:5 Objects from Dedication Offerings

ing contents of this Cerros cache were jade beads, a jade earflare, and some red-orange spondylus shells—highly prized by the Maya. Because of the careful arrangement of these items, we think this cache was probably tied up in a bark-cloth bundle, then reverently placed at the bottom of this deep bucket. Archaeologists discovered a similar bundle in an Early Classic lip-to-lip cache at Tikal.[21]

At Tikal our friend Peter Harrison found another soul-vesting cache vessel revealing how Creation was connected to dedication rites. Discovered under the stairway (Fig. 5:6a) of Structure 5D-46, Jaguar-Paw's palace in the Central Acropolis,[22] this black cylindrical vessel contained spondylus shell figurines, jade, textile (possibly a bundle), wood, and nine imitation stingray spines. A text running around its rim says, "It was made holy, the house of, the holy place of the nine successor lord, Moch-Xok, Six-Sky ???, Jaguar-Paw, Tikal Lord" (Fig. 5:6b). In other words, this building was the palace of the king.

The imagery carved on the side of this cache vessel includes the Quadripartite God plate, the holy heart of heaven, holding its special things, the feathered shell, a bloodletting instrument, and the "kimi" variant of the *way* glyph. This is the sun-marked offering plate that generates the Raised-up-Sky World Tree and Vision Serpents, and accompanies the Cosmic Monster. On the Tikal incised pot, First Father, in his guise as the Maize God, Hun-Nal-Ye, appears next to the deified offering plate. The Double-headed Scepter he holds in his arms is the ecliptic snake with the Paddler Gods emerging from either end. Three hundred years later, Hasaw-Ka'an-K'awil, the twenty-sixth successor of the same Tikal royal lineage, represented this same scene on a set of incised bones as the Milky Way canoe carrying the Maize God to the place of his birth as the Na-Te'-K'an at the center of the night sky (Fig. 5:6c).

The objects placed in these offering plates are the material manifestations of *ch'ulel*, the holy "soul-force" of the universe. Because *ch'ulel* resides in blood, red pigment—both cinnabar and hematite—were often substituted. When heated, cinnabar yields mercury—a liquid mirror that the Maya made offerings of in small bottles or pooled in shells.[23] Hematite, before being ground into red pigment, could be polished and used as mosaic mirrors. And mirrors, both liquid and solid, were portals to the Otherworld into which people could gaze and discern true reality.[24]

Classic Maya artists depicted *ch'ulel* as streams of holy substance represented by jade beads, bones, shells, the color signs *chak*, *yax*, and *k'an* (red,

a. a cache pot from Great-Jaguar-Paw's Palace in its original location with eccentric flints spread around it (photograph courtesy of Peter Harrison)

b. the Maize God holds the ecliptic snake with the Paddler Gods emerging

c. incised drawing on a bone from Tikal Burial 116 showing the Paddlers carrying the Maize God to the place of Creation in Orion

FIGURE 5:6 Dedication Cache from Great-Jaguar-Paw's Palace at Tikal

blue-green, and yellow), zero signs, maize kernels, and mirrors. In addition, glyphic texts add spondylus shells, jade earflares, jade beads, ahaw signs, shell or pearl counterweights, bones, and obsidian to the inventory of things that represent *ch'ulel*. These are exactly the kinds of objects placed in caches—jade jewelry of all sorts, raw jade, eccentric obsidians and flints, spondylus shells, pearls, and precious stones.[25] Sea creatures—sharks, sea fans, coral, and so forth—re-created the Primordial Sea of the Creation. Stingray spines, thorns, obsidian, and flint blades were used to draw the blood that opened the portal in the offering plate. Once this portal was opened, *ch'ulel* from the Otherworld, and from ancestral buildings encased within the new structure, could enter into the new building and bring it alive. These dedication rituals replicated the conditions that existed at the beginning of the world when the Lying-down-Sky pressed against the dark sea and the acts of First Father raised up the sky and let holiness enter the cosmos. Placed in the ground like the festival stews and breads in their *pib,* like the body of the sacrificed First Father in the place of Ballgame Sacrifice, cached things became transformed by the dedicatory act of caching into the source and sustenance of fruitful life.

AFTER THE SPANISH CAME

The ceremonies for the creation and ensouling of sacred objects and houses continued to be complex and interrelated activities even in the centuries following the ninth-century collapse of the Maya kingdoms. Bishop Diego de Landa[26] described the making and renewal of god images in the sixteenth century as a long and very dangerous ritual sequence beginning in the month of Mol. After consulting with a *chilan,* an interpreter for the gods, to choose an auspicious day, the men who served as Chakob, the four assistants, began fasting. Meanwhile, workers, usually belonging to the family who were making the idol, went into the forest where they chose a tree,[27] cut it down, and brought it back to the town. There they built a small hut, fenced it in, and brought in wood, carving tools, and large urns to be used for storing the statues as the work progressed. Here is the shrine place again, the properly prepared house where the divine can enter into things. Incense was burned in front of gods called *akantun,* "erected stone," which were placed at each of the four directions. When all was ready, the workers brought lancets, paper, and bowls for the bloodletting rituals they would conduct in the house. Then the priests and the Chakob closed themselves into the hut with them. During the carving of the gods, the sculptors and ritual practitioners drew blood to anoint the wood and burned incense. None of the participants in the ritual could see their wives during this time and the owners of the new gods were obligated to feed the craftsmen until the work was done.

During the following month of Ch'en, after the carvings were finished and images perfected, the new gods were taken from the hut and placed in an arbor built outside the owner's home. The priests and the sculptors washed off the brazier soot they had used to anoint themselves during the carving. Then the priest solemnly blessed the new sculptures and drove out any evil spirits that might have become attached to the wood. Now completely consecrated, the sculptures were wrapped in cloth and placed in a hamper. After a lecture from the priest about the danger of breaking abstinence and fastings when making god sculptures, the people feasted to celebrate passing through a dangerous time.

The following month of Yax saw the renewal of houses in a festival called *Ok Na*. Once again the *chilanob* read prognostications for an auspicious day for the renewal to begin. New braziers were placed in front of the gods, who were themselves renewed, perhaps by painting them or placing new clothes on them.[28] If necessary, houses were rebuilt or renovated and then the renewal ceremonies were recorded in hieroglyphic writing on the walls of the house. In Landa's time, the months of Mol, Ch'en, and Yax fell between December 5 and the middle of February. The day the Wakah-Chan was raised (February 5 in the Gregorian calendar) fell on 13 Yax in 1547. Thus the time of renewal corresponded to the ancient time of Creation.

THE LAKANDON GOD POTS

Although many modern Maya still clothe and renew the saints and crosses in their churches, only the Lakandon still make gods like the ancient *sak-lak*. Like the ancient plates, each *u läkil k'uh*, "god pot"[29] serves as a brazier (Fig. 5:7) to transmit offerings to the gods. Each consists of a bowl with an upturned face modeled on one side. Although they combine the qualities of the god they serve and the human being who owns them, these pots are neither the god nor a representation of the god. They are instead living beings with all the important anatomical parts of a human being. Male god pots have a vertical strip and female ones are painted with a grid of horizontal and vertical black lines. Their body parts are given to the pot in the following manner:[30]

> Five cacao beans are placed in the bowl of a new god pot to represent the heart (*pixan*), lungs (*sat'ot'*), liver (*tamen*), stomach (*tzukir*), and diaphragm (*bat*). Specific features molded on the head of the god pots are the ears (*xikin*), earrings (*woris u xikin*), hair (*tzotz'ebo'*), jaw (*kämäch*), eye-

A Lakandon burns offerings of copal in the god pots

(photograph © John McGee)

FIGURE 5:7

brows (*maktun*), space between the eyebrows (*chi' u pam*), eyes (*wich*), cheeks (*puk*), and mouth (*chi'*). The front of the god pot is called the chest (*sem*), and the bottom is called the feet (*yok*).

(Davis 1978:73)

Five thin threads are placed in the pot with the cacao seeds to become the intestines of the god.

The god pots are made in a special house called *yatoch k'uh*, "house of god," made in a traditional pattern using a dirt floor, thatch roof, and no walls. Set apart from the rest of the village, the *yatoch k'uh* is dedicated in ways that transform the normal space of human activity into sacred space. Circular designs are painted on certain beams and posts with a red dye made from the annatto tree[31] and called *k'uxu* by the Lakandon. These designs, according to Jon McGee, "are a reminder of the time in the ancient past when the creator god Hachäkyum ('Our True Lord') sacrificed human beings, collected their blood in a gourd, and asked the god Tzibatnah ('Painter of Houses') to paint his dwelling red with human blood."

Even the *balche'* canoe in which the sacred drink is made bears these red marks, to warn that the *yatoch k'uh* is a "meeting place of the gods and humans, where they will sit in your presence, partake of your offerings, and listen to your prayers." And in the sacred space of this house, ordinary tortillas are transformed by clouds of incense into food the gods will eat. The ceremonial tamale *nahwah* becomes human flesh, and the corn gruel called *atole* becomes "sacred water." In addition to food offerings, incense boards (*xikal*) are carved into human form to serve as symbolic human sacrifice, and rubber humanoid figures (*k'ik'*) are ceremonially given life and also burned as a sacrifice to the gods. Finally, in the god house, the annatto dye *k'uxu* become another symbol of human blood offering.[32]

Into this charged space the men of the community who would make the gods came in 1970, along with anthropologist Robert Bruce, for the forty-five-day ceremony—to sleep, to eat, and to mold the raw clay of the earth into the *u läkil k'uh*. All clothing, hammocks, benches, personal possessions, and tools were either made new or scrubbed and refinished for the ceremony, and no nonparticipant could touch them. The women continued to prepare food and leave the men's share on neutral ground, but chile peppers and sexual relationships with women were forbidden, for contact with either would endanger the coming process.

According to Bruce,[33] the participants slept only an hour or two a night during the entire time span. The result was an alternated state where every experience became sharper and brighter. In this growing state of awareness the god pots are made, stage by stage, until they are ready to be brought to life during the last five days. When the new god pots are ready for this transformation, the old god pots are brought forth from their house. Sacred heirloom pebbles are dug out of the thick residue of burned copal in the old pot of the god for whom it spoke and transferred into the new pot that will serve it. These stones,[34] called *u k'anche' k'uh*, "the god's seat," often come from a ruined city of the ancient ancestors of the modern Lakandon. For five days the new pots are fed like human beings, given drink, and awakened by the songs and chants of the participants until at last they have become living beings. When incense burns in the new pots for the first time, they are alive and their faces are turned to the east. The old dead pots have become *u baakel Äk Yum*, "the bones

The bones of god pots resting in a cave with the bones of the dead

(photograph © John McGee)

FIGURE 5:8

of Our Lords," and are turned to the west. Eventually the old bones are taken away and deposited honorably in a cave with the bones of the honored dead ancestors (Fig. 5:8). This god-making ceremony was not held again until 1991.

In his analysis of Lakandon dream symbolism, Robert Bruce gave us critical bits of information that are left out of most of the other descriptions of the ritual. In regard to the interpretation of a dream about painting a god pot, he said, "It should be noted that the painting of the new god pots occurs during the incense burner renewal ceremony, which plays on the allusion to the end of the world and the Celestial Jaguars devouring everyone" (Bruce 1979:133). He went on to say that the Celestial Jaguars (*Nah Tz'ulu'*) and the end of the world are "not quite as fatalistic in Lakandon cosmology as in ours, as *Hachäkyum* may be 'about to destroy the world,' but then be persuaded to stop. . . . The association with the Celestial Jaguars . . . plays on the allusion to the end of the world (or more correctly, the end of one Cycle and the beginning of the next) which is the theme of the incense burner renewal ceremony)" (Bruce 1979:126–127).

So we learn that the renewals of the god pots in 1970 and 1991 were reenactments of the Creation of the world. Today's Lakandon no longer keep the ancient calendar. The date 4 Ahaw 8 Kumk'u would have no meaning to them. Nevertheless, the renewal of their god pots resonates conceptually with Yax-Pak's re-creation of the world in the place of the Lying-down-Sea.

THE LIVING CROSS AND SOUL-FORCE

We believe that the living force that animated all properly dedicated things in the Classic Maya cosmos survived the Conquest and endures today. One accessible expression of this continuity is found in the ancient symbol of the World Tree, which transformed into and merged with the modern Cross of Christ. As we have tried to show throughout our story of the Maya cosmos, the World Tree-Cross as an object had its conjuring houses, its altars, and has today its churches. But it requires none of these things to manifest the soul of god. For the house of the World Tree is the world itself. Here is an unbroken path of dedications, from the Precolumbian Maya to their present descendants. As we saw in Chapter 3, for the Maya whom the padres were trying to convert to the new religion, the leap from tree to cross allowed them to embrace the outward forms of Catholicism without losing the cosmological content of their own beliefs.

Cross-shaped World Trees greeted the first Spanish visitors. When Francisco Hernandez de Córdoba stepped ashore in Kampeche in 1517, he reported ". . . temples of lime and stone, set on pyramids . . . and they saw idols, figures of serpents, and fierce animals, and what appeared to be painted representations of crosses. . . ." At Cozumel, the Spanish saw a pyramid with a three-meter-high cross of white lime in the center of its courtyard. In 1697, Kan-Ek', ruler of the last independent Maya kingdom, planted a stone column called the *Yax-Cheel-Kab,* "First-Tree in the World," in the center of his palace. *Yax-Cheel-Kab* was associated with ancestors and the Creation story.[35] Nearby was another sculpture called *Ah-Kokah-Mut,* perhaps a rendition of the Cosmic Bird that perched in the top of the World Tree. Crosses like these were found throughout Yukatan, where they were associated with rain and called *yax che',* the "first or (green) tree." These crosses are still painted green today.[36]

The Cross-Tree stands, blessed, dressed, and adorned, as the symbol of ancient understanding in contemporary homes and shrines. Crosses have literally spoken to the Maya across the ages, and still do so in the modern world.

In Quintana Roo, crosses became living oracles for the Chan Santa Cruz Maya during the Caste War.[37] The cult of the Talking Cross first appeared the summer of 1850 near the modern town of Felipe Carrillo Puerto. Located at a cenote known as Chan Santa Cruz ("Little Holy Cross"), this particular image of the cross was carved on the trunk of a tree. The first temple the Maya built for this cross was constructed west of the town and called *X-Balam Na*, ("The Jaguar's House"). The cross spoke to the people in several ways—at first through the voice of an interpreter, and later through written messages communicated by the people who served it. The Maya gave the crosses an Indian identity by clothing them in huipil (Fig. 5:9) and fustan, the dress and petticoat worn by Maya women. But the idea of clothing the cross was already an ancient idea, for the Foliated Cross at Palenque wears a necklace and jewelry like the living being it was (Fig. 5:10).

Sometimes the Cruzob carried one of their crosses into battle to guide them to victory, and the losers, in particular the English soldiers taken prisoner at Bakalar, were forced to negotiate with the cross. The negotia-

Teodócio Nahuat Canche with the Cross at the Center of the Earth in Xok'en. The cross wears a huipil and effigies of bread and metal hang from its neck

FIGURE 5:9

the Foliated Cross from Palenque wearing a pectoral and loincloth around its neck

FIGURE 5:10

tion was held inside a house in Bakalar where the cross was kept, attended by boys called "angels." The proceedings unfolded to the accompaniment of drums, bugles, and songs. During the negotiations the oracle cross squawked and demanded more ransom.[38]

In 1901, when the Mexican army finally captured the capital of the rebellious Maya and renamed it Felipe Carrillo Puerto, they retreated into the forest and established four new centers and several independent villages. Each center was protected by its militia led by Maya generals. Señor, where Linda visited Nikolai Grube in the summer of 1992, is one of these. It is only a few kilometers from Tixkakal Guardia, the oldest and most sacred of the four centers. It is there that the Cruzob established a replacement cross for the one the Mexicans burned. Nikolai took Linda to Tixkakal where, accompanied by several of the most important scribes and generals, they entered the Iglesia to see the "Gloria." The cross was there behind the wall that marks off the special space in the east end of the building. It wore a huipil and above it two boughs of plumeria flowers, *nikte'* to the Maya, arched from the four corners of the altar. The door into the inner sanctum had its own arch made of stone and mortar. Nikolai pointed out that the words *Chan Santa Cruz Balam Na*, "Little Holy Cross Jaguar House," were painted on the outer side of the arch. This was the name of the original sanctuary destroyed by the Mexicans.

The cross continues to play a large role in the religious practices of the present-day Cruzob. Crosses are put in milpas to protect them. Lineages have their own crosses, many of which can perform miracles. Of course, the modern inspiration for these crosses comes from the Catholic doctrine taught to the Maya since the Conquest, but the cross shape as the emblem of the World Tree was at the heart of Maya religious thought for at least two millennia before the Spanish ever arrived. The Talking Cross may be Catholic on the surface, but deep inside its voice and soul are Maya.

Even Maya crosses that don't talk are considered to be living beings. In Zinacantan, the cross is always painted some shade of blue-green, *yax*, in honor of the living ceiba tree that is its inspiration.[39] Each cross has an inner soul—a *ch'ulel*. The crosses are never set directly into the ground, but are lashed to a standing pole or set in a cement base. Three crosses are the minimum number that must be decorated for a ritual to succeed. Crosses are added to shrines when a shaman dreams that the Ancestral Gods have appeared and asked for another one. A cross is activated when someone ties the crown of a yellow pine tree onto its vertical shaft, places red geraniums at the crossing point of the arms, and lays a carpet of fresh pine needles in front of it.[40] The pine top always faces away from the town toward the forest, and the geraniums always face inward toward the center, the "navel of the world," *mixik' balamil.*

Although cross shrines must always have three crosses, when one or more crosses are missing, they can be replaced by inserting three pine-tree crowns into the ground and tying on the geraniums.[41] The Zinacantecos call these pine tops and geraniums the clothing of the cross. They are *nichim*, the growing and germinating "buds," like the *sak-nik* blossoms that decorate the arms of the Classic-period Raised-up-Sky World Tree. Zinacanteco shamans speak of the three crosses as three Ancestral Gods "dressed and waiting for candles for their food." These crosses, which are located at the foot and top of the mountains that hold the Ancestral Gods, are doorways to the houses of these important deities. Zinacanteco shamans call the Ancestral Gods *Totil-me'iletik*, literally "Father-Mothers."[42] They have souls.[43] Like the Wakah-Chan of the Classic world, the crosses of today are portals that penetrate into the Otherworld.[44]

The Classic World Trees not only lived themselves but were sources of life. Robert Carlsen and Martin Prechtel bear witness to the continuation of this understanding among the Maya of Santiago Atitlan, who live next to the crystal-blue waters of Lake Atitlán beneath the towering volcanoes that have shaped the highlands of Guatemala. Here is the way they describe Atiteko belief:

> . . . before there was a world (what we would call the "universe"), a solitary deified tree was at the center of all there was. As the world's creation approached, this deity became pregnant with potential life; its branches grew one of all things in the form of fruit. Not only were there gross physical objects like rocks, maize, and deer hanging from the branches,[45] there were also such elements as types of lightning, and even individual segments of time. Eventually this abundance became too much for the tree to support, and the fruit fell. Smashing open, the fruit scattered their seeds; and soon there were numerous seedlings at the foot of the old tree. The great tree provided shelter for the young "plants," nurturing them, until finally it was crowded out by the new. Since then, this tree has existed as a stump at the center of the world. This stump is what remains of the original "Father/Mother" (*Ti Tie Ti Tixel*), the source and endpoint of life.
>
> (Carlsen and Prechtel 1991:27)

This Father-Mother tree stands at the heart of Atiteko religion just as the Father-Mother ancestors of the Zinacantecos stand at the heart of theirs. If properly maintained, the Father-Mother renews and regenerates the world. For the Atitekos, as for the Zinacantecos, this great central concept of life has a physical representation—the main altar of the church in Santiago Atitlán:

> This altar, constructed when the church was without a resident priest and under full *cofradía*[46] control, is dominated by a mountain carved in wood. To either side of the mountain are carvings of *cofradía* members, complete with their staffs of office and shown ascending the mountain. Atop the mountain is a World Tree in the form of a sprouting maize plant. Atitekos believe that as long as the primal ancestral element, as "Flowering Mountain Earth," is "fed," it will continue to provide sustenance. In Atiteko religion, this "feeding" can be literal. For example, some Atitekos will have an actual hole on their land through which offerings are given to the ancestor. In the Tzutuhil dialect, this hole is called *r'muxux* ("umbilicus"). More commonly, "feeding" is accomplished through ritual, *costumbres.* For instance, dancing, sacred bundles, burning copal incense, or praying can feed ancestral form.
>
> (Carlsen and Prechtel 1991:27–28)

The Maya have been feeding the tree at the center of the world for three thousand years. The details of the rituals have changed, but the differences do not belie the underlying continuity in the way they think of the world. It is alive, and all sacred things in it—including things we would call alive and things we would call inanimate—have soul. And as with a child, the souls of these objects must be brought into them and then nurtured for their continued health. They understand that these rituals replicate the Creation of the world. They dance; they make offerings at the four corners or the four sides; they use flowers, maize, copal, candles, and food—whereas their ancestors offered shells, eccentric flints, eccentric obsidians, spondylus shells, stingray spines, jade, red pigments, and all of the other things that were the physical manifestations of the ch'ulel conjured in ritual. Moreover, the ancients brought living creatures from the sea—sponges, coral, sea anemones, small sharks, fishes, fossilized shark's teeth, and others—to re-create the Primordial Sea of the "Lying-down-Sky" time so that they could again raise the sky and order the world. The form they gave the Raised-up-Sky and the central axis of the world was that of a great ceiba tree. That tree is still there both as the cross and as the tree. The Maya continue to re-create the world and to feed the tree and its sprouts so that humanity will continue to prosper. Even the Lakandon, who know they perch on the edge of extinction, bring a new Creation—a new beginning—into being as they urge the celestial jaguars back from their threatened communities and bring their god pots to life. Like the fruit of the Atiteko World Tree, Maya traditions and communities have sprung up all around their ruined past, each sprouting into a newer version of the Classic vision, each rededicating itself to the future by transforming and honoring the past.

(photograph © by MacDuff Everton)

1. The sacred landscape of the valley of San Lorenzo Zinacantan. The three-peaked Senior Large Mountain lies in the center of the distant horizon. Its peaks are the three stones of the cosmic hearth. The wooded hill called the navel of the world lies to its left, just to the right of the tallest tree in the foreground.

(photograph © by MacDuff Everton)

2. Incense rises from people praying in front of the church at Santo Tomás Chichicastenango, Guatemala.

(photograph © by MacDuff Everton)

3. The Great Ballcourt of Chich'en Itza looking toward the North Temple

(photograph © 1982 by Justin Kerr)

4. A rollout of a Classic-period pot showing the birth of supernaturals at *Na-Ho-Chan-Witz-Xaman*, the "First-Five-Sky-Mountain-North." The snake-cords represent the sky umbilicus.

(photograph © by Duncan Earle)

5. The Passions and the Flowers run the Sun-Christ banners along the Path of Fire at San Juan Chamula.

(photograph © by Linda Schele)

6. The crosses overlook the graves of Chamulas at Romería. The planks are doors that are opened when family members talk with their dead.

(photograph © by Duncan Earle)

7. The Ah Itz divines the defeat of the K'iche' in the Dance of the Conquest. The doll that represents the youngest brother sits on the low table.

(photograph © by Duncan Earle)

8. The Passions and Flowers of San Juan and San Sebastián pause to touch their banners during their run around the square at San Juan Chamula.

(photograph © by Linda Schele)

9. The Group of the Cross at Palenque taken from the front of the Temple of the Foliated Cross. The "tree-sea" is visible beyond the Palace.

(photograph © by MacDuff Everton)

10. The head of the Pawahtun from Temple 11 at Copan. A huge ceiba tree rises from the top of the mound behind him.

(photographs © by Justin Kerr)

11a and b. An Early Classic pot showing one of the Hero Twins aiming his blowgun at Itzam-Yeh. A long-lipped being, perhaps a monkey from the previous creation, offers the bird a fruit as he worships it. The cylinder on the side of the bird is marked as a nance tree by *ki*-sprout signs just below the rim.

(photographs © by Justin Kerr)

12a and b. An Early Classic plate with lid showing Itzam-Yeh spreading his wings above the Primordial Sea. The lid handle represents his human nawal, Itzamna, sitting on a water-lily pad.

(photograph © by David Freidel)

13. A line of men and boys playing the Chakob and the frogs wait to approach the mesa in a Ch'a-Chak ceremony at Yaxuna.

(photograph © by John McGee)

14. A Lakandon dips *balche'* out of a canoe to strain it into a pot.

(photograph © by MacDuff Everton)

15. The alcaldes of Santo Tomás Chichicastenango carry their staffs of office in ceremonies before the church.

(photograph © 1987 by Justin Kerr)

16. A figurine of the Maize God sitting on a mountain. Mature maize plants circle his arms.

(photograph © by Linda Schele)

17. Carved bone from Tikal Burial 116. The image shows an ancestor in a cartouche. Bone was one of the substances that carried *ch'ulel* or soul force.

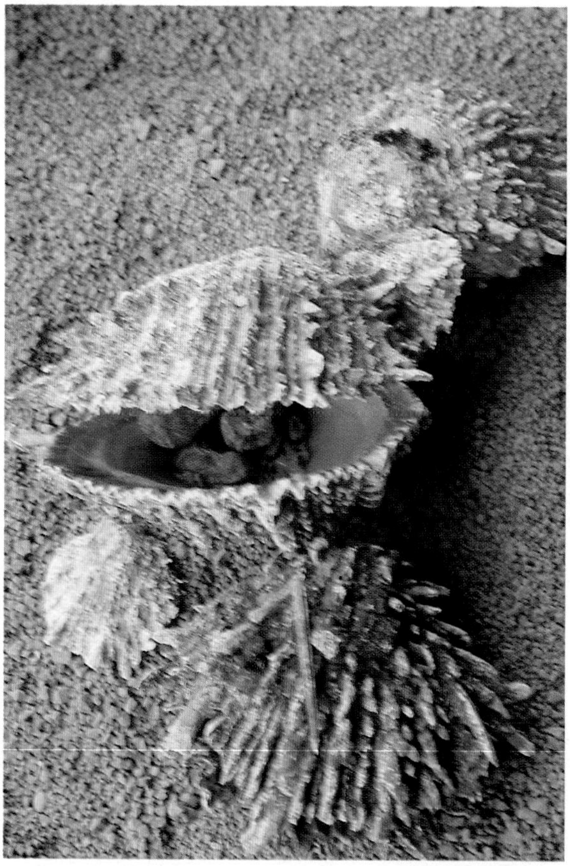

(photograph © by Linda Schele)

18. Spondylus shells from a dedication cache at Copan. Notice the spines and the red color of the shell.

(photograph © by Linda Schele)

19. A slab of jade carved with the image of the World Tree. As part of a building cache for Structure 10L-26 at Copan, it brought *ch'ulel* into the building.

(photograph © by Justin Kerr)

20. The Great Plaza of Tikal with temple-mountains, tree-stones, and the Primordial Sea. The ballcourt is in the lower right corner.

21. A monkey being from the last Creation walks among the people at a Maya festival in Guatemala.

(photograph © by Duncan Earle)

(photograph © by Linda Schele)

22. The ladies of Tekpan dance in the dedication ritual for their new weaving cooperative. Doña Juana sits in the center.

23. A masked Xibalban dances in a Maya festival.

(photograph © by Garrett Cook)

24. The Ah Itz and Kaqik'oxol take a divination together in the Dance of the Conquest.

(photograph © by Garrett Cook)

(photograph © by Garrett Cook)

25. The Ah Itz stands with his red ax that activates the lightning in the blood of shamans.

(photograph © by Debra S. Walker)

26. A open *pib* at a Ch'a-Chak being prepared to receive the food that is to be cooked.

27. The Passions and their Flowers wait to begin their run along the Path of Fire. Notice the different shapes of the silver heads on the banners.

(photograph © by Duncan Earle)

(photograph © by MacDuff Everton)

28. The cenote at Tz'itnip, Yukatan. Caves such as this provided "virgin water" for special ceremonies. They are entries into the Otherworld.

(photograph © by MacDuff Everton)

29. The soldiers of Tixkakal Guardia with tamales at a celebration in 1974

(photograph © by Linda Schele)

30. A Chamula cross with its flowers carved into the wood

(photograph © by Linda Schele)

31. A boundary cross of the Cruzob between Señor and Vallalodid

(photograph © by John McGee)

32. Old Chan Kin prepares a pallet of copal offerings.

(photograph © by Linda Schele)

33. A copal pallet found on January 4, 1993, deposited behind the statue of Bird-Jaguar in Temple 33 at Yaxchilan.

(photograph © by Linda Schele)

34. The west side of Copan Stela C stands before an altar representing the Orion turtle.

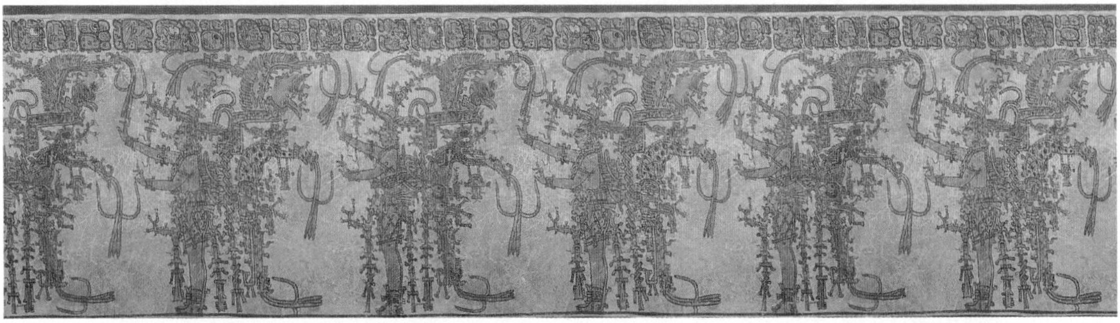

(photograph © by Justin Kerr)

35. Reborn Maize Gods dance with stones of Creation. Their backrack animals carry severed heads of the sacrificed Maize Gods.

(photograph © by Justin Kerr)

36. A scribe sits on a bench above an 8 Ahaw glyph, a Creation reference. He and the monkey are *itz'at*, sages. The gods as sages painted Creation in stars across the sky.

(photograph © by Justin Kerr)

37. A procession of *wayob*, spirit companions of people and gods

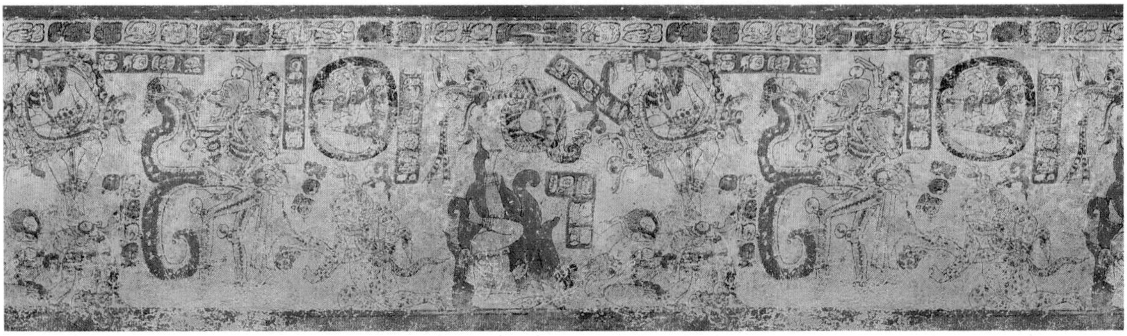

(photograph © by Justin Kerr)

38. A group of *wayob* in a ritual

(photograph © by Justin Kerr)

39. A scaffold sacrificial scene like that modeled in stucco on the Tonina terrace facade

(photograph © by Justin Kerr)

40. The Old Jaguar Paddler God wears the headdress of Three-White-Dog in a scene from Creation mythology.

(photograph © by Justin Kerr)

41. A Lord wields the 18-Rabbit-Snake Battle Standard

(photograph © by Justin Kerr)

42. The Maize God stands in the cosmic ballcourt.

(photograph © by Linda Schele)

43. Duncan Earle sits with Don Lucas outside the cave of Utatlan as they start the copal offering burning.

(photograph © by Linda Schele)

44. Masks on the Temple of the Warriors wear the flower headband of Itzam-Yeh, the Magic-Giver-Bird.

(photograph © by Justin Kerr)

45. Chuchkahawob throw cornshuck-wrapped packets containing pellets of copal into a fire on the altar in front of the church at Chichicastenango.

CHAPTER SIX

DANCING ACROSS THE ABYSS: MAYA FESTIVAL AND PAGEANT[1]

Maya ruins are usually quiet places, steeped in the natural sounds of the forest. The bass roar of the howler monkey, the croaking of frogs in the rainy season, the scream of a hawk, and the song of cicadas might punctuate the sounds of gentle winds and rain or the sharper murmur of human visitors. People who come from mechanized civilizations take a few days to get the engine sounds out of their ears and settle into this timeless peace. But the tranquil modern rhythms of the abandoned royal capitals lull tourist and scholar alike into forgetting that once these were vibrant metropolises whose plazas rang with the voices of the thousands of city folk who made their homes around these urban centers. Here crowds filled the great plazas, their hordes covering the red-painted floors and terraces, their voices lifted in song, their feet dancing across the abyss of drought and death to the green rebirth of life and abundance. Here the miracles of existence were not merely given lip service—they were performed and affirmed by the kings and their people in pageants proclaiming their views of the mysteries of life and death.

Our friend Gary Gossen is as unlikely as we are to imagine these great open plazas as empty, for he has spent years living in Chamula,[2] the community that Linda visited to experience the re-creation of the world. He knows as much as an outsider can about its festivals and ceremonies. Chamula's large population, more than 100,000, accounts perhaps for the elaborate scale of its ceremonial life. More than twenty major festivals mark the rhythm of the solar year—nearly two a month by our way of reckoning. The greatest of them all, the *K'in Tahimol* or Festival of

Games that Linda attended, today wears the appearance of the Christian Carnival, as do similar pageants in many other Maya communities, but beneath its outer guise lies the modern analog of the ancient Maya festival that ended the old year and renewed time in the new year.

This pageant, which takes place in Crazy February (also the time of ancient Creation), involves over two thousand official participants in the dances and rituals that take place in the center and the surrounding community. Of course, these are just the official participants, for everyone in Chamula who can walk or ride—and a huge number of visitors from the outside—comes to witness this five-day extravaganza, to eat and drink heartily, and to remember the true human nature of things. The deafening roar of hand-held bombas and skyrockets; the out-of-tune cacophony of handmade guitars and harps, the plaintive cry of dented brass trumpets blaring before the flashing standards with their flowered flags; the whistles, cheers, jeers, and laughter that encourage the runners charging through clouds of sweet incense as they race around the plaza—these are the music the people of Chamula make as they shout out to their bare hills and the nation beyond that they are still here. As their ancestors did thousands of years before them, they remake the Maya world and renew time itself though the drama of dance and pageant.

The Maya kingdoms of the Classic period boasted populations equal to and often greater than modern Maya communities. The ancient kingdom of Palenque was at least as large as Chamula, and the population of Tikal may have been the better part of half a million souls.[3] The natural amphitheaters of ancient Maya towns and cities must have regularly vibrated with the deep staccato voice of drums and the thud of dancing feet as the ancients renewed their own world, celebrated the history of their kings, and materialized the dangerous beings of the Otherworld in festivals timed by the regularities of the ancient calendar. Although we have long known these rich and complex rituals existed, we could only imagine what they were like. Recently, a new discovery has opened a window into the ancient Maya mind, allowing us to truly get inside their world. In the spring of 1990, our friend Nikolai Grube found this window when he was able to decipher the glyph "dance"—*ak'ot* in the language of the inscriptions (Fig. 6:1).[4]

Nikolai's insight followed upon almost a century of research in which many people had noticed depictions of important lords standing with one heel raised (Fig. 6:2). Nevertheless, it wasn't until 1966 that Michael Coe and Elizabeth Benson[5] finally realized that this position signaled dancing.

FIGURE 6:1 The Glyph for "to Dance"

Yet even with the pose identified, either we dismissed the examples we knew as insignificant or else we did not know how to interpret dance scenes in the larger context of Maya history and religion. Nikolai's decipherment of the *ak'ot* glyph was a crucial breakthrough, for it identified dance as one of the public actions most often depicted by the court artisans of the Maya. Their representations of dancing kings, consorts, and nobles bear witness to the fact that Maya rulers and their courts were, above all things, public performers. Everyone, citizens of the realm and neighbors from other realms, knew that the king's body functioned as a vessel for awesome spiritual forces that could be both inimical and beneficial, but their confidence in that knowledge depended at least partly upon how often and how well the king affirmed his power to control these forces through dancing in the plazas of his city.[6]

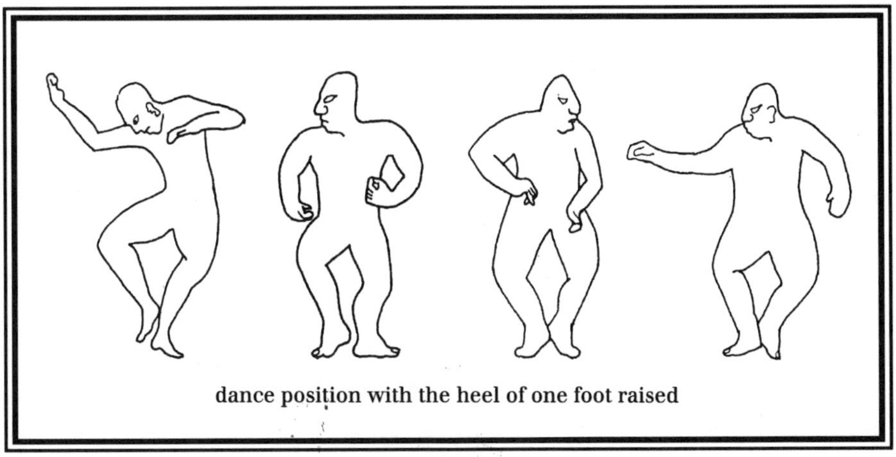

FIGURE 6:2

We can now say that dance was as central to most of the public rituals of the ancient Maya as it is to their modern descendants in highland Chiapas and Guatemala. Just as they did in ancient times, the dances of contemporary Maya serve as a vehicle for expository pageants: the events they reenact bridge the mythological past and political present,[7] the seasons of the year, and the generational cycles of families and communities. Festival, dance, and pageant are midwives to the periodic rebirth of the Maya soul in a people faced with daily hardships and the pressure to conform to national identities as the rural poor and lower caste.[8] A beautiful example of this can be seen in our colleague Gary Gossen's description of the Chamula Festival of Games, portions of which we discussed in Chapter 2:

> . . . the Festival of Games addresses a whole range of normative themes in everyday Chamula life, from sex roles to political integration, the agricultural cycle and world view. Thus, the festival celebrates virtually all aspects of Chamula everyday custom, internal ethnic identity and historical reckoning while acknowledging, without charity, the wider world of the Mexican nation, the state government of Chiapas, even Guatemala, Spain, and foreign tourists.
>
> Because the festival considers these themes so poignantly and so self-consciously, even angrily, it is not an exaggeration to attribute to the event a powerful political statement about who the Chamulas are. They state in countless ways, banal and sublime, that they are not about to be engulfed by Mexican national culture. They reenact once a year just how untenable and immoral cultural assimilation—better said, capitulation—would prove to be.
>
> (Gossen 1986: 228–229)

GODS OR PEOPLE

Ancient Maya art and dance share similar themes, but in the realm of dance the Maya could combine the dynamics of pageants with the mystical transformation of human beings into supernaturals by means of visionary trance. Participants in the pageants of the Classic-period kingdoms transformed into their *wayob* ("soul companions"). These soul companions were depicted in dramatic reality through the masks and the costumes people wore in the dance. Scenes painted on elegant pottery serving vessels from the myriad ritual meals of Classic festivals show humans, both kings and nobles, dressed in costumes. Their human faces

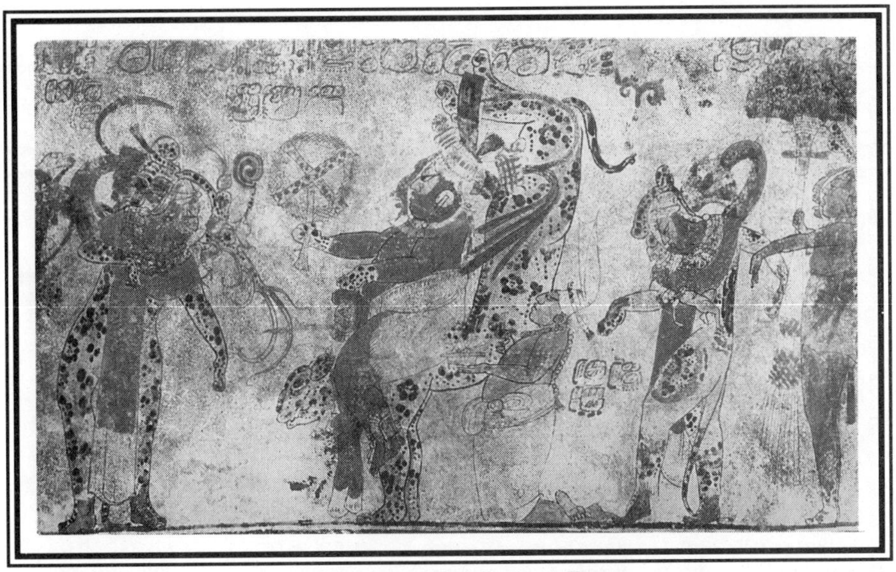

(photograph © Justin Kerr)

FIGURE 6:3 The Jaguar Dance

are shown in cutaway view (Fig. 6:3) inside the fantastic creatures they have become through the transformation of the dance. Some of these *wayob* are recognizable as animals like jaguars and birds of prey, but others (Fig. 6:4) are creatures far stranger than the most imaginative of the monsters conjured up by Hollywood's makeup magicians.

Some of these bizarre creatures show up in the dedication dance

(photograph © Justin Kerr)

FIGURE 6:4 Dance of the Monster Suits

(photograph © Justin Kerr)

FIGURE 6:5

depicted in the Bonampak' murals. There we can identify them (Fig. 5:4) as beings of the before-time who participated in Creation. The masked performers of Bonampak' are anonymous, but those depicted in a ball-game at Yaxchilan are not: they are elite performing for their liege and realm.

Moreover, the boundary between a human dancing as a supernatural and supernaturals materializing in this human ritual was never as sharply drawn as we modern researchers would like. For example, in one dance scene painted on a pot (Fig. 6:5), we see a supernatural stabbing himself in the neck as he dances with other lords who have transformed into their soul companions. One lord dives through bright red fire, while the other dances with huge insects, perhaps bees, in his hands. The blood-splattered belts these individuals wear show up on another pot, where the Hero Twin One-Ahaw dances gracefully on his toes with the White-Bone-Snake wrapped around his body (Fig. 6:6). In this case, we can't determine from the glyphic caption whether the individual is a historical personage dancing in his *way* as One-Ahaw, or alternatively the god One-Ahaw himself dancing with his *way*, the White-Bone-Snake. In still another scene from a Tikal altar (Fig. 6:7), we see named historical lords dressed in the guises of gods kneeling beside the bones of a woman who

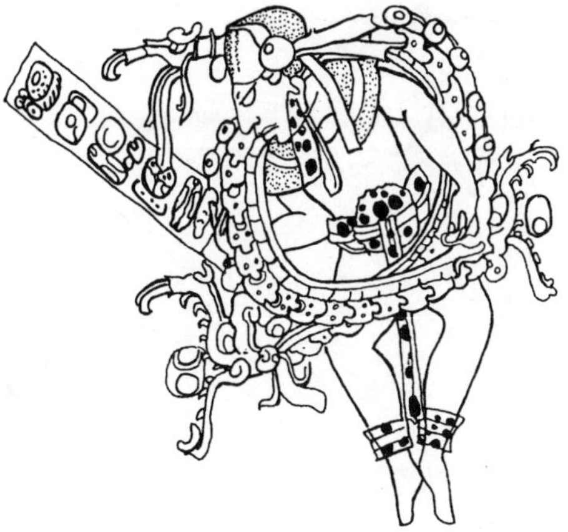

a Hero Twin dancing with the White-Bone-Snake around his neck

Figure 6:6

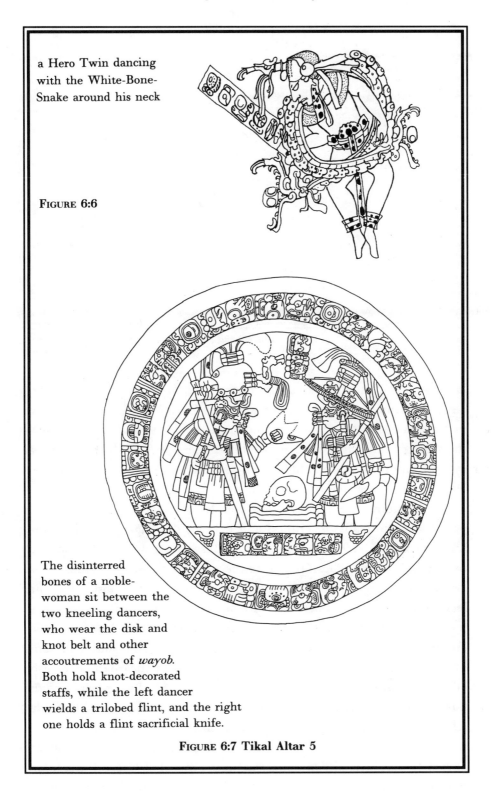

The disinterred bones of a noblewoman sit between the two kneeling dancers, who wear the disk and knot belt and other accoutrements of *wayob.* Both hold knot-decorated staffs, while the left dancer wields a trilobed flint, and the right one holds a flint sacrificial knife.

Figure 6:7 Tikal Altar 5

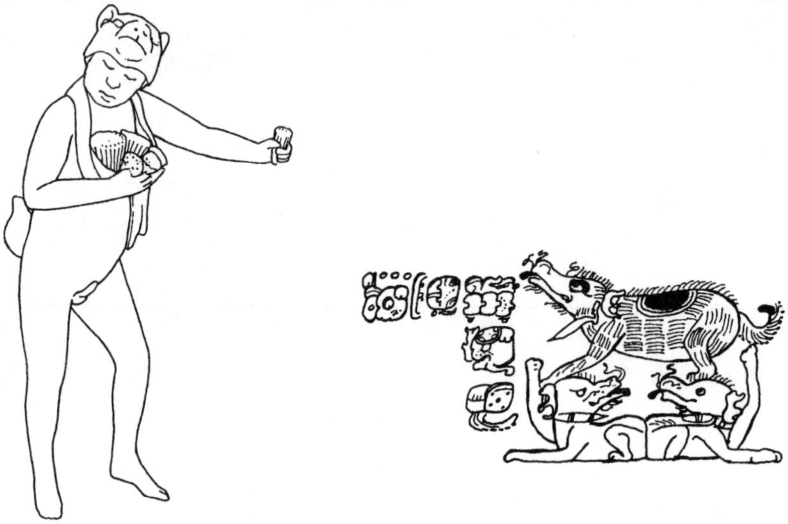

human dancing in his *way* as White-Three-Dog   White-Three-Dog represented as three dogs

FIGURE 6:8

had been exhumed eight years after her death. An extraordinary figurine depicts a human dancing in the costume of White-Three-Dog whom we met in Chapter 5 dancing before the magic house. Yet another pot shows the same god depicted as three white dogs (Fig. 6:8). Which is the real image? All of them are. Through dance, people became gods, and gods became people—if only for a moment.

Commemorated on the serving vessels of ritual meals, carved and painted on the walls of temples and palaces, the pageants of gods and kings display an elaborate, fuguelike pattern of substitution in the symbols of divinity that adorn the performers. There are some constants that label actors beyond a doubt—for example, One-Ahaw almost always has black spots on him—but many attributes combine and recombine in impressive patterns of alteration and transformation that leave the modern researcher breathless. It is as if, in the end, the gods and some exceptional people, like the kings, were aspects or incarnations of each other: as if all of them were just temporary historical expressions of the same cosmic stuff. Each expression, however, was uniquely and individually real, enriching the totality of the cosmos.

This continuity of identity with the sacred and supernatural was not always accomplished by masquerade. In the mural of Room 1 at Bonampak', the procession moves from the Masque of the Monsters to the beautiful formal dance of three nobles of the king, revealed plain in all their human majesty (Fig. 5:1). Similarly, Bird-Jaguar plays ball unmasked on the stairway of Temple 33 at Yaxchilan, but is flanked in the outer scenes by lords wearing monster masks (Fig. 8:13). This dichotomy of the monstrous and bizarre in one phase and the stately and exemplary in another is a compelling feature of many contemporary Maya major festivals. These paired opposites are modes for moving into and out of the sacred time and space of ritual performance.[9]

It is important to realize that Classic pageants were more than just acts of civic pride and piety. They transformed participants into supernaturals,[10] as the paths across the abyss opened on the grand stairways and plazas of their cities. Both gods and humans danced, and through the dance the one became the other. For the Maya, the ambiguity was as it should be. Sorcerers, kings, and nobles transformed into their *wayob* and journeyed into the Otherworld before the transfixed gaze of their people.[11]

DANCE OF THE *WAYOB*

With particular relish, Maya pottery artists painted the walls of cylindrical vessels with the images of *wayob* dancing in fantastic scenes of pageant. One such pageant is depicted on one of the most beautiful ceramic paintings ever found—the polychrome vase from Altar de Sacrificios (Fig. 6:9). These are the *wayob* of the lords of Classic cities like Tikal and Yaxchilan dancing in charged, ritual space.[12] Yax-Balamte,[13] the *way* of a lord from a place called "Four-Sky, Thirteen-Gods," stands on the left, dancing in jaguar-skin pants, mittens, and a head pelt, which he wears as a hat. A personified perforator dangles from his belt in front of the red stain from his bleeding genitals. His partner in dance, Buchte'-Chan,[14] the *way* of a lord from an unknown place, wears pants made from the diamond-marked skin of a strange tailed creature, and swings a living snake, probably a boa, above his head. His body is paunchy, his head bald, and his face swollen with the features of a tortured captive. The text clearly identifies this person as a *way*, but whether this image was meant to be understood as a human being wearing the guise of his *way* or as a human being who has undergone a physical transformation into this being, is unknown. We only know that the particular *way* he is manifesting is always shown with the features of a tortured sacrificial victim.

The other four *wayob* in the group are arranged in pairs. One pair

Six *wayob* in a pageant: the left dancer performs bloodletting, while the right one dances with a snake; two float above holding objects, while the lower left *way* decapitates himself

FIGURE 6:9 Altar de Sacrificios Vase

includes Nupul-Balam, the *way* of the *ch'ul ahaw* of Tikal. He is floating above Ch'akba-God A', who sits on the ground engrossed in the remarkable feat of cutting off his own head.[15] On the other side of the dancing Yax-Balamte, floats a *way* holding a giant death's-head in his arms. His name is Tzak[16]-God A'. Below him, a *way* named for his dragon-deer headdress sits with half his face hidden behind his shoulder.

These beings are the *wayob* of actual human dancers who participated in a real, historical Maya ritual. Their floating positions and impossible acts, like self-decapitation, signal that these dancers have transformed themselves through ecstasy into sacred and very frightening beings. These wondrous and terrifying dances were part of pageants performed by great lords in their urban centers. The records of these great public events were recorded on commemorative drinking cups, serving bowls, and offering platters, some of which were perhaps used in those very rites. These keepsakes sealed the political and spiritual covenants of the Maya's collective vision and accompanied their owners to the grave on their final journeys to the Otherworld.

The artist of the Altar vase painted a similar scene of dancing *wayob* on another pot (Fig. 6:10). These *wayob* include the following creatures: a great bird with a boa around his neck; a death god with flint knives on his knees and elbows; one of the headband Hero Twins holding fire[17]; Ch'aktel-Ix, a jaguar in a bamboo cage; a trumpet-blowing death god; and

Dancing *wayob* in a scene painted by the same painter as the Altar de Sacrificos pot

(photograph © Justin Kerr)

FIGURE 6:10 Dance of the *Wayob*

Nab-Ix, the Waterlily Jaguar who roars from the center of a water cartouche. The dancers in this scene are "Tzak-God A'," our old friend "White-Three-Dog," and a handsome young dancer wearing diamond-marked pants combined with jaguar mittens and headdress. He blows a flute and shakes a rattle as he dances to his own tune. Here is another assemblage of *wayob*—some the same, some different. The cast of characters crossing from the Otherworld changes, depending on the occasion.

Classic Maya lords danced to bare their souls to each other, to the gods, and to their people. Masked and manifesting the eternal gestures of the great *wayob,* or toe-stepping enraptured in vision, wielding as talismans the outward forms of their spirit allies, the dancing lords declared in performance the presence of the true reality behind all other realities.[18]

Of the three of us who wrote this book together, only Joy Parker has had direct experience of trance dancing. She describes it this way.

THE DANCE OF THE SUFIS

(as told by Joy Parker)

I personally experienced this state of "the reality behind all other realities" during the period when I was studying Sufi dancing with a master, Adnan Sarhan. One of the practices of Sufi is to achieve a state of clear-minded, ecstatic trance while engaged in whirling. One moves counterclockwise in the direction of the rotation of the earth, arms outspread, the right hand extended palm downward to draw energy up from the earth, the left hand extended palm upward to release that energy. With practice I was able to whirl for extended periods of time, moving into a timeless space where I experienced a deep feeling that I can only describe as being at the center of true reality. It is perhaps impossible to convey what this feeling was really like, but it filled me with a vivid sense of unity with all animate and inanimate things everywhere and a powerful feeling of wholeness. After long periods of dancing or whirling, I was able to observe the confusions and conflicts of my daily life line up into meaningful patterns. Creative inspiration was also one of the benefits, and ideas for projects, and even their consequences should I decide to pursue them, would simply appear in my mind.

Once I got used to experiencing mental and emotional states through my body, I could call these energies forth by simple, stylized dance movements or even by slow "meditative walks," of which there are several different types in the Sufi tradition. Speed and energy of movement often have no direct correlation with the intensity of effect dance can have upon your psyche. I thought of these meditative walks immediately when David and Linda described to me the slow, stately dancing of modern Maya elders.

There are ways of knowing far more readily accessible to us through the body than through the mind. Modern Maya shamans who diagnose illnesses by paying attention to the messages given them by the "lightning" in their blood are examples of this type of knowledge. I cannot claim to understand precisely what the Maya lords were feeling when they danced upon their pyramids. But I do know that sacred dance raises powerful energies, especially when performed in the company of other, like-minded dancers. It is not difficult to imagine how practiced Maya dancers could shift from normal consciousness into an altered state by means of movement. It has been three years since I stopped Sufi dancing on a regular basis, but even now, when I extend my arms outward and move my body in certain ways when I am dancing at social functions, I will immediately feel myself lifted to the edges of another state of consciousness.

For the Maya, both ancient and modern, dance was public in its performance. It induced visionary trance in which individuals and large groups went into altered states that allowed both transformation and

a. Tzum Stela 3: a king dancing over the White-Bone-Snake, the opening to the Otherworld (drawing by Eric Von Euw)

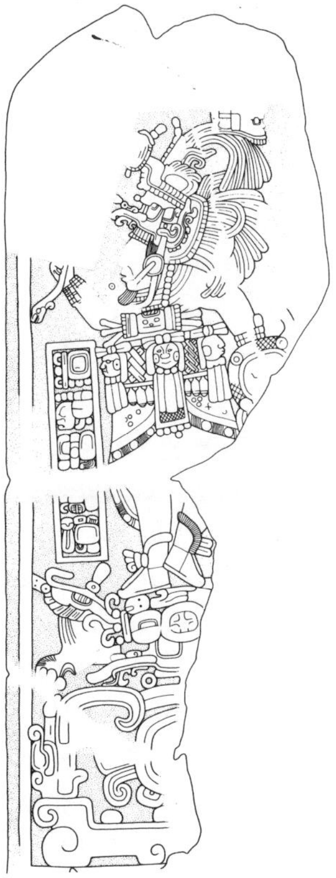

b. Palenque Sarcophagus: Pakal falls down the throat of the White-Bone-Snake along the Milky Way (drawing by Merle Robertson)

FIGURE 6:11

communication with the Otherworld. Just as the real world in which they danced had a geography with its rivers, mountains, and valleys, so the Otherworld had a topography that had been mapped out by those who were strong enough to travel into that land and return alive to tell the tale. The great Maya lords showed themselves dancing quite literally out over the abyss that leads into the Otherworld. For example, the lord of Tzum, a small town in the northern part of Kampeche (Fig. 6:11a), depicted himself prancing on the name glyph of the supernatural place named *Wuk Ek'-K'anal*, the "Seven Black-Yellow Place."[19] Beneath the glyph for this place yawns the mouth of the White-Bone-Snake. This is the bony monster that represented the portal to the Otherworld. It is this abyss that receives Pakal's falling soul as he slides down the road to Xibalba at the moment of his death (Fig. 6:11b). The Lord of Tzum dances atop the same opening presumably because his dance will let him journey to *Wuk Ek'-K'anal* and return without having to die.

DANCING WITH SACRED OBJECTS

The formal dancing in modern Maya ritual often proclaims the sacred obligations of the performers and is not much more than a dignified shuffle.[20] But the lack of scintillating steps only serves to underscore that it is the simple act of performance that designates one as the holder of a powerful status—that and the adornment of the dancer and the objects grasped during the dance. Freidel recalls seeing a prosperous rancher on Cozumel Island dance sedately around an arbor in the little village of El Cedral, holding a severed pig's head in a bowl, followed by an entourage of Maya and mestizo natives. A comic sight, this was clearly a serious moment in the festival, despite the exotic offering, which Freidel later learned marked the ritual obligations of the rancher to the community in the coming year.

The display of objects denoting power and status was one of the hallmarks of ancient Maya festivals as well. But there is a further, and more important, reason why the ancients and their distinguished descendants hold powerful objects when they dance. As we saw in the previous chapter, these objects are actually considered to be imbued with soul force. Evon Vogt,[21] steeped in decades of experience with the Tzotzil Maya of Zinacantan and other communities of highland Chiapas, observed that "in this profoundly spiritual world, the staff of office carried by officials has a strong inner soul which is placed in it by the Ancestral Gods, and like a newborn infant, a newly acquired staff must be baptized to lock the soul in and guard against soul-loss. . . . Furthermore, the living Zinacantecos are constantly in touch with, and receive messages from, the Ancestors via dreams in which the inner souls are in communication. Hence, it is the inner soul of the batons of command that gives them their potent sacred power."

Vogt goes on to observe that the really powerful staffs of the Maya officials are handed down as heirlooms from one holder to the next throughout the annual festival cycles. Imbued with their own souls, these objects accumulate experience far beyond that of any individual human bearer and are regarded in some cases as infallible guides. Ritual actions like prayer and offerings enhance the potency of these objects. As in the symbolic and literal supernatural displays of the masked pageants, the formal staff dances are performances of souls acting in concert. Formal dances are not just declarations of sacred and social status. They are opportunities to amplify the power of both the dancers and their objects, some ancient and some being brought to life for the first time. Dancing with objects played the same role for the ancient Maya. Many of the scenes depicted on the lintels of Yaxchilan show Bird-Jaguar, his friends and family, dancing with the flapstaff called *hasaw-ka'an*[22] (Fig. 6:12), the tree-scepter called *xukpi*,[23] the serpent-footed K'awil, bundles called *ikatz*,[24] and other sacred objects (Fig. 6:13).

And it is clear from the art that these dances were performed with staffs imbued with soul-force, like the ones we described above. The

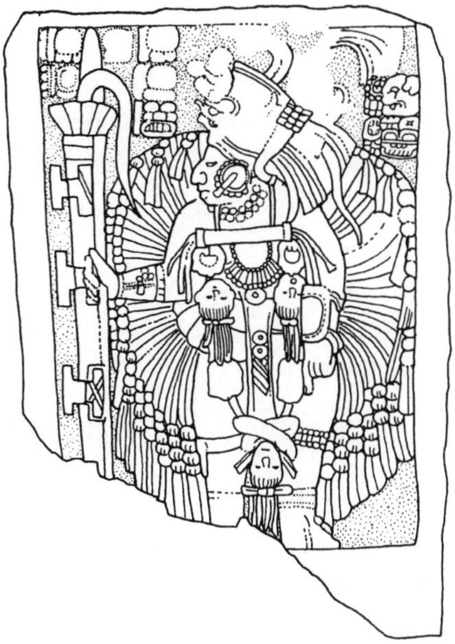

a. Stela 16

b. Stela 11

Dance of the flapstaff called *hasaw-ka'an* at Yaxchilan

FIGURE 6:12

a. Lintel 5: Bird-Jaguar dances with a *xukpi* "bird" scepter while his wife dances with a bundle

b. Lintel 53: Shield-Jaguar dances with a K'awil scepter while his wife dances with a bundle

FIGURE 6.13 Dances with Various Sacred Objects

Manikin Scepter of K'awil along with the Double-headed Serpent Bar are two of the most venerable and popular soul containers wielded by Maya lords. Sacred bundles contained many kinds of objects—various sorts of scepters, eccentric flints, bloodletters, and offerings of shell, red pigment, and jade—all of the things that embodied *ch'ulel.* Modern Maya of Chiapas venerate this soul-force, bundling their most important staves when they are kept at the homes of officials.[25]

THE SNAKE DANCE

The staves, bundles, and scepters of Maya royalty were animated in ritual dance, but sometimes kings danced with real animals. The lintel that gave Nikolai Grube the clue to the *ak'ot* reading depicts Bird-Jaguar of Yaxchilan dancing with a snake. This extraordinary panel was looted from an unknown site we have designated Site R.[26] It was probably located somewhere within the larger kingdom of Yaxchilan. The sahal[27] who ruled the town depicted himself participating in important rituals with his king. In this scene (Fig. 6:14), both men are wearing elaborate

(drawing by Peter Mathews)

Bird-Jaguar and a sahal dancing with a snake at Yaxchilan

FIGURE 6:14 Lintel from Site R

headdresses with *hasaw-ka'an* staffs stuck in them, personified wings, mat and reed decorations, and long feathers arching above and behind them as they dance with living snakes. The text reflects this: "[On] 11 Ik' 13 Mak, he danced with a snake, Bird-Jaguar, He of Twenty Captives, three-k'atun lord (*Buluch Ik' Oxlahun Mak, ak'otah ti chan chan, Yaxun-Balam, Ah K'al Bak, Ox-k'atun Ahaw*)." Bird-Jaguar leans toward his sahal, lifting his left leg in the dance and holding a boa constrictor[28] in both hands. The sahal holds a smaller snake in his right hand and extends his left hand toward his lord, his outer fingers raised in an elegant gesture.

The lords of Palenque also celebrated this Snake Dance. On the piers of House D of that city (Fig. 6:15), a male, presumably Pakal, dances with an ax in one hand and a rearing serpent in the other. A second person, perhaps a woman of the king's family,[29] grasps the lower body of the king's snake, as it rears upward between them. Here, however, the dancers wear the costumes of First Father and First Mother, the deities whose actions enabled the final Creation and the birth of all the gods. We presume these figures represent the king and his consort (or perhaps his

Pakal performs the Snake Dance with a woman. He is dressed as First Father and she as First Mother

FIGURE 6:15 Palenque House D Pier

mother) performing in the guise of the Creator gods. Surrounded by *ch'ulel*, "holiness," they dance on a platform marked by a sun-cartouche and point toward the role of dance in the story of Creation.

THE DANCE OF CREATION

The miracle of dance and its capacity to regenerate life is central to the story of Creation as it was told in the art of the Classic period, but to understand it fully we must return to the story of the Hero Twins in the Popol Vuh that we introduced in Chapter 2. We have told the story of how the Hero Twins faced a series of tests and ballgames with the Lords of Death, and how, after defeating the Lords' every trick, they allowed themselves to be killed in the ovens where maize gruel was made. Their bones were ground up and thrown in a river. This is the part of the story where dance and performance in pageant play a critical role.

Five days after the Twins immolated themselves, they reappeared in the water looking like catfishes, and the next day they emerged dressed in rags like vagabonds. Even though the Xibalbans took them to be country bumpkins, they were soon enchanted by the Twins' skill as dancers and magicians. The Hero Twins danced the Dance of the Poorwill, the Weasel, and the Armadillo. They performed miracles also, like burning down a house and then making it whole again. Most important, they demonstrated their ability to sacrifice each other and bring each other back to life.

Soon their fame came to the attention of the Lords of Death, who promptly invited them to a command performance. Feigning great humility, the Twins told the Lords that they had no home and never knew who their parents were. The Lords told them to perform, so they sang and danced before a whole crowd of Xibalbans. As they danced, the Lords tried out their powers by telling them to sacrifice a dog. To everyone's amazement, the dead dog rose up and walked away wagging his tail in happiness.

For their next trick, the Lords ordered them to burn down a house full of Lords. Miraculously, no one in the house was burned as the house disintegrated around them. The Twins were even able to reconstitute the house as if it had never been burned at all. Amazed, the Lords ordered them to sacrifice a person. They removed the victim's heart, held it up high so the Lords could see, then brought the man back to life. Like the dog, the man walked away overjoyed to be alive.

Thinking that they were asking for something difficult, the Lords next ordered the Twins to sacrifice each other. In the middle of their dance, Xbalanke asked his brother to stretch out on the sacrificial altar. He cut off Hunahpu's head, which rolled away, and then tore out his heart. Xbalanke was left to dance alone before the ecstatic audience. After a while he shouted, "Get up!," and his brother came back to life and joined him in the dance.

Dazzled by all these displays of magic, the Lords were overcome with a yearning for the Dance of Hunahpu and Xbalanke. One-Death and Seven-Death, the two principal Lords, began to cry, "Do it to us! Sacrifice us! Sacrifice both of us!" The Twins began with One-Death, quickly dispatched Seven-Death, and then killed all the other Xibalban Lords who were getting up to leave. But this time they did not bring them back to life. When the Xibalbans saw their Lords die, they ran away in terror to hide in the depths of a canyon. There they fearfully surrendered to the Hero Twins.

After limiting the Xibalbans' power over human beings forever, the Twins dug up their father's and uncle's bodies from the floor of the

Ballcourt and brought them back to life. In this part of the story father and uncle are both called Fathers of the Hero Twins, perhaps in the sense that they were the prototypical ancestors of all humanity. Seven-Hunahpu, their uncle, never fully recovered, for he could not name all his body parts. Leaving their ancestor there in the Ballcourt, the Twins consoled him with the words, "You will be prayed to here. You will be the first resort, and you will be the first to have your day kept by those who will be born in the light, begotten in the light. Your name will not be lost." The Twins then rose into the sky where the sun belongs to one and the moon to the other.

DANCE AND THE REBIRTH OF THE DEAD

This story of Creation, Death, and Rebirth appears over and over on the painted pottery and monuments from the Classic period.[30] Hun-Nal-Ye, the Maize God of Classic-period imagery, was the equivalent of One-Hunahpu and his twin brother Seven-Hunahpu, the fathers of the Twins. In the Classic-period tale, he was the instigator of the last Creation. He was so often portrayed with his arms outstretched and his feet raised in dance that he has become known as the Holmul Dancer after the first image of him ever excavated.[31] Wearing a special backrack representing the cosmos, he often appears on painted vessels dancing with a dwarf[32] (Fig. 6:16). In fact, as Nikolai Grube and Werner Nahm pointed out to us, he can appear alone, in pairs, or in triplets. Moreover, he carries one of three animals in his backrack—a jaguar, a snake or lizard, or a monkey. They suggested that these animals represented the three thrones that were set up in the first act of Creation. We think they are right. The scenes showing one, two, or three images of the Holmul Dancer are showing different moments in the Creation epic.

The story of his rebirth as the maize plant explains how the forces of death were originally defeated, and the method by which the souls of human beings can survive their own journeys to Xibalba after death. The Classic Maya believed the soul traveled down the Milky Way into the White-Bone-Snake,[33] the same path taken by the Maize God in his unsuccessful confrontation with the forces of the Underworld. Depicted on his magnificent sarcophagus deep inside the Temple of Inscriptions at Palenque, Pakal[34] falls dressed in the same net skirt that Hun-Nal-Ye wears in his guise as the Holmul Dancer (Fig. 6:11). The king, defeated

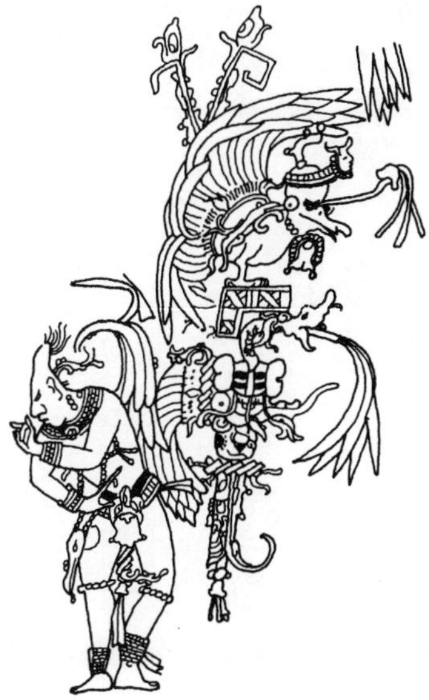

Hun-Nal, the Maize God, dancing with the snake-throne backrack

(photograph © Linda Schele)

FIGURE 6:16

by death, can be resurrected, just as the Maize God was by his sons the Hero Twins. The child of the king enters trance and retrieves his parent through the Dance just as the second set of Twins retrieved their own parent. The ancient Maya believed the souls of all human beings faced the same trials, although the children of common people may not have been as well equipped as their noble counterparts were, by education and access to the holy power of ritual dances, to redeem their own dead.

We now believe this exact act is portrayed on Stela H at Copan, a stela that has long baffled students of the ancient Maya. In their struggle to identify this carved figure, iconographers have given it many different names. We have called it an astronomer priest, a god, the only woman portrayed at Copan, and even a man in woman's dress. Now we know from the inscription and the costume that the stela portrays 18-Rabbit in the guise of First Father (Fig. 6:17) on the day or, more likely, on the night of December 5, A.D. 730. He stands, gorgeous in his net skirt, his shark-shell belt, and his ornate backrack with its image of Itzam-Yeh and the Plate of Sacrifice. Like his prototype, the Holmul Dancer, he has danced into the court holding the Double-headed Serpent Bar with

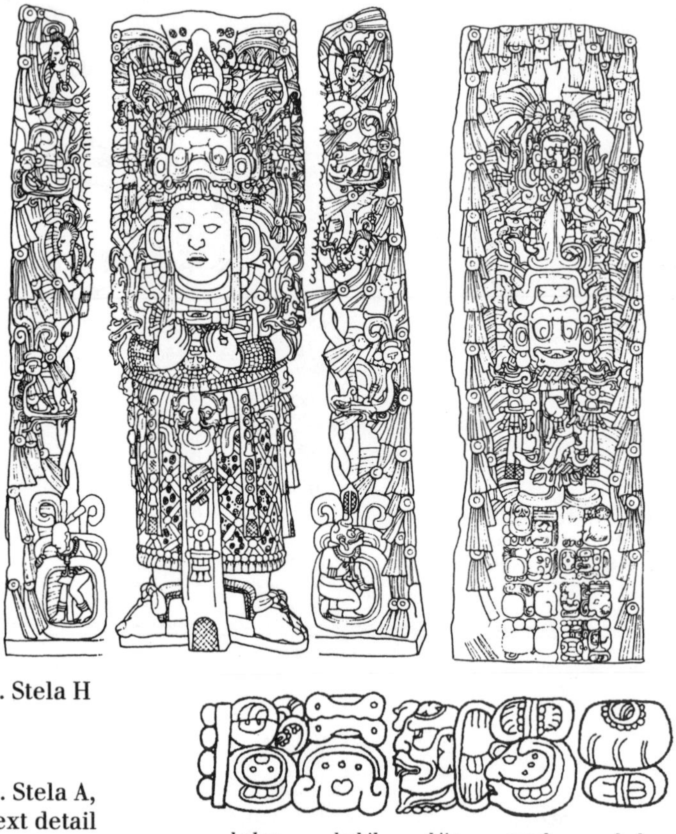

a. Stela H

b. Stela A, text detail

FIGURE 6:17 Copan, Dance of the First Father

K'awils emerging from its gaping mouths. But perhaps the most extraordinary detail occurs on the sides of the stela where the snake cords that are the umbilicus of the sky twist upward. Tiny Maize Gods wielding the Personified Lancet of the bloodletting rite cavort among the twisted strands. The *sak-nik-nal* soul sign sits on the tips of the umbilical snake's nose.

The text on the back of this beautiful stela records that its dedication occurred on 9.14.19.5.0. On Stela A, the big-stone just across the court, another text tells us that on that same day, 18-Rabbit conjured up a Vision Serpent as he commemorated one of his ancestors named Butz'-Chan. We are still trying to understand the verb, which reads *bolon-iplah*, "nine-strengths" or "nine-deaths,"[35] but the action is clearly connected with bones. In fact, our friend Nikolai Grube has read one of the phrase as *susah bak*, "bones were cut." He suggested that this event was the exhumation of bones from Butz'-Chan's tomb to take relics.[36] Alternatively, the actors could have been using reliquary bones that had already been taken from the dead king when he had been buried a hundred years earlier.

That the king would be dressed in the guise of First Father makes sense now in a way it never has before, for First Father's bones were exhumed by his own sons so that he could be reborn. We think 18-Rabbit's dance and the pageant in which he performed replayed that tale of First Father,[37] and the timing seems to support our contention. At sunset on December 5, the canoe configuration of the Milky Way arched over the Great Plaza in its east to west trajectory. At this time of year, the canoe would be just ready to sink, taking First Father to his turtle shell and the hearth of Creation. At midnight, that very hearth was directly overhead, and the hour before sunrise saw the Milky Way rim the horizon, forming the *Ek'-Way*, the Black-Transformer. The date inscribed on Stela A fell sixty days later, on February 3, just two days short of the day when the Wakah-Chan was raised at dawn. This tells us that Stela H signals the change coming from the exhuming of First Father's bones. Stela A celebrated the climax of the raising of the sky on the north-south axis. We think this timing was no accident, and that the pageant involved 18-Rabbit dancing in the role of the resurrected Maize God.

Michael Coe[38] has identified another scene from this resurrection myth. This one is drawn in elegant calligraphy on a beautiful codex-style pot (Fig. 6:18) and depicts the second-born of the Ancestral Heroes, Yax-Balam, as he lifts up a huge plate holding his father's head and jewelry. We assume this is a depiction of the scene that occurs after the defeat of the Lords of Death, where the Twins collect the remains of their father prior to putting him back together again. The baby in the plate is the reborn Maize God held above the figure of his son in the gesture still used today by the K'iche' of Chiniqi to carry the afterbirth to its burial. The second Twin, One-Ahaw, sits atop a coiled fish[39] holding a bundle.[40] In an adjacent scene on the same vessel, the regenerated Maize God dances in front of a woman[41] who sits inside the mouth of the White-Bone-Snake. She hands him the shark-shell device that will go on his belt.

The panel from Temple 14 at Palenque shows the same scene reinterpreted for a king. Chan-Bahlam dances toward his mother, who awaits him extending the K'awil manikin ready for him to take. She wears the

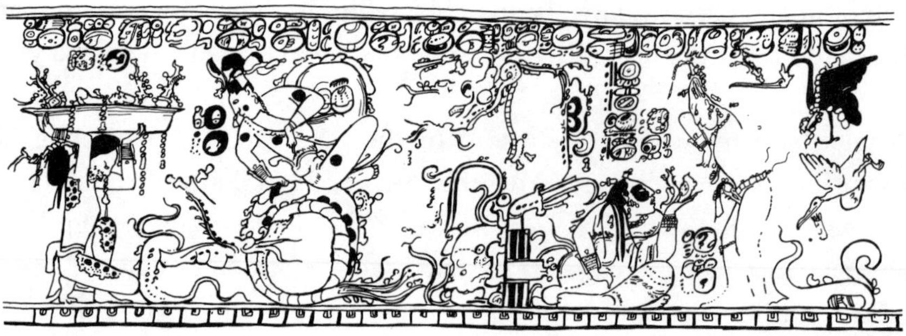

The Hero Twins bring their father his clothes

A goddess dresses the reborn Maize God as she sits inside the White-Bone-Snake

Figure 6:18

guise of First Mother, the Moon Goddess who is named in the text as the person who first performed this action 932,174[42] years earlier. This puts her action in the time of the last Creation. This resurrection panel was commissioned by K'an-Hok'-Chitam,[43] the younger brother of Chan-Bahlam and the next king, who wished to mark his dead brother's rebirth[44] (Fig. 6:19a). His dance of rebirth took place, we are told by the text, at a watery place called "Five-Flower-Valley Lying-down-Sea-Place, White-???-Place" (*Ho-Nikte-Hem Ch'a-K'ak'-Nab Sak-???nal*[45]). This is the Primordial Sea of the past Creation where the sky lay before the gods set the first three stones on the face of heaven.

K'an-Hok'-Chitam commissioned another resurrection scene on House D, pier c, of the Palace at Palenque (Fig. 6:19b). Here, surrounded by *ch'ulel*, the sacred essence of the universe, a Palenque king, maybe Pakal himself, dances across the watery surface of the Otherworld in the net skirt of Hun-Nal, just as the older brother does on Temple 14.[46] Below this watery surface, waterlilies grow from the skeletal heads as if they are seeds.[47]

The resurrected king dances toward a kneeling attendant. In the crook of his left arm, the king holds the same sun-marked sacrificial bowl that held his falling body on the sarcophagus lid. Now instead of holding his lifeless body, it contains a sprouting and emergent World Tree growing

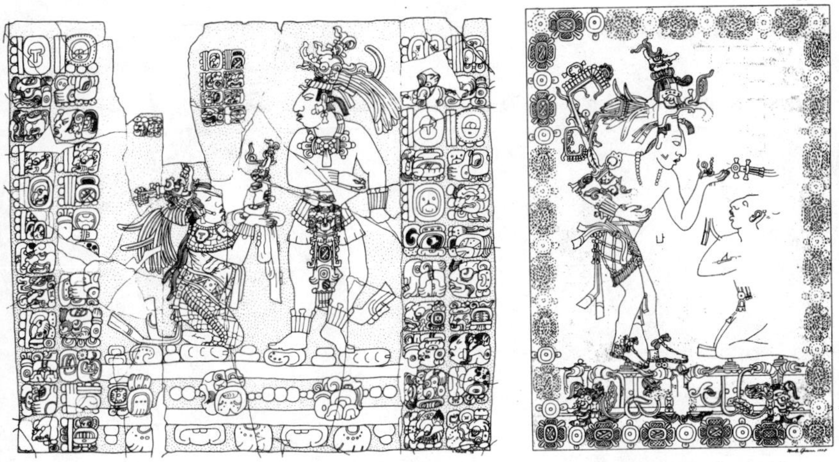

a. Temple 14: Chan-Bahlum dances toward his mother, who plays the role of First Mother dressing her child after his resurrection.

b. House D, pier c: a lord, perhaps Pakal himself, dances in the guise of Hun-Nal-Ye (drawing by M. G. Robertson)

FIGURE 6:19 First Father's Dressing Dance

from the stingray-spine bloodletter. Here is the ultimate symbol of rebirth from sacrifice. In his right hand, the dancing lord balances the shark-shell belt ornament of Hun-Nal-Ye that he has received from an attendant, who perhaps represents Chan-Bahlam, the son who went to the Otherworld to retrieve his father.

The clearest depiction of the Maize God's resurrection occurs on an offering plate (Fig. 6:20). Here, flanked by his two sons, he emerges from the cracking turtle carapace. The torch skull engraved on the side of this carapace names the crack from which the Maize God emerges.[48] In 1992, Nikolai Grube realized the crack is named in a Dos Pilas text as *K'an-Tok-Kimi*, "Precious-Torch-Death." And at least one of these resurrection scenes shows the reborn Maize God emerging from a *k'an* sign. The monster at the base of the First-Tree-Precious on the Tablet of the Foliated Cross is the personified version of this crack. Reborn maize emerged from it to sprout with renewed life. Through this crack the sons gave rebirth to the father who had engendered them in the first place.

This engendering of the father by the sons is the great central mystery of Maya religion. This is the culmination of the Classic-period myth we recounted in Chapter 2. The sequence of events in the Classic myth is a

a. The Hero Twins help their father emerge from the cracked turtle shell

b. Hun-Nal-Ye rises from the crack and a K'an-cross

FIGURE 6:20

Chan-Bahlum and the Yax-Hal-Witznal

the Na-Te'-K'an

Pakal standing on K'an-Hub-Matawil

FIGURE 6:21 Palenque, Tablet of the Foliated Cross

bit different from the later Popol Vuh. In the Classic version, the re-dressed Maize God was carried to the place of Creation in a canoe. There, he emerged from his cracked turtle shell to set the three stones of the hearth and help his companions in Creation draw the images of the constellation on the sky.

Chan-Bahlam narrates the final result of these acts of Creation and the dance of Hun-Nal-Ye in the Temple of the Foliated Cross (Fig. 6:21). The imagery in this temple, along with its complements in the Tablet of the Cross and the Tablet of the Sun,[49] depicts his last visionary communion with his father Pakal and his transformation from the status of heir-apparent into king. This tablet is set inside a small conjuring house (*kunul* or *pib-nail*[50]), like the one we described in the ecstatic dance of dedication in Chapter 4, inside the larger temple.

The tablet inside this *pib-nail* depicts Chan-Bahlam wearing the xok-fish-shell and the netted skirt of the Maize God. The mythological section of the text celebrates the birth of the third-born of the Palenque Triad,[51] the god K'awil, and the first conjuring up of a god from the Otherworld by his mother, the Moon Goddess. This amazing birth and the subsequent conjuring happened in the Otherworld at three sacred locations, which are shown under the feet of the actors and in the center of the image. First, the new king, Chan-Bahlam, lifts the unbundled image of Sak-Hunal,[52] the Jester God, toward the Na-Te'-K'an, the "First Ripe or Precious Tree." We believe the Maya saw this maize tree in the Milky Way emerging from the place of Creation near Orion.

His father holds the personified bloodletter, an instrument that, like the stingray spine, nurtures life out of death. He stands upon a cornstalk being pulled into a shell by a god personifying the white-flower-soul. The shell is called K'an-Hub-Matawil,[53] "Precious Shell of Matawil." First Mother and her sons, the three gods of the Palenque Triad, were lords of Matawil. It was the native place of the Creator gods that generated existence, time, and space.

Chan-Bahlam himself stands upon Yax-Hal-Witznal, the "First-True-Mountain." In keeping with its role as the place holding the substance from which humanity's flesh was made, the cleft summit sprouts maize. This maize takes the form of both leaves and the king himself. Chan-Bahlam is the child who has resurrected his father, but as king he has also *become* the father. All three sacred locations float on a thin band marked as "water." This is the Ch'a-K'ak'-Nab, the Primordial Sea into which the canoe carrying First Father sank. It is the same watery sea portrayed on

Tablet 14 and the pier from House D. This sea is the place of Creation, where First Father raised the sky and where First Mother modeled the flesh of humanity from maize dough and water.

The Popol Vuh tells us that after the triumph of the Hero Twins at the end of the previous Creation, the gods made one last attempt to make beings who would honor, obey, nourish, and sustain them. In the three preceding creations they had tried to fashion nurturing beings from animals, mud, and wood, but none of these materials had worked. Following these failures, they sent Fox, Coyote, Parrot, and Crow to "Broken Place, Bitter-Water Place" to find white maize, yellow maize, and the water from which Xmukane, the mother of One-Hunahpu and Seven-Hunahpu, will mold the flesh and blood of humanity. This time the gods are successful, and corn, or maize, is the stuff of human beings in the world today.

Pan paxil, pan cayala is the K'iche word for the place where the substance of human flesh was found. The expression means "at the split, at the bitter water," and the Yax-Hal-Witznal, as it is depicted on the Tablet of the Foliated Cross, has a cleft in its forehead and sits on the bitter water of the Primordial Sea. The First-True-Mountain is that very special place of Creation that gave existence to all of humanity. Thus, the image of the central maize tree, and the king as the Maize God sprouting from that mountain, represents both the creation of humanity and the regeneration of the sacrificed Maize God. The son, now become the father, stands on the First-True-Mountain from whence came the flesh of human beings, and the father who has passed on to become an ancestor stands on the "Precious-Shell-Matawil," the Otherworldly place where the mother of the gods and her children reside.

In the story in the Popol Vuh, the Hero Twins planted maize seeds in Xmukane's house before they began their own descent into Xibalba to confront death and to resurrect their father and uncle. That maize tree grew in her house just as the Na-Te'-K'an grows inside the *pib nail* of the Tablet of the Foliated Cross. The Twins planted the corn kernels, also known as "little skulls,"[54] left by their father and these kernels grew into a plant that symbolized their own well-being. When they died in the fire pit, the maize plant withered and dried. Their grandmother, Xmukane, harvested the dried ears and replanted them, just as she gave birth to the original twins and life to humanity. And when the Twins were reborn from the river, the maize tree in her house sprouted once again. The imagery in the Temple of the Foliated Cross includes all of these miracles: the death of the father, his rebirth as the tree and in the flesh of his son, and the maize flesh of humanity brought from the First-True-Mountain.

Borrowing from Dennis Tedlock's[55] explanation of "dawning" and "sowing" in the Popol Vuh, Robert Carlsen and Martin Prechtel have compared the modern Tzutuhil concepts of birth and rebirth, of replacement and transformation (*haloh* and *k'exoh* in K'iche'), of the relationship of parent to offspring, and interrelations of the living and the dead to a Möbius strip. These concepts begin on one side of the strip and get to the other without changing sides. Carlson and Prechtel describe this unifying concept as *jal*, pronounced "hal." *Jal* means "transformational change" as these transformations evolve through an individual's life cycles. Traditionally, Maya believed that life arises from death. Consistent with this belief, beginning with death, *jal* is the change manifested in the transition to life through rebirth, through youth and old age, and finally back again into death.

As we saw in Chapter 3, *k'ex* means substitution, as in Don Pablo's substitution of the liquor for the body of the baby as the place where the sickness should go. In Carlsen and Prechtel's analysis, *k'ex* is generational or replacement change. It is "a process of making the new out of the old." It leads to the idea that offspring replace their parents and grandparents, so that a grandchild is named for a grandparent and becomes that person in personality and soul. The child can often be addressed as parent, and "males will address their fathers as *'nuk'jol'* or 'my son.' Likewise a woman will often call her father *'wal,'* which translates as 'child.' " The Maya are even more explicit about the metaphorical relationship of children and ancestors to trees. "When the fruit, the children, eventually drop to the ground splitting open to form new sprouts, these are the grandchildren. Significantly, grandparents will often call their grandchildren *tzej jutae*, which means 'sprouts.' As the grandparents become older, they symbolically assume the position of the old tree. In fact, old people will sometimes be addressed as *Nim Chie Nim Kam*, or 'Big Tree Big Vine,' this being perhaps the most respectful title that can [be] given an Atiteko."

This ever-changing reciprocity and tension between transformation and replacement—the child giving birth to the parent—between regeneration and sacrifice, between the king and his successor, lie at the heart of the imagery of the Foliated Cross and the myth of Maize God and the Hero Twins. The gods engendered humanity, but require sustenance and nourishment from them. The father engenders his son, but relies on his son's courage and skill in dancing across the abyss to dig him out from under the ballcourt floor and bring him the gift of renewed life.

When Chan-Bahlam danced on the tablet in Temple 14 (Fig. 6:19), his mother rose to meet him dressed in the net-skirted costume of First Mother and First Father.[56] His grandmother, Sak-K'uk', wore this costume as she handed the crown of kingship to Pakal the Great on the Oval Palace Tablet. Male or female, dead or living, the costume of the dancing Hun-Nal, the resurrected Maize God, and his wife, the Moon Goddess, were the symbols that united the nobility of the Maya in sacred mystery and linked their regeneration to that of corn, the food of all human beings and the basic substance of life. When their ancestors were clothed in this costume, the ancient Maya, like their modern descendants, regarded them as being of one essence and transcendent of gender, *Totil-Me'iletik*, "Father-Mothers," as the Zinacantecos say it. To the Tzutuhil, the stump of the generative tree at the center of the world is *Ti Tie Ti Tixel*, "Father/Mother," the source and endpoint of all life.[57]

So in the end, Classic-period masked pageant, the visionary ecstatic quest, and the formal dance in which the awesome power of venerable heirlooms was passed in dignified toe-shuffle from one generation to the next were all ritual expressions of the same epic myth cycle, interpretations of the same basic reality embraced by all Maya. Anchored in the daily experience of death and birth, and in the annual planting and harvest of maize, this reality was truth beyond question, requiring only honor and celebration. It was a reality that shaped the opportunities of royal statecraft; one that, over the centuries, was rich enough to allow for amplification and elaboration; but one that, because it lived in the hearts and perceptions of all Maya villagers, endured the collapse of the Classic kingdoms. We are confident that it will continue to endure, transformed and tailored to the ongoing experience of the modern world, still remaining true to its basic premises, as long as there are Mayan-speaking people.

THE CONQUEST AND AFTER

The myths guiding Maya festival endured, as did the context of the celebration of the official status of high lords. So did the manner in which the Maya joined their festivals to the cycles of time. All major Maya festivals of the Classic period were firmly linked to the cycles of days, and regenerated the community by mirroring the cyclic events of actual and mythological history. The *Xibalba Okot* was one such dance. It was performed by the Maya of Yukatan on Kawak, one of four days in the 260-day calendar that can bear the burden of the 365-day year. These are the Year-Bearer[58] days on which the first day of the year, 1 Pop, can fall. Landa described these ceremonies with great detail, and it is from his accounts that we get the following information.

The Yukatekan New Year's ceremony actually began five days before the year ended. At that time an appropriate[59] lord was chosen to preside over the coming year's festivals. At the correct time, the people made an image of the god of the Wayeb days (the final five days of the year). During the years when Kawak was the first day of the new year, this god was called *Ek'-u-Wayeyab* ("Black-his-Wayeb"). The Maya carried this god to a pile of stones at the appropriate entrance to the village—west for the Kawak years—where they dug up the god who had been buried there at the end of the last year's ceremonies. They also made a statue of another god named *Wak-Mitun-Ahaw* ("Six-Hell-Lord") and took it to the house of the chief.

From there they returned to Ek'-u-Wayeyab and prepared a path to bring the Six-Hell-Lord into the center of the village. Priests incensed the god and then cut off the head of a chicken[60] as an offering. They put the Six-Hell-Lord on a staff called *yax-ek'* ("first [or blue]-black"), which was surmounted with a skull and the image of a dead man alongside the head of a vulture. They carried this strange staff-banner[61] with great devotion as they processed toward the center of the village, all the while dancing the *Xibalba Okot*, the Dance of Xibalba.

Cupbearers brought drink for the lords and dancers as they bore Ek'-u-Wayeyab to the house of their ahaw and placed it opposite Six-Hell-Lord. The lords carried offerings to the two gods, incensed them with smoke, and drew blood from all parts of their bodies to anoint yet another statue called Ek'el-Akantun ("Black Upright-stone"), which may have been a black-painted stela mounted in front of the lord's house. After the five days of Wayeb had passed, they took Six-Hell-Lord to the temple and placed Ek-u-Wayeyab in a pile of stones at the south entrance of the village, where he would be ready for the following year's ceremonies.

According to Landa, the Kawak years were considered to be very dangerous, with drought and death the sure outcome for many people. To combat these predicted calamities, the Yukateks made the statues of four gods—Chi-Chak-Chob, Ek'-Balam-Chak (Black-Jaguar-Chak), Ah-Kan-wol-Kab, and Ah-Buluk-Balam (He-of-Eleven-Jaguars). These gods were put into the temple. There they were incensed and given balls of rubber that were burned, and offerings of iguanas, bread, headbands, flowers, and precious stones. When the offerings had been made, the men of the village built a huge arch of wood and filled it with firewood. A singer mounted the huge structure and sang and drummed as men carrying unlighted ocote torches, looking just like the figure of White-Three-Dog in Classic pageant (Fig. 6:8), danced solemnly through the arch and around the pyre until they were exhausted.

After a day of dancing these men went home to eat, but returned at night with all the people of the village in attendance. They danced again, this time with lighted torches that they eventually used to ignite the great pyre. When the fires had burned down, they spread the coals across the ground and the men, barefooted and naked, danced across the path of fire in an act of supreme devotion. Many passed unharmed, while others were singed or badly burned, but Landa[62] says they thought this dance of fire would fend off the calamities that always came with the Kawak years. Just as the Hero Twins threw themselves into the fire and returned unscathed to defeat the lords of death, so the spiritual leaders of the community in Yukatan tested themselves in like manner, dancing across the fires of Xibalba.

In 1543, the Popol Vuh story of the Hero Twins was still being performed as dance pageant among the Q'eqchi' Maya of Guatemala.[63] The Q'eqchi of San Juan Chalmelco danced the story during the accession ceremonies of Ah Pop'o Batz' (Lord Howler-monkey). A hundred years later the Maya were still dancing, as Thomas Gage,[64] an English-born Dominican friar, reported in the account of his travels in México and Guatemala between the years of 1625 and 1637. Gage does not tell us in what town he observed these dances, but his descriptions contain far more detail than the spare descriptions of Landa and other Spanish chroniclers.

He described[65] how each town had several houses appointed to oversee each kind of dance. The men in charge would meet months before the ceremony, drinking *chicha*[66] and chocolate while they planned the dance. The house where these preparations took place was very probably a Popol Nah. For two or three months before the great event, the dance masters from this special group would train other men from the community to do the dance. Gage complained that the silence of the night was broken by "their singing, their halloaing, their beating upon [drums and using as trumpets] the shell of fishes, their wails and with their piping." The dance, when the time finally arrived, lasted for eight days, during which the men drank, sang, and danced from house to house.

Modern Maya festivals continue to commemorate the violent confrontations of the community with its enemies. The Maya dance their history, remembering Cortés and the tragedy of the Conquest in the Dance of the Conquest,[67] which traces the movements of Cortés, Alvarado, and other Spanish intrusions into their world.[68] In this dance the Spanish are represented as Gentlemen cloaked in red satin jackets and knee breeches decorated with gold braid. Men also impersonate Spanish Ladies wearing black and purple veils topped with large black hats[69] banded with gold and sporting peacock feathers.

Montezuma is also present, personified by characters called the White Heads, who represent not only the conflict between Montezuma and Cortés but also the Tlaloc war complex of Precolumbian times.[70] Even the ancient cultural hero called *Quetzalcoatl* among the Aztec and *Waxak-lahun-Ubah-Kan*[71] among the Classic Maya still appears in the dance. His modern name is *K'uk'ul-chon* among the Tzotzil and *K'uk'ulkan* among the Yukatek. From the earliest times in the Maya region, this feathered serpent has been a symbol of the vision rite and sacrifice. At Chich'en Itza feathered-serpent warriors were among the great battle leaders, so powerful that they became legend to their descendants. By Postclassic times myth and history had merged into one, producing stories of a great wise man, Quetzalcoatl, who was driven from the Toltec to Yukatan. Some legends, manipulated by the Spanish who knew them to support their subjugation of the Indians, said that Quetzalcoatl had disappeared in a canoe to the east and that Cortés himself was the returning god.

Later wars appear in contemporary dance as dancers impersonate the Lakandon[72] Indians, who fought with the Spanish against the highland Maya. In the dance-drama of Zinacantan, the death of San Sebastián is combined with the ancient scaffold sacrifice,[73] where the victim was also tied down and slain by arrows, a wonderful example of how aspects of the old religion are incorporated into the new. The Tum Teleche, another dance in the highlands of Guatemala, recorded as late as 1624, was "a representation of an Indian, whom, taken in war, the elders sacrificed and offered to the demon, as is declared and said by the Indian himself, tied to a hitching post, and those who attack him to take his life in four figures which they say were those of their *naguals*: a tiger, a lion, an eagle, and another animal which is not recalled."[74]

In the highlands of Guatemala, the Dance of the Howler and Spider Monkey includes "six impersonators of Pedro de Alvarado dressed in feathered tricorn hats and 'excessively blond European face masks,' six men in deer masks surmounted by deer antlers, and four 'monkeys' with monkey masks and tails who were wearing 'Admiral Dewey naval uniforms and shako hats.' "[75] These dances in the highlands of Guatemala and Chiapas freeze history and incorporate everyone in the pageant—Spanish conquistadors, Aztec emperors, Lakandon enemies, French grenadiers, African slaves and mulattoes, even the expulsion of the Moors from Spain.

Even the Xibalbans are still there. Garrett Cook sent us pictures of gold- and silver-skinned masked dancers. He reported that "an elder costumbrista who was the *chuchkajaw rech tinimit* for one of the outlying aldeas told me that the devils were representations of an evil people of the past called Xibalba. These people he said had no religion, went naked, and didn't bury their dead." Modern dances combine post-Conquest elements with the Popol Vuh myth and survivals of even more ancient warfare and conflict to create a contemporary pageant of history.

Finally, returning to the festival we described at the beginning of this chapter, the Carnival at San Juan Chamula in the highlands of Chiapas, we see a continuation of this pattern. Here officials of the town, dressed in bright-red velvet suits like colonial Spanish lords, carry flower-printed cloth banners tipped with silver heads. People of the prinicipal barrio of San Juan tip their banners with trilobed silver shapes like the trilobed flints of the ancient dance of war and sacrifice. San Sebastián and San Pedro tip their banners with a single silver lancet head resembling the ancient leaf-shaped blades of the war lance. The flower-patterned flags are the body and the silver lance blades are the head of the Sun-Christ, who for five days will confront the forces of chaos and destruction.

Around the red-bedecked Spanish gentlemen whom the Pasiones and Flowers have become run primeval Monkeys. Dressed like nineteenth-century French grenadiers, the Monkeys are survivors of a past Creation. They represent chaos and will banish the Sun-Christ from the world. For five days these Monkeys, along with numerous other officials, will perform the Dance of the Warriors around clay water drums called *bahbinob*. There are two for each of the three barrios, each with its contingent of musicians, advisers, policemen, and other officials. The drummers—six in all—carry their clay drums from one location to another, where they sit in pairs beating their drum in a monotonous rhythm. Carrying the flower banners and led by the game master with his unadorned staff, the Monkeys circle the drums endlessly, in the exaggerated walk described by Gage two hundred years ago. The game master and a beller walk out into the crowds to choose a man by hanging a bell-laden jaguar pelt down his back from his forehead. He is handed one of the banners, and then round and round the new dancer goes, bent slightly forward from the weight of his banner. The game master precedes him and the beller follows, jingling the jaguar pelt in time with the drum. Monkeys carrying their own banners circle behind the warrior who has been drafted to the service of the Sun-Christ. Their endless circling creates a surreal scene of interminable movement and flashing color. When the dancer has circled the drums three times, bells ringing on his back the entire time, an Embracer lifts him from the ground. While the dancer dips the banner to the four directions, the Embracer carries him several paces toward the Pasión and puts him down. Then the beller runs into the crowd and picks another man to circle the drum.

This ritual lasts for five days. During the night, before the fourth day dawns, the tired dancers run up the hill called Calvario, where the Monkeys win the struggle for control and banish the Sun-Christ to the darkness of the Underworld. Then, sometime in early afternoon of the fourth day, the Pasión and his people repair to the central plaza, barrio by barrio, with San Juan, the senior barrio, arriving last, where they help the Sun-Christ out of the Underworld by running the Path of Fire.

By then, hundreds, if not thousands, of people are present—Chamulas who do not hold offices, Maya from other communities, mestizos from San Cristóbal and all over México, and gringos from the United States, Europe, and many other corners of the world. Free Monkeys and other officials have cleared a pathway now lined with an excited, dense crowd pressing always inward to get a better look. The Monkeys hold them at bay with whips made from dried and cured bull penises. The entourage from each barrio is preceded by men who carry thatch from an old house. They lay it along the long path, which stretches in a thirty-foot-wide swathe from the portal into the walled-in churchyard, called an atrium, across the open plaza to a station of crosses on the opposite side (Fig. 9:3).

When all the thatch is laid and the Monkeys have the excited crowd under a semblance of control, the Monkeys fire the thatch and the officials of all three barrios sprint across the fire. This thatch, called the "Path of God," represents the road of the Sun, a symbol of Christ, across the sky. Running through the thick smoke clouds, the men carry the banners of the Sun-Christ back and forth across the flames and coals, east to west and back again, a total of three times.[76]

So the Jesus gods (there are two of them in the Maya Christmas crèche, older and younger brothers), soon to be resurrected on Easter Sunday, voluntarily dance out across the flaming abyss. They follow the Hero Twins of the Popol Vuh who, by submitting to sacrifice and resurrecting their Father as Maize, brought life to all Maya. Across the still waters of the ancient kings, or the hot coals of the modern *Alféreces,* the banner bearers, the Maya dance the terrifying path from death to life as they always have, with grace, courage, and devotion.

Chamula dances may differ in their details from the dances of ancient Yaxchilan or Palenque, but this difference is on the surface only. Modern Tzotzil and Q'eqchi' dancers wear the pink face masks of the Spanish conquerors and the long-tailed coats of French grenadiers, but they dance for their ancestors and the soul-force of the world. Chamulas dance to hold back the forces of chaos and to bring the Sun-Christ back into the world after his confrontation with the forces of destruction and war. The ancient kings danced Creation also and, most of all, danced to awaken their ancestors. After his death, Chan-Bahlam danced and his own mother greeted him and dressed him in the costume of rebirth, just as First Mother assisted her husband, the Maize God, in his resurrection. 18-Rabbit danced in the guise of First Father as he exhumed the bones of his ancestor, just as the sons of First Father dug up his bones from the floor of the Ballcourt in the Otherworld. These ancient Maya dancers were creating the same types of historical symmetries that the modern Maya seek to create in their own dances to preserve the drama of the conquest and the passions of the sacrificed.

Dance was a central component of social, religious, and political endeavors for the ancient Maya. Kings danced, nobles danced, the people danced—and together they created the community. Dance created sacred space, penetrated the portal to the Otherworld, and released the dead from the grasp of the Xibalbans. If blood was the mortar of Maya society, dance was its soul.

CHAPTER SEVEN

FLINT-SHIELDS AND BATTLE BEASTS:
The Warrior Path of Kingship

A PUZZLE IN MESOAMERICAN HISTORY

(as told by David Freidel)

The School of American Research in Santa Fe is a lovely and peaceful place for visiting scholars. Graceful southwestern-style adobe buildings, bright gardens, a magnificent panorama of the Sangre de Cristo mountains, all contribute to uplifted spirits and mental clarity. I attended my first advanced seminar there in the fall of 1977. It was a week of intense conversation with a small group of colleagues—mostly elders and betters—about Maya settlement patterns and archaeology in general.

As we sat in easy chairs around the living room of the seminar building, we periodically got into debates about the nature of the Mesoamerican past and how to explain it. I became caught up in these arguments. We talked about the importance of understanding the native point of view and, more generally, native culture, as important factors in our attempts to explain why Precolumbian people behaved as they did. William Sanders, a great Mesoamerican scholar, eloquently argued that information about available natural resources, population densities, and subsistence technologies were the prime factors determining the basic logistics, and hence the general trends, in social development. Sanders is formidable in debate, ruggedly handsome with a full mane of white hair, flashing eyes, and a rich voice. He mesmerized me even as I disagreed.

After a week of making his case, however, Sanders admitted during one of our social hours that there was a puzzle that he couldn't answer by means of material explanations alone. Why did the Aztecs, who called themselves the Mexica, lose the pivotal battle of Otumba in the war of the Conquest? He described how the Mexica warriors, after the uprising in their capital against the invaders and with overwhelmingly superior numbers on the field, had let the haggard and frightened Spaniards and their Tlaxcalan allies slip through their fingers and escape to safe haven in rebel territory beyond the reach of the Imperial government in Tenochtitlan. Sanders told the story well and the puzzle has stayed with me. What aspect of the Aztec way of thinking about warfare might have contributed to this fatal debacle? The Mexica defeat began, according to all accounts, when Cortés charged on horseback out of his encircled troops, struck down the commanding officer of the Mexica army, and triumphantly raised the captured Imperial battle standard.

Over the years, Sanders and I have disagreed energetically about the nature of Precolumbian Maya warfare. I have argued for the importance of its sacred and ritual dimensions and he has focused on the practical motives of conquest and the acquisition of land and tribute.

The connection between Sanders's earlier comments about the battle of Otumba and my belief in the ritual aspect of Maya war came to me in the spring of 1990. It happened at the Maya workshop in Austin when I was watching Linda Schele present an analysis of an Early Classic Tikal monument called the Ballcourt Marker. The text on the marker refers to the calling forth twice of a war god, the first time during a great battle and the second time during the dedication of the marker itself. It occurred to me that this object wasn't a ballcourt marker at all, but a battle standard that housed a god. When the Maya celebrated the capture of noble enemies, they were probably taking the enemy standards and their gods at the same time. This was one of those flashes of insight that comes out of nowhere. I went to the podium during the break to tell Linda. Another person whose name I didn't catch came up with the same idea and was also up on the stage. Linda agreed immediately and announced the discovery to the audience after the break. The battle standards of the Maya and their Mexican allies presage those of the Aztec.

We have started our discussion of war with the Battle of Otumba because it remains a great mystery as well as a vital turning point in the conquest of the Mesoamerican world by the Europeans in the early sixteenth century. The Spaniards were grossly outnumbered on the field, exhausted, surrounded, and cut off from retreat. How then did they manage so decisive a victory? As a practical matter, there was no way that they and their equally war-torn Tlaxcalan allies could have avoided total defeat; yet they not only did so, but also completely routed their enemy.

The answer given to this riddle by the history books is that the Mexica were barbarians with little battle discipline; that the killing of their general destroyed the Mexica command structure and left the elite officers and their militia in disarray. For us, this is an unsatisfactory answer, for the Mexica were the preeminent fighting force in México.[1] They were not barbarians but Imperial masters over an ancient civilized world. The Mexica officers who clustered around the Spaniards like vultures around a dying animal believed they would have a decisive victory at Otumba, as Cortés himself observed in his letter to his king.[2]

Of course, Cortés was as brave as he was brilliant. His attack against the Mexica lord, against all odds and with death or capture as the likely consequence, was just the kind of audacity that had brought him that far on his campaign.[3] He truly believed, on later reflection, that his success and its immediate effect on the battle were miraculous signs from God and Saint Peter, his patron, and that his cause was just. His men thought that Saint James, whom they knew as Santiago, patron saint of Spain, rode with them that day as he had on other equally dangerous occasions. Both Spaniard and Mexica understood miracles. Later explanations of this battle, from both sides, clearly indicate a mystical interpretation of the events that resulted in the defeat of one world by another.

We also believe that the supernatural played a role in the battle of Otumba and the ultimate defeat of the Mexica. For it was clearly the premise not only of the Spanish but also of the Mexica that their gods were in the melee that fateful day. From this vantage, it was not only the slaying of the Mexica leader that was critical, but also the capture of the Imperial battle standard and its god. Battle standards were obviously as important to the Spanish as they are to modern Americans, as witnessed by the war memorial celebrating the raising of the American flag on Iwo Jima or the raging debate over whether people should go to jail for burning it. There is a crucial difference, however. We think of our flag as a symbol of great potency, but it is not the source of supernatural miracles. For the people of ancient Mesoamerica, however, the battle standard was exactly that. They saw their great standards of war not only as the representation of the state, but as an embodiment of a potent spiritual being whose presence and performance were critical to their success.

We believe that this idea was originally a highland Mexican concept first celebrated at the city of Teotihuacan. The Aztec lords of Tenochtitlan may have acquired it locally from the legendary Toltecs via their reputed Culhua descendants in the valley of México. We can make a case for terrifying battle beasts conjured by the Aztec standards of war because, much earlier, literate Maya kings had embraced some of these foreign deities and added them to their own inventory of supernatural beasts and beings. It is the Maya who, through their glyphic texts and imagery, reveal the secret of these hand-held portals of power.

THE FLINT AND THE SHIELD OF WAR

Our next story takes us from the bloodstained brown hills of México's conquered heartland back almost twelve hundred years to the Maya lowlands and the steamy hot city of Tikal. This immense metropolis floated like an island of red and white mountains in the green haze of the Peten forest in Guatemala. How the first emissaries from the great cities of Teotihuacan found themselves at Tikal we'll never know for sure, but we do know that sometime in the fifth century, these turbaned and tasseled lords brought with them a battle standard embodying a great war beast, a War Serpent[4] whom the Maya came to call *Waxaklahun-Ubah-Kan*.[5]

During the late fourth and early fifth centuries, Tikal was a huge, sprawling city. Its outer ring, made up of scattered households nestled among orchards and gardens, surrounded the central pyramids, palaces, and plazas of the royal dynasty (Fig. 7:1). When the greatest of Tikal's early kings, Great-Jaguar-Paw, and his brother Smoking-Frog decided in A.D. 378 to attack and conquer their rivals at the city of Waxaktun, they adopted rituals and weaponry from their Teotihuacano associates to form a new Maya military institution we call Tlaloc-Venus warfare. It involved the conquest of territory as well as the taking of captives for sacrifice. Most of all, decisions about when and where to do battle became tied to the cycles of Venus and Jupiter. It was a kind of holy war timed by the stars. Tlaloc, the goggle-eyed deity borrowed from the Teotihuacanos, appears on many of the war monuments of the time.[6]

For more than a century before, during, and well after their audacious move to imperial expansion, the Tikal rulers refurbished the center of Tikal, especially the huge compound called by today's tourists the Lost World Group. They used the traditional Maya style of building in combi-

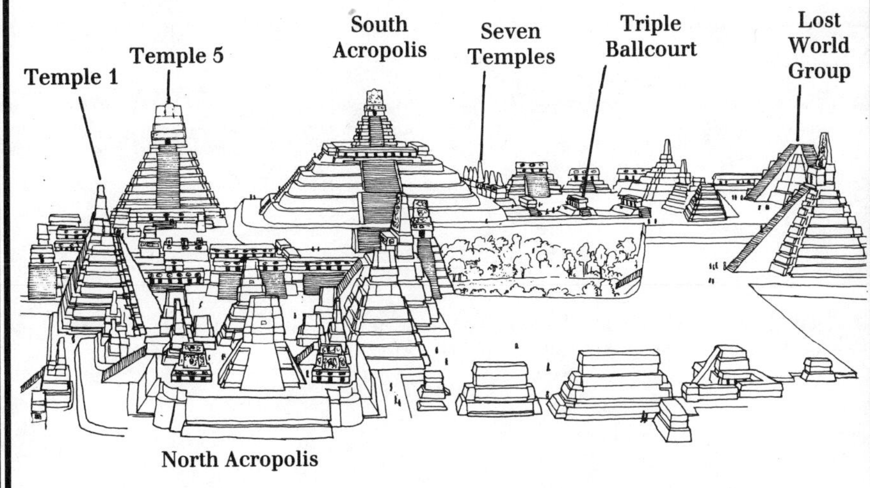

central area of Tikal drawn after Spier (1975)

FIGURE 7:1

nation with a style that characterized the architectural tradition of Teotihuacan. Composed of a framed panel called a tablero set above an angled basal wall called a talud, this style of architecture spread from Teotihuacan along with the Tlaloc-Venus war complex.

Next to this huge newly remodeled group, they constructed another compound, called the Group of the Seven Temples (Fig. 7:1). At the northern edge of its plaza, they built a triple ballcourt with two playing alleys. They covered the outer walls of the seven temples with huge plaster images of shields, flint blades, and the crossed bones of sacrificial victims. This was very probably an area where they trained young warriors and prepared them for battle, or perhaps a place used by a warrior society or cult.

Not far to the south, less than half a kilometer away, vassal nobles of the Tikal king also used talud-tablero decorations to celebrate their participation in the glorious conquest. In the midst of a large and well-appointed compound of houses and temples, called Group 6C-XVI on the survey maps,[7] these lords built a shrine (Fig. 7:2) in the form of a small open-sided stone box, much like the lineage shrines found today among the highland K'iche' Maya (Fig. 7:3). Over the years, it became filled with the ashes of burnt offerings until it was remodeled, at an auspicious time, into a small, elaborate platform in the talud-tablero style. This type of

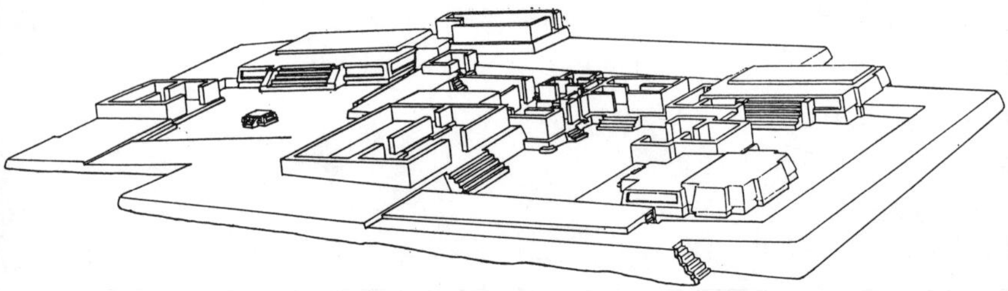

Tikal Group 6C-XVI sub, level 8. The talud-tablero altar is in the courtyard on the left (after Laporte [1989])

FIGURE 7:2

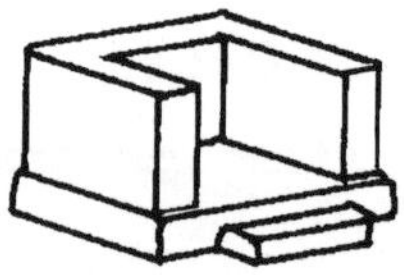

the original box shrine in Group 6C-XVI

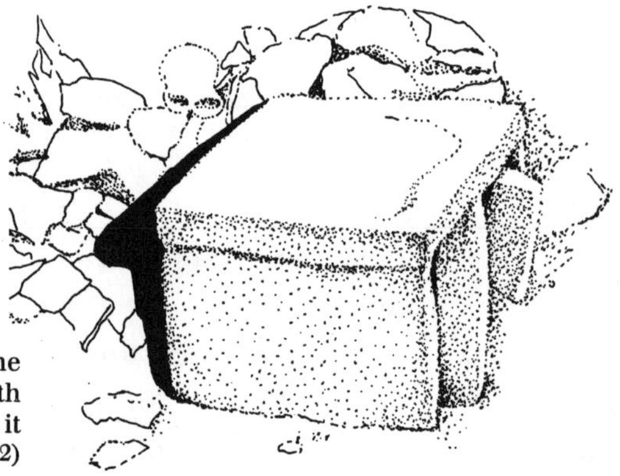

a *waribal* or lineage shrine from Momostenangò with broken pottery sherds around it (drawn after Tedlock 1982)

FIGURE 7:3

a.

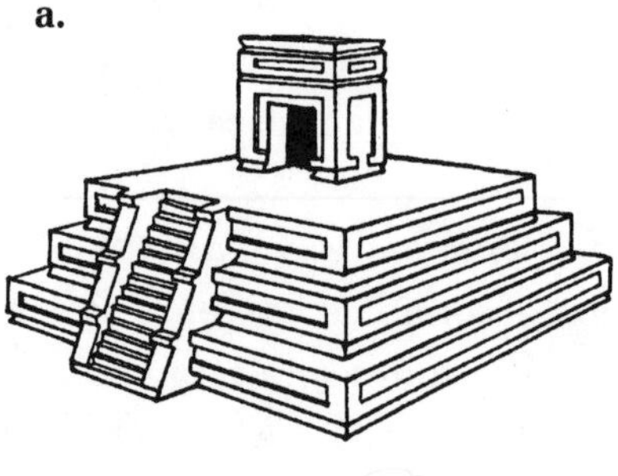

b.

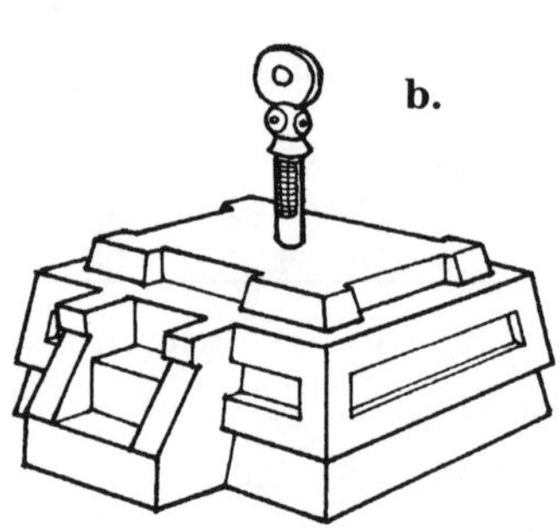

a. shrine at a compound at Teotihuan; b. the shrine from Group 6C-XVI at Tikal (the drawings are not to the same scale)

FIGURE 7:4

shrine emulated the kind placed in the open courtyards of the vast labyrinthine structures called by archaeologists "apartment complexes" in the distant city of Teotihuacan itself (Fig. 7:4a). On January 24, A.D. 414, very near the day on which the Wakah-Chan was raised by First Father, the lineage patriarch of Group 6C-XVI, Ch'amak, "Fox," set a stone replica of a battle standard[8] on top of this shrine to commemorate the lineage's glorious deeds in the war against Waxaktun (Fig. 7:4b).

We have been studying the text carved on the shaft of this effigy battle standard for several years and now understand something of what its makers wanted to remember about that war and its aftermath. They tell us, for example, that the War Serpent, Waxaklahun-Ubah-Kan, came forth at the city of Waxaktun on the day of the battle (Fig. 7:5). They also tell us that thirty-six years later the Serpent came out again, this time to honor the dedication of the stone effigy itself. Most important, we think this effigy is a copy of a standard originally carried by Ch'amak's predecessor[9] into battle alongside his war leader, Smoking-Frog.

In the dedication ritual, Ch'amak planted the effigy standard on its platform altar, then called forth the Waxaklahun-Ubah-Kan to serve as the path along which the companion spirits of important warriors and

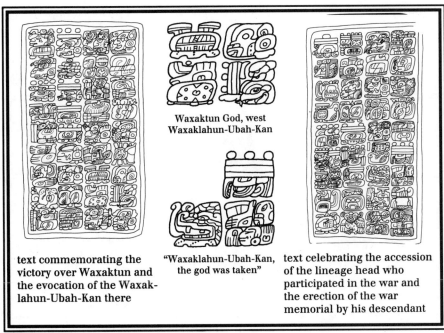

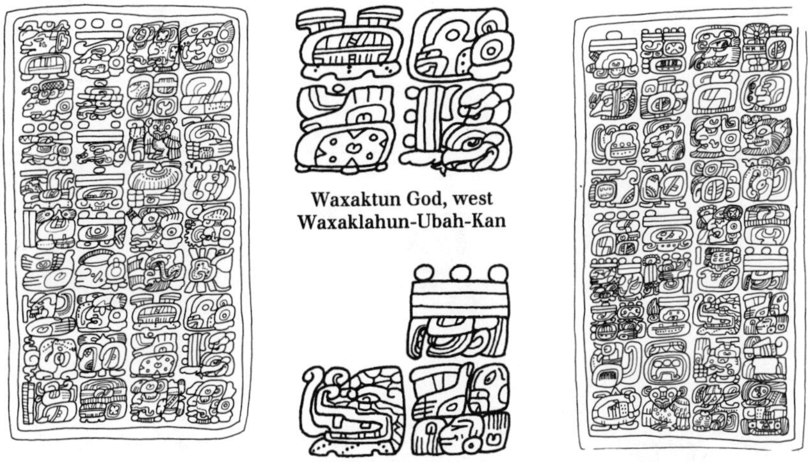

Waxaktun God, west Waxaklahun-Ubah-Kan

text commemorating the victory over Waxaktun and the evocation of the Waxaklahun-Ubah-Kan there

"Waxaklahun-Ubah-Kan, the god was taken"

text celebrating the accession of the lineage head who participated in the war and the erection of the war memorial by his descendant

Figure 7:5

FIGURE 7:6 Banners and Staffs from Teotihuacan

lineage heads traveled when they came to participate in rituals. But this same standard could also be used as an instrument for bringing forth the Teotihuacan god of war that these Maya had adopted as their own.[10]

The Teotihuacanos depicted many objects like this standard (Fig. 7:6) with its feather-rimmed disk, its ball, and its shaft. At Tikal, in a wonderful scene showing the arrival of an embassy from Teotihuacan (Fig. 7:7), a battle standard very like the one raised by Ch'amak stands in front of a Maya talud-tablero pyramid. The cut-shell ornament[11] on top of the staff appears on both Maya and Teotihuacan banners. More important, one of the emblems engraved on the Tikal effigy standard appears regularly on similar banners and shields at Teotihuacan, where it has been identified as a symbol of Tlaloc[12] (Fig. 7:8). The owl and spear-thrower symbol on the reverse side of the Tikal standard mark the shields

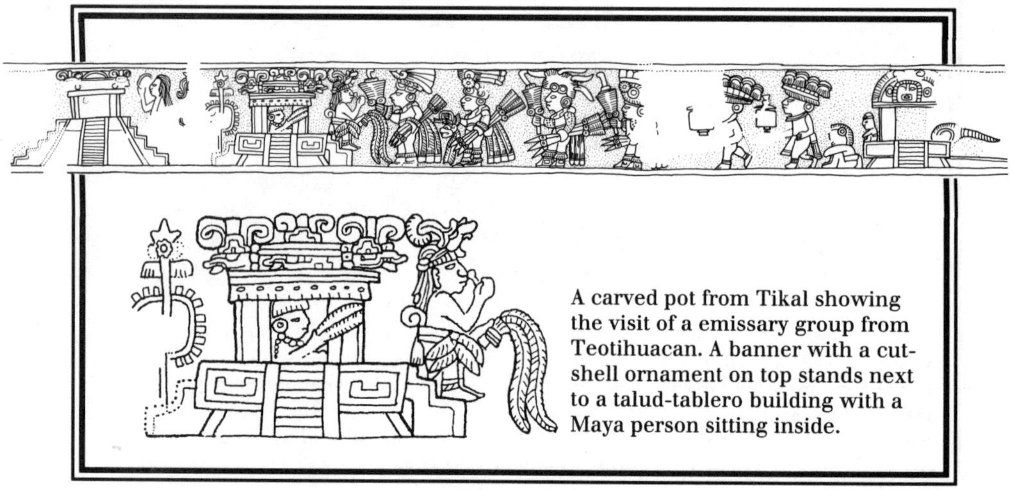

A carved pot from Tikal showing the visit of a emissary group from Teotihuacan. A banner with a cut-shell ornament on top stands next to a talud-tablero building with a Maya person sitting inside.

FIGURE 7:7

The Tlaloc-war emblem from the Tikal effigy banner and from Teotihuacan (the four on the left)

FIGURE 7:8

of goggle-eyed warriors at Teotihuacan (Fig. 7:9). Unlike the Maya, who preferred a side view, the Teotihuacanos liked to depict their war owl front view with his wings, legs, and tail spread behind a shield and crossed throwing javelins.[13] But even with the different approaches to the representational arts, both images depict the emblem.

Moreover, these two signs—the emblem of the War Tlaloc and the owl-javelin-shield medallion—are directly and repeatedly associated with sacrifice and war imagery in the painted murals of Teotihuacan.[14] The owl-javelin-shield combination was a Teotihuacan metaphor for war,[15] just as arrow-shield[16] was for the Aztecs. We think these emblems are symbolizing exactly the same thing at Tikal—Ch'amak's effigy ban-

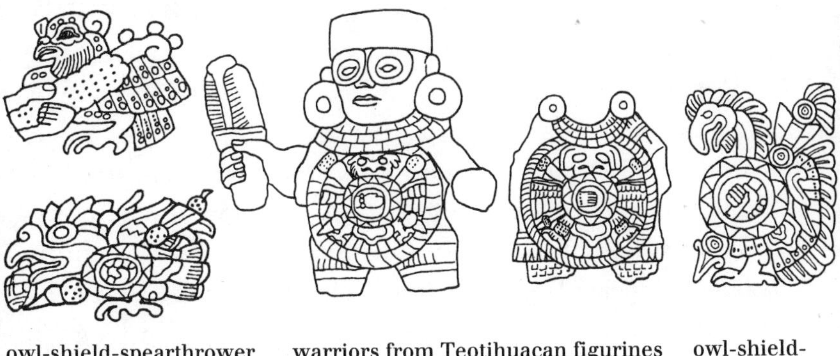

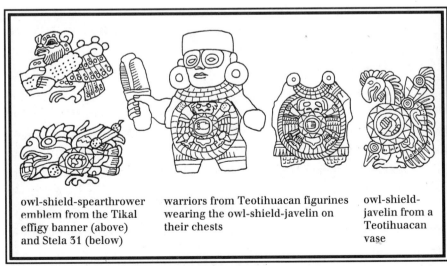

owl-shield-spearthrower emblem from the Tikal effigy banner (above) and Stela 31 (below)

warriors from Teotihuacan figurines wearing the owl-shield-javelin on their chests

owl-shield-javelin from a Teotihuacan vase

FIGURE 7:9

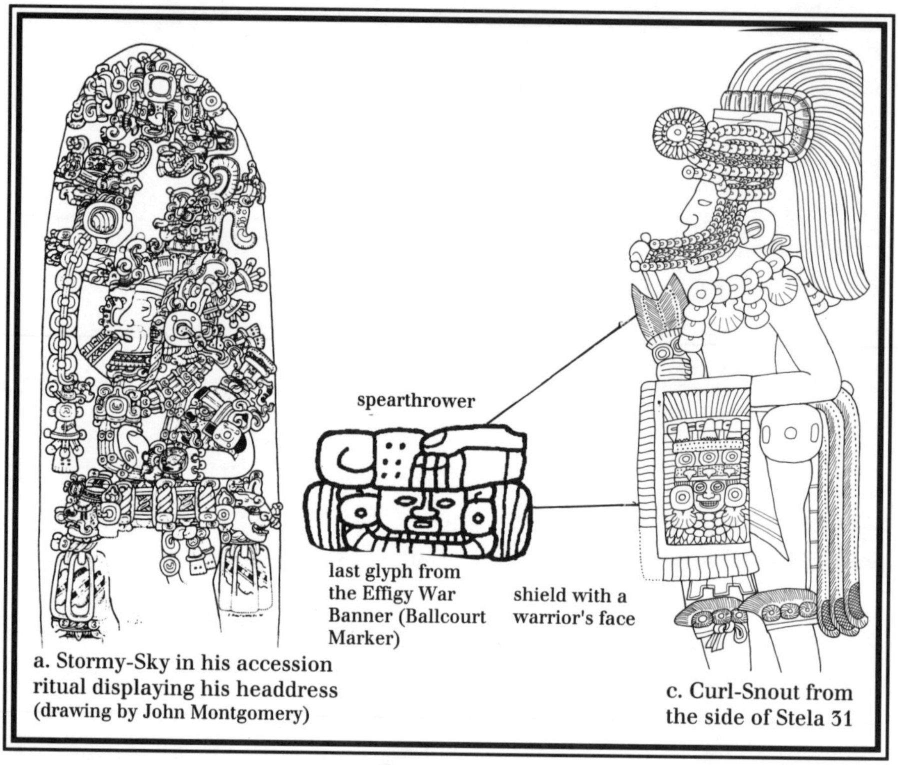

FIGURE 7:10

ner combines the war god Tlaloc with the metaphorical representation of war and battle as a memorial to the greatest victory in Tikal's history.

The Maya, especially those living at Tikal, adopted this spearthrower-owl complex with amazing fervor. Within a few generations, they thoroughly Mayanized the political and religious concepts associated with it and tied them closely to Venus and the concept of war timed by the stars. Great-Jaguar-Paw's grandson, Stormy-Sky of Tikal, depicted himself (Fig. 7:10a) holding up the crown he took as the central act of his accession to the office of the high king in A.D. 439.[17] The headdress represents our familiar friend Itzam-Yeh, but now he is wearing the owl-shield-javelin war emblem (Fig. 7:10b) on his forehead. Stormy-Sky also made very sure that his father, Curl-Snout, dressed as the ultimate Tlaloc warrior, appeared on the same stela (Fig. 7:10c) as if he were standing behind his son on this important occasion. The new king wanted his people to understand that he received the war emblems legitimately from his father, who, in turn, had received them from his grandfather, the conqueror of Waxaktun.

The Maya adopted the war emblem easily, but they very quickly adapted it to their own way of understanding such things. In the earliest examples, they called the war owl *kuh*, "owl of omen,"[18] and wrote it two ways: as a picture of an owl and with phonetic signs that spelled the name in sounds. By doing so, they changed a Teotihuacan owl into a Maya one. They also changed the war emblem in another way. Unlike the Teotihuacanos, the Maya did not like spearthrowers, probably because they were not very effective weapons in the dense forests around their cities. They preferred a blowgun for hunting and a short-shafted stabbing spear for battle. They soon replaced the spearthrower imagery with the large left-shaped flint blade that tipped these spears.

In this way, a borrowed Teotihuacan symbol was converted into a thoroughly Maya idiom. They combined the lance blade, which they called *tok'*,[19] with a shield, which they made by stretching the flayed facial skin of a sacrificial victim over a wooden frame. Their word for this shield was *pakal*. Together, the *tok'-pakal* became the principal emblem of war for the Maya.

We also believe that *tok'-pakal* could refer to the battle standards themselves because of what the text on the Tikal effigy standard tells us. In its last dedicatory phrase,[20] the scribe called the banner itself a "spear-thrower-shield." As our friend Chris Jones[21] first pointed out to us, the shield in this glyph is a very special one indeed. It is exactly like the flexible shields carried by Curl-Snout on the sides of Stela 31 (Fig. 7:10c). The shield in both contexts actually portrays the face of a Warrior as Teotihuacan understood that concept. Curl-Snout, like his father before him and his son after him, was the ultimate Warrior in the tradition brought to the Maya by the Teotihuacanos.

BATTLE BANNERS AT BONAMPAK'

In Chapter 5, we described the feather-fringed battle standards prominently displayed in the house dedication portrayed in Room 1 of Temple 1 at Bonampak' (Figs. 5:2–3).[22] Like the standards at Tikal, the centers of these feathered circles contain images of the supernaturals that resided in them. While dancers in Room 1 at Bonampak' carry the standards in a formal procession, in the mural in Room 2, we see soldiers carrying them (Fig. 7:11) into the heart of a fierce battle. In this extraordinary scene, warriors rush to engage the enemy while standard-bearers hoist their talismans high where they can be viewed and followed by unseen troops

Figure 7:11 Bonampak' Room 2 Battle Scene

behind the vanguard. Amid the blasts of nearby trumpeters, one bearer glances back, as if to exhort his fellows to hurry to the fray.[23] Not surprisingly, these standard-bearers are directly behind the king and his bodyguards, who are confronting their counterparts in a deadly struggle for victory. The tops of two of these standards carry red stone spear-heads—the *tok*' of the *tok'-pakal.*

No battle standards bless the losing enemy side in this scene. One of the hapless enemy lords, however, holds a staff in one hand and protects a box being rushed away by a retreating companion. Both the ancient and the modern Maya use this kind of box to store powerful objects, and it may be that the lord has dismantled his standard to avoid its capture. Certainly, if the shield-parasol part was detachable, it could be stored in such a box.[24] Bonampak' tells us that the great parasol standards not only functioned in battle but also had an important role in rituals before and after battle.

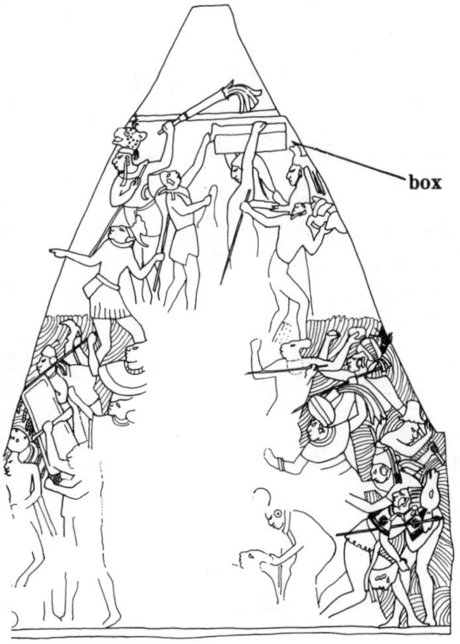

The battle scene from Room 2 at Bonampak'. The battle banners and trumpets enter the scene from the upper left. The Bonampak' king confronts the enemy leader at the center right, whose capture appears immediately below. In the upper left panel, the losers spirit away a box, perhaps containing their banners.

THE FLINT-SHIELDS OF THE ANCESTORS

Texts from Yaxchilan and Palenque identify these flint-shields as objects passed down through the generations from king to king. On Lintel 45, for example, the text says that Shield-Jaguar was the *u tz'akab u tok' u pakal*, the "replacement of the flint, of the shield" of Knot-eye-Jaguar, an earlier king of the city who may have been taken captive by a rival from Piedras Negras[25] (Fig. 7:12).

In the Temple of the Sun at Palenque, King Chan-Bahlam received his flint-shield from his dead father, Pakal, during his accession rituals[26] (Fig. 7:13). The central image on the panel further explains the origin and nature of war in the Maya view. This image is a war shield depicting the face of the cruller-eyed, jaguar-featured god Ahaw-K'in, who was the second-born of what is called the Palenque Triad, the three gods who were the patron protectors of the kingdom. Crossed spears, their blades emerging from the throats of White-Bone-Snakes, emerge from behind the ancient shield. In turn, the shield and spears rest on a great bone throne[27] mounted with the bleeding, severed heads of a jaguar and two flanking snakes held up by two denizens of the Otherworld. On either

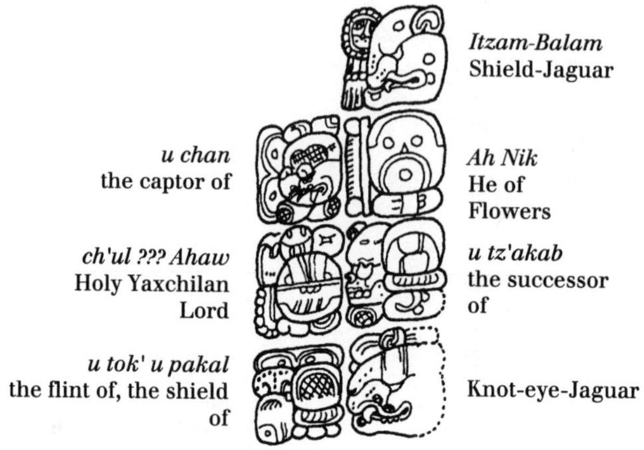

FIGURE 7:12

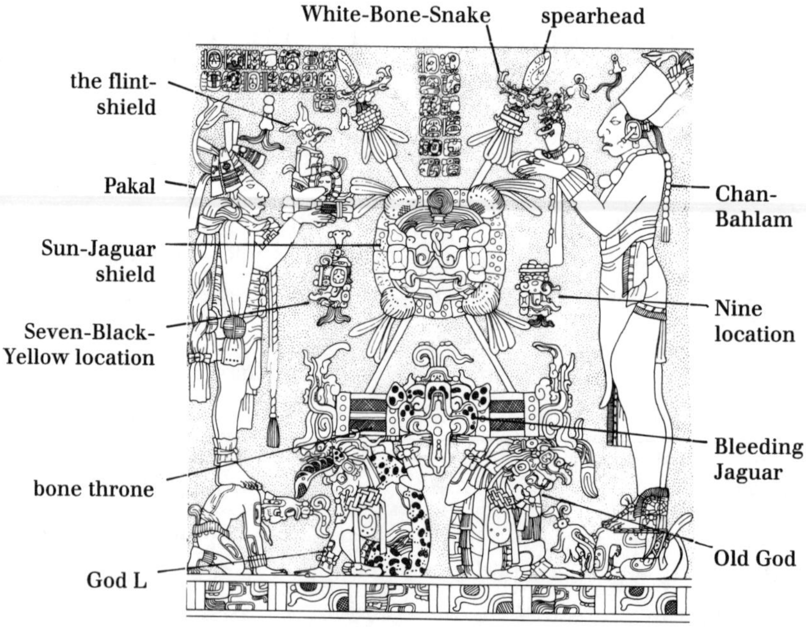

FIGURE 7:13 Palenque Tablet of the Sun

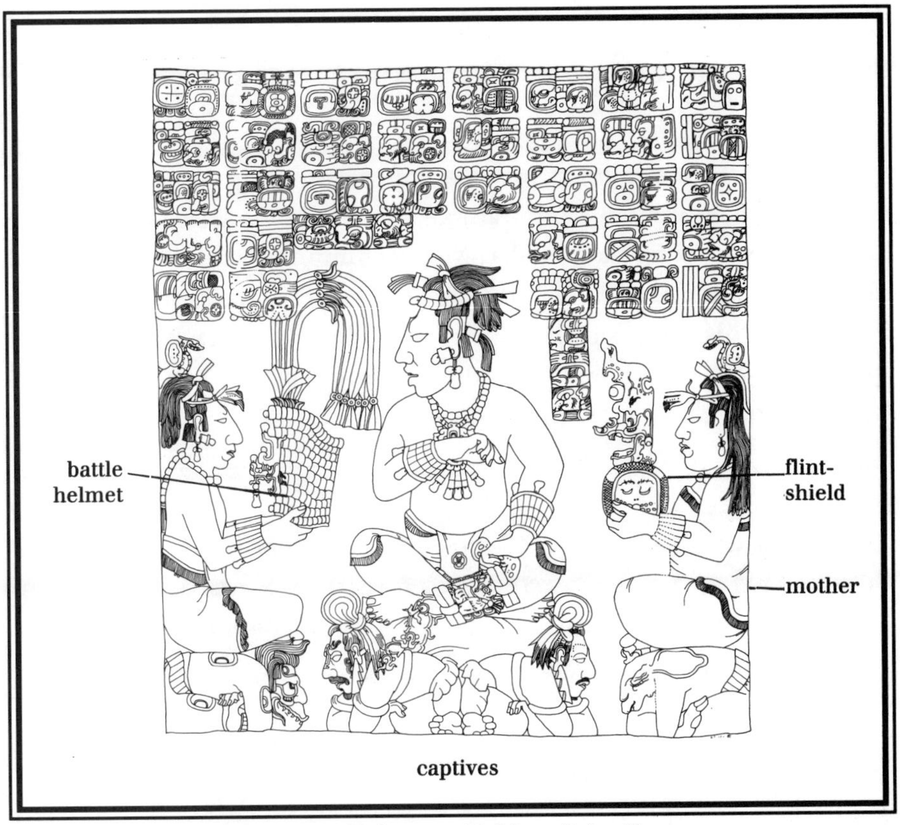

FIGURE 7:14 **Palenque Tablet of the Slaves**

side of the shield floats the glyph for two of the Otherworld locations we discussed in Chapters 4 and 6 (Figs. 4:28, 6:17a). One is the Seven-Black-Yellow Place and the other is nicknamed the "Nine-Place." These glyphs, along with the old gods who hold up the bone throne, tell us that the power of war comes from the Otherworld.

Chan-Bahlam's younger brother, K'an-Hok'-Chitam, got his flint-shield from his mother, while his father, Pakal, gave him the war helmet like the one Curl-Snout wears on the side of Tikal Stela 31.[28] Pakal himself got his battle helmet from his mother, but we know from the Tablet of the Slaves (Fig. 7:14) that sahalob and other subordinate lords also inherited flint-shields and war helmets from their parents and wielded them in battle. Chak-Zutz', a high-ranked vassal lord, served his king well and, like the loyal officer of Smoking-Frog of Tikal, earned the right to display a flint-shield talisman, we presume, through valor on the battlefield. The text certainly speaks of his war exploits and captures. He even sits on two unfortunate captives as he receives his flint-shield from his mother.

THE WAR SNAKE AND THE FLINT-SHIELD

The War Snake, named Waxaklahun-Ubah-Kan, appears with the *tok'-pakal* on Yaxchilan Lintel 25, an extraordinary lintel showing the king's principal wife, Lady K'abal-Xok[29] communing with the founder of his lineage, after Shield-Jaguar had conjured him up through bloodletting during his accession rite (Fig. 7:15). Holding a bowl full of paper splattered with her blood, she kneels gazing up at Yat-Balam,[30] the founder, emerging from the Waxaklahun-Ubah-Kan, the transformed

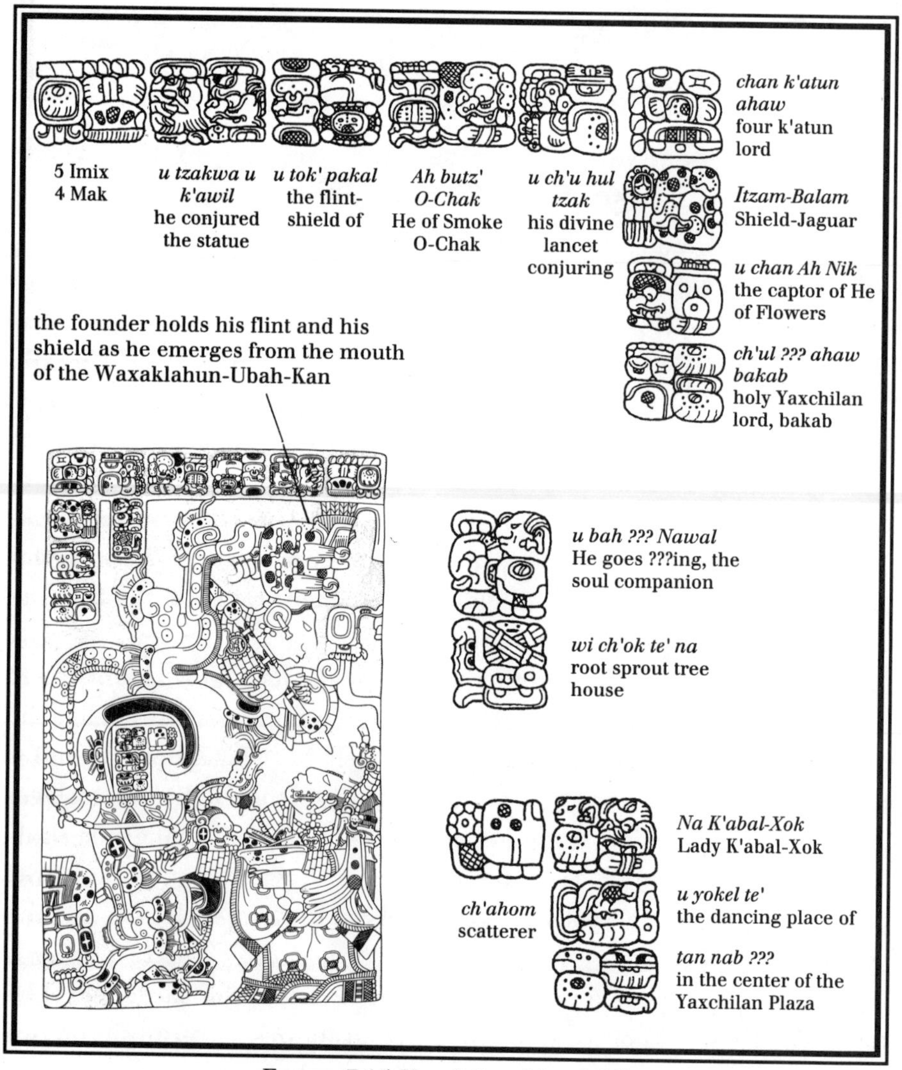

FIGURE 7:15 Yaxchilan Lintel 25

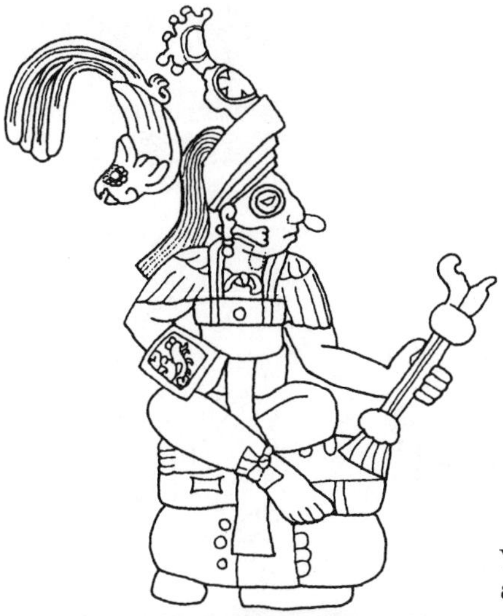

Yax-K'uk'-Mo' wearing the eye rings and square shield of a Tlaloc warrior.

FIGURE 7:16 Detail of Copan Altar Q

war monster originally imported from Teotihuacan. Yat-Balam wears the balloon headdress, the Mexican Year Sign, and the mask of Tlaloc, the other Teotihuacan deity introduced with Tlaloc-Venus warfare. He carries a short double-ended staff tipped with flint blades and a round shield fringed with chopped feathers. His lance and shield represent the *tok'-pakal*. The main text announces that Shield-Jaguar *tzak u k'awil u tok'-pakal Ah Butz' O-Chak u ch'ul hul tzak*, "conjured its k'awil, the *tok'-pakal* of He of Smoke, O-Chak; his holy lancet conjured." The *tok'-pakal* of O-Chak represents the founder himself, conjured up in all his terrifying reality to witness the accession of his descendant. The O-Chak of this passage is a snake-tongued Chak, who may be a special protector of the portal to the Otherworld.[31]

The association of a dynastic founder with the Tlaloc war complex also occurs at Copan. Altar Q, commissioned by King Yax-Pak, and Temple 16 at the heart of the Acropolis, show the founder of that dynasty adorned in the same war imagery that we find at other sites (Fig. 7:16). Apparently, the *tok'-pakal* of Maya dynasties, and presumably their supernatural potency in battle, descended from the founders. At Copan, loyal lords as well as kings could wield these powerful forces, just as they did at Tikal and Palenque. A great noble who served King Smoke-Imix-God K at Copan earned the privilege of erecting his own big-stone, Stela 6, inside

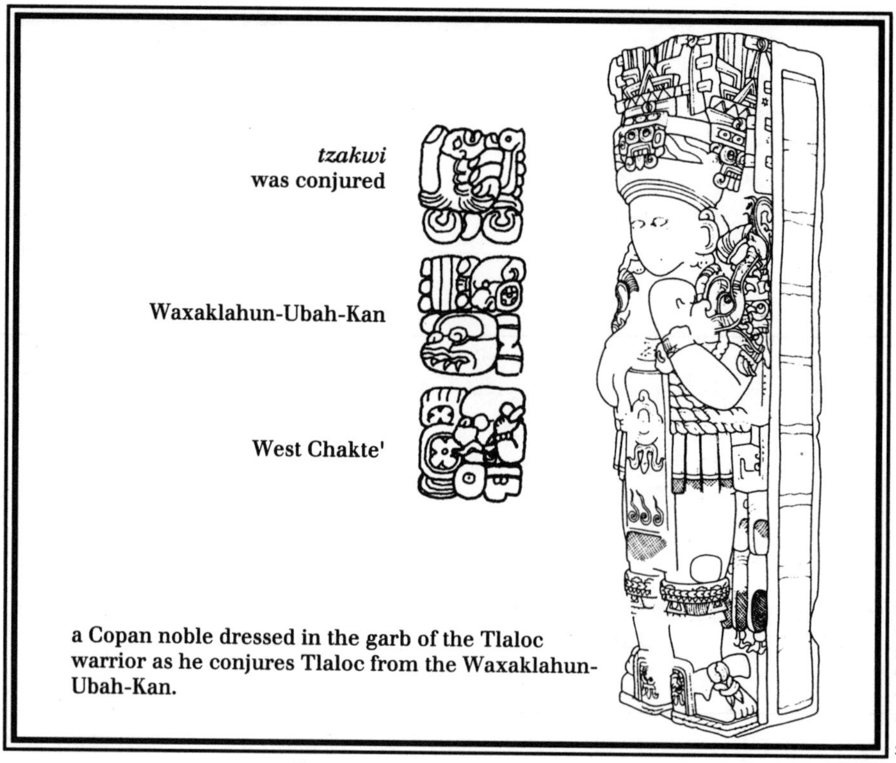

a Copan noble dressed in the garb of the Tlaloc warrior as he conjures Tlaloc from the Waxaklahun-Ubah-Kan.

(drawing by B. W. Fash)

FIGURE 7:17 Copan Stela 6

the confines of his own residential compound.[32] On this stela (Fig. 7:17), the lord declares that he "conjured the Waxaklahun-Ubah-Kan." The image shows him holding the same Double-headed War Snake that graces Lintel 25 at Yaxchilan. Instead of bringing forth the founder wearing the mask of Tlaloc, however, this noble conjures up the Tlaloc war god itself from the gaping mouths of the snake.

BATTLE BANNERS AND LITTERS

Hasaw-Ka'an-K'awil of Tikal, also known as Ruler A and Ah-Kakaw, used Tlaloc warfare to avenge the terrible defeats visited upon his kingdom by his enemies over a span of more than a hundred years. To celebrate his victory over the enemy who had captured and sacrificed his own father,[33] he commissioned two lintels. Each lintel showed him with a different battle beast: the Jaguar Protector who gave him victory over his father's killers and the great War Serpent who symbolized the greatest victory in his ancestral past—Jaguar-Paw's triumph over Waxaktun. These images were carved on the underside of the lintels inside his funerary temple.[34] The lintels convey not only the power of the supernaturals conjured by the vision rites but the fact that there were many battle beasts presiding over war and royal sacrifice.

Lintel 2 of Temple 1 (Fig. 7:18a) shows Hasaw-Ka'an-K'awil wearing the balloon headdress of Tlaloc-Venus warfare adopted at the time of the Waxaktun conquest, and holding the bunched javelins and shield, the original metaphors for war imported from Teotihuacan. He sits in majesty on the litter that carried him into battle, while above him hulks

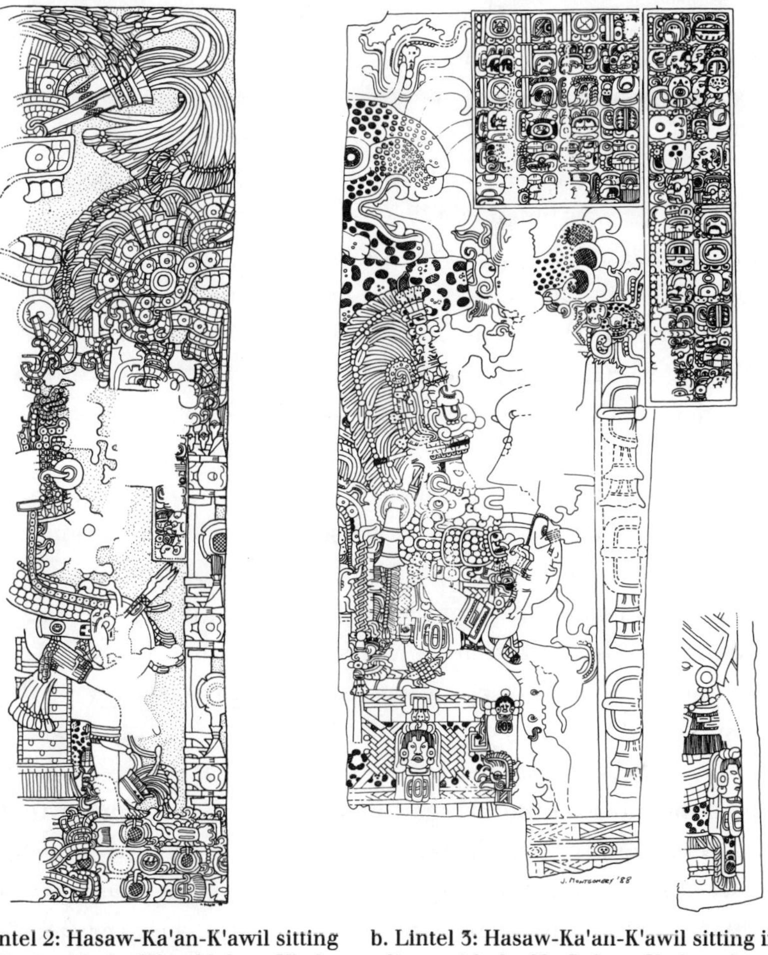

a. Lintel 2: Hasaw-Ka'an-K'awil sitting in a litter with the Waxaklahun-Ubah-Kan arching above him

b. Lintel 3: Hasaw-Ka'an-K'awil sitting in a litter with the Nu-Balam-Chak arching above him (drawing by John Montgomery)

FIGURE 7:18 Tikal Temple 1

Waxaklahun-Ubah-Kan,[35] the great War Serpent. Standing in front of the litter is a standard marked with the trapezoidal Year Signs of the Tlaloc complex and sporting the Waxaklahun-Ubah-Kan at its summit. This image is meant to evoke the Tlaloc standard taken to battle against Waxaktun by his glorious ancestor. The scene also depicts the king in the same battle costume worn by Curl-Snout on the sides of Stela 31, a monument Hasaw-Ka'an-K'awil honorably buried inside Temple 33.

In the complementary lintel over the innermost door (Fig. 7:18b), the king depicted himself in a different litter, this one bearing the huge Jaguar Protector named Nu-Balam-Chaknal.[36] Here the battle standard is decorated with three bundled and tasseled shields, and at its summit perches the bleeding jaguar head we saw on the bone throne in the Tablet of the Sun at Palenque. The massive image of the jaguar looms protectively over the king, reaching out with its great claws to threaten those who come against him.

The text of Lintel 3 confirms its association with warfare. The first event recorded tells us that the flint-shield of an enemy king, Jaguar-Paw of Kalak'mul,[37] was put down.[38] Forty days later, while sitting in the Nu-Balam-Chaknal litter, Hasaw-Ka'an-K'awil conjured a god, probably Nu-Balam-Chaknal himself, by piercing his tongue in self-sacrifice.[39] He did this on the day that he dedicated the final phase of Temple 33, which contained Stela 31, the big-stone memorializing the Waxaktun conquest. This stela, and other sacred big-stones set before the human-made mountains of the North Acropolis, had been broken and desecrated by the Dos Pilas conquerors who had killed his father. Deep in that same building lay the tomb of Stormy-Sky, the grandson of the conqueror and the king who had commissioned Stela 31. Moreover, the Kalak'mul king he captured and killed in the dedication rites for this building was the head of the alliance that killed his father and an even earlier ancestor more than a century earlier. It was a sweet vengeance that still speaks loudly across the ensuing centuries.

Graffiti drawings scratched on the walls of Tikal palaces, depicting the conjuring of supernatural beings from the Otherworld, prove that these scenes were more than imaginary events seen only by the kings. Several of these elaborate doodles (Fig. 7:19) show the great litters of the king with his protector beings hovering over him while he is participating in ritual. These images are not the propaganda of rulers, created in an effort to persuade the people of the reality of the supernatural events they were witnessing. They are the poorly drawn images of witnesses, perhaps the

a scratched drawing of a litter with the Waxak-lahun-Ubah-Kan as the animal spirit

a scatched drawing of a litter with the Nu-Balam-Chak as its animal spirit being

a scratched drawing of a litter with the Sun-Jaguar as the animal spirit being. Notice the bearer below.

Figure 7:19 Graffiti Drawings from Tikal

minor members of lordly families, who scratched the wonders that they saw during moments of ritual into the walls of the places where they lived their lives.

Another graffito at Holmul (Fig. 7:20) depicts a lord on a great rattlesnake litter being carried by bearers. We can't be sure that Maya kings went to war seated on such litters, as the Mexica lords did, but they

a rattlesnake litter being carried by two bearers

(after Merwin and Valliant [1932])

Figure 7:20 Holmul Graffiti

certainly rode them in the great processions that led into the grand plazas where they sacrificed their enemies. These great palanquins were real objects that focused the supernatural powers of the Otherworld for ritual and battle. We also know that these litters could be captured when their owners were defeated.

K'in-Balam, the Sun-Jaguar, of Tikal appears to have been lost in battle against the very enemies who killed Hasaw-Ka'an-K'awil's father, Shield-Skull.[40] That unfortunate Tikal king lost a war to Flint-Sky-God K, the extraordinary lord of Dos Pilas who acknowledged in his inscriptions that he was a vassal lord of the king of Kalak'mul.[41] A remarkable hieroglyphic stairway at Dos Pilas recounts how Shield-Skull's flint-shield was "brought down" by Flint-Sky-God K on May 3, A.D. 679.[42]

Hasaw-Ka'an-K'awil's son and successor appears to have taken vengeance for this particular humiliation by doing the same thing to Naranjo, an ally of Dos Pilas whose dynasty was rekindled by a daughter of Flint-Sky-God K. The Tikal king depicted himself on Lintel 2 of Temple 4 (Fig. 7:21) seated on a litter and bearing the battle standard and body of another K'in-Balam. Here, the protector being is a giant full-bodied form of the god who appears on the shield in the war stack at Palenque (Fig. 7:13). At Tikal, he looms over the king with his arm extended toward a standard decorated with heads of a god, perhaps Chak. The Jaguar Paddler and another god, perhaps his partner, the Old Stingray Paddler, emerge from the eyes of a double-headed monster held by the Sun-Jaguar. These two Paddler Gods conveyed the Maize God to the place of Creation and laid one of the stones of the first hearth.

Most important, our friend Simon Martin[43] has shown us that this particular Sun-Jaguar litter was captured from a defeated lord of Naranjo. In other words, the son of Hasaw-Ka'an-K'awil replaced the K'in-Balam palanquin lost in battle to Dos Pilas by his grandfather by capturing one from a neighboring kingdom.

We can imagine his grandfather, Shield-Skull of Tikal, seated on his Sun-Jaguar litter staring in disbelief as his own black-painted soldiers fell to the triumphant warriors of the upstart kingdom of Dos Pilas. In the moment the Dos Pilas warriors took him captive, they threw down his flint-shield and took possession of his battle litter and its great protective beast.[44] The final outcome was undoubtedly his death, probably with great public humiliation and torture, although the inscriptions do not talk about it directly. The triumphant king of Dos Pilas, however, did keep captive his giant jaguar effigy and the spiritual being who lived inside it.

Figure 7:21 Tikal Temple 4, Lintel 2

In this way the gods of conquered cities were held hostage long after the sacrifice of their human counterparts. In fact, Flint-Sky-God K's grandson, who was himself the conqueror of Seibal, repeatedly called himself "the guardian of the Sun-Jaguar, Holy Tikal Lord" (*u kan K'in-Balam Ch'ul Tikal Ahaw*).[45] As far as we know, he did not attack Tikal and no Tikal king carried that name. We think he was boasting that he guarded the battle beast of Tikal captured by his grandfather, just as the grandson of the captured Tikal king boasted he had taken the Sun-Jaguar of the king of Naranjo.

Perhaps one reason for Hasaw-Ka'an-K'awil's preoccupations with the great war litters in his funerary monument was that they were replacements for those lost by his father to Dos Pilas. The activation of new battle beasts required the sacrifice of suitable victims taken in battle. He avenged the terrible defeats his people had suffered by throwing down the flint-shield of Jaguar-Paw of Kalak'mul, the successor of the man who had helped capture his father.

The most amazing part of his saga of renewal was the name he took for himself. Previously, we knew him as Ruler A or as Ah-Kakaw, based on an inaccurate reading of his name glyph. Recent decipherments by David Stuart, Nikolai Grube, and others[46] have shown that his name was, in fact, Hasaw-Ka'an-K'awil. *Hasaw-ka'an* is also the name of the flapstaffs used by Shield-Jaguar and Bird-Jaguar in their dances at Yaxchilan. This same style of flapstaff, decorated with cloth and other materials along its length, is mounted on the front of the war litters at Tikal. Moreover, when the Yaxchilan kings danced with the *hasaw-ka'an* staffs, they wore war costumes festooned with the shrunken heads of past captives (Fig. 6:11). When the king of Tikal took the name Hasaw-Ka'an-K'awil, he was labeling himself the embodiment of the great war standards that focused his supernatural power and harnessed the war beasts of the Otherworld to his service. We think he chose this name because he rebuilt and ensouled these very objects after they had been destroyed and captured in past defeats.[47]

Waxaklahun-Ubah-Kan (the War Serpent), Nu-Balam-Chak (the War Jaguar), and K'in-Balam (the Sun-Jaguar) presided over war and sacrifice for all Maya kings, not just those of Tikal, Yaxchilan, and Copan. These supernaturals, and there were many others lusting for the flesh of the fallen, were deadly and fearful companions for all who had the power to conjure them. They were especially dangerous on the battlefield where the decisive battles were fought on the supernatural plane—where the key combatants would all be attempting to train their destructive power on the enemy by means of their powerful talismans and ecstatic visions. Military defeat on the savannas of Maya country was the consequence of a king's spiritual failure to hold the covenant and alliance of the gods against the rival supplications and magic of the king across the field. War was a very personal responsibility, anchored in the charisma, breeding, and discipline of the individuals whose responsibility it was to nurture the gods on behalf of their people. This is why their war monuments, declaring war against rival kingdoms and dynasties, bear images of Venus glyphs hovering over the names of enemy lands. These very same monuments celebrate victory over individual kings, nobles, and their battle beasts and *tok'-pakalob.*

In their own way, the Maya thus acknowledged the terrible truth of war as statecraft: the authority of a small number of people over the many who must suffer and die in combat. But unlike our leaders, Maya rulers themselves went to war with the men they sent; and Maya kings and their noble vassals put not only their bodies but also their souls in jeopardy every time they clashed. It is no exaggeration to say that they lived for those moments of truth, those trials of the strength of their spirits. Every major political activity in their lives—the dedication of every public text, image, and building of royal and community importance—required the capture and sacrifice of rival peers. Only in this way could the proper rituals of sanctification be fulfilled, the gods nourished, and the portals of communication opened between the human and the divine.

When thinking about war talismans and gods in the abstract, it is easy to lose sight of the real pain and anguish caused by this code of conduct. Sometimes, however, the Maya ancestors seem to arrange moments of sharp clarity to remind us of the human cost.

A CHRISTMAS ENCOUNTER WITH THE WAR SNAKE

(as told by David Freidel)

The tour I was guiding had just finished its two-day leg at Palenque, where I had been lecturing intensively on the royal family and its history.[48] With my mind on the impending war between my own nation and Iraq, I kept coming back again and again to the importance that combat and capture had to the Palenque kings, the "Holy Lords of Bak." Even their name, *Bak,* has warlike overtones, translating as "bone." It referred both to the name of their kingdom and to the great *way,* the skeletal death god from the hellish side of the Otherworld who was their soul companion.[49] (Fig. 7:22). The ruins of Palenque are studded with reminders of the glories of past battles, including one important monument called House C. On a platform beneath that building, the great king Pakal commissioned six trophy monoliths depicting high-born captives taken in successful battles. These six nobles from formidable distant

Baklel Way, Bonyness Spirit Companion

a skeletal death god

FIGURE 7:22

kingdoms had been born into the Otherworld through sacrifice, making them brothers[50] and eternal companions in the service of the people of Palenque (Fig. 7:23).

In an ironic twist of fate, Pakal's second son, K'an-Hok'-Chitam,[51] shared the same destiny as his father's captives. K'an-Hok'-Chitam came to the throne of Palenque late in life, at the age of fifty-seven, following the death of his older brother, the formidable Chan-Bahlam. Chan-Bahlam had raised some of the most extraordinary monuments in the city, including the magnificent temple Acropolis known as the Group of

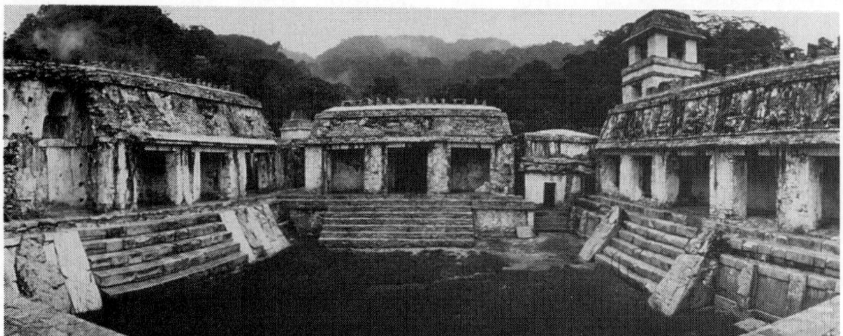

House C and the Six-Stair-Place with its hieroglyphic inscription and six captives is on the right; House B is in the center; and House D with nine additional captives is on the left.

FIGURE 7:23

the Cross. K'an-Hok'-Chitam worked vigorously to preserve the residual power left in his brother's monumental Acropolis. He closed it off with Temple 14 and prepared a magnificent new facade, commemorating his accession to the throne, for the northern face of the Palace, the seat of his dynasty. When everything was completed, and all but the final words of the exquisite glyphic hymn were placed above the image of his throne at the apex of this monument, K'an-Hok'-Chitam came into conflict[52] with Tonina, a rival capital fifty kilometers south of Palenque. With the death of captives taken there, he would be able to finish the text and seal his accession.

Standing before this panel in the museum at Palenque, in the dying light of Christmas Eve, I recounted to my companions the disaster that found K'an-Hok'-Chitam. Instead of taking his enemy, he himself had been taken. A steward of his family actually finished the panel K'an-Hok'-Chitam had mounted in the north side of the Palace. In the end it was the record of the birth and naming ceremony of this steward that surrounded the graceful image of the king seated on a throne he would never see again. I explained that we had known for some time that Tonina had stunned Palenque with this defeat. It was not every day that a king was captured; and when he was, the heart and soul were taken from a city in ways we of the modern world, with our duly elected officials, cannot even imagine. The artists of Tonina commemorated the capture with a carved conquest slab displaying the bound and bent K'an-Hok'-Chitam (Fig. 7:24). It does not take much imagination to guess what happened to him next.

The next morning, bright and early, we piled into our *combis,* the Mexican nickname for the old-style Volkswagen vans, and headed south through the beautiful lush green corridor valleys of Chiapas, dodging the frequent rockfalls and the holes made in the narrow roadbed by landslides. Four hours later our vans bounced into view of Tonina, a mountainside covered with temples and palaces. The city is carved and shaped into a single huge pyramid with massive stepped terraces and faces south into the winter sunlight. When K'an-Hok'-Chitam had seen it over a millennium ago, it had been a bright red and white slash set against the lush green of the surrounding valley and the sacred mountain upon which it was built. I lectured my way up the mountain citadel, my voice squeaky and hoarse from a case of tropical flu and days of nonstop talking at Palenque. We meandered past the ballcourts where K'an-Hok'-Chitam would have faced the humiliation of mock play and final defeat in death,

K'an-Hok'-Chitam bound as a captive at Tonina

FIGURE 7:24 Tonina Panel

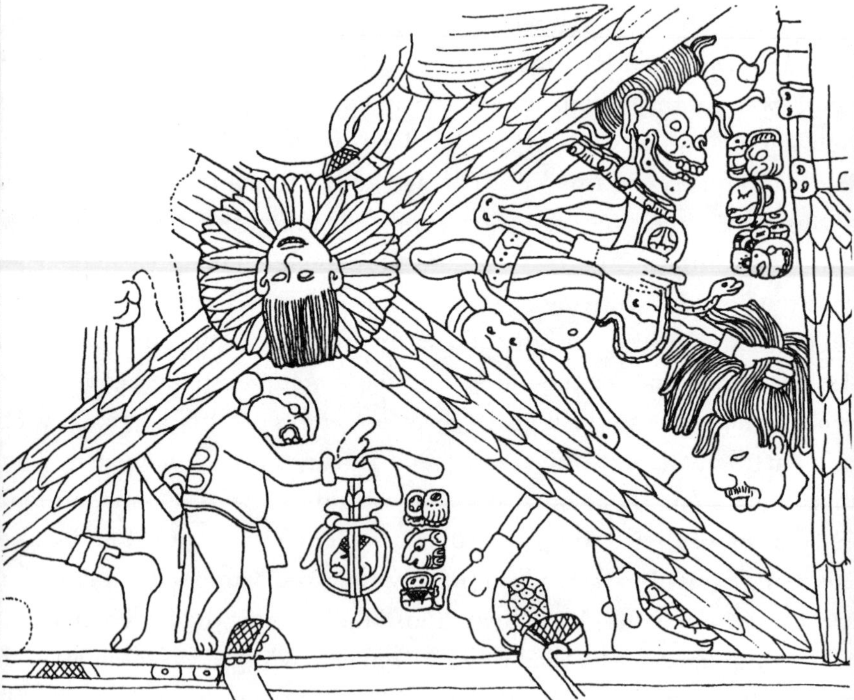

The Tonina stucco wall drawn after a postcard published in México

FIGURE 7:25 The Stucco Wall from Tonina

past the massive stone images of the kings of Tonina, standing on their terraces, looking out over the fertile valley below. Suddenly I heard Gillett Griffin yelling at me from a higher terrace, telling me to stop talking and get up there quick.

I struggled up the steep terrace and found myself standing before a freshly excavated and miraculously well-preserved fragment of a once-huge frieze, uncovered only three weeks earlier by Mexican archaeologists. I stood transfixed before an enormous dancing skeleton, the very being who was the *way* of the kings of Palenque and the constant participant in the sacrificial dance of Chak and the Baby Jaguar at other Maya sites. Cavorting wildly, this huge death god swung the severed head of an enemy by its hair (Fig. 7:25). Next to this image were three identifying glyphs. Although no epigrapher, I still recognized that this title belonged not to a god, but to a human king manifesting himself as a *way*.

My voice returned momentarily and I yelled out what I thought was the reading of the glyphs.[53] As my gaze shifted to the severed head, the only human face on this part of the facade, my skin prickled with recognition. This was not a conventional head, but the detailed portrait of a lord, complete with the high sloping forehead of nobility and a graceful aquiline nose—a king, right down to a fine wispy mustache etched with tiny striations into the wet plaster. The plaster was deeply modeled across the high cheekbones, and the eyes stared at empty space. The slack jaw and extruding tongue were as they must have been at the moment following decapitation. I knew I had seen this man before. Could this be K'an-Hok'-Chitam of Palenque, the moment after he had met his fate?

It ended up that it wasn't the head of the dead king. When I got home, I called Linda and sent her drawings we had made of the glyphs. The glyphs next to the severed head identified him as a lord from Pia, a subregion of Pomona, a kingdom that lay between Palenque and the Usumacinta River. This great plaster scene shows not the fate of K'an-Hok'-Chitam but that of a lord of his nearest rival kingdom.

The real significance of the facade, however, was not the identity of the captive, but the contents of the scene. The huge skeletal dancer was not the only mythological figure represented. Behind him strides a short, chubby rat called K'an-Ba-Ch'o-Xaman, "Yellow (or Precious) First Rat of the North." Linda's guess was that he's the rat who showed the Hero Twins where their father had hidden their ballgame equipment. And sure enough, in a second panel I saw the image of Hun-Ahaw, one of the Hero Twins of the Classic period, his shoulders resting on a bone throne, his legs in the air like an acrobat doing a bizarre dance. There were also images of the great bird, Itzam-Yeh, captives, and other figures on the unexposed portion of the facade.

These bizarre creatures dance within a scaffold frame decorated with feathers and shrunken heads. I had seen quite a few scenes showing this kind of scaffold. Many Maya lords exited to the Otherworld tied to it, and many others sat within it to experience the vision trances heightened by these bloody sacrifices.[54] At the center of this scaffold was a great battle standard, and at the center of its feathered medallion I saw a skull, its tassels marked with the sign of death. The spirit of this Tonina war standard was none other than the Death God himself. I realized later that here was the identity of the skeletal god who dances with the head of the sacrificial victim.

The beings on this facade usually show up in painted scenes on Maya vases that depict the activities of *wayob* and other denizens of the supernatural. By performing here with the war banner and scaffold frame, One-Ahaw locks the sacrificial rituals of war into the great Creation mythology we detailed in Chapters 2 and 6. We think it likely that noble victims like K'an-Hok'-Chitam and the lord from Pomona died in front of this facade, perhaps on this type of scaffold. This scene links war and the sacrifice of captives to the same mythology and cosmology that guided other Maya affairs of state. The same connection can be seen

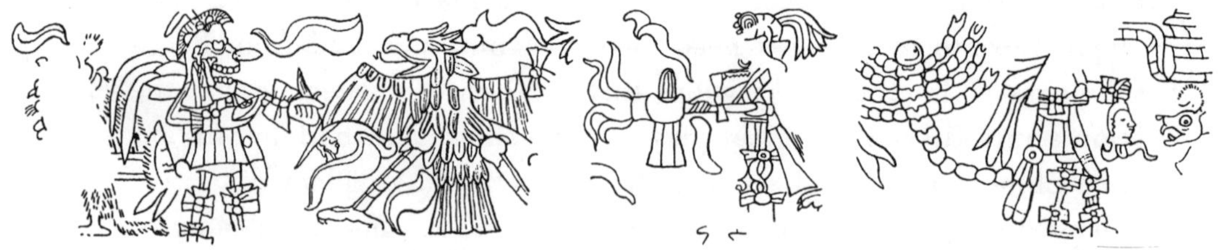

Chich'en Itza, Temple of the Warriors, fragments of the exterior mural on the north

FIGURE 7:26

between the Skeletal Wayob, the images of war, and perhaps signs of zodiacal constellations on the Temple of the Warriors at Chich'en Itza (Fig. 7:26). The same feathered scaffold, Skeletal Wayob, and other supernatural creatures show up on two pots in our color section. The stakes of war may have been control of resources, population, and power, but the Maya did not think of it in that way. To them it was a holy obligation, tied to the beings who created the world. Most of all, it was a battle between the spirit companions of the people who fought it. Victory not only secured wealth and power for the winners, it demonstrated to all that the gods were on their side.

A CHANGE IN THE RULES OF WAR AND ITS CONSEQUENCES

The ability of the Maya kings to call upon and embody powerful spiritual beings did not begin with the appearance of the spearthrower-owl and battle standards during the war between Waxaktun and Tikal. The collaboration of people and gods in Maya warfare likely began a thousand years before the time of K'an-Hok'-Chitam; and it has prevailed into the twentieth century, as we shall shortly see. Nor was the violence of conquest warfare between the great urban centers responsible for the downfall of Maya civilization. Evidence from the ruined ninth-century battlefields, redoubts, and palisades of the great collapsing kingdom of Dos Pilas[55]—and many other Postclassic sites—shows that it was the fading of the great royal cities and their replacement by the violence of town against town and lineage against lineage that savaged the Maya back into a simpler society.

The magical nature of warfare did not change with the fourth-century inception of Venus-Tlaloc war and the victory of Tikal over Waxaktun, but its specific manifestations and political objectives did. Before the fourth century, we believe, warfare was not the wholesale slaughter that it later became. It was fought by rigidly observed codes—the intent not to destroy neighboring kingdoms but to take captives for sacrifice. Long before the Teotihuacan allies of Tikal arrived on the scene, the Maya War Jaguars had prowled for victims on the open savannas of the lowlands.[56] It was the changing of the rules of conduct and the intent of war associated with the new imagery that eventually led to the downfall of

Maya kingship. We believe these new rules opened the way to a "Warring States" phase of lowland society that lasted unabated for five centuries.[57] At the outset, however, the Maya kings who watched the new kind of war in Peten and who embraced the flint-shield, the War Tlaloc, and Waxaklahun-Ubah-Kan as their own might have regarded this change in tactics as rational. They could not have realized that they were speeding their own eventual decline.

We suspect that the harnessing of Teotihuacan military statecraft to that of the Maya unleashed a slow-moving, ever-widening cycle of conflict and destruction. From what we can discern in the painted halls and stone friezes of Teotihuacan, their version of the Waxaklahun-Ubah-Kan served the government[58] rather than particular individual lords or families. In the case of the Maya, this great symbol became identified with mighty and ambitious individuals. The Tikal king Great-Jaguar-Paw, who first used this monster and its flint-shield talisman, took the spear-thrower-shield *tok'-pakal* emblem as his personal title.

The grafting of the symbol of an imperialistic state onto the war practices of individual Maya kings evidently escalated the feuds between their ancient dynasties into a war encompassing all the great kingdoms, a world war—at least in their terms. The ninth-century collapse of the southern lowland Classic kingdoms was preceded by a delicate balance of power that had lasted for hundreds of years. During that time the many kingdoms and dynasties had woven themselves into a treacherous web of marriages and alliances. These grand alliances bought brief respites of support, security, and peace at a very heavy price—the perpetual obligation to engage the enemies of enemies in war. They lived by the tragic old adage we have seen played out in our own world time and again: the enemy of my enemy is my friend. Or perhaps the reciprocal is even more to the point: the friend of my enemy is my enemy. It was not the success of war but the failure of peace that drove the Maya elite to chaos.

THE BATTLE BEASTS AT CHICH'EN ITZA

As warfare and rivalry became endemic in the eighth and ninth centuries, kingdom after kingdom shut down. Central authority collapsed, and in many places, lineages fought each other for the bones of power. Kingdoms fragmented and the people quit building the temple-mountains that had focused the power of the sacred world on their land. Many walked away and many more died in the chaos. There were some cities and regions, however, that survived this time of upheaval.[59] Chich'en Itza was one of them. The lords of this city ruled not as subjects of a divine dynasty of kings but in council. The gods of Chich'en Itza, like those of Teotihuacan, served the state and not any single individual. One such god was K'uk'ulkan, the Feathered Serpent. Although the native Maya chroniclers of the sixteenth century identified this being as a man, a Toltec, and a king, K'uk'ulkan was never mentioned as a man in the contemporary ninth-century texts of Chich'en Itza, nor is any such individual singled out by the artists of the Great North Platform buildings that crown the city in its later history. The images of serpents, scrolled, feathered, and plain, are indeed found there, painted in murals and carved in stone reliefs, but they entwine noble individuals in the manner of the ancient Vision Serpents of the Classic Maya. Like the War Serpents of the southern Maya and Teotihuacan, the Feathered Serpents of Chich'en Itza are also battle beasts. The Feathered Serpent images of K'uk'ulkan are transformations of the earlier Waxaklahun-Ubah-Kan War Snake.[60]

Chich'en's serpent battle beast is also accompanied by a great feather-fringed standard. The nickname for this standard, displayed on many of the city's relief sculptures, is Captain Sun Disk. At the center of this disk is an anonymous lord seated upon a jaguar throne, carrying both a spearthrower and bunched throwing spears—two of the three elements from the original Teotihuacan metaphor for war. This disk is decorated with a fringe of triangular spikes symbolizing the sun and resembling the fringe of feathers surrounding the *tok'-pakal* battle standard of the southern kingdoms. In a battle scene from the Upper Temple of the Jaguars (Fig. 7:27), this disk is depicted literally on the business end of an enormous battle standard, accompanied by many other types of standards. Although this Sun Disk could also represent an ancestor cartouche or a mirror of prophecy, here we see it as the *tok'-pakal* battle standard of Chich'en Itza.

This Captain Sun Disk standard presides over the many battle scenes portrayed in the Upper Temple of the Jaguars. The most spectacular of these scenes shows a battle raging in a village, while up in the sky float noble warriors engaged in mystical combat, their serpent battle beasts emerging from the portals of the Otherworld (Fig. 7:28). Above them all, Captain Sun Disk presides at the center of heaven. The lords of Chich'en Itza may have created a revolutionary government rule by council, but

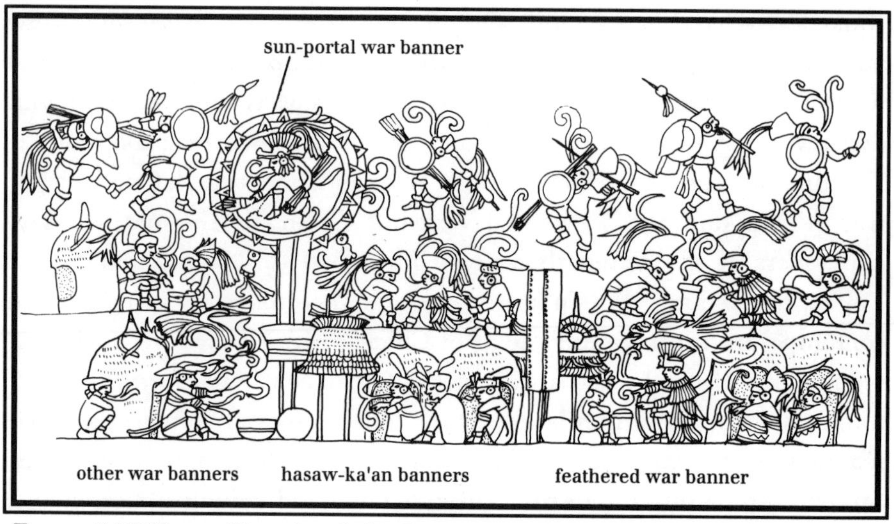

FIGURE 7:27 Upper Temple of the Jaguars, Mural from the Northwest Wall

they still perpetuated the supernatural dimensions of war conjured by Maya and Teotihuacano alike since the beginning of their world.

Escaping from the devastating circle of family feud and military conquest that drove down their southern lowland counterparts, Chich'en Itza and its lords succeeded, for a time, in establishing imperial order in the northern lowlands. They kept the *tok'-pakal* and the Waxaklahun-

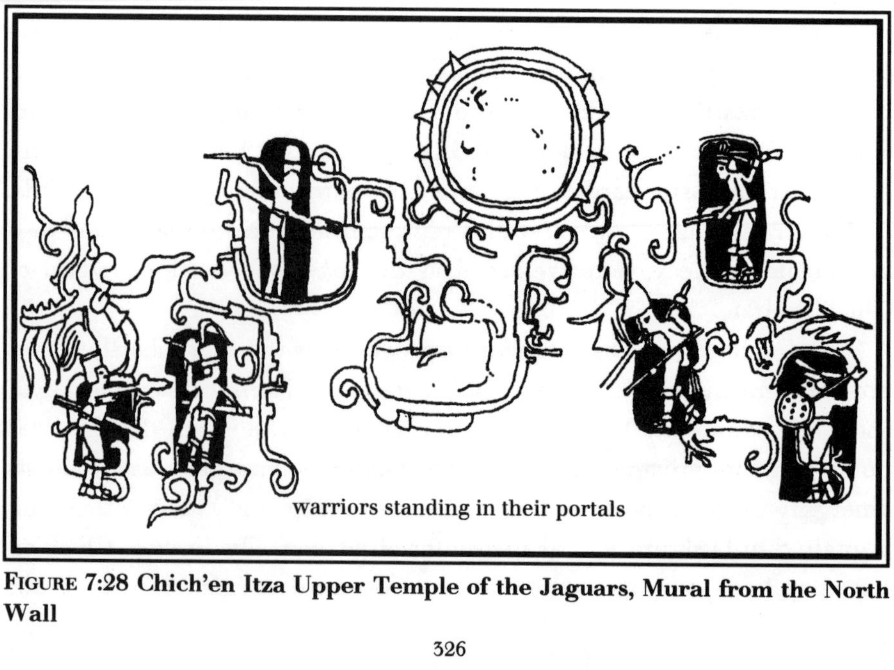

FIGURE 7:28 Chich'en Itza Upper Temple of the Jaguars, Mural from the North Wall

Ubah-Kan as symbols of expansion, but disengaged their meaning from the kingship. Apparently, military power defined in the person and conduct of the divine ruler could not peacefully coexist with the principle of military power vested in gods who embodied the state.

WAR AGAINST THE SPANISH

The powers of the natural and the supernatural worlds continued to be conjoined in Maya warfare long after the demise of the Classic period.[61] In the war between the Maya and the Europeans in the sixteenth century, gods manifested on several battlefields. Although the Spanish accounts of these battles are often mundane, the histories written by the Maya record not only Maya gods but Christian saints and angels. At the battle of Cintla,[62] which took place in the Lagunas de los Términos region of the Gulf Coast, Maya warriors lured Cortés and the Spanish into the ditches of their fields and fired at them from the safety of surrounding stone walls. Defeat seemed imminent until Santiago, the patron saint of Spain, appeared on his great stallion to lead the mired horses of the Spanish cavalry to high and firm ground where they could wheel on the Maya and defeat them. The legend has it that the same Santiago appeared at the Battle of Otumba and saved Cortés from attacking Aztecs.

In their Dances of the Conquest, the K'iche' Maya of highland Guatemala still memorialize the great defeat of their war hero, Tekum Uman, by Pedro de Alvarado, the captain of the Spanish forces.[63] The Spanish account describes the battle in purely material terms. On December 5, 1523, Cortés ordered Alvarado to southern México to suppress a rebellion and begin the conquest of Guatemala. Alvarado took with him 120 horsemen, 300 foot soldiers, 200 Tlaxcalans, and 100 loyal Mexica. This company of 720 men were expected to face 3,000 or 4,000 K'iche'. The battle took place near the town of Xelahuh, today called Quetzaltenango by Ladinos and Xelah by the Maya.

According to Alvarado, this was just another battle among many. In a letter to Cortés, he said that several thousand K'iche' warriors approached his troops while they were taking a break for food and water. They let the Indians close the distance. Then they attacked and routed the Indian army, pursuing them until they were trapped against a mountain. To draw them out, Alvarado's men pretended to flee on their horses and then turned, rallied, and defeated the assembled warriors. He mentioned that one of the K'iche' chiefs was killed, but he did not even record his name.

The K'iche' account[64] is told as if a totally different series of events had unfolded. Their story begins with the entry of Tekum Uman into the town of Xelahuh with 8,400 warriors, including 39 flagbearers and drummers. The warriors prepared themselves for battle with a bloodletting ritual.[65] Tekum Uman was called the Lord of Banners and Staffs. His banner, according to the chronicles, was decorated with gold on the tip and many emeralds (or, more likely, jade). This is clearly the battle standard of the Classic period, rich with the same flashing decoration as its Aztec counterpart. Each Maya lord brought with him 10,000 warriors armed with bows and arrows, slings, and lances, as well as other arms. There were so many warriors they could not be counted.

When his host was assembled, Tekum Uman transformed himself before them. He put on "wings with which he flew and his two arms and legs were covered with feathers and he wore a crown and on his chest he wore a very large emerald [jade?] which looked like a mirror, and he wore another on his forehead. This captain flew like an eagle, he was a great nobleman and a great sorcerer." Tekum Uman had transformed into his *way* and he went to battle as a sorcerer against the magic of the Spaniards who fought in their *wayob*—Santiago and, as we shall see, the Virgin Mary.

The battle began with a skirmish when a chief, "Ah Xepach, an Indian captain who became an eagle," went to fight the Spaniards with three thousand of his soldiers. "At midnight the Indians went and the captain of the Indians who had transformed himself into an eagle became anxious to kill the Adelantado Tunadiú [Alvarado][66] and he could not kill him because a very fair maiden defended him; they were anxious to enter, but as soon as they saw this maiden they fell to the earth and they could not get up from the ground, and then came many footless birds, and those birds had surrounded the maiden, and the Indians wanted to kill the maiden and those footless birds defended her and blinded them." The attackers were paralyzed and blinded by the *way* of the Spanish—the Virgin Mary and the Holy Spirit or perhaps angels who looked to them like footless birds.

The Indians fell back and yet another chief, one who had become lightning, went against Alvarado. "And as soon as he arrived, he saw an exceedingly white dove above all the Spaniards, which was defending them, and which returned to repeat it again and it blinded him and he fell to the earth and could not get up." Three times the lightning warrior went against the Spaniards, and then he too retreated to tell the king that only by killing Alvarado could they win.

Alvarado and his Tlaxcalans charged and routed the Indians before him. After taking thousands captive, and killing and torturing many of them in their search for gold and treasure, the Spanish prepared to go deeper into Maya territory. The next day, February 22, 1524, and 1 Q'anel[67] in the Maya calendar, Tekum Uman himself came against the Spanish in his eagle *way*. "And then Captain Tekum flew up, he came like an eagle full of real feathers, which were not artificial; he wore wings which also sprang from his body and he wore three crowns, one was of gold, another of pearls and another of diamonds and emeralds." Tekum Uman went forward with the intention of killing Alvarado and thus defeating the battle beasts and the *way* of the Spanish. He struck at the great man-beast with all his power, hitting Alvarado's horse and taking its head off in a single blow. According to the K'iche, his lance was not made of metal, but of shiny stone which had a magic spell on it. When Tekum realized he had killed only the battle beast and not the man, he flew upward and came at Alvarado. The Spaniard was ready and impaled the charging king on his lance.

The K'iche', seeing that their king had died on the lance of Alvarado, fled, only to be pursued and slaughtered by the Spanish. So many of the Indians died, their blood made a river. Called Olintepec by the Spanish, this river was named *Q'iq'el*, "blood," by the Indians. The battle of Xelahuh pitted Tekum Uman and his companions who fought in their *way* as eagles and lightning against the Spanish supernaturals—a floating maiden (the Virgin Mary), footless birds (angels), and the Dove of Peace (the Holy Spirit). There is good reason to believe that the Spanish carried their own litters and battle banners into the fray.[68] In the summer of 1992, Federico Fahsen called our attention to a very old statue of the Virgin famous in Guatemala. On the day of Santiago, July 25, a procession brings the statue from the cathedral in Guatemala City to the ruined cathedral in Antigua, the capital destroyed by earthquake in the eighteenth century. Tradition has it that this is the Virgin Alvarado carried into battle with him.

The battle of Xelahuh was fought between worlds and the gods who ruled them. As Victoria Bricker (1981:40) said, "The Indians lost the battle because their magical arsenal was no match against the spiritual arsenal of the Spaniards." In the eyes of the Maya who survived the battle and suffered the rule of the victorious Spanish, the eagles and lightning beasts of their ancestors were grounded, blinded, and immobilized by the battle beasts of the aliens who had invaded their world.

As horrible as it was, however, the conquest of Christian divinities over the old Maya gods was only temporary. Just as the Waxaklahun-Ubah-Kan of Teotihuacan became a Maya god in the fourth century and joined the other battle beasts of those ancient kings, so the Christian supernaturals, the combatants of the terrible battle of Xelahuh, have joined with Maya gods in the modern Otherworld. Tukum Uman became the principal Mundo, one of the Earth Gods that protect the K'iche' people today. He lives, they say, in the cave under the ancient, now-destroyed capital of Kamaar Kah. But the war god of the Spanish also became a Maya god.

In 1906 the K'iche' of Momostenango proved their valor and that of their war god in combat against Salvadoran troops. Led by a famous Ladino military hero from Momostenango, two Indian regiments placed in the front lines broke the Salvadorans sent against them. Priest-shamans performed rituals before and during the battle to protect the K'iche' soldiers from the enemies' bullets. Santiago, the very being who, four hundred years earlier, had appeared on his mighty horse to save Cortés from the Mexica now flew over the heads of the K'iche' wielding his sword in their defense. In an incredible ironic turnabout, we see the patron saint of Spain—brought by the Spanish, albeit unknowingly, to vanquish the battle beasts of the New World—transformed into a battle spirit defending their K'iche' victims.

This Santiago, however, now the patron of Momostenango, no longer speaks Spanish. He speaks only Maya and requires another saint to translate for him into Spanish. A fierce and wrathful being who punishes those who fail him in their duties, he is regarded more as a powerful object vested with irresistible force than as a being with personality. His Maya name means Venus Morningstar,[69] the death star that manifested war for the Maya centuries before the coming of the Christian saint presiding in his church. While the flint-shields of the Maya lie broken or buried in their sacred temples in the forest, the undying beasts of battle follow new talismans and new Maya heroes who wield them in our world today.

The spiritual beings of Maya combat never discouraged their human counterparts from undertaking careful material preparations. Training, logistics, tactics, and strategy served the Maya well. They were very good at war, peasant militia as well as noble officers; and while they fought fiercely among themselves, we don't believe they were ever conquered by non-Maya peoples before the arrival of the Europeans with their horses, guns, and devastating diseases.[70] It is no accident that it took the Spanish the better part of forty years to establish a firm foothold on the peninsula. Ever since, in periodic, armed rebellions, the Maya have resisted the domination of the Spanish conquerors and their Ladino descendants. One of the most extraordinary examples of this resistance is found in Chamula's Festival of Games.

THE BATTLE BANNERS OF CHAMULA

(as told by Linda Schele)

The most dramatic part of the *K'in Tahimol* that I witnessed with Duncan Earle in early March of 1992 was the running of the banners. In between the Dance of the Warriors and the various feasts, when the ritual moves from one place in Chamula to another, the Pasión, the Flower, and the Monkeys of all three barrios—accompanied by other honored office-holders—run to the town square. They form into a huge moving mass of men subdivided by both brazier carriers and barrio-affiliation into three distinct groups—San Juan first, San Sebastián in the center, and San Pedro last (Fig. 7:29). Running counterclockwise, they circle the square, stopping at each cross station to honor the Sun-Christ and to dip the spearheads on their banners to each other, front row to center and center to back. Then amid rising clouds of incense and the cacophony of blaring trumpets, they run to the next station. Around and around and around they run, until on the third pass, they pause at the entrance to the atrium, the walled-in square in front of the church (Fig. 9:3).

Tension builds up around the milling crowd until, with sudden movement, they reverse direction, run back to the cabildo where the civil officials await them on the roof. They stop in front of the building housing the civil government, touch the heads of their banners, and then turn back again in the proper counterclockwise direction. When they reach the southeast corner of the plaza, they stop and lay the banners on the top of the wall surrounding the atrium. Special officials adept in these matters run madly for the gate into the atrium, carrying their smoking braziers with them, knocking aside any gringo too ignorant to get out of the way. When they arrive at the banners, they feed the Sun-Christ by waving the smoking braziers under the clustered heads of each group of banners. Smoke is the food of the Sun-Christ as it was for the gods and

running the Sun-Christ banners at San Juan Chamula

(photograph © Duncan Earle)

FIGURE 7:29

ancestors of ancient times. When the feeding is done, the "Spanish Letter" is read.

Duncan Earle and I were standing in this particular corner during one of the circuits and heard the letter read. I had studied Gary Gossen's (1986:238) and Victoria Bricker's analysis of this document and knew what it represented, although I couldn't understand the words. It is a remembrance of all of the invasions, wars, rebellions, and battles that have peppered Chamulan history since the Conquest—a generic commemoration of war and a reminder of, if not a protest against, the plague of foreign intervention. The war association of the banners and the Festival of Games itself fascinated me. I knew these banners symbolized the Sun-Christ, but they were also the modern descendants of the ancient war banners.

I knew the explanation given for the parts of the banners by Gossen and others. The shaft is the skeleton, the flowery flag is the body, the red and green ribbons are the radiance, and the silver tips are the head of the

Sun-Christ.[71] Each barrio has many flags, but only four, representing the four positions of the Sun-Christ at the rising, zenith, setting, and nadir positions of the sun's circuit, have silver tips. When the runners carry these twelve banners and their lesser companions around the square, they are symbolizing, as Gossen puts it, the totality of the sun.

When I first saw the flying banners I was surprised to find that there were two kinds of tip rather than one. Eight of them carried a single spear tip, which Vicky Bricker has suggested represents the spear that pierced the Sun-Christ's side. When I saw these spear tips for the first time, I made a different connection. While the shape of the silver tip may look like a Roman spearhead, it is also exactly the same leaf shape that blades of the Classic-period Maya spears carried.

The other type had three prongs instead of one (Fig. 7:30). I asked Duncan why there were two types and what they meant. He didn't know, but he asked the nearest Chamula. It turned out that only the banners of San Juan, the senior barrio, carried the trilobed tips. Neither the man he asked nor anyone else we questioned knew the meaning of the three prongs—only that they went on the banners of San Juan. When I later asked Gary Gossen what these tips were, he didn't know either.

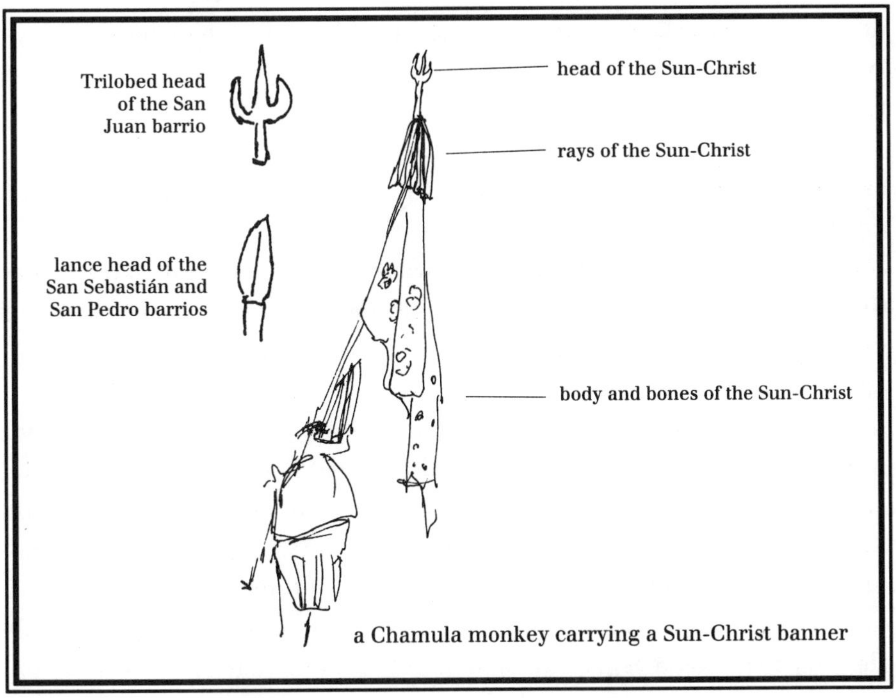

FIGURE 7:30 The Lance Heads of Chamula

When I saw that strange shape it immediately resonated with forms I knew from ancient art. I had seen the three-pronged shape carried by dancers and sacrificers in war and sacrifice scenes carved during the Classic period (Fig. 6:9). I was struck by how much the trilobed tips looked like those ancient sacrificial instruments. But I also remembered the war scene from Bonampak' (Fig. 7:11). The waving flag with its flowers fluttering below the spearhead and its cluster of ribbons reminded me of the spear-tipped standards carried by the soldiers in the mural as they charge toward the enemy. I realized that the Chamula banners were the direct descendants of the ancient war standards, just as the Santiago of the Momostekans was the child of the ancient Tlaloc-Venus war.

WAR AND THE COMMON PEOPLE

The Maya nobility were not the only ones to fight wars. If we can believe the early colonial descriptions of Classic practices, they tell us that commoners also engaged in battle and that warfare also had practical goals, such as the appropriation of arable land and strategic tribute resources. The ritual dimension of Maya war, however, was founded clearly upon two important assumptions. First, war was a sacred activity defined by rules that bound all participants.[72] Second, the needs of Maya ritual life were what gave war meaning. All of the important rituals that took place in the great urban centers required the death of a sacrificial victim. But the death of these valiant war captives was also a celebration of life born out of death, as well as the triumph of humanity over the dreadful diseases and misfortunes that threatened their existence.

Maya peasants no doubt suffered during wars, as they still do today; but they usually managed to endure. David describes how he learned this very lesson while working at the site of Yaxuna in 1986.

I asked Don Emetario to take me and one of my students, Rani Alexander, to see Hacienda Ketelak, the ruined house of the upper-caste Ladino who had owned the land around the site before the Revolution and land reform. The courtyard of the modest two-story masonry structure still contains several tall imperial palms and some decorative carved stone elements on its archways, tawdry remains of rustic elegance. While clearing away some vegetation and debris from the hacienda so that Rani could take pictures,[73] Don Emetario answered our questions about the family who had lived there in the old days. I sensed a certain grim satisfaction about him as he sat in the midst of these ruins, thinking about how rich outsiders, like the ones who had lived here, had been killed at the hands of Chan Santa Cruz Maya insurgents during the Caste War. He told us how the rebel Maya would send messengers into his village before an attack. The people would escape into the woods, the rebels would burn the village, and then the farmers would return and rebuild. As long as they could grow their crops, they could survive, he said with a proud smile. When war destroyed the ability of the farmers to grow food, then they had to leave the area altogether.

While Don Emetario's people stayed near their homes during the nineteenth-century wars, thousands of other Maya farmers, caught in episodes of extensive destruction perpetrated as much by fearful upper-caste Yukatekans as by Maya insurgents, fled into the forests of Quintana Roo or southward into Belize. The "peaceful ones" have settled back into the landscape. David Freidel worked with the descendants of such farmers at Cerros in Belize. In their village, Chinux, they still have a small green sacred cross studded with metal talismans that they brought with them from the north when they fled. In Chinux, only the old people speak Maya anymore. On a day some years ago, the young workers of the village were deeply impressed when the epigrapher Barbara MacLeod showed up to see Cerros and broke into fluent Yukatek. They couldn't respond, but they smiled shyly and declared that their grandparents could still speak this language.

The story is not the same with the rebel Maya, the followers of the *Chan Santa Cruz*, the "Little Holy Cross" of Quintana Roo. Their descendants speak Yukatek primarily and still think about war today. Linda Schele found this out when she went to visit the town of Señor with Nikolai Grube. He has spoken often of the Cruzob preoccupation with war and the imminent end of the world, but she hadn't understood how intense their obsession was until she went there. Every time they stopped to talk to the men of the town, the conversation eventually turned to the end of the world the Cruzob anticipate in the year 2000 or soon thereafter. Nikolai says they have interrogated him repeatedly about the wars and violence around the world they see graphically displayed in news programs. They have televisions in the town, although most of them cannot understand the Spanish commentary, and they watch and wait for the great war in which all societies will fight one another. It will be so bad, they believe, that all the machines will be destroyed and all the armies will be reduced to fighting with machetes and sticks. That is a battle they know they will win. And they will win, they believe, because they alone have kept the faith and listened to the talking crosses. After this victory, they say, a Maya king will rule the land once again.

CHAPTER EIGHT

GAMING WITH THE GODS: Destiny and History in Maya Thought

A CHILD'S BALLGAME

(as told by Linda Schele)

The small whitewashed chapel seemed strangely out of place amid the cracked and tumbled walls of the ruined church and monastery that not so many years ago had housed the public market of the old capital of Guatemala. When Antigua had built its new modern market, the little chapel where the stall-keepers had made their prayers became a laboratory for CIRMA,[1] an organization that supports archaeological, ethnological, and linguistic research of the Precolumbian and colonial history of Guatemala. That day in 1989, the little chapel was hosting a workshop I was giving to forty Maya Indians from the highland communities. Between them, they spoke at least eleven of the modern Mayan languages as their native tongues. That workshop[2] on Maya hieroglyphic writing was one of three that I have given to my Maya friends. In a magnificent irony I did not learn until the following year, these meetings took place in the ruins of the very monastery where the aged Bernal Díaz, one of Cortés's foot soldiers in the conquest of the Aztecs, wrote his personal account of the collision of two worlds in *The Conquest of New Spain.*[3]

We were working on the decipherment of the Tablet of the 96 Glyphs of Palenque, in my own opinion, the most beautiful inscription the Maya ever carved and one of the greatest pieces of literature they ever created. I had copied my drawing of the tablet onto plastic film, and was cutting it up and placing the individual glyphs into a grid lying on the bed of an overhead projector. Cast onto a dry white board that served as our projection screen, the image served as a guide to the Maya, who followed me glyph by glyph, pasting their cutouts onto large sheets of poster paper laid out in the same grid pattern—a column for dates and distance numbers, columns for verbs and verbal complements, and a final one for the actors. I wrote Spanish and Classic Mayan translations of each glyph on the dry white board and they debated the best translations of these expressions into their own languages and wrote them on their papers. While there were whole groups of K'iche', Kaqchikel, and Q'eqchi working together, only single speakers of the other languages, including Chorti, Q'anhob'al, Chuh, Mam, Ixil, Achi, Poqoman, and Jakaltek were present. The participants worked together in groups of related languages, helping each other as best they could.

I went through the first half of the tablet, not only giving them the translations where we knew them but also explaining how we had come to decipher these glyphs in the first place. Then I came to the name of a king whom early Maya epigraphers had dubbed Chaakal, but whose original name was probably Akul-Ah-Nab.[4] He was the father of K'uk'-Balam, the man who had commissioned this beautiful tablet. I went through Akul-Ah-Nab's personal name, glyph by glyph, until I came to one of his titles *Ah Pitzlawal* (Fig. 8:1). I explained how we had discovered that the main part of this particular glyph often accompanies ballgame scenes, and that its two parts read *pi* and *tzi*.[5] Taken together, this gives us the word *pitz*, which in Yukatek means "play ball." I told them about another recent workshop in Cleveland, Ohio, where Terry Kaufman, a linguist with wide knowledge and experience of Mayan languages, had told us more about this title. The second part of the expression, *-lawal*, was formed in a Chol grammatical process in which

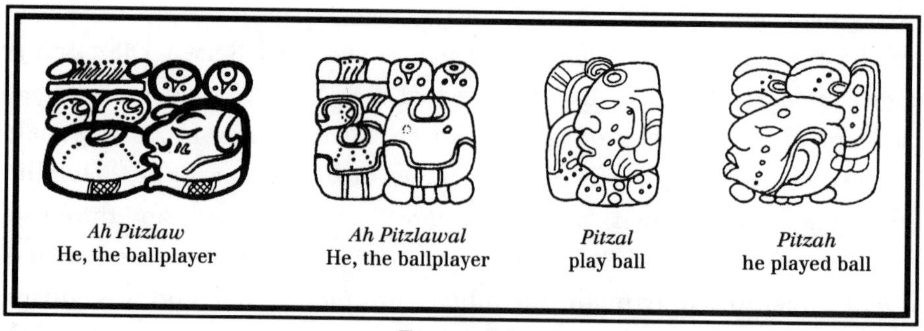

FIGURE 8:1

a verb is derived from a noun and then rederived as a noun. It would be as if we took the English noun "ball," derived a verb like "ballplay" from it, and then rederived the noun "ballplayer."

I looked around the room to see if my Maya listeners understood this complicated argument, especially since it had been delivered in my personal brand of Spanish, learned informally in the field.[6] To my right I spotted a handsome young Q'anhob'al named Ruperto Montejo, who has taken the Maya name Saq Ch'en. He sat with a beatific grin on his face, his hand palmup, moving as if he were bouncing an imaginary ball. I stopped talking in mid-sentence as it dawned on me what he was indicating.

"You know this word?" I asked in excitement. "Do you have it in Q'anhob'al?"

His grin deepened as he explained that *pitz* referred to a children's game played with a grass ball.

"Jugador llamamos 'pitzlawom,' " he said with equal excitement.

"¿Qué me dices?" I asked swiftly, not sure I had understood the Spanish correctly.

"*Pitzlawom* is the word for 'ballplayer,' " he said in Spanish that I understood clearly. A feeling of elation expanded in my chest.

Here it was. A word we had only been able to confirm in Yukatek was still used by Q'anhob'al speakers—and they had the same derivational affix, *-law*, that we had in the glyphic version at Palenque. The *-om* ending of the Q'anhob'al word was something we were also familiar with, in glyphs like *ch'ahom*, "incenser," *kayom*, "fisherman," and *k'ayom*, "singer." *Om* is the suffix that derives instrumental nouns, like fisherman or ballplayer, from their underlying verb.

While I had been talking to Saq Ch'en, I heard laughter and exhilaration bubbling up behind me. Turning, I saw that the Q'eqchis had the same grin on their faces, the same look of excitement in their eyes, and were making the same ball-tossing gesture with their hands. They had this expression in their language too.[7] *Pitz* was their word for "ballplaying." Up to that time, we had not had found this word in those languages, perhaps because it did not describe the kinds of ballplaying Westerners were used to, like soccer or baseball or basketball. Or perhaps those who asked the Maya about their languages had never imagined that there was more than one kind of ballplaying. In the Maya world of today, *pitz* is a word that describes both the old forms of the game and the games children play with balls made from materials that they find at hand.

THE RULES

The game[8] called *pitz* that Saq Ch'en described is a survivor of a gaming tradition already ancient when his Maya ancestors were establishing their civilization twenty-one hundred years earlier. The Olmec, precursors of the Maya and creators of the template of Mesoamerican civilization, also left records of ballplayers. One of these, a portrait of a ballplayer carved around 900 B.C., was found at their ceremonial center of San Lorenzo.[9] This life-sized stone figure kneels in the posture of a player receiving the ball. The ballgame also flourished elsewhere in Preclassic Mesoamerica. At Dainzu, a small center in the highland valley of Oaxaca, devotees of the game carved strange helmeted ballplayers, their limbs askew, on stone slabs set into the side of a temple platform. These figures date from hundreds of years before the first Maya ballcourts were built far to the east in the lowland jungles.[10] Closer to home, portable ballcourt markers and open ballcourts were set up at the major Maya Preclassic city of Izapa located in a low pass of the Chiapas mountains between the modern states of Guatemala and México. And in the valley of Guatemala at Kaminaljuyu, thick-belted ballplayers[11] wearing animal headdresses played the southern highlands version of the game at the same time that towns and cities were emerging in the broad lowlands to the north.

While working at Cerros in the Maya lowlands, David Freidel and his colleagues Vernon Scarborough, Beverly Mitchum, and Sorraya Carr happened upon two of the earliest known formal ballcourts in the region. Initially, they thought they were excavating household mounds[12] in the community surrounding that Preclassic royal capital, but they found ballcourts instead. One of the ballcourts[13] had sloping playing surfaces that flanked a sunken playing field, and both had neat round holes lying in the centers of the well-preserved playing surfaces where the Maya had torn out the central markers.

The masons of this small kingdom constructed these two playing fields right in the midst of the homes of their lords and laborers. Both ballcourts were oriented on the north-south axis that defined the centerline of the community. One was placed at the southern end of this north-south axis and the other was near its center. Here, as at many other capitals, the Maya wanted the sacred power points of the community to lie within their living space, thereby integrating the inhabitants into a magical pattern set down by the shaman king and his associates. The ballcourt was a special place for the ritual enactments of the original ballgame of the Hero Twins. Thus it brought the ancestors and their living descendants together at the place of the original sacrifice and resurrection of the Maize God.

These early ballcourts (Fig. 8:2d) in the Maya lowlands already exhibited the form that the ballcourt would have throughout its history.[14] The playing field of the Maya ballgame was shaped like a capital I. Players bounced the ball off angled benches on either side of the narrow central alley (Fig. 8:2a). The pictorial glyph for "ballcourt" (Fig. 8:2b) depicts the end-on view of these benches. At some sites, one or both ends of the I-shape were enclosed (Fig. 8:2i), while at others they were entirely open. Round or square markers were erected in the center and at both ends of the central alley of most ballcourts. Many had additional markers set into the top of the benches or in the upper panels. At Copan, these markers took the shape of macaw heads, at Tonina they were captives, and in northern Yukatan they were rings with entwined Feathered Serpents carved in sculptural relief.

We are still not sure of the role these markers and benches played in the Maya formal ballcourts nor how the game was scored.[15] Descriptions of the Aztec game called *tlachtli*, however, bear resemblances to scenes of the Maya game that we find carved in relief sculptures on Classic temples and monuments. Durán, the Spanish chronicler who recorded so much of Aztec life, describes the game as follows:

> It was a game of much recreation to them and enjoyment specially for those who took it as pastime and entertainment, among which were some who played it with such dexterity and skill that they during one hour succeeded in not stopping the flight of the ball from one end to the other without missing a single hit with their buttocks, not being allowed to reach it with hands, nor feet, nor with the calf of their legs, nor with their arms.
>
> . . . and at the ends of the court they had a quantity of players on guard and to defend against the ball entering there, with the principal players in the middle to face the ball and the opponents. The game was played just as they fought, i. e., they battled in distinct units. In the center of this enclosure were placed two stones in the wall one opposite the other; these two (stones) had a hole in the center, which was encircled by an idol representing the god of the game. . . . That we may understand the purpose which these stones served it should be known that the stone on one side served that those of one party could drive the ball through the hole which

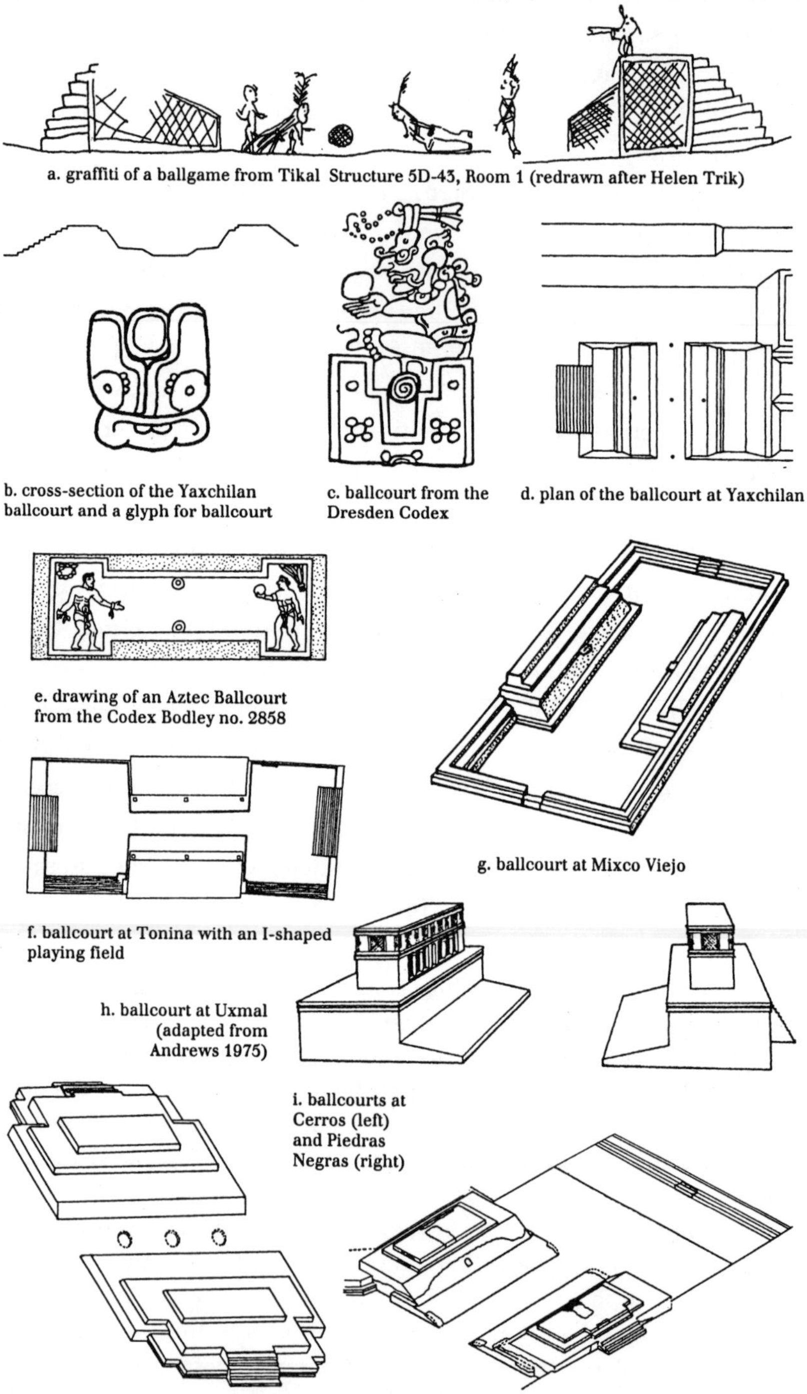

a. graffiti of a ballgame from Tikal Structure 5D-43, Room 1 (redrawn after Helen Trik)

b. cross-section of the Yaxchilan ballcourt and a glyph for ballcourt

c. ballcourt from the Dresden Codex

d. plan of the ballcourt at Yaxchilan

e. drawing of an Aztec Ballcourt from the Codex Bodley no. 2858

f. ballcourt at Tonina with an I-shaped playing field

g. ballcourt at Mixco Viejo

h. ballcourt at Uxmal (adapted from Andrews 1975)

i. ballcourts at Cerros (left) and Piedras Negras (right)

FIGURE 8:2 Ballcourt Forms and Layouts

> was in the stone; and the one on the other side served the other party and either of these who first drove his ball through (the hole in the stone) won the prize.
>
> . . . All those who entered this game, played with leathers placed over their loin-clouts and they always wore some trousers of deerskin to protect the thighs which they all the time were scraping against the ground. They wore gloves in order not to hurt their hands, as they continuously were steadying and supporting themselves on the ground.
>
> . . . They were so quick in that moment to hit with their knees or seats that they returned the ball with an extra ordinary velocity. With these thrusts they suffered great damage on the knees or on the thighs, with the result that those who for smartness often used them, got their haunches so mangled that they had those places cut with a small knife and extracted blood which the blows of the ball had gathered.[16]
>
> (Durán 1971)

Like the Aztec game witnessed by the Spanish, Maya imagery depicts players receiving the flying ball on their hips and sides. The rubber ball of the Maya, however, seems to have been considerably larger than the Aztec version. If its proportional relationship to human players was accurately represented in Maya art, it was larger than a basketball and did not go through a ring, except perhaps at Chich'en Itza and other northern Yukatekan sites.[17]

The hardness of the Maya ball and the danger of receiving it against unprotected flesh was apparently the same as in the Aztec game. A Maya ballplayer (Fig. 8:3) wore cotton padding around his pelvis, cotton padding and a heavy U-shaped protector called a yoke around his waist, thick cloth padding wrapped around his forearm, a single knee pad on one leg, and a calf-length leather skirt over his loincloth. Sometimes they held a small handstone,[18] used perhaps to put the ball into play. A stone object called a palma projected from the yoke in the center of their bodies. Most known yokes are made from stone, but one recovered from Burial 195 at Tikal was made of wood and cut in the typical grooved form shown in Maya imagery.[19] Ballplaying gear also included the headdresses and other symbols of important gods, indicating that players probably assumed the roles of cosmic beings, elevating their play to the level of a great cosmic drama.

As with American football players, the ancient Maya needed all this padding to protect themselves, not only from the impact of the ball but also from their own maneuvers.[20] Images of the game (Fig. 8:4) show them down on one knee, jumping off balance into the air to field the ball

(photograph © Justin Kerr)

FIGURE 8:3 The Ballgame in Play

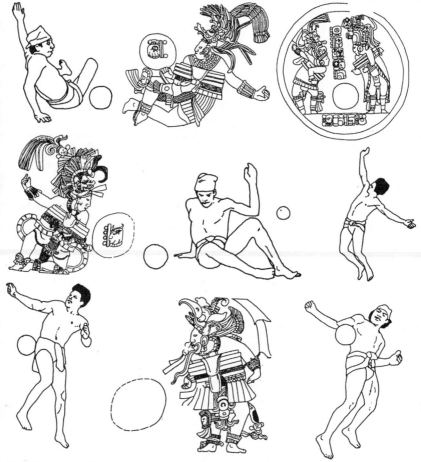

ballplayers from Site Q panels and Cancuen compared to player position in the modern ballgame played in Sinaloa (drawings of modern players after Leyenaar 1978)

FIGURE 8:4 Ballplayers in Various Positions

on the yoke, and throwing themselves under the ball to keep it from hitting the ground. Although we will never be sure of the rules, these depictions suggest that the ball couldn't hit the floor of the main alley—though it is often shown bouncing off the side benches or the stairs.

THE MYTH

While many details of the Maya ballgame remain elusive, we have learned a lot about its mythological and political meanings from Classic-period representations and from the myth of the Hero Twins in the Popol Vuh. We outlined the myth of Creation and the Hero Twins in Chapter 2, but to understand the ballgame, we must now look at that part of the myth in detail. As we told you before, the first generation of Twins, One-Hunahpu and Seven-Hunahpu, were defeated by One-Death and Seven-Death, the chief Lords of Xibalba. The Lords of Death sacrificed them and buried their bodies in the Place of Ballgame Sacrifice,[21] and hung One-Hunahpu's head in a gourd tree.

Blood, the daughter of one of the Lords, was impregnated by spittle from One-Hunahpuh's skull and gave birth to a second set of Twins who are the heroes of the Creation myth. As adults, they became great ballplayers because of an encounter with a mischievous rat who kept interfering with their magical attempts to clear their milpa, the Maya name for their cornfields. Irritated, the Twins caught him and burned off his tail fur. Desperate to get away, the rat offered them a secret. Intrigued, the Twins let the rat go. It led them to where their father and uncle had hidden their ballgame equipment in the rafters of their grandmother's house before they had left for Xibalba. The Twins made a deal with the rat—if he would climb up into the rafters and cut through the ropes that held the ballplaying gear, they would give him and his descendants the right to nibble at the corn and squash in the fields forever. Happy with this deal, the rat scurried up and freed the gear. The Twins donned their newfound equipment and went to the ballcourt to play.

Just as their father and uncle had before them, the boys disturbed the denizens of Xibalba with their noisy ballplaying. Seven days later the messenger owls of the Lords of Death arrived before to summon them before the Lords of Death. Finding the boys away from home, the messenger owls gave the summons to their grandmother, Xmukane. In turn, she sent the words of the gods to her grandsons via a louse who was swallowed by a toad who was swallowed by a snake who was swallowed by a falcon. When this curious array of messengers arrived, they vomited each other out, one after the other, and the Twins learned that they were to face the same trials as their father and uncle.

As the Xibalbans soon learned, these Twins were not the same soft touches as their predecessors. Hunahpu and Xbalanke were accomplished tricksters and were not fooled in the least by the traps laid for them. They detected and eluded every trick and outwitted the frustrated Lords. Outmaneuvered, the Lords challenged them to a ballgame but insisted they use the Lords' ball, the deadly sharp White-Flint-Knife (*Saqih-Tok'*). It was a razor-sharp, round piece of white flint surrounded by crushed bone. Their ball court was the Great Abyss at Karchah.[22]

After an argument, the Twins accepted the Xibalbans' ball but fielded it on their yokes, sending it twisting erratically all over the court in a trajectory that was clearly not that of a rubber ball.

"What's that!" yelled the Twins. "Death is the only thing you want for us!" With that they began to leave.

"Don't go!" shouted the Xibalbans, "We can still play ball, but we'll put yours into play."

The Twins relented and the Lords then suggested that they all place bets on the outcome. The loser would give the winner four bowls of flowers, one of red petals, one of white, one of yellow, and one of whole flowers. With the stakes set, the Twins deliberately let themselves be defeated, for the first and only time in the story.

Believing that the Twins could not deliver the flowers to them the following morning, the Lords placed them in Razor House, a place inhabited by a host of animated knives. To save their lives, the Twins gave the knives the flesh of all animals forever, quieting their furious gnashing and cutting. This accomplished, they asked the cutter ants to bring them some flowers from the gardens of the Lords of Death. The next morning the Twins give the Xibalbans their own flowers as payment of the bet.

That day and each one thereafter, the Twins played the Lords to a scoreless tie and spent the night in a different house. They escaped from Cold House by shutting out the cold by the force of their wills. In Jaguar House they kept off the hungry cats by giving them bones. In Fire House they again survived by the strength of their wills, getting only slightly singed where others would have been consumed.

Bat House,[23] however, was their undoing. To escape from the Death Bats swooping down at them, the Twins climbed inside their blowguns. Unfortunately, as morning approached, Hunahpu began to get impatient and stuck his head out of the blowgun to see if it was dawn. A giant bat dove down and snatched away his head.

Left alone with his brother's lifeless body, Xbalanke had little hope that he would be able to avoid a similar fate at the hands of the Lords in the morning. Casting about desperately for a solution, he called out to all the animals and asked them to bring him samples of their food. Seeing the coati's squash, he had a brilliant idea. He carved it into an image of his brother's head and attached it to Hunahpu's shoulders. The new head miraculously spoke and the body rose and walked like a living being. In the meantime, Xbalanke made a deal with a rabbit, who agreed to wait near the ballcourt and imitate the ball when it landed in the rough outside the court.

The next morning, he took his reanimated brother to the ballcourt to play against the Xibalbans who were now using Hunahpu's head as the ball (Fig. 8:5). Xbalanke received the head on his yoke and sent it flying into the trees where the rabbit was waiting. Pretending to be the ball, the rabbit bounded off with the Lords in close pursuit. Meanwhile, Xbalanke retrieved his brother's head, replaced it on its rightful body, and then

the God of Zero strikes the head
of One-Ahaw as the ball

FIGURE 8:5 The Ballcourt Marker from La Esperanza ("Chinkultik")

yelled to the Xibalbans that he had found the ball. Play resumed with the squash carved in the likeness of Hunahpu until it hit the court and burst into a thousand pieces, seeds flying everywhere. Furious and humiliated, the Xibalbans realized they had been tricked and defeated once again.

The next part of the story we have already told in Chapter 6. The Twins incinerated themselves in an oven, their bones were ground up and thrown in a river, and five days later, they emerged with the features of catfish men. After disappearing for a day, they returned as vagabonds who used dance to defeat the Lords of Xibalba. With this victory, the Lords of death, disease, calamity, and misery were restricted to Xibalba forever. No longer could they walk the world of humanity in their real forms, but only in their avatars of disease and sickness. The Twins went to the Ballcourt and brought their father back to life, but because he could not name all his parts, they left him in the ballcourt.[24] There the Maize God stayed to receive the prayer and worship of the human beings born in the next Creation.

HOW THE MYTH RELATED TO LIFE

By now it should be plain that we regard the tales of the Popol Vuh as more than delightful stories to entertain K'iche' Maya children. The saga of the Hero Twins, their twin ancestors, and their mother and grandmother comprised the basic and pervasive Precolumbian Maya myth cycle that explained many important things about their universe and how it came into being. This myth embodied their concepts of justice, proper behavior, and how to defeat evil. It also told them what would happen after death—how to confront it and win rebirth and everlasting life. The Popol Vuh stands alongside *The Iliad*,[25] the *Ramayana* of Hindu religion, and the Gospels of Christianity as the charter for a way of life that blossomed out of oral tradition into written words. The backdrop against which the Christian myths played out was the Old Testament, Hebrew prophecy, and the events of the Roman Empire. For the Greeks, the mythic underpinnings of their world began on the battlefields of the Trojan War. For the Maya, the confrontation with death, evil, and disease took place in the ballcourt. The roles of the players, the conflict between teams, the winning and losing of wagers, the consequences of defeat, and the recovery of the dead First Fathers from the floor of the ballcourt, all provided the basic metaphors defining destiny and history for the Maya.

For the ancients, ballplaying as ritual was a fateful game destined to end in sacrifice. The uncertain outcome of actual sporting matches gave their pageantry a gladiatorial feeling. Ballgame sacrifice was often the ceremonial climax of the struggles between neighboring kingdoms out to master each other on the field of battle.[26] Historical accounts of the sorcerous combat of the ballgame became a favorite subject on the stairways and buildings celebrating the victories of the winners. We see (Fig. 8:6) just such a ballgame in progress at Ox-Nab-Tunich, a place somewhere within the larger kingdom of Kalak'mul,[27] one of the major combatants in the great wars of the seventh century. This image shows a victorious lord from this northern Peten city, wearing the images of Tlaloc and the Waxaklahun-Ubah-Kan War Serpent. He plays against an enemy dressed in a bird headdress and wearing the flower ear ornament of a captive. While trying to return the huge ball flying at his face, the captive has slipped and fallen to the ground in a tangle of feet. He certainly lost the game and probably lost his life as the final outcome.

In the myth, the odds were clearly with the Lords of the Otherworld. Yet while knowing the probable outcome of the game, the wagering of life against death, or even bowls of flower petals, allowed for the hope of victory as a historical outcome. The Twins won by playing tricks on their dangerous enemies—and the greatest of their tricks was to submit knowingly to defeat and sacrifice in order to win the larger game.

FIGURE 8:6 Kalak'mul Ballgame Panel

The great popularity of the ballgame among Maya of the Preclassic and Classic civilization's makes sense in light of its central role in cathartic rituals. Like the gymnastics of the ancient Greeks, it must have embodied qualities of physical prowess, virtue, and devotion. On another level, it was a celebration of resistance against the inevitability of fate. Its playing unified the great cycles of time and Creation with the lesser cycles of human life and history. Through the ballgame, human beings even defied the finality of death by reenacting the feats of the Hero Twins who outwitted the cosmic forces arrayed against them. The Twins and their Fathers took the ballgame from the gods, and with the game, they gave humanity the gift of history.

The Lords of Death tried to stack the game against the Hero Twins to assure their own victory. Maya kings often played games against captive lords that were just as clearly stacked. As in the myth, a courageous enemy caught in this nightmarish situation might have been able to turn the tables and win, if not the game, then at least his honor. Aztec annals of sacrificial combat record a few unexpected victories by feather-armed captives over fully armed warriors. There's no reason to assume that the same outcome wasn't possible in the Maya captive ballgame—although it must have been very rare. But most of all, the ballgame rituals symbolized the combat between good and evil and the larger struggles of ordinary people for justice, happiness, health, and prosperity. The essence of the metaphor was that life was a game, and that the ultimate stakes were the rebirth of the ancestral dead in the afterlife. Without ancestors to help the living, they had no chance.

THE PLACES WHERE THE GAME TOOK PLACE

We have come to understand that while the ballgame took place in earthly kingdoms, the ballcourts themselves opened into the Otherworld. Moreover, different ballcourts opened into different locations in the supernatural world, and often ballcourts reestablished the time and space of a past Creation. Sometimes the location of the ballcourt is designated in a hieroglyphic text, sometimes in the name of the ballcourt, and sometimes in the imagery on its buildings. The eighth-century Yaxchilan ruler Bird-Jaguar specified several of these Otherworld locations on a series of thirteen panels his artists set into the upper tread of the stairway

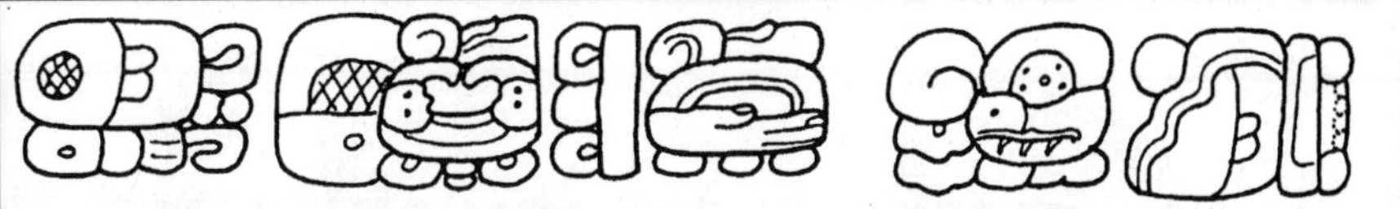

a. it happened at the Black Hole, the Six-Shell-in-hand-Place

b. he entered the road at the ballcourt

FIGURE 8:7 Locations of the Ballcourt

that leads into the interior of Temple 33, one of his most important temples. The images on these panels depict both a critically important ballgame in his lifetime and the mythological sacrificial events that gave that historical event meaning.

The central panel in this series tells us that the ancient mythological game took place in the "Black-Transformer (*Ek'-Way*) at Six-Shell-in-hand" (Fig. 8:7a). The Black-Transformer is the name of the great Otherworld portal created when the Milky Way rims the sky, while "Six-Shell-in-hand" is a place where the gods acted in the last creation.[28] The text (Fig. 8:7b) goes on to say that the ballcourt itself was where one "entered the road" (*och bih*). Thanks to our new understanding of Creation, we know that one enters the road at the south end of the Raised-up-Sky, which is the Milky Way when it arches across the sky in its north-south path. So we learn that the mythological ballgames described in this text were played at a place that existed in the previous creation down the Black-Transformer where the road enters the Otherworld. This is, of course, exactly where One-Hunahpu and Seven-Hunahpu played their own ballgame before they were summoned to Xibalba.

The panel to the right of this center scene says that "he played ball, it happened at the chasm, ball at the Six-Stair Ballcourt" *(pitzah uti hom*[29] *pitz ti Wak-Eb*) (Fig. 8:8). *Hom* is the word for "ballcourt" in the K'iche' of the Popol Vuh and the word for "grave" in modern K'iche'. In Yukatek,

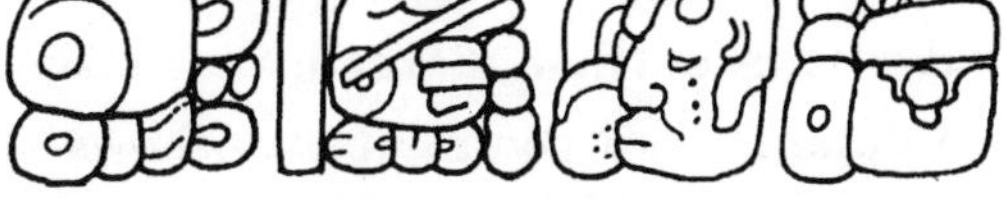

it happened at the chasm, ball at the Six-Stair-Ballcourt

FIGURE 8:8

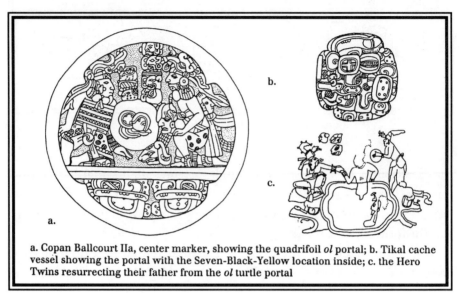

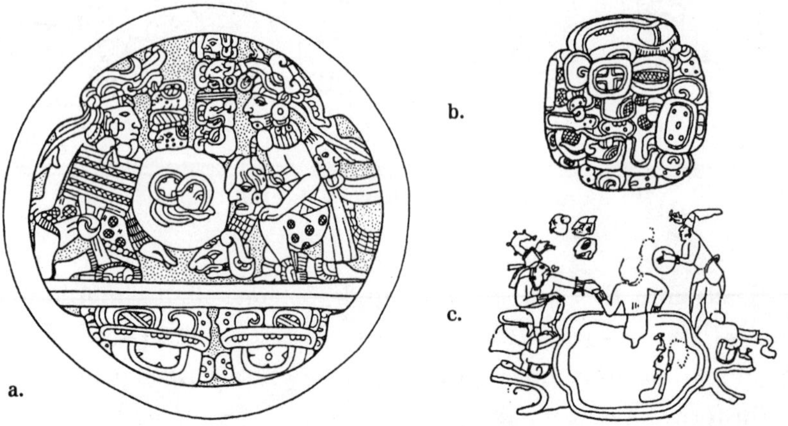

a. Copan Ballcourt IIa, center marker, showing the quadrifoil *ol* portal; b. Tikal cache vessel showing the portal with the Seven-Black-Yellow location inside; c. the Hero Twins resurrecting their father from the *ol* turtle portal

FIGURE 8:9 Portals and the Valley-Place

hom means "chasm" or "abyss," which again describes the ballcourt where One-Hunahpu and Seven-Hunahpu were killed in the Popol Vuh account.

The markers from the Copan ballcourt identify yet another supernatural location. The action takes place inside the quadrifoil cartouche (Fig. 8:9a) we identified with the *Ol* portal in Chapter 4. This shape was used from Olmec times[30] onward to represent the gateway to the Otherworld. Moreover, other locations in the Otherworld often appear inside the portal (Fig. 8:9b), as if these portals could shift positions from one place to another. Most of all, this quadrifoil cartouche shape is also the crack in the turtle's back from whence First Father emerges after his triumphal resurrection by his sons. Thus the ballcourt was not only a place of sacrifice; it was an entry portal to the time and space of the last Creation.

The "chasm" name on the Yaxchilan panel goes on to say that the "ball" was "at Six-Stairs" *(pitz ti Wak-Eb*[31]*)*. This place apparently corresponds to the inscribed stair shown on the central panel (Fig. 8:17) and, as our friend Sandy Bardsley first realized, to an actual stairway on Structure 5 at Yaxchilan.[32] The central panel shows a bound captive bouncing down the Six-Stair-Place like a ball, so that the Six-Stair-Place was a place of sacrifice in the ballgame ritual.[33] Another example of this kind of place is the false ballcourt found on the south side of Yax-Pak's

Temple 11 at Copan. Its dedication inscription calls it the portal of Yax-Pak (*yol ch'ahom Yax-Pak*), so the sacrificial victim who rolled down the stairs was on his way to the Otherworld. And this is more than just words. These six stairs lead down from a terrace—marked as the surface of the primordial waters by huge conch shells—to the plaza below where three ballcourt markers complete the symbolism of the portal. Captives were very probably rolled down these stairs to their deaths.

A text fragment from Anonal, a site in the Dos Pilas region, also refers to such a place (Fig. 8:10a) in the kingdom of Seibal. At Seibal, the last king of Dos Pilas built a stairway made of six stairs flanked by two panels that showed himself playing ball. To make sure we understand that his stairway was the site of the game he played, his artist depicted him playing on top of the glyph, *Wak-Ebnal*, "Six-Stair-Place" (Fig. 8:10b). These Six-Stair-Places have been found at many sites. Only at Copan do they have ballcourt markers, but we are sure that everywhere they occurred they were the sites of special ritual ballgames in which captives met their fates. The Six-Stair-Places were often inscribed with dynastic histories or the celebrations of victories and defeats in warfare.

These very special war memorials were related to another type of ballcourt called *Ox-Ahal-Eb.* When we started working on the name of this place several years ago, we looked up the word *ahal* on the advice of our friend Nikolai Grube. We found that in Yukatek, it means "to dawn," and "to wake up," and it is used in phrases like *ahal kab,* which

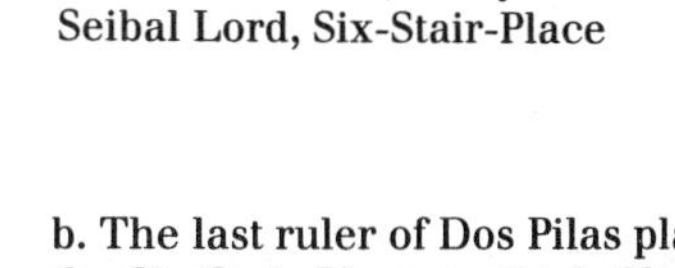

a. Anonal Panel, "Holy Seibal Lord, Six-Stair-Place

b. The last ruler of Dos Pilas plays ball at the Six-Stair-Place on the half-period 9.18.10.0.0. This panel was mounted at one end of a stairway with six steps.

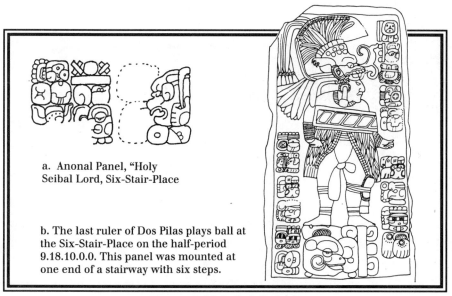

FIGURE 8:10 Seibal Stela 7

a. decapitated himself, the Maize God, the first conquest

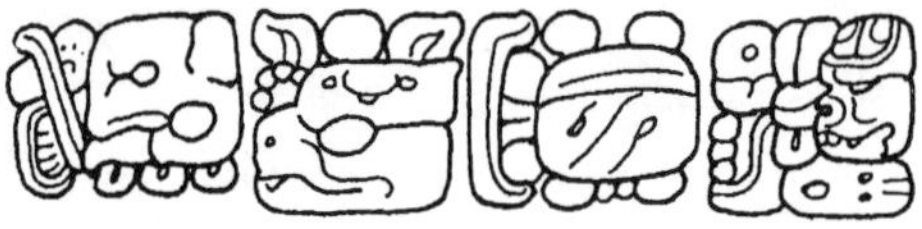

b. decapitated himself, Tz'a-Chan, the second conquest

c. decapitated himself, First Maize Seed Earth Lord, three conquests

d. the severed head of the Maize God lying in the sacrificial plate
(drawing courtesy of Donald Hales)

FIGURE 8:11 Yaxchilan Hieroglyphic Stairs 2

a. Yaxchilan HS 2, Step 7, he conjured at the Three-Conquest-Ballcourt, the step of . . .

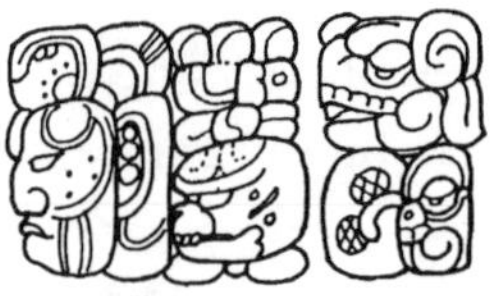

b. Naranjo HS, he played ball on the Three-Conquest-Stair

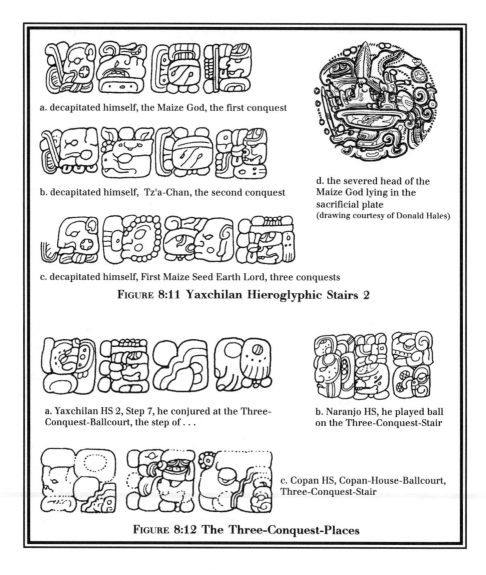

c. Copan HS, Copan-House-Ballcourt, Three-Conquest-Stair

FIGURE 8:12 The Three-Conquest-Places

means "to create the world." Sometime later, Jorge Orejel, one of Linda's students, discovered *ahal* in an old Cholti dictionary with the meaning "to conquer or defeat (*vencer*)." Linda resisted his suggestion that this name marked the stairs as places that recorded military conquests until Nikolai reached the same conclusion from his study of a war memorial stairway at Naranjo. The name, then, is best translated as "Three-Conquest-Stair." It marks the places where the sacrifice of war captives within the metaphor of the ballgame took place.

The Three-Conquest-Stair and the Six-Stair-Place also show up in the mythology of Classic-period texts, although we don't understand the myth as well as we do that of the Hero Twins. The central panel of the Yaxchilan ballcourt sequence records all we know of these three victories. As we have already seen, they took place in the Black-Transformer so that we know they involved supernaturals. Second, they took place in the distant legendary past, at least 1,400 years before the historical date in the text.[34] The text talks about three sacrificial decapitations, and as unlikely as it seems, each was a self-decapitation—*ch'akba*,[35] as the Maya called it. We have pictures of just such improbable actions committed by *wayob* in several pottery scenes (Figs. 6:7, 6:15).

At least one of the victims depicted at Yaxchilan was the Maize God (Fig. 8:11a), whose severed head is shown lying in an offering plate. We have seen other offering plates showing exactly the same image (Fig. 8:11d). We are still unsure about the identities of the other two victims of self-decapitation, but for our purpose the important thing is the way that these three sacrifices are described. They are called the "first conquest" (*u natal ahal*),[36] the "second conquest" (*u katal ahal*), and "three conquests" (*ox ahal*). The ballcourt that Bird-Jaguar dedicated on October 21, A.D. 744,[37] carried this name—it was the "Three-Conquest-Stairway, the step of Bird-Jaguar" (*Ox-Ahal-Eb yeb Yaxun-Balam*) (Fig. 8:12a). Whatever the meaning of these three acts of self-sacrifice, this type of stair is named after them.

At Naranjo in the northeastern Peten, this type of Three-Conquest-Stair appears in a text on an inscribed stairway of four steps. One clause declares that a Naranjo lord, defeated in war by the king of Caracol to the south, "played ball at the Three-Conquest-Stair," or in Mayan, *pitzah Ox-Ahal-Eb* (Fig. 8:12b).[38] The text of the Hieroglyphic Stairs of Copan (Fig. 8:12c) records the name of the ballcourt there as the "Copan House Ballcourt" in one phrase and as the "Three-Conquest-Stair" in another.[39] The temple above the stairway sports the symbolism of Tlaloc warfare in keeping with its militant name.

The ballcourt of the Classic period, therefore, was both a portal into the Otherworld where the game regenerated the time and space of the third Creation, and a war-related place where captives died in a special game. These Six-Stair-Places and Three-Conquest-Ballcourts often displayed texts imparting the dynastic history of the local kingdoms. Both types of ballcourts and ritual playing fields were crucial implements of political and religious performance for the ancient Maya. We will look at four exemplary ballcourts to investigate these functions.

FIGURE 8:13

THE ANATOMY OF FOUR BALLCOURTS
Bird-Jaguar and the Ballgame at the Three-Conquest-Ballcourt

The Spanish never described the complex rituals surrounding the Maya ballgame so we have to reconstruct them from the representations the Maya put in their art. Our best source is the Yaxchilan ballgame panels we described above, because they tell us not only where the game took place but also about the pageantry that accompanied its play. The larger central panel (Fig. 8:13) records the dedication of the Three-Conquest-Ballcourt and it is flanked by six panels on either side. The dates on some of these panels place the depicted games during the lifetimes of Shield-Jaguar and 6-Tun-Bird-Jaguar (Bird-Jaguar's father and grandfather, respectively), telling us that the scribes and artists who

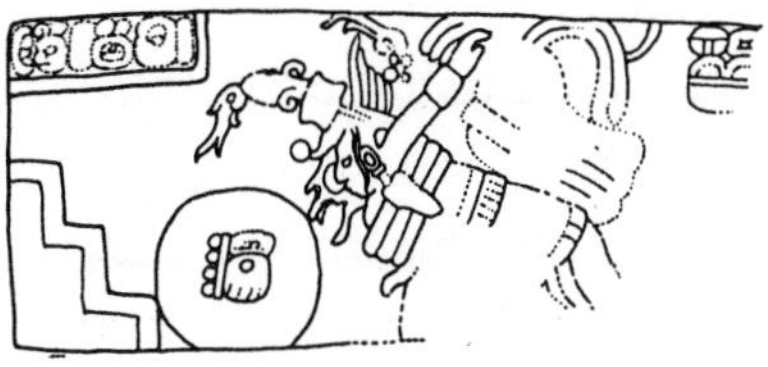

Panel 4

Panel 5

Panel 6

Yaxchilan Hieroglyphic Stairs 2, mounted on the substructure of Temple 33, depicts the ballgame played to sanctify the dedication of the Ballcourt and the Six-Stair-Place. Panels 1-3 and 11 show women, while Panels 2 and 3 depict the Vision Rite that opens the portal so that the game can be played. Panel 2 depicts Lady Pakal; Panel 6 is Shield-Jaguar; Panel 7 is Bird-Jaguar, while Panel 8 is his grandfather, Six-Tun-Bird-Jaguar; Panel 10 is a sahal named K'an-Tok; Panel 11 is Mi-Kimi, a lineage sprout.

Panel 11

Panel 12

Panel 13

created these panels were assembling on this step a history of famous game rituals performed in the city. We believe that the outer panels also record a sequence of activities that would have been common to all ballgames.[40]

The sequence of activities on the left begins with a woman throwing out a ball bearing the glyph 13-*Nab.* Images of Maya balls often have this glyph on their surfaces, but the number may be nine, twelve, thirteen, or fourteen. We do not yet understand what this glyph means in the context of the ballgame. The next two panels depict two other women holding Vision Serpents in their arms. A White-Bone-Snake with its mouth gaping toward them sits on the left of one woman and the right of the other, as if to show us that they are both sitting inside the Maw of the Otherworld. The text next to the first woman names her as Lady Pakal, who was the paternal grandmother of Bird-Jaguar.[41] Since she was long dead at the time of the action, we surmise that important forebears, especially female ancestors, were conjured from the Otherworld to sanctify and witness the game.[42]

Panels 4 and 5 to the left of the central panel, and Panels 9, 10, 12, and 13 on the right show ballplayers positioning themselves to catch the ball on their yokes. One ball is glyphically inscribed with 9-Nab, and the rest are inscribed 12-Nab. Most of the name phrases of these players are too badly damaged to read, but the player on Panel 10 is K'an-Tok', a high-ranking noble known as the first sahal[43] of the kingdom and a major figure in the historical tales Bird-Jaguar inscribed on his buildings. Mi-Kimi, the player on panel 12, is a *ch'ok* or "sprout" of his lineage. We believe that these secondary ballplayers are all subordinate lords or members of the royal family who played in the game. We also surmise that these side panels depict the same game as the main panel, which features Bird-Jaguar, because none of them have additional dates. So, in this ballgame "the team" was composed of very high ranking nobles and members of the king's family. These lords were engaged in a ritual of building dedication (probably the main ballcourt itself), which most probably would have begun with a battle to take captives and concluded with the sacrifice of those captives.

The most important feature of the secondary players carved on the panels are the masks they wear. As we saw in Chapter 5, these ballplayers have transformed into the same supernatural dancers who cavort around the seated Maize God in Room 1 at Bonampak' (Figs. 5:4 and 8:14).[44] Since the same fearsome dancers reappear in pottery scenes (Fig. 8:15), we surmise that the Yaxchilan ballplayers have just come from or will soon go to the same sort of dance. In the form of their *wayob* or spirit companions, they play not as human beings, but as supernatural beings. The ballgame, as well as the dance, charges the space of the ballcourt and transports the players into the Otherworld.

The three center panels focus on the role of the king in the drama of the game. In the wide center scene (Fig. 8:16b), the current king, Bird-Jaguar, dedicates *Ox-Ahal-Eb* the through his own play. His father, Shield-Jaguar, plays on his right (Fig. 8:16a) facing the viewer as he hits a ball against the step in the chasm (*hom*) at the Six-Stair-Place (*Wak-Ebnal*). Shield-Jaguar is dead, and therefore he plays in the Otherworld with a ball made from the body of a bound sacrificial victim. He also occurs on the south marker of the main ballcourt. The image shows him sitting in the middle of an ancestor cartouche, an *ol* portal opening from the Otherworld. Thus, the markers of the real ballcourt at Yaxchilan confirm the location of the game displayed on these panels.

On the opposite side of the central panel, Bird-Jaguar's grandfather,

a. masked dancers from the Bonampak' murals

b. masked players from Yaxchilan

FIGURE 8:14 Masked Performers

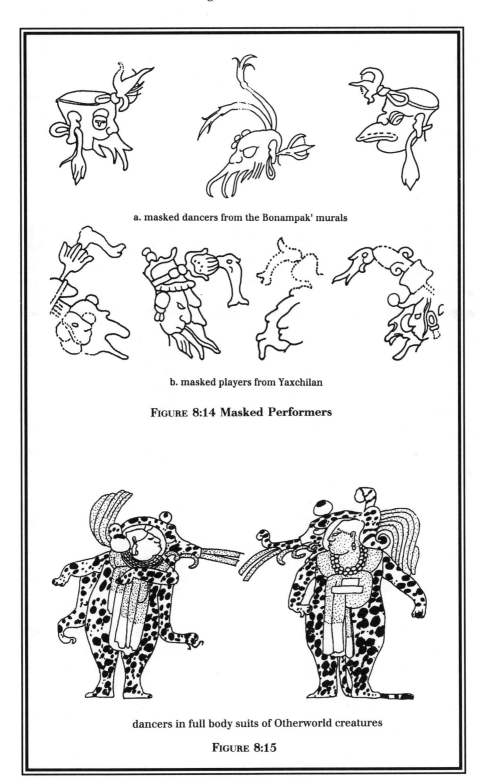

dancers in full body suits of Otherworld creatures

FIGURE 8:15

a. Panel 6, Shield-Jaguar plays down the chasm at the Six-Stair-Place

b. Panel 7, Bird-Jaguar plays ball at the Three-Conquest-Ballcourt Stair

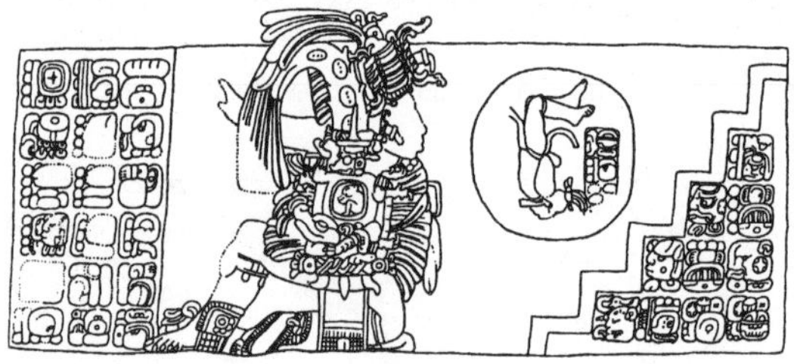

c. Panel 8, Six-Tun-Bird-Jaguar plays ball against a stairway with six steps

FIGURE 8:16 The Three Central Panels of Yaxchilan HS 2

Six-Tun-Bird-Jaguar, plays with his back to the viewer as if he and Shield-Jaguar are reciprocal images of each other—that is, as if they are one player seen from the front and back (Fig. 8:16c). Six-Tun-Bird-Jaguar wears the Cosmic Monster on his backrack as he, too, fields a captive rolling down a Six-Stair-Place. One way of looking at this triptych of kings is to see Bird-Jaguar in the middle of the game, and his father and grandfather in the two end zones facing each other and their descendant, the reigning ahaw. Bird-Jaguar (Fig. 8:16b) also prepares to catch his captive on his yoke. He wears the net skirt of the Maize God, First Father, and has a fish on his back, reminiscent of the Hero Twins who first resurrected with catfish faces. Next to him stand two dwarves, one wearing the shell earflare of Chak and both wearing Venus or star signs attached to their arms.[45] The captive who bounces down the stair is called a *te-tun ahaw*, "stela lord" (Fig. 8:16b).

The balls used in these rituals at the Six-Stair-Place and at the Three-Conquest-Ballcourt really were tightly bound captives. This is confirmed by the ballplayer panels of La Amelia. Like those at Seibal, these panels flanked a stairway that celebrated the victory of the last king of Dos Pilas. This king, like Bird-Jaguar, wears a huge fish on his back (Fig. 8:17a). On both panels, he dances above a jaguar who crouches in the position of a captive. One panel records the dedication of the stairway on August 16, A.D. 807, while the other says that on July 10, A.D. 804, he threw a 9-Nab ball named The-Guardian-of-the-Jaguar.[46] This is the jaguar under his feet. Captives really did roll down the *Wak-Ebnal* to suffer and die in mock play.

The historical, mythical, and cyclical aspects of these ballgame rituals were carefully conjoined. The hieroglyphic stairs pictured in the center of Bird-Jaguar's ballgame sequence record three self-decapitation rituals in mythic time. Furthermore, these sacrifices are called "conquests," thus linking them to battle and the unfolding of fate in warfare. Those mythic decapitations and conquests have their parallel in the decapitations of real-world victims. But if the mythology here involved the game of the Hero Twins, as we believe it did, then the human victims of ballgame sacrifice could also look forward to the same resurrection as the Hero Twins and First Father. The Twins emerged from the river of Xibalba

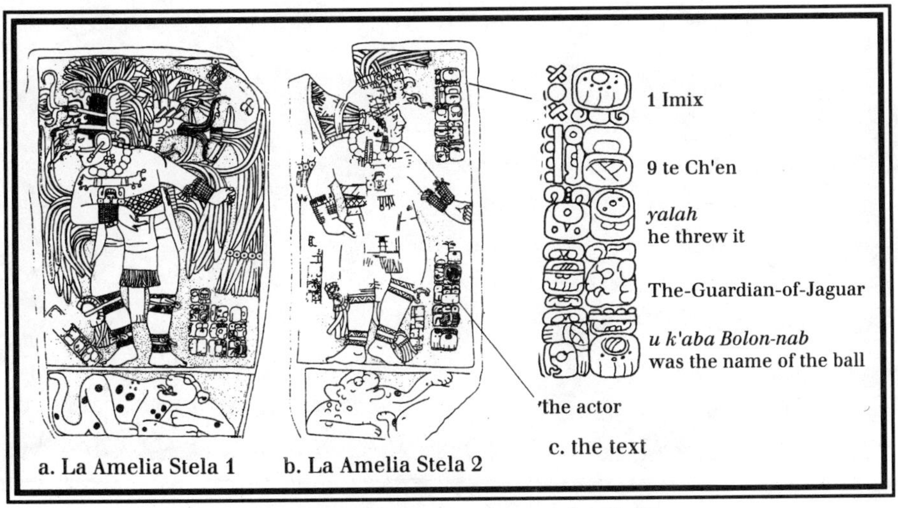

FIGURE 8:17 The Captive as the Ball

disguised as fish men after being burned to bone, ground to powder, and thrown into the water. Bird-Jaguar and the Dos Pilas kings wear huge fishes on their backs. First Father emerged triumphantly from death as the beautiful dancing young Maize Lord. Bird-Jaguar wears the net skirt of the Maize God and is surrounded by lords playing as his supernatural companions. Ballgame sacrifice renewed the basic covenant between gods and people. In these ballgame panels, the lords of the Maya declared that the defeat of their enemies and their own historical triumphs were like-in-kind to those of the Hero Twins. It was no happenstance that Bird-Jaguar framed his own game with those played by his father and grandfather, nor that he linked his building dedication to the original decapitation sacrifices of mythic time, for these parallels harnessed the force of Creation to his dynasty's victories in war. Other lords harnessed the same mythology to their service.

A Creation Ballcourt at Copan

The reverence Copanecs felt for the works of their ancestors is especially evident in their Ballcourt (Fig. 8:18), the powerful portal they

FIGURE 8:18 The Copan Ballcourt

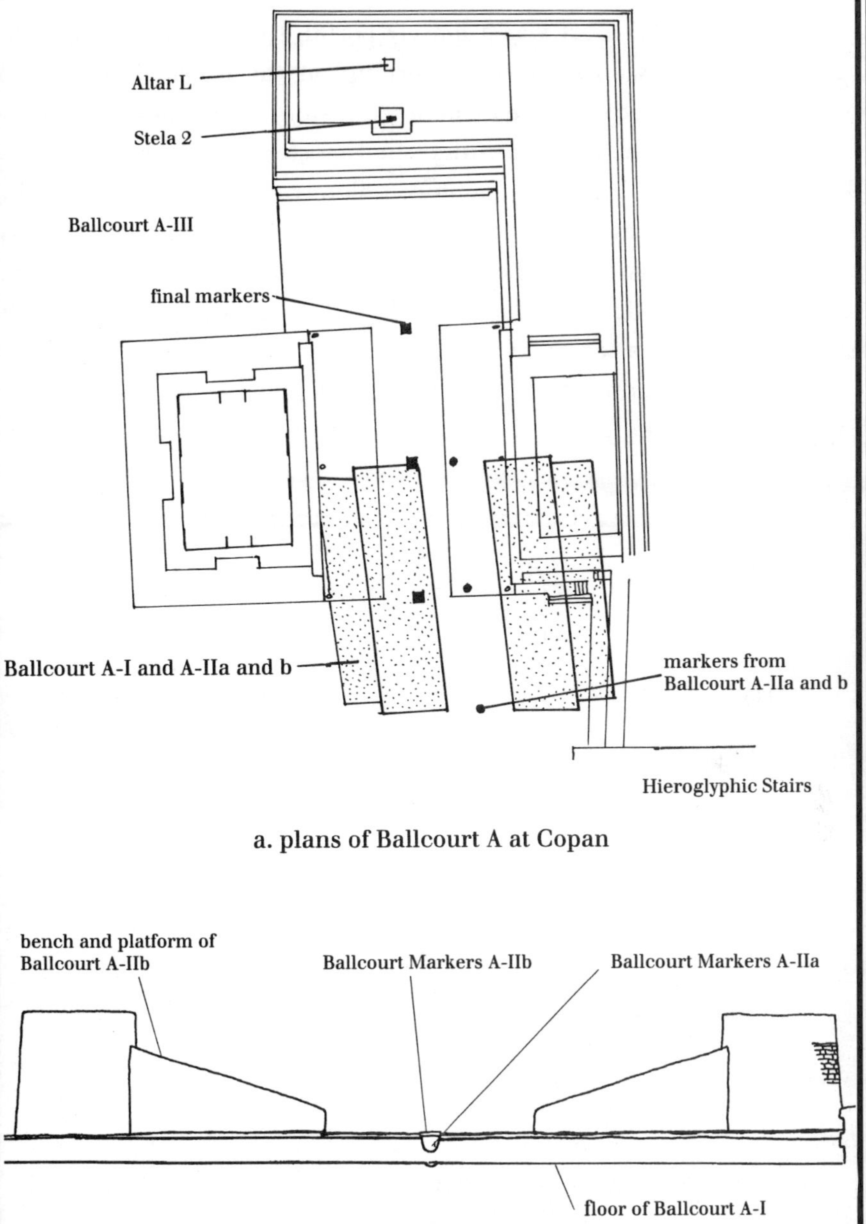

b. section of Ballcourt A-II

FIGURE 8:19 The Main Ballcourt at Copan

(adapted from Stömsvik 1952)

rebuilt and maintained for over four centuries. If, as we believe, kings thinned the membrane separating the Otherworld from this world through rituals reenacting myth and the founding acts of their ancestors, then the Copan Ballcourt was a master portal, the largest *Ol* of all.

Archaeologists, who have been working on the Copan Ballcourt for generations, have reconstructed a detailed history of its phases.[47] Initially built sometime in the fourth century, the first stage was remodeled twice (Ballcourt A-Ia and b), before being completely rebuilt, probably during the sixth or early seventh century (Ballcourt A-IIb). This second phase was itself remodeled at the beginning of the eighth century (Ballcourt A-IIa), and rebuilt for the final time (Fig. 8:19) in A.D. 737. This is the version that tourists see today.

From the beginning, the Ballcourt was closely tied in meaning and function to the Temple of the Hieroglyphic Stairs, located just to the

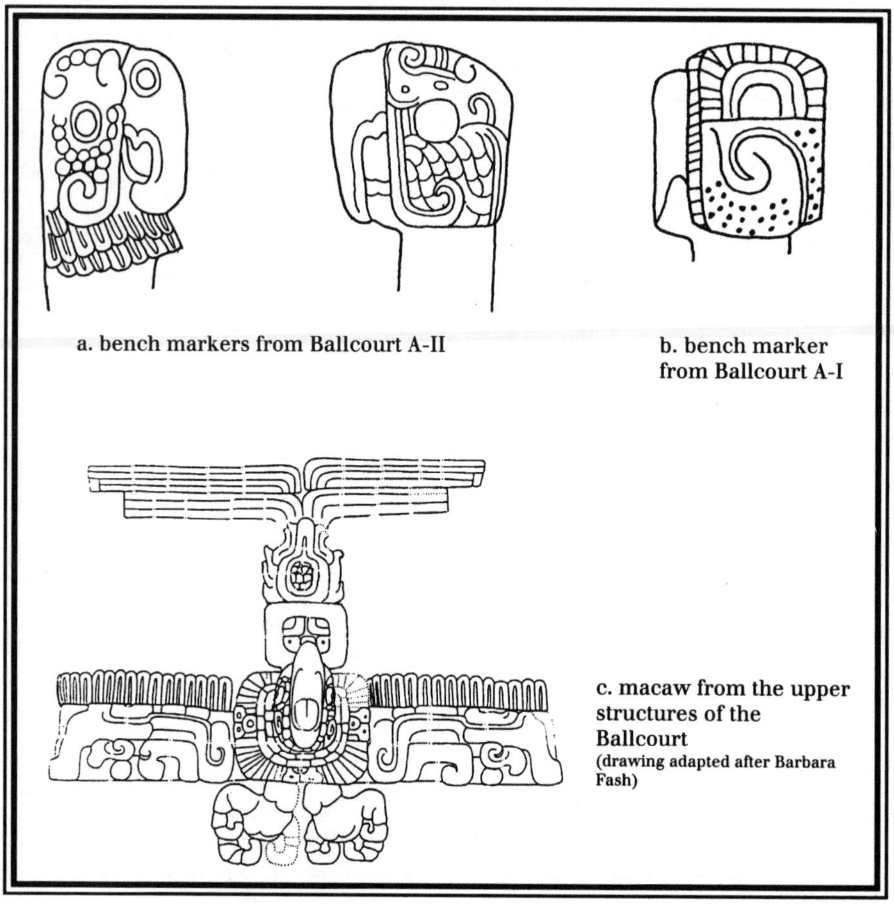

a. bench markers from Ballcourt A-II

b. bench marker from Ballcourt A-I

c. macaw from the upper structures of the Ballcourt (drawing adapted after Barbara Fash)

FIGURE 8:20 The Macaws from the Copan Ballcourt

the macaws on the bench and
the Ballcourt buildings at Copan

FIGURE 8:21

southeast. All phases of the Ballcourt had the same essential form—an I-shaped playing surface flanked by inclined benches and buildings on the east and west. Bench markers in the form of macaw heads from at least three of the earlier stages of the Ballcourt (Fig. 8:20a–b) were deposited as offerings in and around the Hieroglyphic Stairs.[48] Knowing as we do that the Maya put things into the construction fill of pyramids as part of their ensouling, the Hieroglyphic Stairs held part of the soul essence of its companion Ballcourt. Sixteen other macaws with their wings spread wide and their tails upright perch on the medial molding[49] of the Ballcourt buildings (Fig. 8:20c).

Both of these sets of macaws (Fig. 8:21) represented Itzam-Yeh, who became known as Seven-Macaw,[50] the prideful bird of the Popol Vuh who thought he was the sun. Most Classic-period images of this bird show him toothless and grotesque after his defeat by the Twins. The Twins shot him in the jaw with a blowgun pellet, making his teeth hurt so badly that he finally had them pulled. With the loss of his teeth and beauty, people saw him for what he was and his greatness fled.[51] But the macaw that flies above the Copan Ballcourt still appears in all his glory as the majestic, self-declared sun. When the historical champions of Copan and their adversaries walked onto this playing field, they entered the mythic time and space of Seven-Macaw and the Hero Twins, before the outcome of that contest was certain. Today, tourists by their thousands still gaze upon Itzam-Yeh, the bringer of magic, who still hovers above the silent, empty playing alley where Creation was prepared.

In this extraordinary Ballcourt, Copan lords played—sometimes against players of their own city and sometimes with battle captives destined to lose the contest of life. These lords faced the uncertainty of the contest and prevailed, just as their ancestors had prevailed for generations since the beginning of their history. The contest between destiny and risk was no symbolic pose. King 18-Rabbit (Waxaklahun-Ubah-K'awil[52]) dedicated the markers placed in the final stage of the Ballcourt's construction on January 10, A.D. 738.[53] Just a hundred and thirteen days later, he died as a war captive at the hands of Kawak-Sky of Quiriguá, who we think was a relative with ambitions to take his throne. In one of the great ironies of Maya history, the king named for the great War Serpent perished at the hands of a rival who also called upon this supernatural being in war.

Although the markers from his final Ballcourt are no longer readable, 18-Rabbit commissioned an earlier set that did survive in remarkable condition, perhaps because they were not used for very long.[54] These thin capping markers (Fig. 8:22) covered an older set of worn markers. The scenes on all three known sets of playing alley markers[55] are framed with

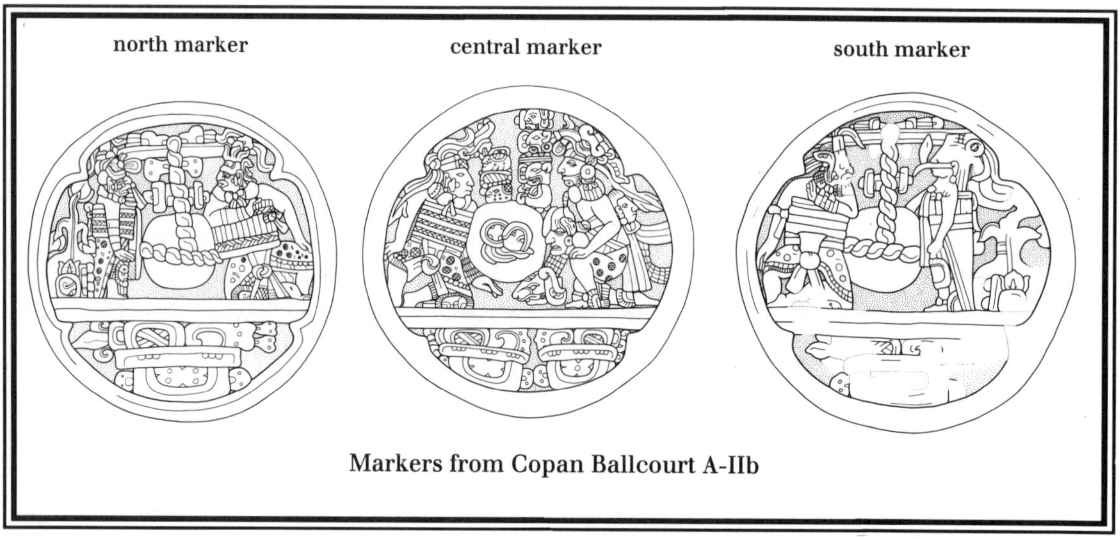

Figure 8:22 (drawings by B. W. Fash)

the quadrifoil portal called *Ol* that pierces through to the watery Otherworld, offering humans a view not unlike that of a glass-bottomed boat.

These windows into the Otherworld reveal (Fig. 8:22) pairs of supernatural ballplayers engaged in the game. Sun-marked sacrificial plates rest under the ground line as if underneath the floor of the Ballcourt where they play—like offering caches. Generally, such plates were used in sacrifice and, when cached, they put *ch'ulel*, "holy spirit," into buildings and places. Moreover, this specific sun-marked plate is located where the ecliptic crosses the Milky Way. It was the offering plate used by First Father to generate the World Tree of the center and it also held his head when he was decapitated by the Lords of Xibalba.

Here, however, the plate of sacrifice is inactive. The generative stingray spine is missing. Its presence is crucial, because it is the stingray spine that brings the Raised-up-Sky World Tree, and its earthly analog, the maize tree, into existence. The generative power of this plate is locked inside, because Hun-Nal-Ye and his brother have yet to be released from their graves under the floor of the Ballcourt. Like the bench markers displaying Itzam-Yeh, with his magnificent beak intact, these images point to a moment in the Creation before the Hero Twins have confronted and defeated their frightening adversaries, in the time before the Fourth Creation.

Each of the three scenes from the Ballcourt A-IIb markers depicts sets of two players flanking a ball. On the north and south markers, the ball is suspended from a house rafter by a twisted rope[56] from which the rat in the Popol Vuh myth will soon cut it loose. A bulbous tree rests on the ground line at the outside of the north and south markers. The north tree has "Nine Successions" (*bolon te tz'ak*) and the south has "Seven Successions" (*wuk te tz'ak*). These, we think, represent the maize trees planted by the Twins in the house of their grandmother so that she will know their fate. The positions of the players on the outer markers mirror each other: the outer players stand upright in profile, while the inner ones kneel front view in the position of play.

The players on the north marker are square-eyed, mirror-headed beings we think might represent Hunahpu and Xbalanke, just as they find the ball and learn to play. They have the "nine successions" tree in their scene. On the south marker, the kneeling player on the left wears the deer ear that usually distinguishes One-Artisan and One-Monkey, the first sons of One-Hunahpu. The standing player wears a downturned half mask associated with moon gods.[57] Since in the Popol Vuh myth Hunahpu

FIGURE 8:23 **Copan Structure 10L-26 and the Hieroglyphic Stairs**

and Xbalanke rose into the sky where the sun belonged to one and the moon to the other, this image might well represent the Twin who owned the moon about to play one of his older brothers.

The central panel depicts the myth's final confrontation scene. Here we see Hunahpu, named in the adjacent glyphs as One-Ahaw, his Classic-period epithet,[58] playing a Lord of Death, who attacks the ball from the right with a hand stone. The White-Bone-Snake headdress this lord wears and the hand over his lower jaw identify him as both the god of the number zero and the god of sacrifice by removal of the lower jaw.[59] This dreadful god strives to defeat the crafty Hero Twin and lead him into a trap. And we know he was successful, for this is the very god who smashes Hun-Ahaw's head with his hip on a Ballcourt marker from the city of Chink'ultik in the western borderlands of Chiapas (Fig. 8:5).

This imagery, commissioned by the ruler 18-Rabbit, whose name floats just above the ball, shows the primordial contest where the Xibalban Lords of Death matched their powers against the sons of First Father. In one sense, this was the eternal conflict between death, disease, and fear, and the forces of balance and light represented by the ancestral gods. The names given the lords in the Popol Vuh—Pus-Master, Blood-

Gatherer, Skull-Master—reveal this. This imagery also represented the mystery of life and sustenance reborn out of death.

The final act of this drama was played out on the magnificent Hieroglyphic Stairs (Fig. 8:23) that rise from what used to be the south end of the old Ballcourt. These stairs rise straight up out the portal to the Otherworld. This structure was raised, after 18-Rabbit's untimely death, by his grandson, Smoke-Shell. Smoke-Shell remodeled the old pyramidal platform, resurfacing the lower terraces, adding new ones to the summit, and putting a new temple on top of the pyramid. To reach that temple he built a stairway of over seventy steps and had the history of his dynasty carved on its risers. The resulting text is the longest inscription to survive from the Precolumbian world.

Barbara Fash[60] has shown that the upper temple was festooned with images of Tlaloc warfare. Sitting in majesty at intervals along the stairs are five figures of ancestral Copan kings dressed in the costume of Tlaloc-Venus war. It is as if Smoke-Shell was preoccupied with the imagery of war and conquest in order to compensate for the terrible harm done to the prestige of his dynasty when 18-Rabbit was taken captive and decapitated—possibly in ballgame sacrifice—at Quirigua. The inscription[61] records the accessions, deaths, and deeds of all the successors of the

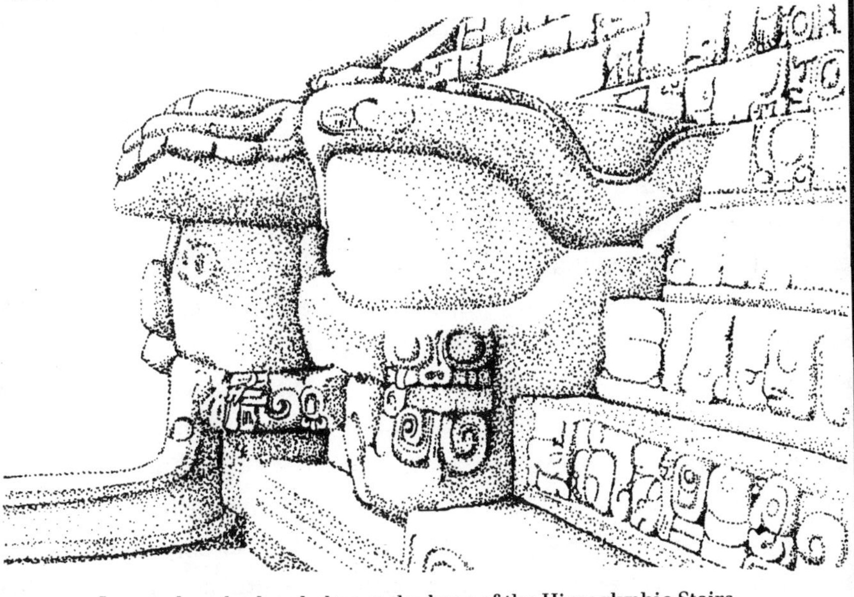

Inverted snake-head altar at the base of the Hieroglyphic Stairs

FIGURE 8:24

founder Yax-K'uk'-Mo', as if to put 18-Rabbit's death in perspective by saying that his defeat was but one small event in a three-hundred-year-long history of glory and accomplishment. This history would continue in defiance of that catastrophe by the time-honored means of war and the ballgame.

So the history of the kings rises up from the plaza at the south end of the old Ballcourt. To affirm the sacred and prophetic nature of the dynastic history, the Copanecs placed an altar at the bottom of the stairs and carved it into the shape of a great Vision Serpent with its gullet open toward the sky (Fig. 8:24). Copan's history emerges from its throat as a materialization of the holy power of the Otherworld. It rises cloaked in the military symbolism of the Waxaklahun-Ubah Serpent and the *tok'-pakal* talisman of war.[62] This is, in fact, an *Ox-Ahal-Ebnal*, "Three-Conquest-Stair-Place."

There is an amazing image carved on the upper surface of this altar in the region representing the palette of the serpent's mouth. But to understand this image, we have to go to Pakal's sarcophagus at Palenque on the other side of the Maya world. The scene carved on the lid of Pakal's coffin depicts what happened to him when he died. Framed by the sky, he falls down the throat of the White-Bone-Snake (Fig. 8:25a). Behind him the Raised-up-Sky Tree, the Milky Way, soars upward with the Double-headed Serpent Bar of the ecliptic entwined in its branches. Itzam-Yeh perches on the topmost branch. The dead king, his jade ornaments and loincloth blown upward by the violence of his fall, sinks into the Maw along with the Quadripartite God, the personification of the sun-marked plate of sacrifice. The crossed-bands that usually mark the contents of this plate are here replaced by a "kimi" sign that is also a variant of the *way* glyph.[63] This sign is meant to tell us that Pakal is being transformed, even as he falls. This is further symbolized by the net skirt, the garment of the Maize God, that Pakal wears. In dying, he becomes First Father, who was called to Xibalba to face the Lords of Death in the ballgame. Losing the contest of wits, he was killed and buried under the floor of the ballcourt. But his sons soon came to release him to new life (Fig. 8:25b).

This same resurrection is actualized in the mouth of the snake at the base of the Hieroglyphic Stairs (Fig. 8:25c). The lower jaw and teeth of the White-Bone-Snake represent the maw to the Otherworld. In the form of a great U-shaped mouth, they frame the scene at the bottom and sides of the altar. Portal within portal, the bony maw cradles the magic turtle

a. Palenque sarcophagus lid: Pakal falls as the Maize God
(drawing by Merle Robertson)

b. the Maize God's rebirth from the crack in the turtle's back

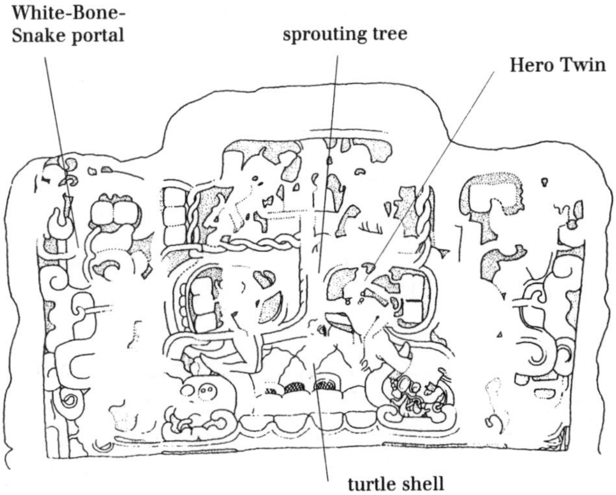

c. top of the altar at the base of the Copan Hieroglyphic Stairs

d. the First-Tree-Precious

FIGURE 8:25 The Maize God in Death and Resurrection

carapace out of which First Father was reborn. The turtle's head is on the left end of the carapace, while a waterlily god emerges from the other end. The Maize God emerges from this crack, not in his human form, but as the maize plant itself, the First-Tree-Precious we have seen at Palenque (Fig. 8:25d). This Copan tree is flanked by the kneeling forms of the Hero Twins as they attend to its growth. Above this altar the Copanecs performed the ritual offerings that would bring about this rebirth.

With this imagery, the cycle of the Ballcourt myth begun on the alley markers is completed. Just as First Father is resurrected in the maw of the snake, so the deeds of Copan's holy lords rise upward, telling the world that their dynasty will also triumph in the end.

A Three-Conquest-Ballcourt at Tonina

The war and sacrificial associations of the ballcourt are nowhere more dramatically illustrated than at Tonina,[64] a kingdom on the opposite edge of the Maya world. A small, elegant panel records the dedication of that ballcourt on 9.13.4.6.12 7 Eb 0 Mol (July 7, A.D. 696). It was called the "Seven-Black-Yellow-Place Three-Conquest-Ballcourt (*Wuk Ek'-K'anal Ox-Ahal-Ebnal*), the ballcourt of Mah K'ina Baknal-Chak, Ch'ul Tonina Ahaw" (Fig. 8:26). Thus, it was a Three-Conquest-Ballcourt where ballgame sacrifice took place and also a portal to a supernatural location in the Otherworld—in fact, the same location where we saw the Tzum lord dance in Chapter 6 (Fig. 6:17).

The markers along the upper benches of this ballcourt reflect the war associations of the ballcourt with particular poignancy. Carved rectangular flexible shields like those carried by Tlaloc warriors hung down the walls on both sides of the court (Fig. 8:27a–d). The text on each shield

was dedicated, the Seven-Black-Yellow-Place
Three-Conquest-Ballcourt, the ballcourt of
the Lord of the Tree, the Ballplayer
Great Sun Lord Baknal Chak
Holy Tonina Lord

FIGURE 8:26 Tonina Glyph Panel

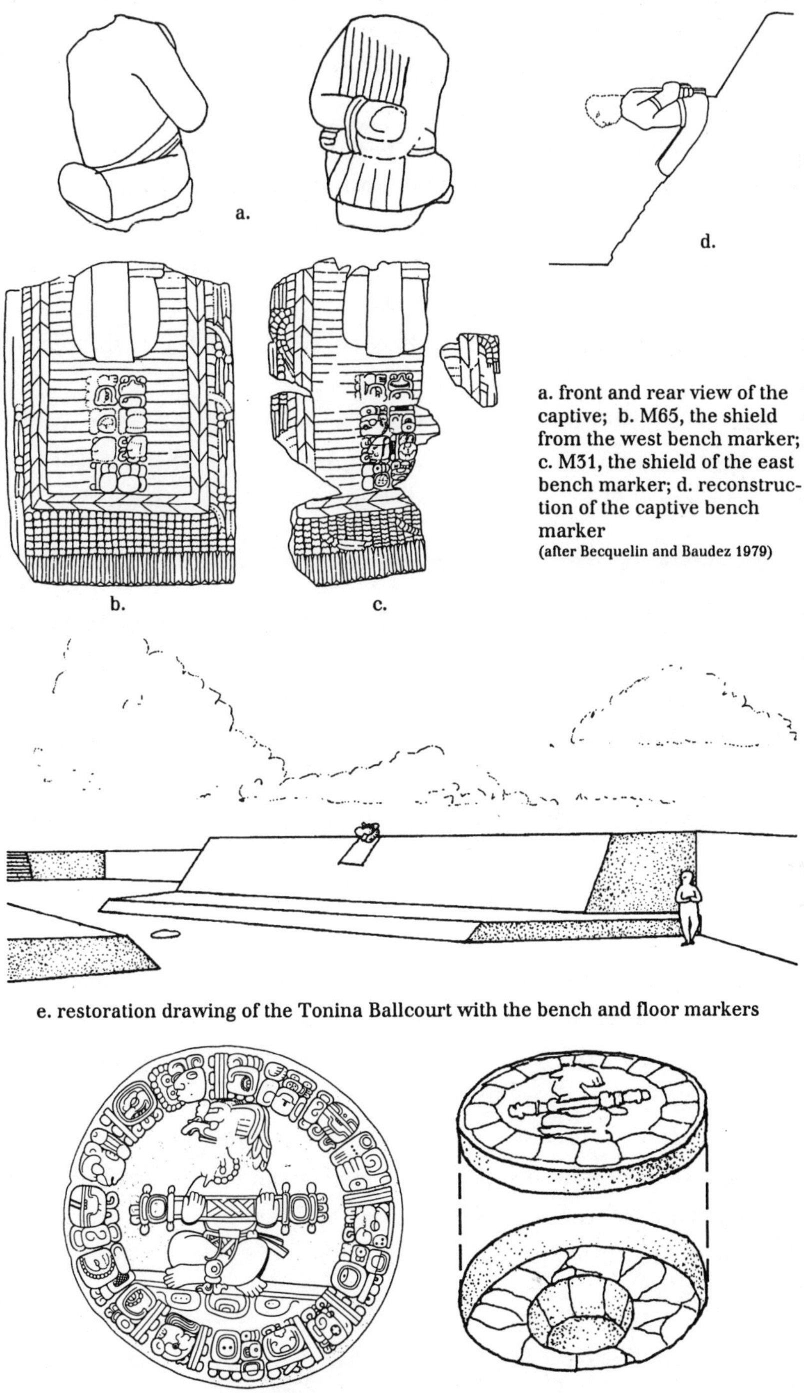

a. front and rear view of the captive; b. M65, the shield from the west bench marker; c. M31, the shield of the east bench marker; d. reconstruction of the captive bench marker
(after Becquelin and Baudez 1979)

e. restoration drawing of the Tonina Ballcourt with the bench and floor markers

f. the central ballcourt marker and the offering hole under it
(drawing of the marker by Peter Mathews)

FIGURE 8:27

named its owner as "the vassal of the Ballplayer, Holy Tonina Lord" (*yahaw Ah Pitz Ch'ul Po' Ahaw*).[65] A captive with his wrists tied behind his back knelt on top of each shield, arranged so that his body cantilevered out over the ballcourt to function as the target of the ball. This ballcourt and its grisly sculptures were dedicated by Tonina ruler Baknal-Chak about fifteen years before he captured K'an-Hok'-Chitam of Palenque.[66] That unfortunate king may have played a losing game with his captor in this very ballcourt before being taken to the sacrificial terrace above, where he was beheaded and reborn into the Otherworld. Nowhere is the imagery of ballgame sacrifice so dramatically and poignantly represented as in the Tonina Ballcourt.

The single marker on the court playing floor (Fig. 8:27e) even further reinforces the role of the ballcourt as a portal to the Otherworld. Added about sixty years after Baknal-Chak dedicated the ballcourt,[67] this little marker covered a hollow place that symbolized the portal itself (Fig. 8:27f).[68] The text here records two events—the death of a man named Six-Sky-Smoke on September 5, A.D. 775, and his entombment two hundred and sixty days later on May 22, 776.

The image shows Six-Sky-Smoke, who is called both a firstborn child and a sprout of the lineage,[69] sitting on a platform marked as a cosmic partition.[70] In his arms he holds a bar. On this bar, like the one carried by Yax-Pak on Copan Stela 11 (Fig. 4:9), the expected serpent heads are replaced by the glyph for the "white-flower" soul *(sak-nik)*. Copan Stela 11 has a picture of this bell-shaped flower blossom with a "white" glyph attached below. This young man sits cradling in his arms the ecliptical path to the Otherworld marked with the symbol of his very soul. It is this soul that enters a child from the *Ol* portal at birth. Here the soul seems to be leaving the world through the ballcourt portal. On Copan Stela 11, the dead king stands atop the *Ek'-Way*, another name for the portal. Presumably, the "white-flower" soul, like the *ch'ulel* of modern Zinacantecos, is indestructible and will return via another newborn sometime in the future. One of the principal entry and exit points for these soul journeys was the ballcourt.

The Great Ballcourt at Chich'en Itza

By far the largest and most impressive ballcourt ever built by the Maya was the Great Ballcourt at Chich'en Itza, fashioned in the twilight of the

Classic kingdoms in the far northern lowlands where ballcourts were far more rare than in the south.[71] Even in modern times, it is a place of legends. Today guides love to embroider stories of champions while tourists suffering under a tropical sun gaze in awe at the steep walls and the gray, weathered stone reliefs of sacrificial ritual. Here is a worthy American comparison to the imperial amphitheaters of the Roman world.

Paul Sullivan tells a wonderful story of the Great Ballcourt in modern times. In 1935 the officers of the Chan Santa Cruz Maya were hosted by the famous archaeologist Sylvanus Morley at the ruins of Chich'en Itza. When these inheritors of the great nineteenth-century Yukatek Maya rebellion first came into the playing alley, they discovered the natural echoes it produced and conversed with the legendary native king and the ancestors, *chilankabob,* they thought buried there under the city. Here, according to Sullivan's informants, is how the dialogue went:

> "Well, hello!" said the captain. "Well, I came to visit you, Mr. King, here in the town of Chich'en Itza. Because since long ago when the world was settled we have known that you are here, your Majesty. We came to visit you. We came to greet you. We came to do our duty, here in the town of Chich'en, here where your office is. Here we are conversing thus. God made you lord. We have come to visit you." Sargeant Chaac spoke for the king: "Good. You, Don Concepción Cituk, if it is you, I am happy. Come and visit me here in the heart of my town, here in the town of Chich'en Itza. I am very happy with all that is said [about] how it happened that the revolution happened here in the town of Chich'en. . . ."
>
> With the King of the Maya and the hidden *chilankabob* speaking to them from beyond, the officers also made special note of the many feathered serpents adorning the temples of Chich'en Itza—especially the great serpent heads at the base of the Castillo—beings which, though stone, also live and wait the day they may stir again and feast on human flesh. This will come to pass, Mayas say, should they fail to defend themselves against the encroaching dominance of the foreigners; only if they fail to make human war will these creatures stir to make apocalyptic war. It was the mission of the officers to keep the enchanted beasts still.
>
> (Sullivan 1989:86–87; brackets original)

When the Chan Santa Cruz officer spoke of revolution in Chich'en Itza, he was no doubt referring to his own people's rebellion in the Caste War. Yet his comment was also in keeping with the spirit in which the Great Ballcourt was originally built. Its imagery and meaning derive directly from the symbolic tradition of its smaller counterparts in the southern

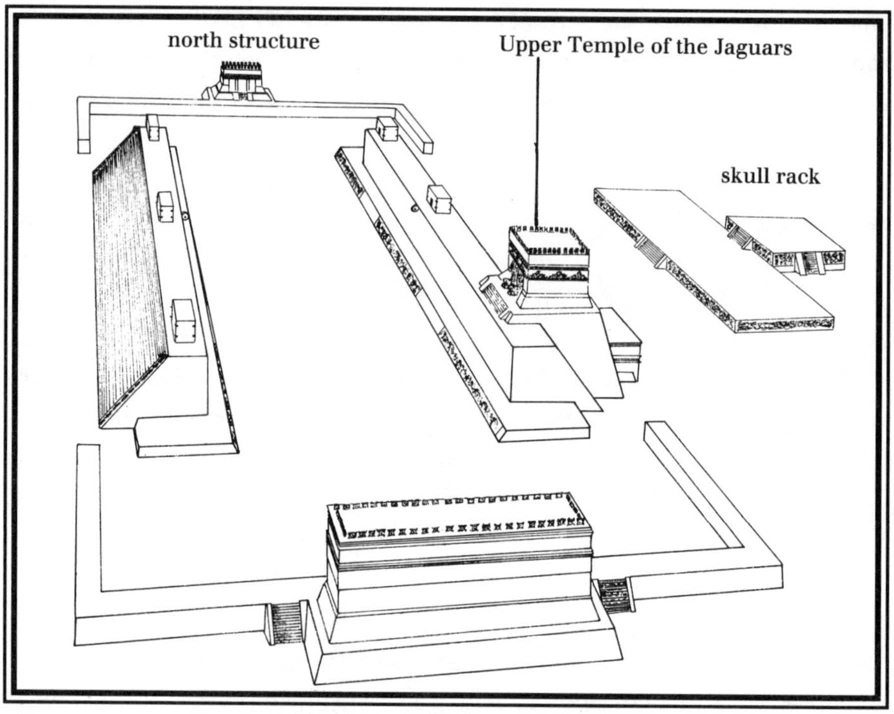

(drawing by Kathryn Reese Taylor)

FIGURE 8:28 Great Ballcourt at Chich'en Itza

lowlands. While the scale is larger and the proportions of the parts different, the components are the same (Fig. 8:2). The playing field has the standard I-shape, and the size of the inclined benches is greatly reduced in proportion to the tall vertical side walls with their ring markers.[72] In a city of many smaller ballcourts, this unique edifice signaled the final rendezvous of royal history with destiny, the closing and most stupendous act of the Classic Maya. For in the city of the Itza, we see not kings or dynasties, but a council of nobles joining the ancient mythology of the ballgame with a new way of ruling themselves that had grown to maturity in the north.[73]

The huge sidewalls of this ballcourt had massive outside stairways that ran their length and led up to wide areas on top (Fig. 8:28). Along this summit stood five small buildings, in all probability free-standing portal arches.[74] The larger Upper Temple of the Jaguars paralleled the midpoint and end panels decorating the playing benches flanking the alley below. Even from afar, people would have heard the noise and music of ceremony on the court and imagined the play while they watched smoke billow from these six summit temples. Both the northern and southern ends of the playing field are enclosed by low walls. The southern end of the playing field rises until it reaches a broad colonnaded building, and the northern end extends to a smaller temple.[75]

Instead of the three alley markers of the southern lowland ballcourts—placed at the center and the ends of the east and west benches—the Great Ballcourt has relief scenes carved on large panels set at the same three points. These panels depict a familiar ceremony: Two teams of seven players each (Fig. 8:29) face each other across a ball with a skull engraved on it. The player to the immediate left of the ball holds the severed head of the captive whose unfortunate corpse kneels on the other side of the ball in the position of a player. His neck stump spurts seven streams of blood shaped like serpents. The central stream, cascading around the victim, transforms into scrolling waterlily plants from the Otherworld. This is a beautiful symbolic rendition of the regenerative power of the ballgame, even in the moment of sacrifice and death.

The players who back the two main protagonists wear not only the gear of the ballgame but the headdresses and face ornaments of Maya warriors. While many experts have interpreted this scene as an engagement between Mexican Toltecs and Maya,[76] we believe the contest pits Maya against Maya.[77] For one thing, some of the reputed Toltecs wear the Jester God diadem of Maya holy lords[78] and all of them wear costumes and emblems long employed by their southern Maya cousins both in the war and in the ballgame.

The military trappings of these players are reinforced by murals in the Upper Temple of the Jaguars. Inside this two-roomed vaulted building, the capital's finest artists painted bright and detailed murals recording the battles of the Itza. We believe these are the very engagements that forged their hegemony over the north—they are the foundation wars. Moreover, the murals are loaded with images of the venerable Maya flint-shield battle standard merged with the spiked Sun Disk favored in highland México,[79] as well as the Waxaklahun-Ubah-Kan War Snake transformed into the Feathered Serpent of its original Teotihuacano conjurors. Together these divine symbols of conquest war express the Itza's god-given right to create a Maya empire.[80]

Towering high above the Ballcourt, the outside of the building displays even more explicit military symbolism (Fig. 8:30). The columns holding up the doorway are huge feathered rattlesnakes, their tongues flickering outward seeking prey as they dive earthward. Below these heads on the balustrades, screaming bird deities, perhaps Itzam-Yeh himself in yet

another manifestation, hold a god, or perhaps a *way*, in their open mouths.[81]

The decorations on the upper walls add more imagery to the overall effect. Entwined Feathered Serpents circle the building in three bands alternating with friezes. Their entwining replicates the pattern on the ballcourt markers and the pattern of the twisting umbilical snakes of the heavens. In the lowest of the intervening friezes, jaguar battle beasts prance between round and tasseled shields like those carried by Feathered Serpent warriors in the murals inside the sanctum. The central frieze has tail-to-tail Feathered Serpents meandering outward to the corners between droplets and jars, symbols of offerings.[82]

The most articulate images of this structure are the freestanding sculptures that line the edge of the roof, forming a low roof comb. Here are shields surmounted by crossed throwing javelins: the ancient Teotihuacan metaphor of war, once transformed into the flint-shield of the southern lowlands and now returned to its original form.[83] It is as if the Chich'en rulers wished to show that they not only understood the Classic Maya symbols of conquest war, but also the history of the

(after Marquina 1950)

FIGURE 8:29 The East Central Panel of the Ballcourt

Mesoamerican alliances that had spawned them. Like a giant standard, the Upper Temple of the Jaguars boldly and explicitly proclaimed the Great Ballcourt as the place of the Flint-Shield and War Serpent. The lords of Yaxchilan, Tonina, and Copan named their ballcourts in writing, but the Chich'en lords used only conspicuously displayed visual symbols of war and sacrifice to name theirs. Nevertheless, Chich'en's Ballcourt is no less a Three-Conquest-Stair place.

Like the lords of Copan, the Itza lords also decorated their Ballcourt with intricate reliefs depicting their mythology of Creation.[84] At the northern end of the playing alley, the lords placed a small masonry building called the North Temple. Although this little building is dwarfed by the Great Ballcourt, the sheer walls of the playing alley create a foreshortened arrow that aims directly at this temple, whatever your vantage. The stairway leading up into this little temple is flanked by images of the World Tree, Itza style (Fig. 8:31), to signal its cosmic import. The piers that stand in the front door are covered with a wide net pattern filled with the same vines that sprout from the necks of the decapitated players in the panels of the Ballcourt.

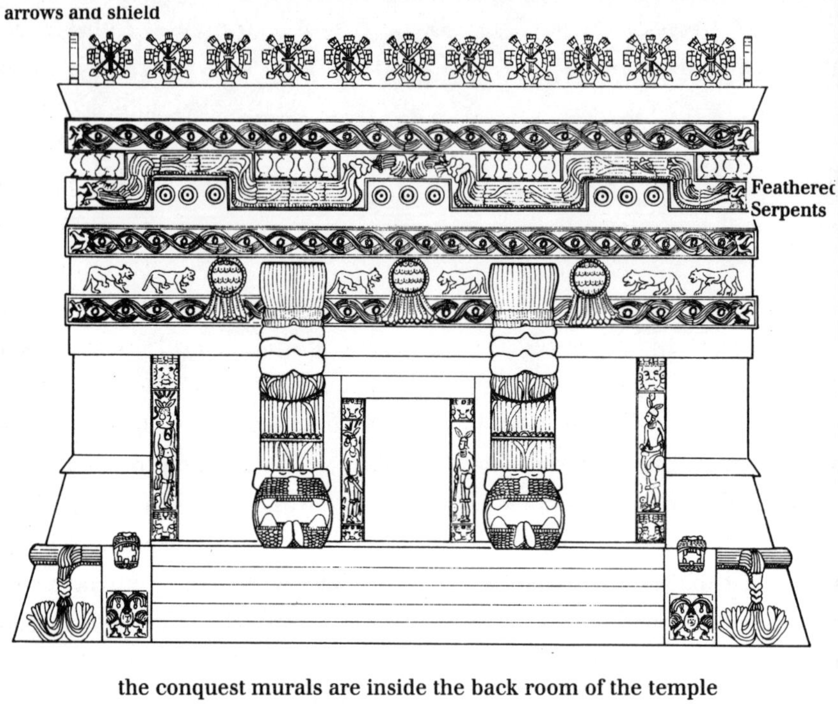

FIGURE 8:30 The Upper Temple of the Jaguars

(drawing by Kathryn Reese Taylor)

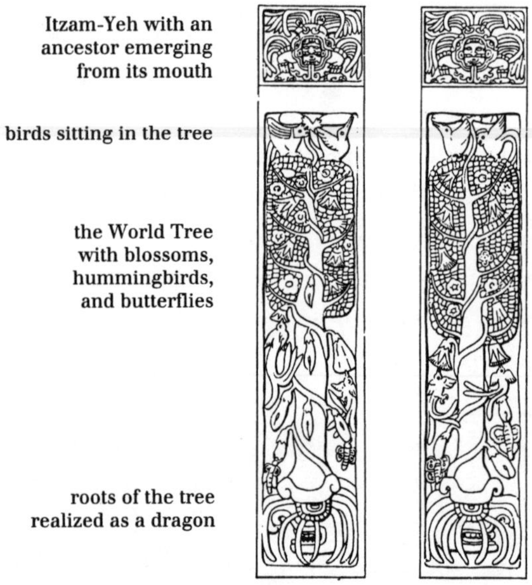

FIGURE 8:31 Balustrades from the North Temple of the Ballcourt

(drawing by M. A. Fernandez in Marquina 1950)

An extraordinary set of relief scenes covers the interior walls, similar to Chan-Muwan's great narrative murals at Bonampak'. These reliefs,[85] however, are devoid of text. They display their richly detailed message solely through narrative images once brightly painted to enhance their vividness for the privileged few who saw them in Chich'en's heyday. These images tell a story about the mythical underpinnings of political power and the role of ballgame sacrifice in the lives of the city's leaders.

The scene in the single chamber (Fig. 8:32) moves from wall to wall, wrapping itself around the viewer. As at Bonampak', the composition uses registers to show different events in this very complex series of ritual and historical scenes. The focus of the building's north wall is the historical accession of a ruler in front of the city's council of rulers.[86] A row of warriors carrying spears and spearthrowers stand in the top portion of the scene. In the middle row the councillors sit on either side of the acceding lord, and more warriors stand in the lower row. In the middle of the scene, just below the jade-shirted ruler, sits a ball and the empty throne he is soon to occupy. Just such a throne sits in the door of the Lower Temple of the Jaguars and inside the temple encased by the Castillo. It is the jaguar throne of Chich'en Itza.

This jade-shirted man is the living counterpart of the being lying prone in the lowest band of the north wall. This giant human wears a cloak of the same jade-sequined material and lies on his back with a Double-headed Serpent flowing out of his belly like eviscerated entrails. One head moves toward the man's feet and one toward the head.[87] The heads on the bifurcated, feathered body have gaping jaws that emit huge flint blades of the kind used in heart sacrifice at Chich'en. The vine tree created by the sacrifice of ballplayers fills the rest of the lowest band as the serpents entwine with the Pawahtun earth-bearers. The sacrificed victim is First Father and the Double-headed Serpent Bar is the ecliptic snake emerging from the place of Creation where he was resurrected. This Ballcourt was dedicated on November 18, A.D. 864,[88] when this particular area of the sky was directly above the playing field at midnight. The image here is a picture of the sky on that night, and the ruler whose accession fills the north wall was the avatar of this god.

The west wall of the inner room was filled with the ballgame itself. Several scenes represent preparation for the game. In the lowest register birdmen dance around the corner from the north wall toward the water-lily plant now become the World Tree. This tree rises from a monster head. Beyond them, ritual actors hold a staff and a perforator. Above

FIGURE 8:32 Reliefs from the North Temple of the Ballcourt at Chich'en Itza

west wall

them is a smoking censer next to a great bird and two warriors who all lean over a kneeling man. In the upper register, a player accompanied by a feather-draped man carries the ball toward seated people. Purification rites that prepared the players for the game are depicted in the two registers on the vault of the north wall. These include scenes of penis perforation, induced vomiting, what is perhaps a steam bath, and a blowgunner shooting birds out of a tree.[89]

On the opposite, eastern wall we see events that occurred after the game. In the upper register a dying player with a snake stream of blood spurting from his neck collapses back against a ball while the victor leans over him. In the final scene below the sacrifice, a bundled, dead body receives homage from a set of warriors.

Taken in its entirety, the entire program portrays the preparation rites, the playing of the game, and the sacrifice at the end as the framework that gives government by council its supernatural and mythological

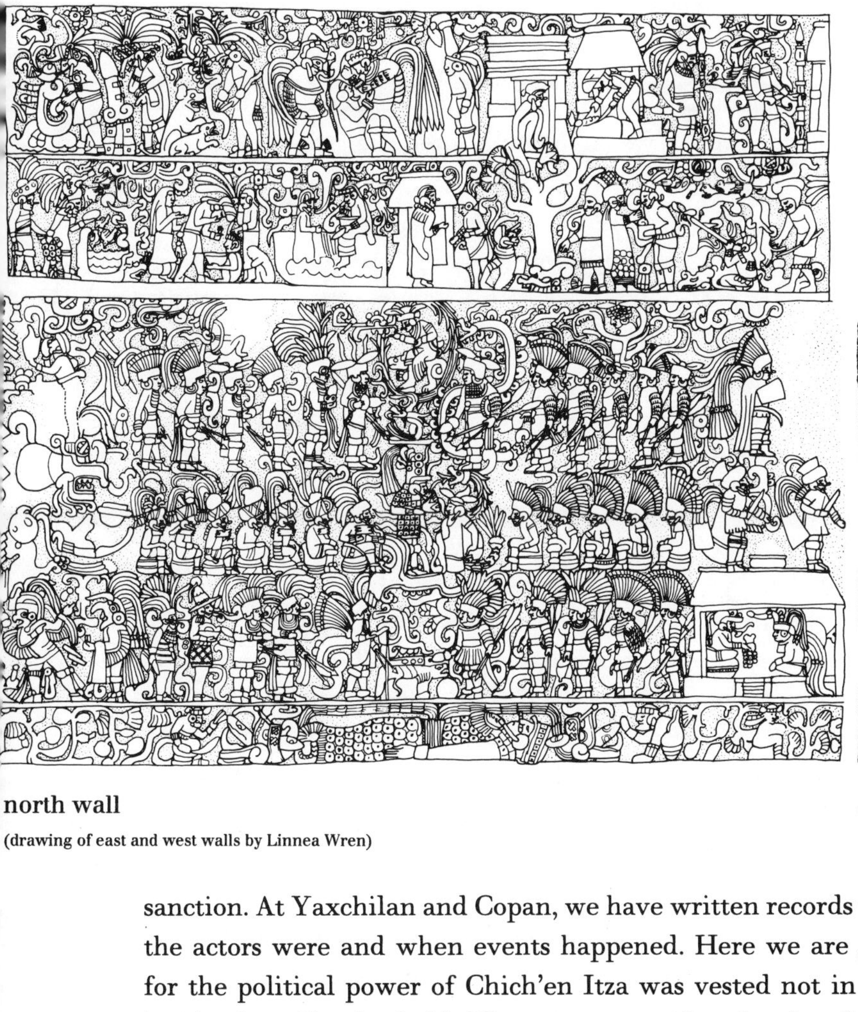

north wall

east wall

(drawing of east and west walls by Linnea Wren)

sanction. At Yaxchilan and Copan, we have written records telling us who the actors were and when events happened. Here we are given neither, for the political power of Chich'en Itza was vested not in an individual but in the office he held. The scenes carved and painted on the Upper Temple of the Jaguars and in the North Temple represented the political and mythological power base for all the men who acceded to high office, robed in the jade shirt and entwined in the Feathered Serpent. These rulers lost their identity as individuals and gained the power of empire.

The Mayaness of the Great Ballcourt

Since Charnay, a nineteenth-century French explorer, first suggested the Great Ballcourt at Chich'en Itza was built by invading Toltecs, it has been seen as the product of invaders from central México.[90] Although many major scholars, including J. Eric Thompson,[91] have opposed the idea that Toltecs invaded northern Yukatan and ruled from Chich'en Itza, most scholars have emphasized how different the Great Ballcourt is from the courts in other Maya cities. The basis of their arguments has been the traits it shared with central Mexican and Gulf Coast ballcourts of the Postclassic period. In contrast, we have emphasized the striking similarities of the Chich'en Ballcourt to the courts of the southern kingdoms.

All of the ballcourts we have discussed in this chapter are alike in the followings ways: in their relationship to the spatial organization of their respective cities, their associations with council houses and pyramids, their emphasis on war and captive sacrifice, their association with founding events and dynastic history, and finally, their relationship to primordial time and space before the 4 Ahaw 8 Kumk'u Creation date. They all contain portals to the Otherworld, whose power could be accessed by human ritual and pageant.

In the light of all these similarities, it becomes clear that the Great Ballcourt of Chich'en is a quite orthodox expression of Maya ideology with at least a millennium of tradition behind it. The Itzas created in the Great Ballcourt an exposition of power drawn from the same source of myth and ritual as their cousins in the southern kingdoms. Like the Three-Conquest-Ballcourts of the south, it emphasizes military themes in the war murals of the Upper Temple of the Jaguars. It was also a Creation Ballcourt. Its images of First Father and the dedication date on the Great Ballcourt stone put the place of Creation overhead at midnight.

In spite of its roots in the Maya tradition, the Great Ballcourt of the Chich'en Itza did represent a major departure from the tradition that had prevailed in the south. As the Cruzob captain suggested, the Itza were revolutionary in the way they addressed political power. The lords of Chich'en disengaged their Ballcourt and its symbolism from individual destiny by banning inscriptional history from its walls. The North Temple portrays not the accession of a particular king on a particular date—but rather kingship in the abstract, imagery that would be usable by everyone who held the office. The Itza lifted their Great Ballcourt out of the linear time of historical consciousness and returned it to the cosmological frame of cyclic time.

The lords of Chich'en Itza turned their backs on dynasty and inscribed history. By doing so they disengaged themselves from the restraints inherent in the southern definitions of kingship. In this way they were able to revitalize an ancient, dying tradition. Their restructuring of the kingship let them establish an empire and brought them into the larger, non-literate tradition of central México and the Gulf Coast. We suspect that Chich'en Itza and its contemporary, Uxmal, gave the Terminal Classic and Postclassic cultures of Mesoamerica a new way of using the ballgame in their own political strategies—one that was more Maya in its origin.

ENDGAME

After the demise of Chich'en Itza, we don't have a lot of evidence for the ballgame in Yukatan. Diego de Landa, the first bishop of that region, mentions it in his *Relación*[92] only once—as a pastime for unmarried men and boys when they gathered at the men's house. For reasons we still do not know, these northern Maya gave up this way of thinking after the fall of Chich'en Itza. In the highlands to the south, it was another story. In the land of the K'iche' people, whose Popol Vuh version of the Hero Twins myth was preserved on paper, no self-respecting lord built a town without a ballcourt. Every town, from Utatlan, the capital of the K'iche', to Iximche', the capital of the Kaqchikel, to Mixco Viejo and dozens more, all have at least one ballcourt. When Linda Schele visited Iximche' with Kaqchikel friends from Tekpan, the town the Spanish established to replace Iximche' after they destroyed it, she saw that each of the two ruling lineages had its own ballcourt. Playing the ballgame and the Creation mythology that underlay it was critical to the political and religious institutions of these highland peoples, just as it had been for the lowlanders for a millennium before the Collapse.

The invasion of Alvarado and his army changed everything. Although eyewitnesses reported the ballgame in remote parts of México during the ensuing centuries and even into modern times, the game was only rarely seen in Maya country after the Conquest. It is as if the Spanish realized the pivotal role of the ballgame and discouraged its play. The only colonial description of a Maya ballgame came from Rafael Landívar, the Jesuit friar born in Guatemala in 1731. He described it this way:

> Nothing, however, provides a more amazing spectacle than a large company of Indians given to play. They first gather a thick gum, discharged by a tree, which gets its name from its elastic properties, and by rolling it together form a large ball which freely bounces high into the air. The crowd then forms a large circle into which the large ball is first tossed and

> it is not permitted for anyone to touch it with his hands when once it has been thrown, but rather he must hit it with his hips, or elbows, or with his shoulders, or knees. Then as soon as the ball is tossed into the middle of the field, the whole crowd excitedly bounds over the plain darting this way and that. One hits the rubber ball with his elbow, another drives it back with his hip, one thrusts his head in its way as it comes down, another with his knee quickly sends it back again into the sky or darting back and forth strikes it with one hip and then with the other. But if at any time the ball should alight on the broad surface of the ground the grounded ball must be retrieved with the elbow or the knees and lifted from the level plain into the air. For this reason you will see the Indians at this point of the game rolling all over the ground until they have raised the fallen ball with their elbows or knees. But if someone should venture to strike the ball with his hands while it is in the air and carelessly disregard the strict rule, he is reprimanded and suffers the loss of the game.
>
> (Leyenaar 1978:40)

So the conquerors did not completely suppress the game; they merely sealed the political fate of the players in their colonial regimes. If we are right about the essential function of the ballgame as ritual, the highland Maya lords no longer had defeated adversaries to bring to the courts—for they were themselves the defeated—and they had no opportunity to challenge history for they had been subsumed into the historical agenda of their conquerors.[93] Modern Maya play ballgames, but they are the soccer and basketball games of the Western world. *Pitz*, the Classic-period word for ballplaying, is now used only for the games of children, as our Maya friends revealed that day in Bernal Díaz's study. *Hom*, the old K'ichean word for "ballcourt," is now the word for "grave." The portals still pierce into the Otherworld, but those that once opened through the ballcourts are now located in modern Maya cemeteries. There, as in the ancient myth, the souls of the dead are trapped on the other side of their graves. They can be freed from the chains of death and brought back to the land of the living only through dance and the pageants of Maya life.

THE PATH AT CHICH'EN

(as told by David Freidel)

On my last day in the field in the summer of 1992, I took seventeen Maya men from Yaxuna village on a tour of the site of Chich'en Itza. It was Sunday, the only day when my friends could get in free. David Johnston, who works on the project, drove our friends up the bumpy road in the back of our big truck. I expected a pretty good group of young workers from the project, but I was surprised and delighted to see Don Emetario, Don Bernardino, and other elders of the community pile out along with the younger men. My shaman friend, Don Pablo, was presiding over a Ch'a-Chak that day, out in the western part of Yaxuna's land, so he couldn't come. Most of the men hadn't been to Chich'en since the new tourist facility at the gate went up a few years ago. They don't usually have the time to go or the money to pay the entry fees. Chich'en has become a place for foreign tourists rather than Maya pilgrimage.

As we passed through the gleaming metal turnstiles, Don Emet told me he hadn't been here for thirty years. It was gently drizzling as we moved briskly off toward the Great Ballcourt, but we didn't mind because the weather was keeping the crowds of tourists small. We must have made quite a sight, a parade of Maya men in American gimme caps; a smaller group of American Spanish teachers and their Yukatekan companions; and Phil Hofstetter, friend and filmmaker, recording it all on videotape.

By the time we got into the Ballcourt I was waking up to the challenge I faced. With the writing of this book, my perceptions of this Ballcourt had changed dramatically. I only hoped that, as I explained its meaning to my Maya friends, I could do it justice. I began by telling them the Maya name for ballcourt, the Three-Conquest-Stair. As I wondered how best to explain that name, I suddenly realized that I had to start by relating the Classic story of the Creation. I had mentioned parts of this story before while working in the ruins of Yaxuna, but now I had to encompass the whole to make sense of the Ballcourt. Sweating with effort and the summer heat, I marshaled my somewhat halting Spanish to the task. I told them about the Milky Way as it cycled through the night sky on Creation Eve. I talked about the First-Three-Stone place, and linked it to the way they lay out new fields and the hearthstones of their kitchens. As the story tumbled out of me, I began to use Maya words where I could, reaching to build a bridge between the world of their ancestors and their modern world by way of mine.

As I spoke, gesturing where words failed, they watched me carefully. Over the years I had talked to some of them about these things, especially Don Emet, but now I was trying to tell the story of this book. They could tell that I felt strongly about what I was saying. They knew I was talking about Maya beliefs because I was using Maya words. But I think it must

David Freidel talks to the men of Yaxuna in the Great Ballcourt of Chich'en Itza

Figure 8:33

have been hard for them to grasp that I was talking about their traditions, that this was a story of their ancestors. It was, after all, tumbling from the lips of a too-tall, light-haired foreigner who spoke imperfect Spanish earnestly and Yukatek Maya not at all. Perplexed or not, they concentrated and listened and tried to make sense of it. Surrounded by American, European, and Japanese onlookers, I tried to breach the chasm between their world and mine, and between all of us and the world of their ancient ancestors. (Fig. 8:33)

We stopped for a time at one of the large panels showing the decapitation sacrifice of a ballplayer. Pointing out the bloody snakes and flowering vines flowing from the neck, I explained that this place was not just for playing ball, but for remembering and re-creating the death and rebirth of First Father, whose name was Hun-Nal-Ye. We spent some time going over just who this ancestor was and the metaphor maize played in their thinking today. As they nodded their heads, agreeing with me that maize was indeed the grace of life, I felt that some of them had grasped the central ideas of what I was trying to say.

We wandered up the court to the North Temple, where I pointed out the Tree of the World carved in shallow relief on the balustrades flanking the small stairway into the temple. There is a grotesque head nestled in the roots, and flowers and birds adorn its branches. I explained to them that the skull represented First Father's, which the Lords of Xibalba had hung in the tree, and that the white flowers were symbols of the soul. Then we moved on to the Lower Temple of the Jaguars where I told them the story of the aged ancestors who hold up the doorways. On his back, the old male god bears a turtle shell, the starry place of First Father's rebirth. The side of the doorway presents an image of First Father emerging out of a cracked mountain monster.

The gentle drizzle that had been accompanying our progress turned into a hard rain, and we hastily bunched into the temple to look at the processional rituals carved inside. These figures ringed the room much like their own circling of the altar during Ch'a-Chak ceremonies. They nodded in recognition. When it stopped raining, we went back outside and passed the grinning stacked heads of the great skull rack, another remembrance of the miraculous tree of First Father and the necessity of sacrifice.

Stopping at the Venus platform, I described the dances of war and sacrifice that would have occurred here and talked about the great War Serpent depicted on the platform, the *Waxaklahun-Ubah-Kan*. I also explained the meaning that the star Venus had to the ancients. When we got to the cenote of sacrifice, I described the portals to the Otherworld. Don Emet said, "That's the way to hell!" He pointed down into the cenote. "Unhappy young girls still kill themselves by jumping into cenotes," he said, "and they go right to hell."

We ended up finally at the summit of the Temple of the Warriors, where I tried to explain the concept of Itzam-Yeh, as we stood before the great screaming battle-bird images that decorate the building. "You know what *itz* is," I said and I started to list some meanings. One of the young men lit up with recognition. "Yes, *itz*. It's sap, nectar." He did know what I was talking about. We spoke of the rain, the *itz* of the sky, and looked about us hopefully in this year of drought. I felt that the path of understanding I was trying to build was shaky but promising.

We went inside the sanctum and up to the table-throne in the back of the temple. In the twenty years I had been visiting this site, I had never seen Maya men standing next to it. Suddenly, in a flash of insight, I understood its function. The table top came up to the middle of their chests, the same height as the altars these men build for Ch'a-Chak ceremonies in their own villages; and the top is just about the same size. Of course, I thought, that's why the throne top is held up by small dwarf

The men of Yaxuna pose around the Chak Mool at the Temple of the Warrior on July 26, 1992

FIGURE 8:34

figures with their arms in the air. Just as the Ch'a Chak table makes a model of the sky, so does the altar inside the temple. And like the sky monster inside Temple 22 at Copan, it is held up by sky-bearers—this time dwarves instead of Pawahtunob.

By now I was pretty exhausted, talking and listening nonstop for more than an hour. I looked out over the restored grandeur of this famous urban center and thought about the odd twist of destiny that had brought me here trying to teach the descendants of its makers about it. I realized how much it mattered to me. Even though my first attempts were clumsy, at least I was on my way. The fates of Maya people, of poverty and marginality, of genocide and exile in a vast and indifferent world, loomed like the ancient gods of death on this gray afternoon.

My friends wandered out to the front of the temple and one of the American Spanish teachers with us called them all over for a group photo. With a boyish whoop Don Emet jumped over to the famous statue of *chak mul* and sat down majestically with a grin, blissfully ignoring the large sign forbidding anyone to touch the statue. The others gathered around and I yelled, "I want that picture!" For just a moment, Chich'en Itza was theirs again (Fig. 8:34). Not mine to explain, not for the tourist strangers wandering in droves across its broad expanse, but theirs to enjoy. Don Emet holds the hope of the kings and Hero Twins in his heart. He and all our Maya friends who have shared their world with us stand prepared to defy the fate that tried to reduce their people to obscurity, and we will stand with them.

The Maya cosmos is a place that is still alive today. The Maya still play ballgames; still dance; still stand prepared to battle for their cultural autonomy; and still nurture their gods with holy objects, food, and the places they make. Their reenactment of Creation occurs in their fields, their homes, and in their places of worship, as it has been done from the beginning. Looking up into the sky, we showed you how the Maya book of Creation is inscribed in its stars. *Bey ti' ka'an, bey ti' lu'um*, "As is the sky, so is the earth," says John Sosa's shaman teacher in Yalkoba. And so it is: the book of the sky reflects the daily creation and renewal of the earth. The Maya people who have befriended us, who teach us even as we try to teach them, want to be part of the modern world around them. They want their children to live better lives, with good health care, good education, and opportunities for gainful employment. But they also want to speak their languages, to practice the ways of their parents and grandparents, and to bring to the world, openly and proudly, their insight into its workings. We think the people who want to know about the ancient Maya should know their descendants. Theirs is a great tradition, a gift for us as we look forward to a future in which we too face a changing world that we must make anew.

AFTERWORD

A FINAL WORD ON TREES FROM LINDA

When last we wrote a book together, David wrote the epilogue and had the final eloquent word in our combined effort. This time I get the last word and I have given much thought to what I want to say. The evening I opened the first file and keyed in the first words of this book, my colon ruptured and I found myself poised at the entrance to the road, hovering just above the gullet of the White-Bone-Snake, ready to fall down the Wakah-Chan after my old friend Pakal. A skilled surgeon, heavy doses of antibiotics, and a lot of time pulled me back from the brink, and gave me time—a third try, if you will, at the business of life. The first time I had stood at the edge of the Ek'-Way and looked down the black hole was in 1982 when I first said the words "I am an alcoholic" and admitted that I needed help. Both instances had the effect of reminding me of what was truly important in life.

During the last year of writing this book, a series of trees conspired to remind me of what I had begun to learn in those traumatic experiences. The first of these trees spoke to me in January of 1992, when I was guiding a group of people through Tikal's history and magic. My friend Merle Robertson was leading a second group, which had merged with us for our five glorious days at Tikal. That morning her group was late, so our guide parked us beside a tree at the place where the dirt road in the park forked. One fork led into the ruins and the other led up to the Tikal Inn.

Most of the people in our van took advantage of the interlude to shop at the nearby stalls set up by highland Maya who had come down to Tikal to take advantage of the steady supply of tourists. However, one of our group, Harriet Gillett, a retired physician and an inveterate bird-watcher, had other interests. She noticed a nearby tree heavy with white blossoms and surrounded by a raucous sphere of birds and bees. She climbed out of the van with her binoculars around her neck, and walked over to take advantage of the unexpected opportunity the morning had provided. Our local guide, Francisco Florián, who knew the forest and its creatures in an unusually intimate way, joined her, explaining that the birds came to the tree only early in the morning.

The sounds and the odd sight finally drew my attention and I too disembarked from the van and edged closer to the buzzing center of the action. I stared at the screaming birds as they fought for positions among the flowers and the hovering drone of thousands of bees. How beautiful, I thought, and then my gaze happened to settle on the trunk of the tree. It had thorns and it bulged just above the ground. It was a young ceiba tree. I already knew that the ceiba was the model for the sacred World Tree of the Maya, but I had never seen one in flower when I knew what I was looking at. I was really excited because normally you can't see the blossoms even if you're there when the tree is in bloom. The fully mature trees are hundreds of feet high (Fig. 9:1) and the blossoms are very small.

"It's a ceiba," I chirped and began looking for a branch low enough to see one of the blossoms up close.

Joyce Livingston, a retired teacher, did the logical thing. She bent over, picked up a fallen branch, and held it out for me to see.

I was too excited and full of myself to listen. She tapped my arm more insistently and still I didn't hear her. Finally, in frustration, she grabbed my wrist and raised her voice.

"Will you look at these?" she said, waving the branch, and finally I did.

What I saw stunned me, for in her hand (Fig. 9:2) lay a perfect replica of the earflares worn by the Classic Maya kings. Suddenly I understood the full symbolism of so many of the things I had been studying for years. The kings dressed themselves as the Wakah-Chan tree, although at the time I didn't know it was also the Milky Way. The *tzuk* head on the trunk of the tree covered their loins. The branches with their white flowers bent down along their thighs, the double-headed ecliptic snake rested in their arms, and the great bird Itzam-Yeh stood on their head. I already knew as I stood under that young tree in Tikal that the kings were the human embodiment of the ceiba as the central axis of the world. As I stood there

A mature ceiba at Tikal

FIGURE 9:1

The flowers of the young ceiba at the entrance to Tikal

FIGURE 9:2

gazing at the flowers in Joyce's hand, I also learned that the kings embodied the ceiba at the moment it flowers to yield the *sak-nik-nal*, the "white flowers," that are the souls of human beings. As the tree flowers to reproduce itself, so the kings flowered to reproduce their world.

When I returned to Austin, I went to the Plant Resources Center of the University of Texas Herbarium and spoke to Dr. Carol Todzia about what I had seen. She helped me find out about the life cycle and natural history of the ceiba and exactly how unusual the little encounter was. Not only are ceibas so high that you usually can't see the blossoms, I learned that they do not necessarily bloom every year. In fact, it can be as much as ten years between such flowerings. I also learned that ceiba blossoms open at night and that the fruit bat is one of its principal pollinators. The blossoms remain open only in the very early hours of the day and then close against the rising heat. If we had come an hour later, the birds would not have been there to draw Harriet's attention. And finally, the ceiba flowers in January through the first week or so of February. If I had come at any other time of year or a little later in the day, that baby tree's flowers would not have been there to see. Finally, I am realizing only now, as I write this, that the tree blossoms in the month just before Creation day on February 5, so that it was in flower when First Father raised it into the sky. It bears fruit in the dry season around March 21, the equinox, just before the first planting of maize. The myth of Creation maps onto the life cycle of the ceiba just as surely as it does onto the movements of the Milky Way.

I thought the ancient Mother-Fathers were done with their gifts to me after that glorious little revelation, but they had just begun. In the midst of my research[1] into the life cycle of the ceiba, Creation broke open and I was diverted by the heavens for the next three weeks. When next I met a tree, it was late February and I was with Duncan Earle during our trip to the Festival of Games at San Juan Chamula in Chiapas, México. I described portions of this trip in several earlier chapters.

We had arrived in San Cristóbal de las Casas late on the last Thursday of February, spent one day orienting ourselves in the town, and gone out to Chamula to meet Duncan's friend, the outgoing Pasión. He was the principal official for the barrio of San Sebastián during the five days of the festival. On Saturday, we had made our way back to Chamula for the afternoon activities where the incoming Pasión of each of the three barrios gives his *compiral*, a feast in which he feeds most of the community.

Maya festivals, like many such activities in the world, are characterized by long hours of negligible activity punctuated by moments of furious action. During one of those slow interludes, Duncan and I wandered into the central square of the town and sauntered around the edges of the open market that occupied the middle of the plaza (Fig. 9:3).

As we walked, I noticed one of the crosses that ring the plaza at regular intervals and moved closer. I thought I knew what to expect, but I was wrong. The cross was more than ten feet high and mounted in a concrete base along with another one just a few inches shorter. Brown, dry pine tips almost as high as each cross were tied to the back of the crosses, which were painted a bright pastel blue-green. All this was exactly as I had read about it, but when I looked more closely, I noticed the image of five-petaled flowers cut into the arms of the cross and picked out with a bright silver paint.

I was fascinated. No description I had ever read described flowers on the crosses of Chamula. Duncan and I began walking from cross to cross and I noticed other things. There were angular drawings of pine and a leafy vine on the vertical shafts and between the flowers. I began drawing

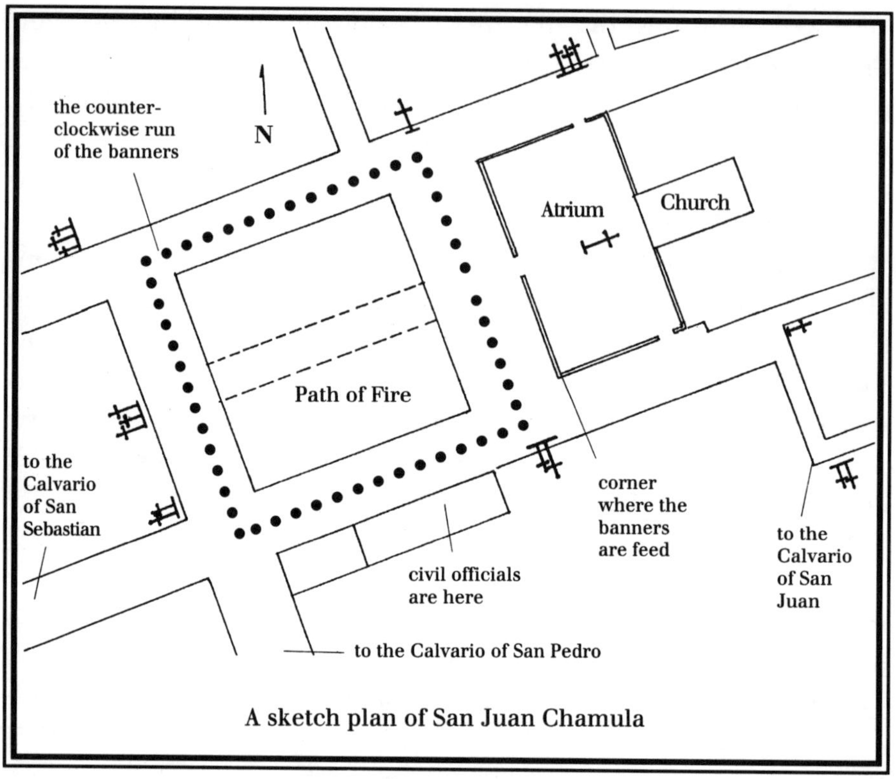

FIGURE 9:3

them in my book, marking where they stood in relationship to the plaza. I found that each cross was a little different from its neighbors, and that most of them were dated so that we could began a seriation of them right there. At first, Duncan was bemused as the archaeologist came out in me, but he was soon as excited as I was.

As we neared one set of crosses on the north side of the plaza, we passed a house holding a saint, and we were invited to enter. I liked the idea of sitting down for a while, so we went in, looked around, and I sat down on the wooden benches that ringed the walls. Soon after our eyes adapted to the dark interior, Duncan came and asked me if I didn't want to buy our hosts a couple of soft drinks. I agreed and found myself buying a case. As I downed my tepid orange soda, I had Duncan ask if they minded if I drew the altar in front of the saint's compartment, which resided behind a hanging wall of bromeliads. Like the other Chamulas I had met, they were fascinated with my drawing and soon there was a crowd pressing around me. I took the opportunity to ask them about my drawings of the crosses outside.

I pointed at the flower patterns on the end of the arms (Fig. 9:4).

"Nichim," they replied.

"Flowers," Duncan translated.

I had him ask them in Tzotzil if there was anything more specific.

"Yes," Duncan replied. "They say, these are the flowers of the arms." He pointed at the flower on the end of the cross bar. "And this is the head flower . . . and the foot flower is there on the bottom."

"And this one in the center?" I asked.

"Oh, they say that's the heart, *yonton*."

So I learned that the cross is a living being that, like a Lakandon god pot, has anatomy analogous to a human being. Later, David Stuart showed me a paper that Bob Laughlin (1962) had published thirty years ago showing that the Tzotzil use flowers as a statement of holiness. Victoria Bricker (1973:88) has even said: "The adjective 'flowery' (*ni-chim*) is used in contexts in which Christians used the term 'divine.' " The flowers marked the crosses as divine beings.

And then, ten days later, when I showed my drawings to Gary Gossen and Victoria Bricker, both of whom have written definitive descriptions of the Chamula Festival of Games, I asked them why no one had mentioned the flowers on the Chamula crosses. Gary said he had just never paid any attention to them. When I was talking with him, we

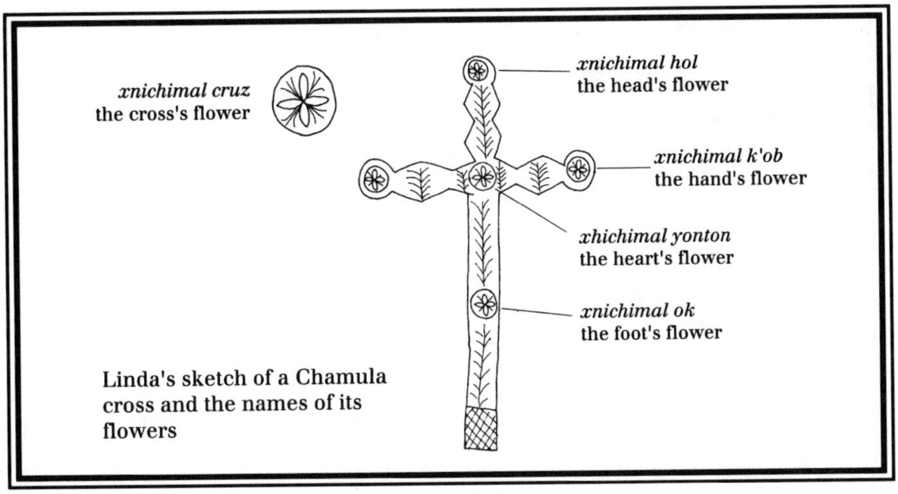

FIGURE 9:4

happened to be looking through Vicky's book, *Ritual Humor in the Highlands,* when I noticed the drawing of the Festival of Games her Chamula informant, Mariano López Calixto, had made for her. He had made sure the flowers were on the cross when he drew one in a scene of the Dance of the Warriors (Fig. 9:5). I realized that for the Chamulas, crosses are trees in blossom and always dressed. They are like the ceiba I saw six weeks earlier in Tikal. They are the Wakah-Chan World Tree with its blossoms carved in the conjuring house Chan-Bahlam put in the Temple of the Cross at Palenque. They are the Wakah-Chan in the sky.

But the Father-Mothers were still not done with me. There was one more gift that awaited me, this time delivered by Duncan on the Monday before we went to Calvario in the night.[2] He had a friend who had rented a car and volunteered to take us anywhere we wanted. It was late in the morning when we finally set off and we didn't have much time because I had promised to meet another friend for a late lunch. We piled into the car and headed for Zinacantan so that I could see the navel of the world that David described in Chapter 3. After seeing the churches in the center of the town, Duncan asked where the navel was, and an official pointed

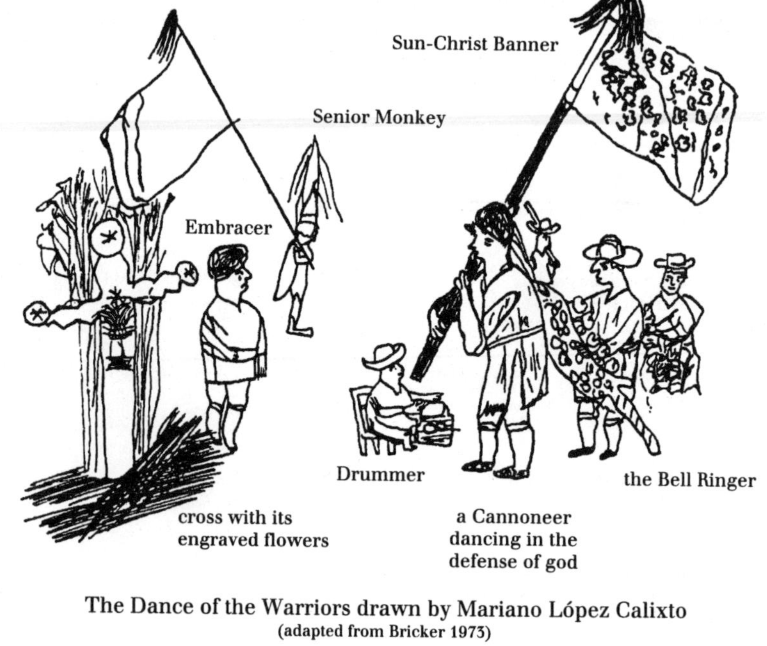

The Dance of the Warriors drawn by Mariano López Calixto
(adapted from Bricker 1973)

Figure 9:5

up toward the road we had driven down to a pointed little hill topped by a stand of pine. It was bigger than the little bump that David had described, yet it still seemed small and a little off to the side to be the center of the cosmos. David was right. You can't always judge the importance of things by the yardstick of our world.

Duncan urged us back into the car and we headed back toward San Cristóbal, to my lunch appointment, I thought. Then, he suddenly turned off a rutted dirt road that circled the outskirts of San Cristóbal.

"I want to show you something. It's only fifteen minutes away," he assured me.

Twenty minutes later we were still zipping around tight mountain curves as I repeatedly glanced at my watch. I was deep in my gringa mode because I thought I was going to be late. We raced through the mountains, and finally Duncan slowed and began aiming the car toward the side of the road.

"What's wrong?" I asked anxiously.

"Nothing. We're there," he answered.

"Where?"

"Romería."

I looked around at a bare hill, with the stain of orange-red earth broken by scraggly clumps of dry, brown grass. There were a few poor houses on the other side of the road and litter thrown carelessly about. I was confused.

"Where?" I asked again as he drove off the road and into the hard, rough ground of a field.

"There." He pointed up to the crest of the hill where a line of huge crosses poked their needle-thin silhouettes into the steel-gray, overcast sky.

We got out of the car and, pummeled by an uncomfortably cold wind, we walked into the strangest, most haunting place I have ever been.

The crosses, each painted green, incised with its proper flowers, and dressed in straw-brown pine boughs, marched across the crest of the hill like a row of sentinels thirty feet high. As we made our way up, carefully picking our way among the graves and their tiny, knee-high crosses, Duncan described the Day of the Dead, when thousands of Chamulas came to Romería to have a picnic and share food with their dead. Around us lay old shoes, an occasional fork or broken plate, and everywhere the little green flowery crosses at the heads of the graves, sometimes in two or threes, sometimes fallen and rotting on the red-clay earth.

All across the hill lay huge wooden planks the size of doors. Some had been carefully arranged on top of the graves, others lay askew, the individual planks broken apart.

"What are these?" I asked in voice barely above a whisper. Sound seemed so out of place there that I felt I might be disturbing the souls lingering nearby if I talked too loud.

"These are doors," Duncan answered. "The Chamulas believe they open on the Day of the Dead to let their ancestors come back to share food and drink with their descendants."

We left soon after that, but the image of Romería has been sitting just below the surface of my mind ever since. For me, Romería and the tomb of Pakal in Palenque merge into a continuous stream of thought and understanding that flows across thirteen hundred years of history. That history has often been a trying one, punctuated with momentous and sometimes devastating events. We had thought the trauma of these events—the Classic-period collapse, the Spanish Conquest, and five hundred years of colonial domination—had cut the umbilical cord linking the ancient world of the Maya to their modern descendants, as in the Yukatek story. We were wrong.

The dead of Romería are buried at the foot of flowery crosses that are the direct analog of the Wakah-Chan that covers the body of Pakal. Both crosses are alive; both have flowers at the ends of their arms. To the ancients the flowers were the soul; to the moderns, they are divinity. Pakal lay in a coffin connected to the floor of the temple-mountain above him by a hollow tube that was a Vision Serpent. His descendents conjured up his soul through that snake so that he could aid them in unfolding the future he had helped make. The modern dead rest on the summit of a low hill that is the natural analog of a temple-mountain. They lie below a door that opens so that their descendants can come and share food and drink with them, and secure their ancestors' assistance as the future continues to unfold. The dead in both worlds remain a vibrant force in the world of their descendants, who live in a cosmos alive with the power of *ch'ulel.*

The time between August 6, 1991, when I stood near Waqibal in Momostenango and knew I was at the center of a world more than three thousand years old, and April 13, 1992, when I write this epilogue, has been the most productive period of my life. When we began writing this book two years ago, I wrote in what I now call the voice of authority, as if I truly knew something about the Maya world. In the process of writing this book we have become uncomfortable with that voice, especially when it is wielded by those who speak from the world of the Western scholar. We have become instead witnesses, who are privileged to go into that Otherworld of the Maya and come back to our own people to report what we have seen. All of us have much to learn from the Maya and their cousins who lived in these continents for many millennia before the European peoples happened upon them and changed their lives forever.

NOTES

Foreword: The Orthography

Page 16

1. Cognate sets are groups of equivalent words with the same meaning that are found in different but related languages, and that are derived from an original word in a mother language common to them all. Historical linguists use these cognate sets to determine what phonetic changes have occurred in a language family and the sequence of changes that resulted in the development of daughter languages from their mother language. A great many of these cognate sets allow the vocabularies of such mother or protolanguages to be reconstructed. These kinds of studies allow us to know that *kan* ("four") in Yukatek is the cognate of *chan* ("four") in Ch'ol. Knowledge of cognate sets in vocabularies and in the phonetic patterns of all the Mayan languages are also fundamentally important to the use of the dictionaries and the reconstruction of the vocabulary of the Classic-period inscriptions.

Chapter I: Worlds Apart, Joined Together

Page 39

1. Vogt has discussed this important symbol in a number of contexts, but he has summarized his views on the syncretism of Christian and Precolumbian Maya cross imagery in a recent paper entitled "Indian Crosses and Scepters: The Results of Circumscribed Spanish-Indian Interactions in Mesoamerica." We quote from his conclusions in this article:

> the ancient Maya "cross" (world tree) and the Spanish cross were indeed startlingly similar in form but quite different in basic meaning. The Maya "cross" symbolized the world tree, the *axis mundi*, and was personified (which surely indicated that it had an inner soul) and clothed (in jewels and mirrors), and had a spiritual relationship to supernatural ancestors. The Spanish cross was basically a structure on which Christ was crucified, but, at least in some Catholic thinking, was related to the tree of life in paradise (Nuñez de la Vega [1988]), a concept that is similar to the Maya world tree. After syncretism occurred, the ancient Maya meaning continued, but some new meanings were added, or at least strongly emphasized by the Maya: the cross as a boundary (between Nature and Culture, between town and woods); the cross as a guardian; and especially the cross as a symbol of collective identity for families, lineages, hamlets, towns. Further, I suspect that not only in Yucatan (Farriss 1984:314), but in Chiapas and Sonora as well, the plethora of crosses are manifestations of drawing ideological support and symbols from Christianity to oppose Spanish rule and bolster Indian power.
>
> (Vogt n.d., parens original)

Crosses and trees played central roles in the Maya vision from the earliest material manifestation of their symbol system until today.

2. Determining the language spoken and written at any particular site is difficult in the best of circumstances. The most reliable evidence comes from specific spelling or grammatical forms that occur only in Yukatekan or Cholan. Several of these kinds of spellings identified Palenque, Yaxchilan, and Copan as Cholan-speaking, while the spelling of the month Pax and other grammatical features identify Caracol and the Nah Tunich region of southern Belize as Yukatekan. No such unequivocal evidence has been found at Tikal, but the phonetic spelling there seems to be far more productive in Yukatekan than in Cholan.
3. The lowland Maya of the colonial period did work to maintain long-distance contact, exchange, and communication. See the absorbing account of Maya resistance to the Spanish given by Grant Jones (1989).
4. The Popol Vuh, or the Book of Counsel of the K'iche', was written down around 1550–1555 by a member of the K'iche' elite from Santa Cruz K'iche' in the highlands of present-day Guatemala (Edmonson 1971:vii). The Community Book of Utatlan, the original from which the Popol Vuh was derived, was likely a codex or set of codex books written in Maya glyphs. Edmonson notes in his preface to his 1971 translation that there are clear indications that the original also included almanac pages used for divination, comparable to the pages of the four surviving Maya codices, three of which are probably from Yukatan and all of which date from the Postclassic period or the centuries just prior to the Spanish Conquest. There are several English translations of the Popol Vuh. The Edmonson edition has the advantage of a running K'iche' transliteration for those interested in checking the Mayan. The translation by Dennis Tedlock (1985) is a flowing and accessible narrative that incorporates the insight of a native K'iche' speaker knowledgeable in the ways of his people, and Tedlock's own training as a K'iche' day-keeper. Most of our summaries of the Popol Vuh stories are taken from the Tedlock translation.
5. Michael Coe has written extensively on the subject of the Ancestral Hero Twins (1973, 1978, 1989). In 1986 at his public talk at the Symposium of the Blood of Kings exhibition in Fort Worth, Texas, he first proposed that, from the Late Preclassic period onward, the stories of the Hero Twins functioned for the Maya as the *Iliad* and *Odyssey* did for the Greeks
6. Our intention here is not to review the entire history of research in our fields, but rather to acknowledge the most important contributions to the particular approach we are taking in this book. However, we recommend Michael Coe's (1992) interesting and highly entertaining recounting of the decipherment of the Maya writing system.
7. In his unpublished history of research, George Stuart (n.d.) evaluated the remarkable contribution of Charles Etienne Brasseur de Bourbourg. When he became fascinated with the Maya, Brasseur de Bourbourg focused his attentions on searching for, copying, purchasing, and collecting all the old manuscripts he could find. He found the Yukatek Motul dictionary, which is one of the oldest and most important dictionaries of Conquest-period Yukatek. He borrowed and copied the Popol Vuh and then he published the K'iche' along with his own translation. He also published the *Annals of the Cakchiqueles*, and he transcribed the first dictation of the Maya drama called the *Rabinal Achi.* In pursuing his interests, he learned Nahuatl, K'iche', and Kaqchikel. And finally, he discovered the major part of a codex, the third one to be found, in the possession of a Spanish scholar and published it under the auspices of Napoleon III. Many of his interpretations have not stood the test of time, but his contribution in collecting and publishing primary sources was fundamental to the field.
8. We rely on a history of research written by George Stuart in 1985 for a book he was writing with Linda Schele and David Stuart, called *Ancient Maya Writing.* We never finished the book, but George published his history of research in 1992. It remains one of the most detailed and insightful accounts available to us. His research was less concerned with who discovered what detail when than with when resources were published and who had access to them. An equally interesting history has recently been published by Michael Coe (1992) in his *Breaking the Code.*

   Among the early sources he cites are the descriptions of Peter Martyr, a witness who chronicled the Spanish Conquest; Diego López de Cogulludo, who wrote *Historia de Yucathan*, which was published in 1688; and an account of the conquest of the Itza published by Villagutierre Soto-Mayor. A guide to these and other early sources is published in Volumes 13–15 of the *Handbook of Middle American Indians* (Wauchope 1974–1975) and many of the most important have recently been translated into English, annotated, and published by Labyrinthos Press (Culver City, California). The Universidad Nacional Autónoma de México is presently engaged in a truly remarkable program of publication concentrating on these earlier sources: these include both transcription and facsimile additions. Both these publication series are providing an invaluable resource to a wide range of scholars in the field.

9. Maudslay's amazing plans, drawings, and photographs were published between 1889 and 1902 in the four volumes of the Archaeology section of the *Biologia Centrali-Americana* (Maudslay 1889–1902). Maudslay, along with his photographer, Henry Sweet, and his artist, Annie Hunter, created a primary resource of incalculable importance. The standards of drawing and photography, as well as glyphic commentary, were so high that we still struggle to match them today. Not until Ian Graham's *Corpus of Maya Hieroglyphic Writing* was Maudslay's work equaled in accuracy and expanded to incorporate major new bodies of texts and images. We still use his work today as a major resource.

 Teobert Maler, the other great recorder of the period, was born in Rome of German parents. He came to México with Maximilian's army and stayed after the fall of the regime. Like others before and after him, he became fascinated with the Maya and found himself sketching in Palenque. Taking a huge heavy camera and glass plates with him, Maler traveled through the Maya region from 1877 to the turn of the century, photographing and mapping and giving special attention to the inscriptions and images of the sites he visited. His meticulous journals and photographs were published in the Peabody Museum Memoirs beginning in 1901 (Maler 1901–1903, 1908–1911). Since many of the monuments he photographed have since been lost, stolen, or destroyed, his work preserved an invaluable segment of written history and religion of the Maya. Maudslay and Maler left us an unparalleled heritage that would not exist had they not committed their lives to the ancient Maya. Furthermore, even when the monuments have survived time and looters, they cannot be studied effectively without photographs and drawings that can be taken into the libraries and homes of Mayanists in distant lands.

 Sylvanus Morley not only directed many of the early excavation projects, but most notably for us, his two great resource works, *The Inscriptions at Copan* (1920) and *The Inscriptions of the Peten* (1938) are still used today and have played a vital role in the process of decipherment. More recently, Merle Greene Robertson (1982–1992) has spent her life recording through drawings, photographs, and rubbings the art of Palenque, Chich'en Itza, and many other sites. Without these pioneers, and their later successors, Ian Graham and the other photographers and artists who record the words and images of the ancient Maya, we could not work.

10. Our understanding of the ancient Maya world was also enhanced by the pioneering work of three outstanding scholars. The first was Ernst Förstermann, a German philologist and librarian of the Royal Library of Dresden. If you can imagine working at your desk with an original painted Maya book sitting in front of you, you can picture the great privilege that Förstermann enjoyed. The extraordinary copies he had made of the Dresden Codex (two hundred in number) in 1892 are today our primary source of study of that work because the Allied fire-bombing of Dresden during World War II caused severe water damage to this most precious of books. Förstermann had not only the original Dresden codex at his disposal but also reproductions of the other two codices then known, as well as the ongoing publication of the Classic monuments by Maudslay and other ethnohistorical documents that were just then coming to light. During the last part of his long life, in a virtuoso performance that has rarely been equaled, he revealed the workings of the Maya calendar, interpreted the glyphs associated with it, and read or correctly identified many of the actions and actors shown in the codices.

 His colleague in this great endeavor was another German, Eduard Seler. Seler not only worked with ancient Maya religion and writing but combined these studies with parallel work on the Aztecs and other Mesoamerican cultural traditions. Discovering many connections between the Aztecs and the Maya, he identified gods, described ritual, suggested the nature of Maya religion, and proposed a number of decipherments that are still accepted today. Paul Schellhas, another friend and colleague of Förstermann, worked with the same material and published the first systematic study of Maya gods, giving them the alphabetic designations we presently use.

In many ways, the contributions of Förstermann and Seler are still not appreciated as they should be because only a few of their original German articles were translated into English, under the supervision of the great Harvard scholar, Charles Bowditch. Bowditch (1904) published the most important articles by both authors along with other interesting works by German researchers in a Smithsonian Institution Bulletin. Seler's collected works, *Gesammelte Abhandlungen zür Amerikanischen Sprach- und Alterthumskunde* (Seler 1902–1923), were translated under Bowditch's supervision by A. M. Parker, Selma Wesselhoeft, and others, with J. Eric Thompson overseeing the work and making slight editorial changes. The translations were copied by mimeograph, without illustrations, and distributed to a few interested scholars and to the libraries of institutions then involved in Maya research. Until the advent of copy machines, however, only students and scholars at those institutions had access to these translations. Fortunately for the latest generation of scholars, these translations are now being published, with the illustrations, by Labyrinthos Press (Seler 1990). Bowditch also translated and published, with the Peabody Museum at Harvard, Schellhas's study of the gods in the codices. Bowditch did a great service to the field by facilitating communication among scholars around the world. Working as editor gave Thompson the opportunity to absorb much of Seler's data and interpretations, information that he would eventually use in his own work. As we shall see, Thompson engendered a vision of Maya religion and life that still absorbs and fascinates us today.

Another giant of those early decades of study was Herbert J. Spinden, a man who became a leading epigrapher and the exponent of an incorrect correlation between the Maya and Christian calendar that gave us the modern myth of an astronomers' conference at Copan. His earliest work remains the most influential. As a doctoral student in anthropology at Harvard, he spent the years 1906 to 1909 studying the Maya art and architecture of the Classic period, searching for patterns. Using the drawings and photographs from Maler and Maudslay, the expeditions of the Peabody Museum, and other publications, Spinden undertook the first systematic study of iconography and meaning in Maya art with remarkable results—often underappreciated by modern researchers. He identified many of the main themes of Maya art and invented most of the nicknames we use for the images today. Most of all, he instinctively realized he was seeing a record of both historical action and religious belief. We latecomers sometimes find that our most recent discoveries were found first by Seler, Föstermann, and Spinden during the youth of the field.

11. Brasseur de Bourbourg was one of the historians and collectors who assembled invaluable archives of material from the colonial period. Juan Pío Pérez, a Yukatekan scholar who published several early documents in Stephens's collection, *Incidents of Travel in Yucatan*, was another early historian. His work was soon supplemented by others. Among these, Daniel Brinton, Ralph Roys, France Scholes, William Gates, Susan Miles, and Adrián Recinos stand out as particularly important because they not only found and preserved critical documents, they translated, analyzed, and published them so that others could have access.

George Stuart (n.d.) also cited the work of Don Crescencio Carrillo y Ancona of Izamal, Yukatan, and Carl Berendt, who both continued the work of Pío Pérez and assembled his own enormously important collections of colonial documents. The Berendt Collection of the University of Pennsylvania is still one of major importance. Daniel Brinton, a physician and professor of American archaeology and linguistics at the University of Pennsylvania, published many of the documents preserved by Pío Pérez, Carrillo y Ancona, and others, thus making them available to others for study. William Gates, who founded the Maya Society in 1920, assembled a remarkable collection of documents that were to become the foundation of the Maya library of Tulane University. He ended up at Tulane as the first head of the Middle American Research Institute, which was to become a major player in the years ahead. Gates also brought in Ralph Roys, Oliver La

Farge, and Frans Blom, three men who would make unparalleled contributions in archaeology, ethnohistory, and ethnology. Frans Blom became an important archaeologist in Chiapas; Roys published English translations of the *Chilam Balam of Chumayel* (Roys 1967), the *Ritual of the Bakabs* (Roys 1965) and other critically important documents (Roys 1943, 1957, 1962); and La Farge studied several major Maya communities in western Guatemala. The Gates collection, in original and photographic copies, eventually became part of the collections of Tulane, Harvard, Princeton, and Brigham Young universities, the Library of Congress, and the Newberry Library. See Robert Brunhouse's (1975) informative and entertaining biographies of these and other Mayanists during this early era of the field.

In 1941, Tozzer published an English translation of Landa—still the principal source we use because of the extraordinary notes he wrote to accompany the text. France Scholes, working with Roys and others (Scholes and Roys 1948; Scholes and Adams 1960), translated and published other vital documents, especially the papers of Paxbolon, which contain a history of the rulers of Akalan-Tixchel, a town visited by Cortés on his journey to Honduras. Perhaps one of the most extraordinary publications is Susan Miles's (1957) study of the Poqom Maya from early dictionary and documentary sources.

Adrián Recinos, a Guatemalan who served as ambassador to the United States from 1928 to 1944 and to the United Nations after that, translated the major historical documents of several Maya groups in Guatemala, including the Popol Vuh, the Annals of the Kaqchikels, and others, in an anthology of documents (Recinos 1957). Perhaps most important to English-speaking scholars was the publication of the Popol Vuh (Recinos 1950) and the Annals of the Kaqchikels (Recinos and Goetz 1957) in English translations by the University of Oklahoma Press. Robert Carmack (1973) published a listing of the sources available on the K'iche', including translations of several important but previously unpublished documents.

The resources above were not, of course, the only documents that have been published since the middle of the nineteenth century. There are many scholars we have not mentioned, and many more who are working today, both interpreting early documents already known and finding others still waiting in libraries, archives, and private hands around the world. In great part, these documents have provided the basis of our understanding of Maya social and religious structure, and of Maya ritual practice and belief at the time of the Conquest. They have also been invaluable to the process of decipherment. There are doubtless many more documents of this nature waiting to be found.

12. Alfred Tozzer (1907), lived with the Lakandon, the only non-Christian group of Maya to survive into the twentieth century, and wrote an invaluable comparison of Lakandon and Yukatek life-styles and beliefs. Oliver La Farge, Charles Wisdom, Robert Redfield, Alfonso Villa Rojas, Michael Mendelson, Maud Oakes, Rafael Girard, and many others recorded and published descriptions of the rituals, pageants, and practices of many twentieth-century Maya communities. Oliver La Farge, for example, published observations of several communities in Chiapas, Veracruz, and western Guatemala (Blom and La Farge 1926–1927), of a Jakaltek community (La Farge and Beyer 1931), and of a Q'anhobal town (La Farge 1947). Charles Wisdom (1940, n.d.) and Rafael Girard (1966) accumulated major ethnological materials on the Chorti of western Guatemala, including descriptions of religion, ritual, and social interactions. Wisdom compiled a huge still-unpublished dictionary of Chorti. Redfield and Villa Rojas (1934 and Villa Rojas 1945) assembled the same kind of data on towns of Yukatek-speakers in the northern lowlands. Mendelson (1956) worked at the Tzutujil town of Santiago Atitlan, and Oakes (1951) did her research at Todos Santos. Ruth Bunzel wrote on Chichicastenango (1952), and this list is far from exhaustive.

Thomas Gann, a British doctor who dabbled in archaeology throughout British Honduras (now Belize), described and photographed whatever modern Maya rituals he happened to witness on his many expeditions to various ruins. Gann's "archaeology" expeditions bordered on looting, but his penchant for detailed description of what he saw and how he found the material he excavated, as well as his attention to the Maya communities he encountered, make his many "popular" books extremely valuable records of a life-style that has been, or is being, lost. His books are far more than just travel adventures.

Since the fifties, there have been many more people who have contributed to the mass of information we have on Maya communities and their rituals. In the context of our narrative, we have drawn extensively on select scholars who have written recent major monographs and articles on contemporary and ethnohistorically documented Maya cosmology. Our attempt is not to synthesize all the work of our many colleagues and intellectual ancestors on Maya thought, ancient and modern, but rather to illustrate certain central connections and continuities that support the principle of an enduring understanding. Our Bibliography will, we hope, show our awareness of the many people who have contributed to knowledge of the Maya, even if we do not draw regularly on all of their insights and evidence.

These ethnographies—along with the dictionaries and other linguistic materials that accompany them—compose some of the most valuable resources we have available to us today. Moreover, since the earliest of these observations and studies dates from the last third of the nineteenth century, they also represent a record of the process of adaptation and transformation in Maya communities over the last hundred and thirty years.

13. Thompson's masterpiece in this approach was published in *Maya History and Religion* (1970). In this amazing work, he outlined his ideas about the Putun Maya at the end of the Classic period; summarized what was traded lowland to highland and vice versa, using archaeology and ethnohistorical records as his sources; put together an unmatched collection of comparative mythology from all Mayan languages, and detailed his understanding of how Maya religion worked. We disagree with some of his conclusions, which we found confusing, but quite frankly we had lost sight of the extraordinary resource of information he assembled in this remarkable book.

14. Thompson released his *Ancient Maya Civilization* in 1956. Morley preceded Thompson in 1946 with an enormously popular book titled *The Ancient Maya.* In this work, Morley used the colonial texts, both about the Maya and the indigenous Chilam Balam books, along with ethnographic information, to craft an interpretation of the Precolumbian Maya. He proposed an Old Empire—what we now call the Classic period—and a New Empire, the Terminal Classic and Early Postclassic periods. He further argued for massive migrations of people out of the southern lowlands into the northern lowlands during the ninth-century collapse in the south. From the ashes of the Old Empire rose the New Empire. His analyses were brilliant, articulate, and influential for several generations of archaeologists. In basic contrast to Thompson, Morley regarded the Maya of the Contact period, with their kings, dynasties, nobility, wars, and intrigues, as the proper guide to interpreting the Classic Maya. Sadly, he never knew the glyphs he loved to study could confirm his basic intuitions in this matter.

15. Tatiana Proskouriakoff of the Carnegie Institution of Washington convinced her colleagues in Maya studies that the glyphic texts on carved stone monuments pertained to historical individuals in a series of articles (Proskouriakoff 1960, 1961, 1963–1964, 1964) on the kingdoms of Piedras Negras and Yaxchilan.

16. Thompson articulated his views on Itzamna as the apex of the Maya pantheon and as a deity of nearly monolotrous focus for the Classic elite in *Maya History and Religion* (1970). Karl Taube (1992a, 1989) reviewed Paul Schellhas's (1904) analysis of the Maya deities and discussed Thompson's hypothesis. Taube, following Nicholas Helmuth (1987), supported the notion that God D, Itzamna in the Postclassic and colonial sources, was head of the Classic pantheon of gods. During the Classic period, his name was written Itzam or Itzamhi.

17. Taube (1992a:31) points out that Seler and Fewkes had correctly identified the codex portrait of this god with the Itzamna of the colonial sources.

18. David Stuart (1987) first found the syllabic spelling of this god's name and read it as *k'awil,* a word that means "sustenance" in Yukatek and "idol" or "embodiment" in the Poqom languages and Kaqchikel.

19. At the heart of the God D—Itzamna—problem is the meaning of the god's name and its distribution in the codices and Classic-period imagery. Much of the evidence has been assembled by Taube (1992a:31–41, 1989), but we think a summary of the high points is warranted here. First of all, the name glyph that has been read as Itzamna in the codices appears only beside the old anthropomorphic deity given the designation God D by Schellhas (1904). This glyph does not appear—to our knowledge—with any of the saurian or crocodilian images. We can surmise,

therefore, that the association Thompson made with iguanas is based entirely upon his interpretation of the name. Using previous scholars' work along with his own research, Taube (1992a:36, 1989:2) has shown that Itzamna Kab Ain was probably glossed as a "whale" in colonial Yukatek and that in Classic and Postclassic imagery, the name referred to a great crocodile that symbolized the surface of the earth.

Itzam remains the problem, for if it does not mean "iguana," as the great Yukatekan linguist Barrera Vásquez (1980:272) asserted, then what does it mean? The glyphic name is composed of three signs—a small flower or shield image, a square-eyed head, and a phonetic *na.* We have known for some time that the first two signs must represent *itzam,* by recording the sounds either as *i-tzam* or *itz-am.* To our knowledge Floyd Lounsbury, in his work on the codices during the late sixties, was the first to suggest that the "shield" sign should be read as *itz,* based, he thought at the time, on its being the head of a caterpillar. The iconic identification is surely wrong, but the reading appears to be correct. In Classic-period inscriptions the *itz* sign can appear alone or with the head of God D attached to it, so that the god's portrait is the personified variant of the phonetic sign. The Shield-Jaguar of Yaxchilan and the Shield-God K of Naranjo and Dos Pilas all appear with this head variant in their names. Their names were probably pronounced Itzam-Balam and Itzam-K'awil, respectively.

Using this *itz* reading for the first sign, Freidel and Schele (1988a) proposed that the second sign in the name, the square-eyed head, reads *am.* They supported their proposal by showing the use of this head on small eccentric obsidians and jades that were used to cast prognostications. These stones are called *am* in Yukatekan sources.

Combining *itz* and *am* with *na* gives the expected name Itzamna, but we also note that in the Classic period the *na* sign was not present. Instead, the name of the god closes with a *hi* sign to give *itzamhi* or *itzamih.* It is interesting that this same set of *hi* signs stands in the place of an expected *na* sign in the *yitah* glyph for "sibling" so that there may be some regular process of correspondence registered in these differences. While we cannot explain the difference between the Classic and Postclassic versions of the name, we think it clearly points to the *na/hi* as being separate from the main part of the name. Furthermore, since *itzam* appears in many different god names, just as *k'awil* occurs in many different people's names, it must record some common property that these gods share.

In trying to determine the meaning of *itzam,* we can presume that it is a either a self-contained root or a root combined with an affix giving *itz-am* as the form. We also know that *-am* is the agentive affix in Yukatek; it forms nouns of agency from other nouns and verbs. In other words, *itzam* may be "one who does the action of *itz*" or "one who makes *itz*" or an *itz*-er. If this is the correct analysis, then *itz* is the critically important component. In Yukatek (Barrera Vásquez 1980:272), it means "milk, tears, sweat, hardened or thickened resin or gum from trees, bushes, and some herbs." It also means the wax that melts down the side of a candle and the wax from honeycombs, resins and gums used to dye cloth, rust, juice, as well as all body fluids like semen, sweat, and tears. In Proto-Cholan (Kaufman and Norman 1984:121), *y-itz* is "pitch, sap, resin" and there is an equivalent root *iitz* reconstructed for proto-Maya. *Itz* is the magic stuff brought forth in ritual and as secretions from all sorts of things—living and (to us but not the Maya) inanimate. Barrera Vásquez (1980:272) used this meaning in exactly the way we propose, for he also glosses *itz* as "wizard, sorcerer, enchanter, witchcraft, enchantment." *Itzam* he took to be composed of *itz-a'-am,* which he analyzed as follows (translation ours): "*itz* is a morpheme whose significance is related to ideas of knowledge, magic, occult power; *-a'* means water and *-am* is the actor, so that *itzam* means water wizard, he who has and exercises occult power in the water."

We agree with Barrera's logic, but we see nothing in the word *itzam* that indicates the presence of an extra *a'* to refer to "water." We believe *itzam* is simply "magician" or "wizard." In her analysis of early Poqom (Poqoman and Poqonchi') sources, Susan Miles (1957:751) described *Ah itz* as "the ancient counterpart of the modern *Ah itz,* consulted by individuals of every rank for private divination, curing, and witchcraft. It was the only office open to women, and probably had prerequisites similar to those for modern brujos of illness, cure, and dreams." For the Kaqchikel also, *itz* was the word for "wizard, sorcerer" (Coto 1983:268), and although like *way* in many modern languages, it has negative connotations, *itz* is also enchantment and sorcery in K'iche'.

Itzam, therefore, appears to be a general term for a person who manipulates the magic world—in other words, a shaman. Barrera Vásquez (1980:272) pointed out that *na* means "first," "house," and in many languages, "to know," especially through dreaming. His suggestions may explain the presence of the *na*, but since it is replaced by *hi* in Classic-period texts, we feel it may represent an additional inflectional or derivation suffix on the basic term for "magician" or "shaman." This means that the presence of *itzam* in other god names is simply to register that they too operate in the magical world as *itz*-ers. Furthermore, God D, the Itzamna of the Yukatek sources, the Postclassic codices, and the Classic pottery painting, was the principal shaman, the "first wizard" of the Maya cosmos.

The *itz* glyph, which is very probably a flower ornament rather than a shield, is worn on the headband of the god Itzamna as a phonetic and iconic clue to his name. The same *itz* headband is worn by the sacred bird that sits atop the World Tree of the Center. This shared feature led Nicholas Hellmuth, who first identified the Classic version of God D (Helmuth 1987:303–312), to associate the bird with God D. His suggestion seems to be correct, for on the Blowgunner's Pot (Fig. 2:7), this bird is shown "entering the sky" as it lands in the tree. The glyphs following the verb name it as *Itzam-Yeh*. *Yeh* or *ye* also occurs in First Father's name, Hun-Nal-Ye, and, like *k'awil* and *itz*, it has many meanings relevant to these contexts. In Cholan (Kaufman and Norman 1984:137), it is "to take in the hand" and "to give." In Yukatek (Barrera Vásquez 1980:973), it means "to put forward, to present or offer something." *Itzam-Yeh*, perhaps, names the bird as the "Wizard Giver."

20. See Nancy Farriss (1984:294–295) for an interesting discussion of Max Weber's notions of the rational evolution of religion as applied, or perhaps more accurately as misapplied, to the Maya case.
21. Farriss describes this category as follows: "Defined by the Spanish clergy as superstition and corresponding to what later taxonomies have come to call magic, this level involves the manipulation of highly discrete and localized supernatural forces for the benefit of the individual and his family" (Farriss 1984:296).
22. Farriss describes the middle level as "corporate or parochial cults with their patron deities or saints; they are still tied to a particular group but are less particularistic than the magical. Although these corporate cults might shade into semiprivate devotions to family patrons, I place them in the public, collective sphere of religious activity . . ." (Farriss 1984:296).
23. These arguments are developed in her Chapters 9 and 10 (Farriss 1984).
24. Evon Vogt also questions this position of Farriss's. In a review of her book he writes: "I have reservations about labeling the cluster of curing ceremonies, maize-field rites, and house dedication rituals as 'magic' with all the connotations of that concept. It might be preferable to consider them as 'domestic rituals' (Vogt 1976). Further, the deities invoked in these ceremonies are not always highly discrete and localized. For example, the h-men in Chan K'om regularly invoke 'the great *Chaac* for the protection of the maize plants.' (Redfield and Villa Rojas 1934:346) This deity symbolizes rain, thunder, lightning, fertility. These concepts are hardly discrete and localized, but rather represent some general phenomena in the Maya cosmos" (Vogt 1986:42).
25. In the spring of 1971, David Freidel wrote a graduate term paper for Evon Vogt and Jeremy Sabloff on shamanism in the Maya area. It was the beginning of his own journey to this book. He sat in the basement lounge of the Peabody Museum a few days after handing in the paper, and asked a question of his mentor, the great Tatiana Proskouriakoff: Did it made sense to her that the modern Maya peasants retained shamanism from their peasant forebears? And if there were archaeological remains of crystals, divining stones, and other paraphernalia of shamanism in the Precolumbian record of the elite, could shamanic beliefs and rituals be said to be the conceptual foundation of Maya political power? She answered David that it did make sense but that it would be difficult to prove. But as encouragement, she pointed out the documented descriptions of ecstatic trance by the Chilanob given in the ethnohistorical descriptions of Yukatan, a clear case of shamanic practice by Maya elite.
26. Further, as John Sosa shows in the case of Yalcoba village, the ceremonies of devotion to these gods and the cosmological principles shared by shamans and farmers alike, in their fields and in

their homes, are replicated in Farriss's second tier. There these principles are expressed by the entire community's devotion to its saints (Sosa 1988, 1990).

27. William Hanks (1990), a trained knower in the Yukatek cosmology of Oxk'intok' town, provides a detailed discussion of the spatial organization of Ch'a-Chak ceremonies in which three altars are raised, representing three powers and temples in heaven. There is a fourth power and temple, Ah Kin Tuus (High Priest Deceiver), in the west, but he is carefully avoided in ceremonies of this kind, for when he is aroused, evil flows into the world. The triadic principle is also an ancient one in Maya cosmology, as we shall see in Chapter 2.
28. In the Book of Chilam Balam of Chumayel (Roys 1936), there is a passage called the interrogation of the chiefs, in which local lords are required to answer a series of riddles in the "language of Zuhuya." For many years, Mayanists speculated that this might be a foreign language introduced by conquerors in the northern lowlands. What it actually is, as Roys understood, is an examination on the ritual knowledge of such lords, designed to test their ability to conduct the kinds of ceremonies that shamans today conduct. We speculate that "Zuhuya" is just a way of spelling "suhuy," which means "untainted" or "virginal" in Maya and refers to the pristine quality of the water and other special materials used in offerings to the gods. Here it would refer to the untainted or pristine quality of the acts and concepts given in the riddles.
29. Ch'a-Chak ceremonies were not always performed as overtly as they are today, but the basic cosmological ceremonies show evidence of continuity through the period of European domination. Nancy Farriss herself points out the time depth of the Ch'a-Chak ceremony in Yukatan: "The one quasi-communal rite of which we have any record [in the colonial period, eighteenth century] was the Cha-Chaac ceremony to avert drought. A few images preserved as family heirlooms might be taken from their hiding places and brought to a milpa, where a small group of men—certainly no substantial portion of the community—would gather to make their offerings" (Farriss 1984:292–293; brackets ours).
30. This description and explanation of Yukatek cosmology is transmitted by John Sosa (1990, 1988) from his shaman teacher. The design of the altar and the arrangement of the materials in Ch'a-Chak ceremonies carried out at Yaxuna corresponded well with the descriptions given by Sosa for Yalcoba, to the northeast. Don Pablo confirmed many of these terms and concepts in his own sacred work in Yaxuna. There is certainly variation in the particular terminology and interpretations of shamans in contemporary Yukatan. Freidel recalls his colleague José Aban of neighboring Yaxkaba observing in 1986 that Don Pablo was a good practitioner but that he did things differently from the shamans of Yaxkaba, only a few kilometers away. William Hanks (1990) provides very detailed accounts of the Ch'a-Chak ceremonies of Oxk'intok' and their spatial-cosmological import. There, the shaman does not have the "hanging platform of the sky" portal. Instead, spirits are called down to specific positions within the fourfold/center pattern to participate. The variations are important, but they operate on the basic theme of cosmic order and communion with the supernatural forces.
31. As we discussed in Note 19, *itz* refers to special liquids and essences that include morning dew, the holy water sprinkled by ritualists with an aspergillum, and semen. *Itz* can also refer to the nectar of flowers. John Sosa (1990) reports that his informant, a contemporary Yukatek h-men, explains *yiitz ka'an* as the blessed substance of the sky which flows through the portal represented by the hanging sky platform on the shaman's altar. On the altar, the dripping wax of the votive candles, also food for the gods and spirits, is *yiitz kab*, "wax," and this represents the flowing liquid of heaven. This notion of the blessed substance of the sky is also an old idea, noted by the historian Lizana and mentioned by Taube (1992a:34).
32. The seminal comparative work on Maya houses is Wauchope (1938).
33. Evon Vogt (n.d.) made this observation of the mirrors on ancient and modern crosses.
34. Many religions have a cabalistic use of numbers to signify powerful and important ideas. Judaism, for example, still employs this practice.
35. Nancy Farriss (1984:324) makes this point in her detailed analysis of colonial Yukatek Maya cosmology: "The Maya expressed the dual sacred quality of food—as something the divine powers give and receive—in their use of the Spanish term *gracia* as a synonym for the all-important, and sacred, maize."
36. In all, there are three focal positions in the ar-

rangement of Maya ritual—the center, the altar and offerings, and the practitioner who opens the portal. In modern Yukatekan ritual, the center position is demarcated by the cross, while in the Classic system the center was represented by the World Tree in its various manifestations. The altar/offering container has many manifestations, but in Classic-period imagery, its principal representation was the Quadripartite God, who is the personification of a large offering plate made with flaring sides. Known from many archaeological contexts, this plate, called a *lak* (Justeson n.d.; Houston and Taube n.d.) in the nomenclature of the time as well as in modern languages, often carries a *k'in* sign, meaning "sun," "day," and "festival." Some of these plates also have two small loops reading *te',* the word for "tree" and things made of wood, such as the gourd offering bowls used by Don Pablo. The plate when it is used in lip-to-lip cache vessels often has the quadrifoil shape of the portal or a field of *ch'ul* incised on the top. The plate functioned as a portal into the Otherworld through which sustenance was dispatched to the gods and ancestors and through which they in turn journeyed into the human world.

The function of the offering plate as a portal was enduring. In the Postclassic Dresden Codex, offering bowls are depicted resting on the ground in front of a stela, which Landa called an *akantun/akanche'* ("standing-up stone/standing-up tree") (Fig. 1:5). Called a *Yax Am Che'* in the Dresden, this sacred tree was erected during the New Year's ceremony in which the new Year Bearer was put in place. In these scenes the Year Bearer brings offerings to the Yax Am Che'. These plates hold offerings of fish, turkey, incense, and most important, maize bread, called *wah* by the Maya (Love 1991; Bricker 1991). Bruce Love (1991) documents the continuous use of sacred breads, with various shapes and meat additions, from Precolumbian and colonial times up to the present. Don Pablo places various kinds of *wah* on his altar after they are baked in a nearby underground oven called a *pib.* The altar supporting these offerings is called a *ka'an che'* ("sky tree"; *che'* is "tree" in Yukatek, while *te'* is "tree" in the Cholan languages), and it is conceived by the modern Yukatek shaman as a portal through which sustenance travels in both directions.

The last analogy concerns the person of the shaman. For the Classic Maya, the most potent shaman was the king himself, who signaled this function by adorning his loincloth with an image of the World Tree and wearing the Cosmic Bird in his headdress (Schele and Miller 1986:76–77; Schele and Freidel 1990:90–91). He was the earthly embodiment of the *axis mundi* and the instrument for opening the portal. In modern thinking on the matter, Maya shamans create a conduit to the supernatural through meditation, prayer, and trance, particularly when they are in close proximity to the altar and the cross.

37. John Sosa (1990) has discussed the altar of the Yukatek-speaking h-men, "doer," in terms of its cosmological metaphors. The Yukatek Maya called their altar a *ka'an che',* "sky tree." In Classic Maya theology, the World Tree of the Center was called *wakah-ka'an,* "Raised-up Sky," in its Yukatekan form (Schele and Freidel 1990:66–67). The name comes from the action of separating the sky from the earth performed by First Father a year after creation began. As we shall see in Chapter 2, the erection of the central axis not only separated the sky from the earth, but it imparted rotation to the star fields around the polar star. Over seven hundred years later, the three children of the first parents were born. Two of these were the sun and Venus, who dance across the heavenly house created by star fields. John Sosa (1990, personal communication 1991) informs us that circles of vines arching over the altar are called *speten ka'an,* the "platform of the sky," and that they metaphorically create a "hole in the sky." Sosa describes this "hole" (which is coincidentally *hol* in Yukatek) this way: "The shared qualities of 'perforation' at the 'edge of Yalcoba' [entrances to the community] as well as at the 'edge' or 'center' of the world, seem to be a function of the Maya belief in a flat world, since such a conception requires some sort of 'hole' to explain how the sun, moon and stars can seemingly pass through its plane at the horizons. At the center, the realm of humans and that of deities are joined at *kumuk' lu'um,* 'the center of the sky' where the h-men believes *u hol gloriyah,* 'the door or hole of heaven' to exist. The conduit itself, then, which achieves this joining is *u yitzil ka'an,* 'the liquid substance of the sky,' and corresponds to *u yitz kib'* of the ritual table [candle wax on a stick piercing the altar directly below the circle of vines], while *u hol gloriyah* is equivalent to the *speten ka'an* hanging platform." (Sosa 1990:139–140; brackets ours). Not only is the modern altar a portal to the world of the

gods, it is the current transformation of a specific kind of ancient portal that was, in Classic belief, brought into being in First Father's act of creation.

38. This name for the h-men's altar in contemporary Yukatan was supplied by social anthropologist John R. Sosa (1990), based upon his fieldwork with modern ritualists. *Ka'an che,* with the likely same meaning of "sky tree" or "sky of woody substance," is reported by the distinguished Yukatekan scholar Alfonso Villa Rojas (1945:51 and 57) as the name of little gardens raised upon wooden frames among the Maya of east-central Quintana Roo. The gardens were used for onions, parsley, and other minor additions to staple fare. Villa Rojas's (1945:Pl. 5c) photograph of one of these gardens shows it to be about the height of a person and supported by six posts. In these Maya communities, the altar is called *zuhuy mesa,* "untainted or holy table." There is no doubt that in the case of Sosa's observations, the name of the altar is intentionally metaphorical and pertains to a cosmology in which the upper elements of the altar represent the sky and its portals. The linkage to raised gardens is more problematic, but illustrates the care we must take in dealing with the linguistic homophonies and metaphors of other people.

39. William Hanks (1990:334) gives "elevated wood" as the English rendering of the term *ka'an che* when used in reference to the altar. Normally, *ka'an* means "sky" or "heavens" in Yukatek. Clearly it can also mean "elevated." This ritual usage of the term *ka'an* combines and collapses both the given terms of the Classic World Tree, "raised or hoisted up" and "sky," *wakah ka'an.*

40. See Barrera Vásquez (1980:748) glosses *tab* as "cord or rope with which the Indians tie and carry cargos on the back."

41. William Hanks (1990: Fig. 8:4) shows a quite different spatial and numerological patterning for the Ch'a-Chak ceremonies of his teacher in Oxk'intok'. Nevertheless, his comprehensive analysis of spatial patterning and cosmology shows important common structures underlying such variability in performance and expression. We deal with his schemata more in Chapter 3.

42. We work within the theoretical framework of Structuralism (Freidel and Schele 1988a; Freidel and Schele 1988b; Schele and Freidel 1990) as do many of the scholars we draw on in this book (e.g., E. Z. Vogt 1976; G. Gossen 1986). We believe it is possible to explain variability in cultural expression among historically related peoples in a region, and, alternatively, through time among sequential phases in a cultural tradition, as derived from basic structures of thought. Structuralism is a controversial approach to interpretation and there are numerous other ways of making sense of human experience. We find that it works for the Maya and we think it works because the Maya think about things like cosmology in rational and structurally ordered ways. We further believe that the empirical, material evidence of the ways that the Maya think support this contention.

Chapter 2:
The Hearth and the Tree

1. Dennis Tedlock (1986:79; brackets ours) explains the "mountain-plain" concept as follows:

> Through their *tzih* ["word"], their *naual* ["spiritual essence"], and their *puz* ["sacrifice"], the K'iche' creators "carried out the conceptions of the *huyub tacah* . . . , literally 'mountain(s)-plain(s).' " *Huyub takah* is what might be called a pair of antonymic synecdoches for the earth as a whole; at the same time, the two nouns together are the principal K'iche' metaphor for the human body, sometimes phrased in modern prayers as "my mountains, my plains. . . . " Indeed, the *Popol Vuh* itself, elsewhere in the passage under consideration, makes it clear that the creators conceive both the earthly and human forms at the same time. . . . Confronted with a world apparently containing only a sky and a featureless sea, they make manifest an essence concealed in the sea, namely the *huyub takah.*

2. The full series of events and insights are detailed in the *Workbook for the XVIth Maya Hieroglyphic Workshop at Texas.* David Freidel provided the first impetus, but the main set of discoveries unfolded in Linda Schele's graduate seminar on "Origins" in February 1992 at the University of Texas. She was the focus of the discoveries, but her students, especially Matthew Looper, Khristaan Villela, Kent Reilly, and Barbara MacLeod, were central contributors. Dr. Robert Robbins of the Department of Astronomy

of the University of Texas also participated in the debate and investigation. In the middle of February, preliminary copies of the workbook section setting out the discoveries was sent by Federal Express to ten other scholars in the field. Anthony Aveni, Dennis Tedlock, Barbara Tedlock, David Kelley, Nikolai Grube, and Werner Nahm all responded and added their own insights and cautions to the growing discoveries. Dr. William Gutsch helped us check the astronomy at the Hayden Planetarium of the American Museum of Natural History in New York and Von Del Chamberlain did the same at the Hansen Planetarium in Salt Lake City, Utah.

Duncan Earle guided Linda through the days of the Ch'ay K'in, the "Lost Days," at San Juan Chamula in the state of Chiapas in México so that she could experience the Chamulas' re-creation of the world for herself. And finally, when Linda Schele presented the new interpretations of Creation to a large audience at the 1992 Maya Meetings in Austin in March, many people in the audience spoke up to add their own insights and cautions to the growing understanding. We are especially grateful to Victoria Bricker, the Tedlocks, Evon Vogt, Terrence Kaufman, John Carlson, Karl Taube, José Fernandez, John Fox, Garrett Cook, and others for their generous participation and sharing of ideas. This new understanding of Creation is the result of a true collaboration among many different people who gave their ideas freely and generously. And we also acknowledge that we do not present the final word here, for the details will change as others contemplate and check our ideas. We are reasonably confident, however, that we have found the "big" picture of Maya Creation.

3. The Aztec counted this as the fifth creation, while the Popol Vuh counts three creations before the present, making this the fourth. We don't know the count used in the Classic period.
4. Dennis Tedlock (personal communication, 1992) informed us that he now reads Tziiz as Coatimundi as did Edmonson (1971:4).
5. At the 1992 Texas Meetings, Terry Kaufman, the most highly respected of Mayan linguists, pointed out to Schele that the name might just as well read Hun-Ye-Nal as Hun-Nal-Ye as Schele herself, Mathews, and Lounsbury (1990) had suggested. He pointed out that *ye nal* reads "the tooth of maize" and refers to a single kernel of corn. We think this idea is a good alternative, but will retain the original reading because so many portraits of this god have the name Hun-Nal, without the *ye* sign. This leads us to suspect the first two components should be Hun-Nal, "One-Maize."
6. That the Maya thought of the calendar in this way is confirmed by the long count recorded on Step VII of Hieroglyphic Stairs 2 at Yaxchilan where eight cycles above the bak'tun are recorded with their coefficients still set at thirteen to give the long count 13.13.13.13.13.13.13.13.9.-15.13.6.9 3 Muluk 17 Mak (October 21, A.D. 744).
7. MacLeod (1991b) isolated a substitution pattern at Chich'en Itza where the crossed-planks replace the phonetic spelling *ha-l(i)*. She showed this reading to be productive in several important contexts and suggested that the Creation verb reads *hal*, a term that appears in Yukatek and Cholti as "to say, to manifest itself, to say his name in order to be known," and in some derivations as "to make appear."
8. Mayer had sent it to Nikolai in Schele's care so that it could be displayed at the Texas Meetings in March 1992.
9. Schele realized from this text that the second glyph, which represented the side view of a closed fist, had a phonetic complement of *ho*, suggesting that the word ended in *oh*. She simply went to the dictionaries of Yukatek and Chol and looked up every consonant combined with *oh*. When she found *k'oh* as "mask" and "image," she remembered that Nikolai Grube and Barbara MacLeod had suggested *k'oh* as the reading of a glyph for "mask" on a looted lintel from the Yaxchilan region. She went to the lintel and saw that their *k'oh* was exactly the same glyph as the one in the Creation phrase. That sealed the reading as *k'oh*, "image or mask." See the 1992 *Workbook for the XVIth Maya Hieroglyphic Workshop* for the full argument.
10. Brian Stross (1988) first identified the glyphs on these bundles as *ikatz*, a Tzeltal word for bundle and burden of office.
11. Based on its occurrence in the glyph that introduces distance numbers, J. Eric Thompson long ago read this sign as *hel*, a Yukatek word for "change." In his study of successor titles, Berthold Riese (1984) suggested a reading of either *hel*, *hal* or *tz'ak*. He pointed out the frequent presence of *ka* as a phonetic complement to the sign and associated it with successor expressions in the royal genealogy of the Chontal of Akalan-Tixchel. Nikolai Grube (personal communication 1987) has supported the *tz'ak* reading by identi-

fying the "nine successor" title as Bolon Tz'akab, "Nine (or many) Generations," a deity name documented in colonial sources. Barbara MacLeod prefers the *halhi* reading for this verb, giving it the meaning of "was manifested." Since this particular version of the "successor" glyph most often occurs with a *ka* complement, we prefer it for this context also. In Yukatek (Barrera Vásquez 1980:873), *tz'akah* is glossed as "existir (to exist)" and "hacer existente (to make or create existing or being)." It also is used for "grade, step, degrees of relatedness, and knots (as on a string)" and "measure of a milpa." In Chorti (Wisdom n.d.: 735) *tz'akse,* a derived form, is "regulate, arrange or adjust, put in order." The sense of the verb may refer to both meanings "to bring into existence" and "to put in order."

12. This location consists of the sign for "black" followed by *u* (the third person pronoun) and T606, a glyph David Stuart (n.d.) read as *tan* in the T1.606 "child of mother" expression. He marshaled convincing evidence that this relationship glyph is *huntan,* a term in Yukatek meaning "to care for greatly." *Tan* means "in the middle of."
13. While we have the images of seven gods and a list of seven names, we still don't know which names refer to which of the gods.
14. The verb shows a deer hoof in a hand. We have the same verb recording an action occurring to human kings when they were seven and thirteen years old. Since First Father was about eight years old at the time, he may have undergone the same kind of event as the human kings, but until we have a decipherment, we have no way of knowing what the action was.
15. At the end of March 1991, Schele first saw the great ceibas of Copan in fruit, and to her astonishment the trees, huge in size, shorn of leaves by the dry season, and heavy with fruit looked exactly like this image. The presence of fruits on this tree places the time of this mythical event in the dry season of the year. The date 13.0.1.9.2 just happens to have fallen on February 5, 3112 B.C., the beginning of the dry season, which usually lasts well into May. The Calendar Round date of the pottery scene is not 13 Ik' end of Mol, but instead 1 Ahaw 3 K'ank'in. If, as Dennis and Barbara Tedlock suggested to us at the 1992 Texas Meetings, this is the last major event before the present Creation began, that date would have fallen on 12.18.4.5.0 1 Ahaw 3 K'ank'in or May 28, 3149. Zenith passage at Palenque falls on May 11, but May 28 might well have been acceptable as the beginning of the rainy season.
16. Robiscek and Hales (1982:56–57) first identified this scene as the shooting of Seven-Macaw from the Popol Vuh myth. Also see Coe (1989) for a magnificent Early Classic pot depicting the same scene.
17. The act of perching on a tree is an act of special meaning to modern Yukatek Maya as William Hanks's shaman teacher in Oxk'intok', Yukatan, explained to him. In trying to explain how an evil spirit afflicting a household could be identified as being in "the south corner . . . where it makes its shrine (nest)," Hanks gives the following:

> As a man *kul* "sits" in his home, a bird "alights" on its *k'uh* "nest," and a saint resides in a *k'uh* "shrine," the evil [spirit] *k'u* "alights, perches" in the south corner. This stationary aspect is what makes it possible for DC [the shaman] to locate the thing on one day and perform a . . . ceremony only several weeks later, without fear that it has moved. There is a cultural premise that all animates, including spirits, occupy stable places from which they occasionally move.
>
> (William Hanks 1990:344; brackets ours]

The act of Itzam-Yeh or First Father's alighting or perching on the ceiba tree declares their places in an action that is homophonically and metaphorically like a shrine. The texts from Palenque and the pot call this action *och,* "enter," because each being enters a shrine in the sky. These sky places are their regular places, so that when each "entered the sky," each was manifesting in his proper place.

18. Floyd Lounsbury (1976) first recognized that this Creation event was recorded in a couplet construction, the most common discourse pattern used in Mayan languages. In the first half of the couplet, the scribes said that the event was 1.9.2 after the 4 Ahaw 8 Kumk'u event. In the second, they used the proper Calendar Round date, 13 Ik' 20 Keh, and a different way of talking about the event to contrast and complement the first half of the couplet.
19. This dedication phrase consists of the God N verb which Barbara MacLeod (1990a, 1990b) has read as *hoy.* Her suggestion has recently been confirmed by a pot in Justin Kerr's archives. The text on this pot has two glyphs, one with the syllabic signs *ho* and *yi* and the other the standard *yuch'ib* ("drinking vessel") glyph expected in Primary Standard Sequences. This text is the two-glyph

reduction of the PSS known from a large number of pots. It has the phonetic spelling *hoyi* replacing the God N dedication verb.

In Chorti (Wisdom n.d.:468), this term is glossed as "make fitting, make proper," while in Chol, Tzeltal, and Tzotzil, it means "to circumambulate." To move successively to the four main house posts to make offerings and thus through the four directions is often part of Maya house dedication ritual. As we shall see, this meaning has particular relevance to the context of the Creation events. *Wakah-Chanal* combines the number six with the glyph for "sky" and a combination of signs recently read as phonetic *nal* (Schele, Mathews, and Lounsbury 1990). Nicholas Hopkins (personal communication 1978) long ago suggested that *wakah* functions in this context not as the numeral, but as the rebus for the verb *wakah*, which is glossed in Yukatek (Barrera Vásquez 1980:907) as "to raise, lift, to straighten, to erect, to set upright." The equivalent term in Chorti (Wisdom n.d.:755) is *wa'ar*, "standing, stood up, erect or erected, perpendicular." With the *-nal* suffix, this is the "Raised-up-Sky-Place."

The second part of the name is composed of the number eight, phonetic *na*, and the head of GI. This same head occurs in the introductory glyph to the Primary Standard Sequence as the substitute for the mirror sign Grube and Schele (1991) have read as *tzuk*. This substitution demonstrates a phonetic value of *tzuk* for the GI head when it functions phonetically. However, we do not know if it carries the same value when it stands for the name of the god. Unfortunately, there is ample evidence that the portrait glyphs that stand for the names of gods have independent phonetic values.

These two glyphs are then described as the "holy name," *u ch'ul k'aba* of the house of the north. The decipherment of the *ch'ul* was made independently by several people in 1988, including John Carlson, David Stuart, and in published form by William Ringle (1988). The *k'aba* was read the year before by Judy Maxwell at the 1987 Texas Workshop on Maya Hieroglyphic Writing and by Nikolai Grube in independent studies of dedication phrases. Finally the object whose holy name is "Raised-up-Sky-Place" is recorded as *yotot xaman*, the "house of the north."

20. The recognition of the names of these sanctuaries began in the 1984 Texas Workshop on Hieroglyphic Writing, when Schele identified a glyph on the Tablet of the Foliated Cross (a combination of T4 *na*, a tree sign, and *k'an*) as the name of the image in the center of the panel—a maize tree rising from the forehead of a K'an-cross Waterlily Monster. Once the naming and dedication patterns were recognized through the work of David Stuart, Nikolai Grube, and Schele at Copan, the correlation of *Wakah-Chan* with the World Tree at the center of the Cross became evident (Schele 1987g:139–140; Schele and Freidel 1990:472). *Wakah-Chan* is the name of the *pib nail* (the small sanctuary building inside the temple), but it is also the name of the tree image in the center of the panel.

21. We think this complex and powerful image of the cross contains one of the fundamental covenants of Classic Maya belief. It is not merely an icon of what happened, but a declared relationship between humanity and the gods. Freidel believes that the cross itself is an embodiment of First Father, who was Raised-up-Sky (see our earlier discussion of the saint-cross-idol conceptualization). He points out that the kings were embodiments of the World Tree (Schele and Miller 1986:77 and Fig. 1:4). At Copan, Smoke-Imix-God K went even further by portraying himself on Stela I in the mask of GI (whose image may be either First Father or his first-born son) wearing the sun-marked bowl of sacrifice as a helmet. While both the king and the First Father could embody the Wakah-Chan, it does not follow logically from this observation that the original Wakah-Chan Tree represents First Father. Instead, Freidel bases his interpretation on the statement that First Father "entered the sky," which he suggests means "manifested in his place." His place was certainly the Raised-up-Sky, for he is the Raised-up-Sky-Lord.

Schele does not accept the identity of Hun-Nal-Ye GI with the World Tree, but argues instead that he was the principal actor in the events that led to the World Tree being erected and the sky being lifted. She takes the title Raised-up-Sky-Lord to reflect his role in Creation.

22. John Carlson has also worked on the directional glyphs painted in this remarkable tomb. We find the intercardinals especially interesting in relationship to the organization of the cosmos. Northeast is recorded as *wak-nabnal*, "six (or stood-up-ocean-place)," a location that seems to be directly in contrast to another location, *ch'a-*

nabnal, "lying-down-ocean-place," which is recorded on the panel from Temple 14 at Palenque. Thus, the "lying-down-sky" and "raised-up-sky" opposition has its complement in the "lying-down-ocean" and "raised-up-ocean." The southeast is recorded with a dragon head over the sky sign in what seems to be a direct reference to the Cosmic Monster. The glyphs associated with northwest and southwest are more difficult to analyze, but one has *yax,* "first," as a prefix and the other has six or "stood-up." While we do not fully understand these eight directions, we can identify them as the partitions that radiate from the center and were generated in the act of Creation.

23. The head variant of *tzuk* is also the name of First Father in the Palenque Tablet of the Cross text, coupled with his other name, Hun-Nal-Ye, the resurrected Maize Lord. David Freidel thinks that this face inscribed on the Raised-up-Sky tree does not simply name it as a partition but rather signifies the severed head of First Father hanging on this World Tree. On the second form that the World Tree takes, the *Na-Te'-K'an,* First-Tree-Precious, the Maize Tree on the Tablet of the Foliated Cross at Palenque, the tree wears a full-frontal version of the *tzuk* mask and two profiles of this mask occur on the stalk. Presumably, there were four such masks altogether on the maize plant. Four *tzuk* masks may denote *kan tzuk* (Barrera Vásquez 1980: 298), "four partitions." However, we also know that the *Na-Te'-K'an* is Hun-Nal-Ye, the Maize Lord and First Father. Freidel argues that transfer of the *tzuk* mask from the trunk of the Raised-up-Sky tree to the face of the Maize Tree denotes the transformation of First Father from sacrificed head to resurrected Maize Lord.
24. This concept of the partitions of the world is especially central to ritual in the town of Momostenango in Guatemala. The priest-shamans, who are called mother-fathers, in Los Cipréses, a rural canton within Momostenango, make a four-part ritual circuit of the great directional mountains that bound the Momostekan world. On 11 Keh, they visit Kilaja, the sacred mountain located in the east; thirteen days later they cross the axis to visit Sokop, the west mountain. Thirteen days after that, they visit Tamanku, the mountain of the "four-corners of the sky" (*kajxukut kah*), the K'iche' word for south. Then thirteen days later, they go to the mountain called Pipil, located at the "four-corners of the earth" (*kajxukut ulew*) or the north. As Tedlock relates, the K'iche' call this circuit "both the 'sowing and planting' (*awexibal tikbal*) of the town and the 'stabilization' (*chak'alik*) of the town. The K'iche' term *chak'alik,* which in everyday usage refers to the firm placement of a table on its four legs, here refers to the firm placement of the town within its four mountains, so that it will not wobble or tip over during a revolution, earthquake, landslide, or other catastrophe" (B. Tedlock 1982:82).

    This circuit has its analog in the Classic period also. The Río Azul tomb has the signs for the cardinal directions associated with four glyphs consisting of the head used for the day sign Men (which reads *am* as a phonetic sign), phonetic *ah,* and a variable element assigned to each direction. East is *ah k'in* ("he of the sun"); west is *ah ak'bal* ("he of the night"); north is *ah uh* ("he of the moon"); and the south is *ah Lamat* ("he of Venus"). The sun-night and the moon-Venus contrast occur in distance number introductory glyphs at Palenque and elsewhere as paired oppositions derived from these axial associations. Furthermore, Stela A of Copan lists the directions in these axial oppositions—north-south, east-west—in conjunction with the opening and closing of a cruciform vault which has its legs oriented along these axes. A throne discovered at Copan during the 1990 season uses the same oppositions to place the person seated on the throne at the center of the vertical and horizontal directions. Crossing the center along the axis was a alternative ritual circuit to circumambulating counterclockwise or clockwise through the direction. The counterclockwise pattern is used in Landa's New Year's ceremonies and at Zinacantan today, while the clockwise circulation was used in the 819-day count of the Classic period.
25. Between the tropics of Cancer and Capricorn, the sun will pass through the zenith position on two days during the year. These zenith passage days and nadir passage nights occur on different days at each site depending on their latitude. We never see this phenomenon in the United States, Canada, or Europe.
26. This is a major difference from modern Maya cosmology. In Yukatan, the portal of heaven opens at the zenith of the sun and that is where *itz* is showered down on the altar and the supplicants. The connections to the sun are there in the

ancient materials: Seven-Macaw declares himself to be the sun and he is certainly derived from Itzam-Yeh of the Classic period. Furthermore, many of his earliest images at Kaminalhuyu and on Early Classic pottery show him with *k'in* signs on his wings. His face regularly has a mirror in its forehead and a squint-eye, which have been taken as sun-god traits in the past. We know they derive from the *tzuk* "partition" head on the trunk of the World Tree, because Itzam-Yeh, like the World Tree, is associated with the divisions of the world.

27. Elizabeth Newsome has associated the stelae of the Great Plaza at Copan with the Creation myth as it is recounted in the Chilam Balam of Chumayel. She also pointed out a remarkable resemblance to the Creation story at Palenque. Here is her analysis:

> This genesis myth opens with the events that precipitated the destruction of the past creation: the theft of the seeds of various food plants of the world from the gods of the thirteen layers of heaven. Chaos swiftly follows the assault on the gods of heaven; they are seized by the nine gods of the levels of the underworld, and the narrative relates that the Bakabs, the four giants who support the earth, consequently caused a flood to destroy the world. As the primordial ocean rose to consume the land, the sky was in danger of falling. To prevent this calamity, the four Bakabs were set at the corners of the world to lift the sky above the sea. The giants then undertook the first step in forming the cosmos for the dawning of a new era, one which would make the raising of the sky a permanent feature of the world when it was created anew. They began a procedure of raising Trees of Abundance at the quadrants of the world, beginning with the red tree placed at the corner of the eastern sky, and proceeding clockwise through the quadrants of the north, west, south and center. The final act of this phase of creation was to raise the *Yax Che',* or green tree, at the center of the world.
>
> (Newsome 1991)

She went on to describe how five gods were set up—four at the cardinal points and one above the whole world. In addition there were "four lights" that were denizens of the "four layers of the stars." These were associated with Lahun-Chan, a god of the Morningstar Venus in the Dresden Codex. Thus she associated Creation as Yukatek Maya recorded it with the arrangement of the stelae in the Great Plaza of Copan and with the Creation passages at Palenque.

MacLeod (1991b) has associated other passages in the Chilam Balam of Chumayel to the setting of three stones. These links suggest that the post-Conquest Chilam Balam of Yukatan, even with the overlay of Spanish and Latin terms from the Christian liturgy, still reflects the Precolumbian Maya conception of Creation and the world with the core definitions remarkably unchanged.

28. Nikolai Grube and Werner Nahm (in a letter circulated on January 18, 1991) provided convincing substitutional and structural evidence supporting a decipherment of the round disk sign as *pet,* a word meaning "round" and used as a numerical classifier for things round, plazas, milpas, and other geographical areas. They pointed out that *pethal* means "to make round" in Yukatek and Barbara MacLeod suggested that this sense could be extended to the sense of turning. Either or both may be what the Maya intended. The horizon from the viewpoint of a person is round and the sky has the sense of a round, domed shape. But in addition, the constellations circle around the north pivot (today near Polaris, the North Star) in a round pattern so that the turning or rounding idea is very appropriate.

The *ki* sign attached to the sky glyph is more problematic. *Ki* has no productive meanings except in Chorti where it is the word for "heart." Since "heart of heaven" and "heart of earth" are frequently used references in Maya religious terminology, it is an appropriate interpretation here, but because we have found it only in Chorti, it must remain a tentative one. One tentative association may be with the Yukatek word *kil,* meaning "pulse" and "to pulse."

There is one other connection that supports this tentative identification of *ki* as "heart." The World Tree as it is represented in the Temple of the Cross has a Double-headed Serpent entwined in its branches. These serpent heads are regurgitating a motif consisting of a triangle of dots infixed into a round shape with a braided ribbon hanging from it. This motif regularly appears in the mouth of Itzam-Yeh, the bird on top the World Tree.

This same motif appears over the mouth of a special version of the god Chak found in early texts at Yaxchilan and on Early Classic censers. In one example from Yaxchilan, this god's name is spelled with the head variant of *cha* holding this motif in its mouth. Since the spelling is phonetic a *ki* sign is required and the only candidate is this motif. Given this possibility, the motif may read phonetically as *ki* and thus be the reference to the *ki* in the *chan-ki* glyph at Palenque.

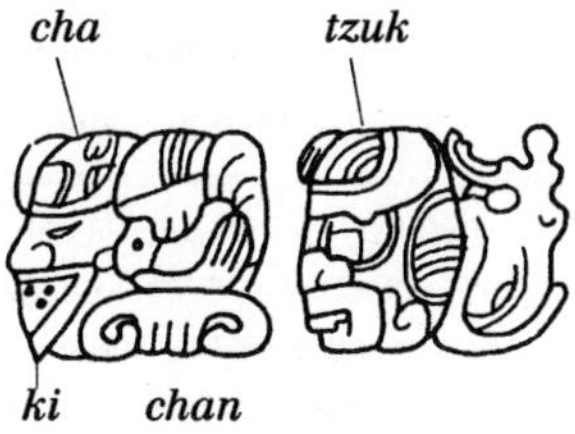

Yaxchilan Lintel

drawing from a clay box

image from an offering bucket

Chak gods from Early Classic inscriptions and offering vessels

This little bundle also occurs in the chest cavities of sacrificial victims so it may actually symbolize a heart taken in sacrifice.

The other thing that Itzam-Yeh carries in his mouth is a cord personified as a snake. This may represent the *xtab ka'an*, "the cords of the sky," that some modern Yukatekan shamans regard as a means of drawing in the water of the sky the way that people draw up water from a well or from a cenote with ropes. These also may relate to the snake-headed umbilical cords that emerge from the supernatural place called Na-Ho-Chan. Piedras Negras Stela 40 shows another kind of personified sky cord rising out of an underground stone house, that is, a well, and more specifically out of the mouth of an ancestor manifesting in a plate. The cord rises from the mouth of the ancestor, out of the well, and into the sky, where it turns and starts downward again. The ancestor is clearly "harvesting" blood/water from his mouth, which, in good Maya ecological thinking, goes up into the sky along the cord, where it can then fall as rain for the descendants. The living king kneels at the opening of the well, scattering into it in sacrificial offering. As he gives, so he receives from his ancestor.

29. Readings of the Classic glyph for "north" and "south" as "zenith" and "nadir" were first formally presented by Victoria Bricker (1983). The concept was challenged by Moore (1982) and Closs (1988) and defended by Bricker (1988) and Coggins (1988). The pertinent evidence was summarized by Tom Jones (1989) and reviewed by Schele (1992). It turns out that the concepts of "up" and "top" were associated with "north," while "down" and "base" were fitting to the concept of "south" in Classic period imagery, but these are not "zenith" and "nadir" of the sky.

30. It turned out later that Scorpius was not the constellation that Bruce Love associated with the scorpion in the Paris Codex. He had information from a Yukatek informant that the scorpion was a huge pattern occupying a good portion of the sky around the constellation Gemini. We proceeded blithely for the next two months, presuming the constellation was our Scorpius, a conclusion we still support. *Sinaan*, the word for "scorpion," is glossed in the Motul dictionary, a very early Yukatek source, as "scorpion," and also, "the scorpion sign in the sky." Furthermore, as Peter Keeler pointed out to us, Redfield and Villa Rojas (1934) list *sinaan* as the constellation Scorpius for the villagers of Chan K'om in Yukatan. Nikolai Grube informs me he has received the same identification from numerous Yukatek speakers in the Cruzob villages of southern Quintana Roo. Duncan Earle and I were told by Chamulas who were standing around with me as I drew our scorpion constellation that they call the same constellation *tz'ek*, their word for "scorpion." This information was volunteered, rather than elicited. Combined with other evidence from the Dresden Codex and Classic-period imagery, we are convinced that our constellation Scorpius was also called the "scorpion" by the ancient Maya.

31. This was the LAILA (Latin American Indian Literature Association) meeting held in San Juan, Puerto Rico, in January 1992.

32. This was an article on worldview and religion called for the catalog of the exhibition *Die Welt*

de Maya, that was organized by the museum at Hildesheim, Germany.

33. David Stuart circulated this decipherment in a 1989 letter.
34. This identification makes eminent sense once it is made. Images from many Late Preclassic sites, including Izapa, Cerros, Tikal, Kohunlich, and Waxaktun, show heads identified as the sun or other planets hanging from these double-headed serpents. As we thought about it, we realized that this was also the skyband of the codices and that almost all the symbols directly attached to double-headed serpents and to skybands are planets or constellations along the ecliptic.
35. Barbara Tedlock has paid special attention to modern K'iche' astronomy and helped Dennis apply her knowledge to the Popol Vuh. They had been sending Schele papers for years, which Schele had passed on to her students. Matthew Looper especially had absorbed all the information in their papers and held it ready for this remarkable time of discovery. See B. Tedlock (1985, 1992) and D. Tedlock (1985, 1992, n.d.) for their understanding of how the astronomy worked.
36. See D. Tedlock (1985:330, 360) for the evidence he used for these identifications.
37. At the 1992 Workshop on Maya Hieroglyphic Writing, Barbara Tedlock added that the K'iche' call Orion *Je Chi Q'aq'*, "dispersed fire." The three hearth stars are called the *Oxib' Xk'ub'*, "three hearthstones," and Nebula M42 is *Q'aq'*, "fire."
38. See Aveni's study (1980:28–40) of Aztec constellations for a full discussion of the evidence. He also presents evidence that the Aztec called Scorpius *citlal colotl*, "Star-scorpion."
39. See Mary Miller's (1986:30–38) study of Bonampak' for Lounsbury's arguments.
40. See Redfield and Villa Rojas (1934) for this identification.
41. There are two kinds of peccaries found in the Maya region and there are at least three documented words for peccary—*ak*, *chitam*, *chakwo*. Matt Looper has shown that the *ak* peccary occurs in the phonetic spelling of *chak*. This example has the same trifoil shape in its eye.
42. This was Richard Berry's 1987 *Discover the Stars*.
43. Schele (1990a) showed that the glyph for "cenote" or "hole" was the glyphic version of the skeletal snake heads that frame the portal to the Otherworld. Barbara MacLeod (personal communication 1989) used the occurrence of this sign in the month glyph Wayeb to suggest that it was read *way*, the term for "sleep, dream, animal spirit companion, and to transform into your *way*." She suggested that the month read *wayeb* as "resting place" of the year and pointed out the presence of *u* prefixes and *ya* suffixes as phonetic complements. We accept her reading and suggest that the portal was the "Black-Transformer," the place where change occurs.
44. In her thesis on the Late Preclassic bird deity, Constance Cortés (1986) identified this stela scene as part of the Popol Vuh myth. Magdiel Castillo Baquero, a student of Schele, first patterned iconographic information in a way that allowed Schele to see how this particular scene fits into the motion of the Milky Way on Creation day of August 13. Most important, Castillo first identified the floor plan of Pakal's tomb in the Temple of Inscriptions at Palenque as a symbolic ballcourt. He used the four niches in the wall as the ends of the I-shape and the sarcophagus area as the playing field to complete the I-shape. When he (personal communication 1990) first suggested this, Schele was skeptical, but now we know that Pakal fell down the road to Xibalba in the guise of the Maize God. The Maize God was resurrected and left in the Ballcourt by his sons, where he was to be worshiped forever by humanity. To make the floor plan of a tomb resemble the I-shaped plan of a ballcourt makes real sense.
45. This was Burial 116 under Temple 1 in the Great Plaza. This king is also known as Ruler A and Ah-Kakaw.
46. This interpretation occurred to Nikolai Grube and Schele when they were giving a workshop on Creation to thirty Maya in Antigua, Guatemala, during July 1992. When Linda told David Freidel, he said that couldn't be right, because human babies normally emerge from the birth canal facedown and away from the mother. David's wife, Carolyn Sargent, is a medical anthropologist specializing in birthing practices. She settled the matter by pointing out that while human babies first emerge facedown, they rotate at the shoulder while they come out so that by the time they are fully out of the canal, they are in the faceup position of the Maize God on this pot.

47. Karl Taube sent a letter to Schele after the 1992 workshop predicting that Chak would be one of the actors in Creation. When Schele (1992), following a suggestion by Werner Nahm, showed that Virgo was represented both as a peccary and Chak, Taube pointed out in his letter that K'eq'-chi' has *chakow* or *chakwo* as "peccary." Here Chak stands in the place of the Stingray Paddler who sits in the eastern end of the canoe, where Gemini, the other peccary constellation, is. I think this may be a play on the Chak-peccary substitution, but it also confirms Taube's idea that Chak was a player in Creation. In August of 1992, he pointed out to us a scene of two Chakob cracking the turtle shell from which the Maize God emerges. Their lighting axes made the rebirth possible. See his discussion of caves, lightning, and origins (Taube 1986) for a full discussion of these concepts.
48. In Yaxuna village, David has seen the ceremony in which the chief helper to the shaman officially takes the role of principal Chak. He ties a headband onto his head, often a white one but sometimes red, with a small leafy twig bound over the forehead and dangling outward. This little ceremony shows an uncanny resemblance to the act of royal coronation of Holy Lords in the Classic period as given in glyphic texts and accompanying images.
49. Susan Milbrath (1988) has identified this configuration as critically important at another time of year. She observed that when the Gemini-Orion-Milky Way junction is at zenith at midnight, it is in late November, so the sun is at nadir. She pointed out that this configuration can happen only in the tropics. At Chich'en, it takes place on November 22 and at Palenque on November 12. Now we also realize that this configuration positions the place of Creation—that is, the Copulating Peccaries, the hearth of Orion, and the K'an-cross of the Na-Te'-K'an opposite the sun. We have not yet checked to see if particular dates and monuments in the Classic period refer to this configuration, but we wouldn't be at all surprised if we find a cluster of monuments that refer to it and to Creation iconography.
50. Kent Reilly (personal communication 1990) has found the same cross at the center of four symbols he tentatively identified as the four directions on the Humboldt celt, an image of the Olmec culture.
51. Barbara MacLeod (1992) first made these associations with the Popol Vuh and these titles.
52. These are the dates in the 584,285 correlation that we favor. The days would change to August 11 and February 3 in the 584,283 correlation that many other researchers favor. The latter correlation has the advantage of being consistent with the modern calendars used in the highlands of Guatemala.
53. Matthew Looper associated their demise under a falling lintel with the Milky Way sinking into the earth atop them after they set.
54. Dennis Tedlock (1985:251–252) discusses this use of "dawning" as a reference to "creation" and "birth" in his commentary on the Popol Vuh. However, the verb *ah* in Yukatek also means "to dawn" and "to create." Barrera Vásquez (1980:3) glosses *ahal* as "to wake up," and *ahal kab* as "to create the world" and "to dawn." *Ahi kab* is "from the beginning of the world." This *ahal* term has been identified in the inscriptions of Palenque associated with the Creation myth and at Yaxchilan with the pre-Creation myth of the ballcourt. The fact that the stones of Creation were at zenith at dawn on the Creation day was no happenstance.
55. Floyd Lounsbury (personal communication 1992) suggested that the August 11 or 13 date may have been chosen precisely because it was a zenith passage and then cast backward to the mythic past where they wanted to set Creation. If he is right, then the myth may have locked to the celestial and solar clocks somewhere in the region between 14° 30" and 15° 30" north latitude. Other evidence gleaned from the orientation of La Venta suggest the myth and its timing were already in place there by 1000 B.C. The planetariums showed us that the hierophanies we are describing went from zenith at sunset to zenith at dawn at La Venta in that year. Moreover, working with Kent Reilly, we found that the alpha star of the Big Dipper was on the horizon at sunset on August 13 and that the beta star of the same constellation was in the same position at sunrise on February 5. The timing of these alignments to the two Creation days seems more than simply fortuitous.
56. Karl Taube suggested to us that these represent the intestines of the Maize God taken in sacrifice. The scenes may then represent the death of the Maize God in New Year's rites.

57. Several other people, including Bruce Love on the one hand and Victoria and Harvey Bricker on the other, have studied this zodiac and come to different conclusions about how the pictures match the constellations. Several of these studies as well as other important papers on the subject are published in *The Sky in Maya Literature*, a volume edited by Anthony Aveni (1992). Schele assumed the scorpion matches our constellation Scorpius, while Bruce Love places it elsewhere. In the summer of 1992, Nikolai Grube and Schele asked the Maya participating in a workshop we gave on Creation what they called the constellation of Scorpius. Several of the K'iche' told us it is *sinaj*, their word for "scorpion." Later they asked the same question of the people of Tixkakal Guardia and Señor in southern Quintana Roo. The answers there were confused and unsure, but the most knowledgeable man they talked to, Agapito Ek' Pat, told them that *Sinaan* was high in the sky (which it is there); that it occurs in a band (it is in the ecliptic); that it is far away from the *Ak Ek'* (which is either Gemini or Orion); and that it disappears in November. All of these characteristics fit the behavior of Scorpius.

    Each of these three different proposals for the zodiacal identification have a well-researched and thoughtfully derived body of evidence to support them. We will all have to test them over the coming years and eventually the correct one will accumulate the necessary corroborative evidence to be accepted by the field. For our purposes, an agreement on the exact identification is not necessary. The recognition that the Paris Codex is a zodiac and a picture of Creation is enough.

58. The Hauberg configuration corresponds to the pattern of the Milky Way just before dawn. Schele tried the date 9.0.15.11.0 (May 1, A.D. 451) for Tikal Stela 1 and found the same configuration at midnight.

    At the 1992 Advanced Seminar of the Texas Meetings, Michel Quenon and Richard Johnson checked my conclusions and tried as many other possibilities as they could. They came to the conclusion that the pattern that Dave Kelley had found was the most workable one. They also found a set of Xultun stelae that represented constellations as small beings held by the protagonist. These monuments show rulers holding a jaguar manikin extended out from their shoulders in their right hands and a Chak with his foot turning into fish-monster head clasped tightly to their bodies in their left arms. The rattlesnake seems to have been removed, but its absence is signaled by the holding of the jaguar (Capricorn) extended out from the body.

    Most important, Michel Quenon was particularly fascinated by the tree in the backrack of the Hauberg Stela. It has figures with severed bodies falling headfirst down the tree. He noticed that the lowermost figure has a sun sign in his helmet, while the middle one wears a long-beaked bird headdress. He proposed that these three beings represents the configuration of the sky at sunset on the same day. The sun sank through the western edge of Pisces (the bat), while Venus and Jupiter hovered 12° and 16° above the horizon in the constellation of Aries. In the Paris zodiac, Aries corresponds to a bird and the middle figure wears a bird headdress. The particular pattern we propose for the Paris zodiac may be incorrect, but the supporting evidence seems to be accumulating in a very strong pattern.

    In additional work on the Paris Codex zodiac, Johnson and Quenon (1992) have discovered that the row of green numbers written below the column of days was a correction device for the zodiac. They pointed out that the thirteen columns of days are organized in five rows. Each column is separated from right to left by twenty-eight days so each row of thirteen dates covers a span of 364 days (13 x 28 days). To return to the beginning date of 12 Lamat, one reads through all five rows for a total of 1,820 days. They calculated that an error of 6.28 days accumulates in the position of the sun relative to the constellations during each full run of the table. Three runs of the table gives a total error of 18.64 days. They noticed that a row of green numbers written below each column of dates is separated from the red numbers in the top row by a factor of twenty. They suggest this was a correction device that was used every fourth run of the table. This correction brings the sun back to approximately the same position that it held on the beginning date of the table. Switching back and forth between the green and red numbers would thus have allowed the use of the table over a long period. I checked it over a period of five hundred years and found it to be very accurate.

59. The paper he sent me, "A Stellar City: Utatlan and Orion" (Fernandez 1992), was presented at the International Symposium, "Time and Astronomy at the Meeting of Two Worlds," in Warsaw, Poland, April 27 to May 2, 1992.

60. Just before the 1992 workshop, Peter Dunham sent us research on umbilical cords he had done several years earlier. He described some of his conclusions in a letter he sent to Freidel on March 8, 1992.

> Three of the enclosures relate to the Milky Way and *kuxan sum*, or celestial umbilical cord in the world center. As I mentioned, there is a suite of myths among the Yukatek, with reflections elsewhere, that describe a heavenly road connected to and sometimes contiguous with a "living rope" filled with blood that descends from the sky, often through the heart of a giant ceiba at an important site. In some versions, the cord was connected to the ancient Maya rulers, supplying them with heavenly blood, obviously a metaphor for divine celestial right. The rope was severed (in some cases by the Spaniards), depriving the rulers of their sustenance and legitimacy and leading to the decline of the Maya. This complex is also associated at times with a giant cenote and network of underground caverns, again joining significant places. Bits and pieces of this mythic body survive today in various forms and are embedded in colonial texts and I believe even Classic Maya art.
>
> The impression I was left with by these tales and interviews I conducted with elderly men familiar with them is that the underlying conception is that of a sort of giant cosmic machine or organism. It consists of a route in the heavens—the Milky Way—along which the major celestial bodies circulate, imbuing it with power (in fact, the planets and sun do keep to the same band of the sky defined by the seasonal wobbling of the Milky Way, and indeed they often pass through or along it; a Mexican myth even describes such an association). This route articulates with another one—the *kuxan sum*—that transcends the void and transports the essence of celestial power (blood) to the earthly realm, where it animates life and those ordained by the cosmos. The living rope passes through the world tree at the center of creation. It may be connected, perhaps through the tree's roots and the cenote (the cenote is a sort of the underground equivalent of the tree), to subterranean routes that may somehow recirculate this divine celestial power through the realm of the dead and back to the world of the heavens. In many instances, due to natural transformations and the disruptions of the post-contact era, the various components seem to have been conflated (years ago, Alfonso Villa Rojas told me he, too, felt they had been merged and confused in recent times).

The symbols associated with the Milky Way and the ecliptic change from group to group on the surface, but the underlying way of thinking about the constant and repeating patterns in the sky is the same.

61. David Freidel describes his argument as follows: Both the conjunction points of the ecliptic with the Milky Way were the heart of heaven. A riddle given in the Chilam Balam of Chumayel, a Conquest-period Yukatekan book of myth and historical lore (Edmonson 1986:174), goes: "And so this is the heart of God the blessed, as they are told then: it is cordage." The word used for cordage here is *k'aan*, a very close homophony to *k'an*, the "precious, yellow" symbol of the Na-Te'-K'an. The Na-Te'-K'an is the second form of the World Tree, the Maize Tree, First Father as Hun-Nal-Ye, which means not only One-Maize-Revealed but One-Maize-Kernel, according to a conversation with Vicky Bricker at the Austin Workshop in 1992. The word used for "heart" in the Chilam Balam riddle is *puksik'al*. Barrera Vásquez (1980:673) pointed out that *puksik'al* referred to the physical organ, while *ol* refers to "heart" in the positional sense. *Puksik'al* is the word for "heart" that can apply to the heart of the maize plant where the stalk comes out; that is, to the kernel. *U puksik'al ixim* is "el meollo y corazón del maiz; cuando sembrando se corrompe, de cualquier meollo sale la caña" (Barrera Vásquez 1980:673). At the zenith conjunction the ecliptic divides the Milky Way into a waterlily plant to the south. This is the body of water that the canoe of the east-west form of the Milky Way floats upon and sinks into. The Maize Tree is above, emerging from the blossom of the waterlily, a slight expansion in the Milky Way at this point. The roots of the waterlily descend to the south-sky nadir position. This root "cord" may be related to the umbilicus stories of post-Conquest Maya in which the sky umbilicus of the created world is identified with the Milky Way. Iconographically, the waterlily plant resembles the umbilicus attached to the placenta in a human birth.

62. Karl came to Austin in August 1992 and shared these insights with us. They are contained in his dissertation on the New Year's ceremonies (Taube 1988b).

63. There are twenty day names in the 260-day calendar. Twenty goes into 365 eighteen times with a remainder of five so that each New Year advances five in the twenty day names. These four

days are called the Year Bearers. Every four years, New Year falls on the same day name, but it is combined with a different number. To go through all thirteen combinations of day and number for all four Year Bearers takes fifty-two years.

64. Anthony Aveni (1980:32, 36) points out the interesting possibility that the Aztec called Gemini *citlatllactli* or "ballcourt." If this same identification is someday found in Maya sources, then the ballcourt where the Maize God stayed to be worshiped may be where the copulating peccaries are. Gemini is immediately adjacent to the three stones of Creation and the turtle carapace from which the Maize God was reborn. Since we have also speculated that the Milky Way emerging from the turtle constellation in Orion is the Maize Tree, this area of the sky is the logical place for the Maize God to reside so that he can be worshiped.
65. Girard (1949, 1966) compiled one of the most detailed and eloquent ethnographies of a modern Maya people. His work was somewhat premature, however, in its attempts to link modern Chorti beliefs and practices with the ancient Maya. He was correct in his assumption that the connections were there, but he was often wrong in his evidence because the decipherment had not yet occurred to give him access to the ancient world. Our descriptions are drawn from his 1966 work.
66. There are two Pasiones, an incoming one and an outgoing one, for each of the three barrios. There are also six Flowers distributed in the same way. The Pasiones care for the heads of the Sun-Christ banners during the year and are the principal officials. The Flowers care for the poles and the banners attached to them. See Victoria Bricker's (1973) and Gary Gossen's (1986) analysis of the Festival of Games for the full details.

Chapter 3: Centering the World

1. See Evon Z. Vogt's *Tortillas for the Gods: A Symbolic Analysis of Zinacanteco Rituals* (1976) for a comprehensive discussion of space, time, and cosmology in this community.
2. At the 1992 Symposium of the Texas Meeting on Maya Hieroglyphic Writing, Evon Vogt informed us that the three peaks of the Senior Large Mountain are called the three hearthstones. As the Zinacantecos look at the mountain, it is the peak on the right that contains the corral for the animal soul companions. We know now that the name "three hearthstones" associates these mountains with the Classic-period place of Creation and the ancestors.
3. In Tzotzil, the word for shaman is *h'ilol*, "he who sees."
4. See Schele and Freidel (1990) for a full discussion of Structure 5C at Cerros.
5. See Freidel and Schele (1988a:558–559) for discussion of *am* as a Precolumbian Maya name for divining stones. These fourfold *am-tunob* appear in the opening incantations of the Ritual of the Bakabs (Roys 1965), a colonial-period Yukatekan book of shamanic ritual and magic.
6. The word *vaxak* actually means the number eight in Tzotzil and other Mayan languages. That the fourfold divinities are called by the number eight here is intriguing. As we have shown, in the text of the Tablet of the Cross at Palenque, First Father created a house of eight partitions in the north sky when he raised the sky. See also Schele (1990f).
7. The neighbors of the Zinacantecos are the people of the *municipio* of San Juan Chamula. Like the Zinacantecos, the Chamulans speak Tzotzil Maya. The Chamulans are a thematic source for our insight into Maya cosmology, here and throughout our book. Gary Gossen's paradigmatic article on the Chamula Festival of Games (1986) is a major inspiration for our general interpretive approach in this book and will be cited periodically. His 1974 monograph on Chamulas in the Land of the Sun is a full-scale ethnography viewed through the lens of cosmology.
8. Contemporary Maya have many beliefs concerning hot and cold foods, and medicines as elements that affect the health of people. This is called by scholars "humoral" medicine after the ancient Greek theory of humors. While the Maya may well have acquired some medical beliefs through

contact with Europeans, it remains to be seen whether their understanding of heat and cold as supernatural forces is adopted, indigenous, or—we think most likely—syncretic.

9. Arthur Miller (1974) reviewed the ethnographic and archaeological evidence for this magical cord and for umbilical-cord myths and images in contemporary and Precolumbian Maya thought. Peter Dunham (1980) also supplied us with his own review of the matter as well as a summary of information he had gathered in the field. In this widespread story of the floating, living cord, it has been cut and its life-giving properties ended. However, several features of past and contemporary Yukatek shamanic practice imply that the cord that links this world and the Otherworld can be rejoined through ritual. Arthur Miller notes that in the Ritual of the Bakabs, ". . . a reference is made to four snakes which form the world. These snakes are in the next paragraph compared to (among other things) 'arbors' and to umbilical cords: 'He would be beheaded by his mother, by his progenitress. . . .' Here is a reference to snake, i.e., personified umbilical cord, being cut by the mother, the cutting a necessary concomitant to every childbirth" (A. Miller 1974:181).

   Miller noted the image of the umbilical cord in the Paris Codex zodiac, but no one knew then that the Maize God was First Father, that First Father was the primary actor in Creation, that the Paris Codex showed Creation, or that the snake-headed umbilical cord was the ecliptic. These twisted cords did carry sustenance from one world to the other.

   We observed in Chapter 1 that the Yukatek shaman's altar is a leafy arbor called the *ka'an te',* the "sky tree." The word for "snake," *kan,* is homophonous with *ka'an,* the word for "sky." Further, radiating from the peak of the altar-arbor are thick vine cords called *xtab ka'an,* the "ropes of the sky." These are the modern analog of the ancient snake cords that represented the umbilical of the ecliptic. John Sosa (personal communication, 1991) tells us that in past times old men stood at the ends of these ropes of the sky and threw buckets of water toward the altar during the climax ceremony of the "taking of the rain gods," *ch'a-chak.* He also says (Sosa 1990) that the Yukatek shaman's altar is the place where *u y'itzil ka'an,* the "liquid substance of the sky," flows through from the center of the sky to the center of the earth at the climax of the ceremony.

10. There are subtle variations in the way that shamans and their anthropological pupils from different Yukatekan communities describe the cosmos. John Sosa (1990:132) states that a cardinal direction is associated with "its face" or "its side," *u tan* in Yukatek. The intercardinals are placed at the corners and the major east and west sides are such that the sun always rises and sets between the corners in the course of solstitial cycles. The northern and southern sides are consequentially defined by these corners as given in the annual Sun cycle. William Hanks (1990:-299–300) says, "It is clear that the cardinal points are defined as regions containing specific, named places . . . 'winds, spirits' are described as brothers and . . . 'people' who originate from these places, and their qualities depend upon their provenance."

11. Actually, the Maya have no single, simple way of expressing the notion of the spatial cosmos. William Hanks (1990) argues that the spatial framework is so pervasive that it is embedded into the practice of speaking the language as a whole. According to another expert colleague, John Sosa, modern Yukatek Maya say "center of the sky, center of the earth" to convey an encompassing axial principle (Sosa 1990); the K'iche' Maya speak of "heart of sky, heart of earth" in the same fashion. We have glyphic parallels to this duality of the spatial cosmos in Classic-period texts.

12. William Hanks (1990) describes the process of laying out fields in Oxk'intok' in great detail. He states: "The first prayer should be performed before even beginning the perimeter cut, outside the protected *haál* 'edge' of the field, at the corner where one will start work. . . . The perimeter cut made, another offering protects the farmer from snakes during the felling. The intended consequence of these prayers and the thoughts they express, all of which constitute *primisia* 'first fruits ceremonies' (even though they start long before the fruits exist), is to *wá'akuntik yuntzil* 'stand up the lords' at the five cardinal places of the milpa. . . . This is the same term used to describe how one 'stands up' the liquid offerings . . . and the semantic equivalence is a straightforward result of the iconic

equivalence of the two acts in Maya culture. The offerings embody their recipients" (Hanks 1990:363).

13. See William Hanks (1990:299).
14. Here is a situation where different Yukatek shamans have different interpretations, according to their anthropological observers. John Sosa's informant regards the *balamob* as protectors, not jaguars. Sosa (1990) points out that the modern Yukatek term for jaguar is *chak mul*, "red/great paw." Hanks (1990), on the other hand, refers to the *balamob* as jaguars in such a fashion that we surmise his informants also thought of these deities as jaguars.
15. This is a variation on the Classical term for these deities, "Pawahtun." According to John R. Sosa (1990), his Yukatek shaman teacher explained that the *babatunob* are the greatest of the *chakob*, the rain gods. As Sosa observed (1988: Chapter 6), the term *babatun* is subject to varying interpretations, including "he who breaks the flow from precious stones," the interpretation given by Barrera Vásquez. The word *babah* means "to pour, spill or vomit" in the Cordemex dictionary (Barrera Vásquez 1980). This convergence of divinities suggests that to this h-men, being a *babatun*, an earth-sky bearer, pertains principally to the quality of being a fourfold deity rather than to a particular distinctive category of such being. One of the names of the *balamob*, according to Sosa's informant is "the clay dwarf."
16. William Hanks puts it this way: "As a man *kul* 'sits' in his home, a bird 'alights' on its *k'u'* 'nest,' and a saint resides in a *k'u'*, 'shrine,' the evil *k'u* 'alights, perches' in the south corner. This stationary aspect is what makes it possible for DC [his shaman informant] to locate the thing on one day and perform a *heetz lu'um* ceremony only several weeks later, without fear that it has moved. There is a cultural premise that all animates, including spirits, occupy stable places from which they occasionally move. From its place it lops off a corner of the yard as its path, all the way past the front door, itself a metonym for the (first-level) family sleeping quarters" (Hanks 1990:344; brackets ours). We have good epigraphic evidence that the Maya scribes of Precolumbian Yukatan, at Chich'en Itza, made the same connection between "nest" and "shrine" from the analyses of dedicatory texts from the Temple of Four Lintels by Ruth Krochock (1988:100). The connection with "alights, perches" makes fundamental sense of a venerable Maya iconographic composition that is the representation of a bird "perched" on the world trees of the four corners and the center. Rather than the bird itself being the image of importance, it would serve to convey this idea of a spirit alighting or perching on its corner, or at the center.
17. The Yukatek expression for this act is *liik'sik y iik'al* (Hanks 1990:345). Another Yukatek word for "something stood up" is *wakab* or *bakab* and the gods of the four corners or edges can be called *bakabob*. God N, Pawahtun, of the ancient Maya pantheon, is now called *babatun* by John Sosa's Yukatek shaman teacher (Sosa 1990) and he is also a *bakab*, a "stood-up one." Traditionally, students of the Maya have emphasized the idea that the Pawahtun or Bakab beings hold up the world and the sky by its corners. It is likely that the primary importance of these and other beings that pertained to the four corners and the center is that they manifested in that spatial pattern. That is to say, a portal bringing forth one of these beings was necessarily part of a pattern of portals defining the four corners and the center. In this way, an act of "centering" defined, simultaneously, edges and an ordered domain of space surrounding that center. The triadic pattern was equally formulaic. Evon Vogt (1976; n.d.) describes how Zinacanteco cross shrines have to be prepared so that there are a minimum of three crosses before the portal can be opened. If there are only one or two wooden crosses available, then extra ones are made of pine boughs for the occasion.
18. The passage is a paraphrase after William Hanks (1990:342–349), one of the many detailed Yukatek shamanic ceremonies he describes.
19. This modern metaphor of binding up a sacred space, discussed by John Sosa (1988) and William Hanks (1990) is paralleled in the Classic Maya word for dedicating or ensouling a place, *hoy*, which Barbara MacLeod (1990) deciphers as "to make proper" and also as "to circumambulate." As we discuss in Chapter 2, this is one of the formative acts of First Father in the Creation of the world.
20. Kent Reilly (1988, 1989, 1991, and extensive personal communication 1987–present) is the source for many of our views on Olmec spatial order, cosmology, and kingship and we draw on that work for our discussion. His research and interpretations are presented in his dissertation in

progress, *La Venta and the Olmec: The Function of Sacred Geography in the Formative Period Ceremonial Complex.* The use of Classic Maya text-image correlations as a source of analogy for earlier symbolism is a major new consequence of the success in glyphic decipherment. Implicit in Reilly's arguments is the premise that the Maya inherited major concepts and their symbols from the peoples of the Formative Ceremonial Complex, the Olmec civilization. The empirical evidence to support direct diffusion between Olmec and lowland Maya peoples is still circumstantial but it is gaining in quality and quantity. For example, Richard Hansen (1989, 1992) has discovered evidence at the Preclassic center of Nakbe in Peten, Guatemala, that the earliest Late Preclassic monumental masks on architectural facades are of finely shaped monolithic stone block mosaic suitable for a thin veneer of stucco. This technique of monolithic construction of images is characteristic of the Formative Ceremonial Complex as seen in Middle Formative contexts at Chalcatzingo (Grove 1987) and Teopantecuanitlan (Martínez Don Juan 1984) in highland México. Subsequently, Late Preclassic Maya masks on architecture increasingly become the product of thickly applied and modeled stucco over crude stone armatures that merely outline the images. The prospects are there that the Maya tradition of decorating architecture began with a diffusion of existing Olmec practices (Reilly 1991a; see also Freidel 1979). Working closely with David Grove, Kent Reilly has forged a remarkable new understanding of Olmec symbolism. They have addressed the problem of what is Olmec and what is not by unifying Olmec style under what they call the Formative Period Ceremonial Complex, a body of symbol, myth, and ritual which was shared by many different peoples speaking different languages and coming from different ethnic traditions during the period from 1500 B.C. and 400 B.C.

21. At the architectural level, the cleft of the Olmec is like the cleft between living mountains, mountains called *witz* by the Maya of later times (Stuart 1987). The cleft in the *witz* mountain of the Maya certainly functioned as a portal to the Otherworld. At the level of smaller artifacts, Kent Reilly (conversation with David Freidel, June 6, 1991) suggests that the cleft element functioned as a representation of ground stone, iron mineral, mirrors. Olmec people used such mirrors as headdress and body ornaments. The cleft elements decorating Olmec headdresses sometimes flank central trilobate elements, and this composition seriates into the Late Preclassic crown of kingship in which U-shaped signs continue to flank trilobate diadems. The trilobate element is an Olmec maize sprout that becomes the Maya Jester God diadem of rulers (Fields 1991). The U-signs mark polished stone in Classic Maya iconography, and the Jester God diadem is glyphically depicted as a bright mirror, generally commensurate with Reilly's idea that the sprouting rectangle represents a mirror to the Olmec.

22. The circumstances of discovery are described in Blom and La Farge (1926–1927:45–46).

23. When Schele checked the astronomy of Maya Creation at the planetarium of the American Museum of Natural History, she discovered that the most perfect timing of the events on August 13 and February 5 occurred at 1000 B.C. Since we had set the planetarium for the approximate latitude of La Venta, we knew that the astronomy worked perfectly for that site. A month later, Kent Reilly brought Schele its precise latitude, 18° 06" north, and we ran the two days in *EZ-Cosmos 3.0.* Following hints from earlier work by Marion Hatch (1971), Kent knew to look at the Big Dipper. She had tried to associate its meridian transit and horizon orientation on summer solstice, but to get her ideas to work she had to presume the astronomy was set at 2000 B.C. We followed her lead and found that in 1000 B.C. alpha Ursa Major touched the horizon (at azimuth 349.72) exactly at sunset on August 13. On February 5 at sunrise, beta Ursa Major sat on the horizon at azimuth 343.92. The exact timing of these alignments seems unlikely to be coincidental.

24. La Venta was the hub of a well-populated region of farmers and fisherfolk during the first half of the first millennium B.C. Today, the island is a base of operations for oil production in the area and many of the huge stone monuments that decorated the ancient center have been removed to a park in the city of Villahermosa.

25. The fluted form of the pyramid was discovered, surveyed, and reported by Morrison et al. (1970). The resemblance to the form of a volcano is not an artifact of erosion. It was designed that way by the original builders. See Robert Heizer (1968) for a discussion of the identification of the La Venta pyramid as a volcano.

26. Reilly (1989) originated these interpretations, which rely heavily on a careful reexamination of archaeological evidence (Drucker, Heizer, and Squire 1959) from Complex A.
27. Kent Reilly in his work on the Olmec and Maria Elena Bernal-Garcia in her work on the nature of sacred space and the city in the Mesoamerican world have contributed substantially to our conception of ceremonial space. Freidel (1979) identified the pervasive distribution of the triadic architectural design in Late Preclassic Maya centers (see also Schele and Freidel 1990a: Chapter 3). Following this long-term interest in the triadic pattern, the newly deciphered importance of the three-stone place in the Classic Maya creation myth as discussed in Chapter 2, and the centering value of the three stones in contemporary Yukatek cosmology, Freidel arrived at the interpretation of the triadic design at La Venta. In conversation, Kent Reilly said he had himself observed this pattern and its significance in light of the Classic creation myth and proposed it to Schele. It is tough to come up with original interpretations of the Olmec around Reilly. Our work also parallels the conclusions of many other researchers who have worked on other cultural traditions in Mesoamerica so that we are tapping into the patterns that were fundamental statements of reality and experience for this broader cultural tradition. As we can see in the work of ethnographers, these ideas and definitions still live today among the peoples of modern Mesoamerica.
28. Olmec rulers and their craftspeople used serpentine, along with even more precious greenstones such as jade, to fashion celts inscribed with magical symbols and other power objects necessary to cleave open their portals into the Otherworld. The hundreds of tons of this precious stuff cached at this center testify to the wealth, labor, and political support commanded by Olmec rulers. Olmec kings, and the Maya rulers who came after them, did not build their centers through coercion or brute force, but rather through the willing work of their subjects. Thousands of farmers, fisherfolk, and craftspeople, backed up by an equal number of families providing food for these workers and preparing their neglected fields, created the beautiful architectural programs and the mighty stone carvings that adorned them to cement their relationship with the divine. This relationship was exemplified by their king, who invoked the powerful, nurturing, counseling gods and ancestors in the plazas for the good of all.
29. These capping mosaics are decorated with double-merlon designs, so nicknamed because they have two angular bumps on their lower edge. The four double-merlons surround a central rectangle, forming the four corners and the center of sacred space. The double-merlon, Reilly deduced (1991a), derived its shape from the cross-section of a sunken court. The two bumps are the enclosing walls. Symbolically it signifies a portal to the Otherworld. Combined with the central rectangle in the mosaic designs at La Venta, the four double-merlons create the Olmec version of the quincunx.
30. Kent Reilly (1989) has detailed the argument for his identification of these symbols as water plants and of the surface of the court as the surface of water. He cites the initial recognition of this motif as vegetation to Rosemary Joyce. His contribution was to identify the type of plant as a waterlily pad. His deductions fit well into Mesoamerican concepts of ritual and city space. The Maya, for example, used the word *nab,* "ocean" or "lake" (Schele and Grube 1990a), as "plaza." Courts are regularly associated with water imagery at many sites and architectural dedicatory caches often included marine animals and objects in dedication caches. The Templo Mayor in Tenochtitlan, the capital of the Aztec, symbolically sat in the Primordial Sea (Broda 1987) and was considered the center of the world (Matos Moctezuma 1987), and at Cacaxtla, the dancing figures of Structure 1 cavort atop water bands. Teotihuacan also represented the city in the murals of the Tlalocan as a sacred mountain with a cave from which water emerged. Maria Elena Bernal-Garcia (n.d.) has discovered that Mesoamerican conceptions of founding a city and creating sacred space within the community require that both a mountain and water be present. If they were not available in the natural landscape in which the city was built, they were created in the architecture of the city and sometimes, as at Teotihuacan, the mountain and water were there in both the god-made and human-made landscapes.
31. Kent Reilly (personal communication 1991) informs us that the erosion patterns documented on this mosaic pavement could have been caused by a combination of human use and the effect of long-term exposure to rain and other environ-

ment elements. The other pavements were, in contrast, in pristine condition with a mirrorlike polish on the surface blocks. They had clearly been buried very soon after they had been laid. We think this practice reflects their function as part of the process of spiritual activation of the space. They did not have to be seen or used. They had only to be there.

32. While Reilly's evidence for the identification of the central ally as a ballcourt is still tentative, it matches up with long-held suspicions that the Olmec played the ballgame. Archaeologists have long wondered if the pictures of Olmecs wearing knuckledusters and the images of colossal, helmeted heads have ballgame associations. The ballcourt has always been a central idea in Maya cosmology. For them, the ballcourt was a place where the Hero Twins, the founding ancestors of humanity, battled for life in a game of wits against the gods of death and disease, winning the immortality of rebirth that their descendants witness in the unfolding of the generations. Straddling both this world and the Otherworld, the ballcourt is the place of confrontation, skill, sacrificial death, trickery—the uncertainty intrinsic to our concept of the word *sport.* It very well may have played a similar role for the Olmec.
33. David Joralemon (1976:37, 41) first pointed out the four- and five-part patterns in Olmec art and specifically on these celts. He also identified this complex of images as God I, whose associations he identified as warmth, water, and agricultural fertility. Building on this foundation, Reilly (1991b) went on to identify God I as the deification of the center and the prototype to the Maya World Tree.
34. Three is a sacred number to both the Olmec kings and the Maya. Three points mark the sprout of the maize seed rendered as a diadem jewel of kingship worn on Olmec headdresses. Maize is a symbol of the source of creative force and sustenance for both the Olmec and the Maya of the Middle Preclassic period. See Virginia Fields's (1991) study of the Jester God diadem and its Olmec antecedents. See Freidel (1990) for a discussion of the Jester God diadem through Maya history.
35. We infer this sequence of events from the imagery on the Tablet of the Foliated Cross at Palenque, one of the accession monuments of king Chan-Bahlam. The central image is of the Maize Tree, the First-Tree-Precious, sprouting from the K'an-cross. As discussed in Chapter 2, we believe that the K'an-cross represents the near-zenith conjunction of the Milky Way and the ecliptic in the night sky, when First Father as the Maize Lord manifests as the northern half of the Milky Way. The existing iconographic substitution patterns we are aware of support the idea that the First-Tree-Precious is an expression of the Maize Lord, the reborn First Father. Flanking the Maize Tree is king Chan-Bahlam dressed in the beaded skirt and shell loin ornament worn by First Father as the Maize Lord. The king is standing upon the *Yax-Hal-Witznal,* the First-True-Mountain, whose clefted head sprouts maize plants. In this image the king depicts First Father as born from the mountain of abundance where humanity was shaped of maize dough. This interpretation is confirmed by a passage in the accompanying text, which glosses "it happened at First-True-Mountain, White-Flower-Born, First-Tree-Precious" (Schele 1992:163).
36. Schele and Grube (1990a) first identified the word for "court" as *nab.* Their argument is based on the iconography of the lower register of many monuments, one a glyphic spelling on Stela M at Copan, and on Stuart and Houston's (1991) identification of toponyms. *Nab* is a word for palm of the hand and the measurement made from the end of the thumb to the tip of the first finger. It also means waterlily, lake, ocean, and other bodies of water. The Maya, like the Olmec, conceived that the plaza corresponded to the surface of the Primordial Sea.
37. There are a number of "flying Olmec" figures and some might be swimming rather than flying, according to Kent Reilly (personal communication 1991). A sensation of swimming is documented for trance states among many peoples across the world. Most specifically, a segment on A&E's *Footsteps of Man* featuring David Lewis-Williams's work with the Bushmen of the Kalihari shows them making swimming motions as they lie in the sand after collapsing in deep trance. These sensations have been reported by many other groups around the world and seem to be part of the physiological response many human beings have during trance experiences. See J. D. Lewis-Williams (1986) for a more detailed discussion of these phenomena.
38. Many contemporary Maya peoples have supernaturals who have the word mountain, or a refer-

ence to mountain, in their names. Among the Maya of San Antonio, in the Toledo district of Belize, J. Eric Thompson (1930:57) reported more than half a century ago a belief in *witz-hok'*, gods of the mountains and the plains, and he reports the name *witz-ailik* from Oliver La-Farge's work among the Chuj—again Mountain-valley. Garrett Cook (1986) discusses the *mundo*—"world"—divinities of the K'iche' Maya of the highlands of Guatemala. This is clearly a related concept. Mountains, and their human-made counterparts, not only embody supernaturals, they also constitute the home of ancestral gods among many Maya of the highlands (Thompson 1970). We can document this belief in at least some Classic Maya cases: the Temple of the Inscriptions at Tikal is glyphically termed a "sleeping house" (Houston and Stuart 1989b). As we discuss in Chapter 4, this is still the term for lineage shrine among the modern K'iche', and such places are certainly the abode of ancestral souls.

39. Maya make plaster, or mortar, from limestone, using a simple technology that requires only abundant firewood to drive off moisture, creating quicklime. When water is added again, the material is slaked lime, which, when exposed to air, chemically returns to the original calcium carbonate of the parent stone (L. Roys 1931). The Yukatan peninsula is primarily a limestone formation geologically and most Maya enjoyed ready access to the limestone to make plaster.

40. Richard Hansen (1992, personal communication 1991) has already discovered and excavated an early Late Preclassic (400–300 B.C.) pyramid at Nakbe with the Principal Bird Deity represented in huge five-meter-high masks. Middle Preclassic period (1000–400 B.C.) pyramids have been found at Nakbe and at Río Azul, where one has stucco sculpture. Hansen is confident he will find sculptured Middle Preclassic pyramids at Nakbe during the next season and perhaps at nearby Tintal and Guido soon thereafter. These pyramids are massive in scale, rivaling any constructions from the Classic period and competing seriously in scale with the largest buildings ever constructed at Teotihuacan. Hansen's work has shown this Maya activity to have been highly developed and ongoing by 500 B.C. and very likely by 700 B.C. to 600 B.C. If even earlier structures are found, then the earliest Maya centers of scale and stratified organization may be contemporary with the last phase of Olmec civilization. The present assumption is that the first Maya centers are at least contemporary with Chalcatzingo and other regions using the Formative Ceremonial Complex. The size and scale of Maya architecture and population density appears to dwarf anything contemporary with them, and most everything that followed them in subsequent history.

41. This strategy for interpreting earlier events on the basis of later ones is called the Direct Historical Approach in archaeology. The method requires clear evidence of continuity in the form, function, and meaning of culturally defined ideas and artifacts, that is to say, a cultural tradition. Many archaeologists regard the Direct Historical Approach as an unreliable way to interpret past events. Many others, however, use and defend this method. We are confident that the method works for the Maya because, in addition to the formal continuities in artifacts, we have specific continuities in concepts defined in words over long periods of time. This book, taken as a whole, is one defense of the existence of the Maya cultural tradition spanning the Precolumbian and post-Conquest eras.

42. Recall that *witz* is the word for "hill" and "mountain" in many Mayan languages, and most important, it is used as the word for "mountain" in the writing system (Stuart 1987).

43. Richard Hansen (1989, 1992, personal communication) observes that the early great plaza and pyramid complexes at Nakbe in Peten, around 600 B.C., were designed with sloping surfaces so that water would run off them into undetermined reservoir areas. He suggests that, contrary to popular understanding, this jungle environment was and is a "dry season desert" lacking vital drinking water sources on the surface during part of the year. He hypothesizes that one practical reason for building centers was to generate drinking water reserves. If these people had the concept of *nab*, as we believe they did, their plazas would have been both metaphorically and literally a source of their sustenance.

44. Grube and Schele (1991) found an example in which a God C variant head replaced a gourd sign that had long been accepted as phonetic *tzu*. This God C variant directly replaces the mirror sign in several contexts, including the introductory glyph to the Primary Standard Sequence in pottery. Taking this substitution as a clue, they identified a whole set of contexts where the gourd and God C variants appear in the spelling

of *tzuk*, the word for "partition." In many contexts, the *tzuk* head has an infixed or appended *ku* sign, which functions as a phonetic complement. Most important, they realized that this is the God C sign that appears on the base of the World Tree, on the noses of Mountain Monsters, in skybands, and alternating with stone signs, earth signs, and many other similar locational signs on benches, in framing bands, and so forth. They realized that these contexts combined a type of location symbol with *tzuk* as a reference to the great partitions of the Maya cosmos.

45. *Tzuk* is also the word for "belly" and "stomach."

46. Juan Antonio Valdés (1987) hypothesizes that the triadic theme represents dynastic kingship for the Preclassic Maya. Richard Hansen (1989, 1992, personal communication 1990) generally agrees and identifies the triadic architectural design as a Late Preclassic innovation accompanying elaborate stucco and masonry decoration. Freidel (1979) also identified the triadic arrangement as a feature of the Late Preclassic lowland Maya interaction sphere between emerging elite centers. In contrast, Hansen sees Middle Preclassic Maya design as focused upon a single pyramid on one end of a support platform or raised plaza. As we have seen, the triadic theme is certainly present in Middle Preclassic contexts at La Venta. There is good reason to believe that the lowland Maya of that time also used it. Only future investigation and a larger sample of Middle Preclassic public architecture in the Maya region can settle this issue of timing. Certainly for the Maya, the triadic theme had to do not simply with lineage descent and dynasty but with the Creation of the cosmos, the actions of the founding family, and the creation of humanity.

47. *Ch'ulel* is the word for the inner soul or holiness that resides in all living things, in powerful objects, in sacred places, and in the many energy-laden objects in the Maya world. See Chapter 4 for a full discussion of this concept.

48. Barbara Fash (Fash and Fash 1990) first recognized a *popol nah* in Structure 10L–22A at Copan. It is identified by the same *pop*, "mat" signs. Like this threshold building at Waxaktun, it defines a space of assembly and ritual. A dance platform opens up in front of it. In her research on the function of the *popol nah* as a category of Maya building, Fash has observed that they were community houses where councils of high nobles met with the king and where dances were preserved and taught for performance. We think her research applies equally well to the *popol nah* at Waxaktun, identifying it as a place where the community interacted with the king—where dancers were taught the dances and then performed them in the great pageants vital to Maya politics, community, and religious life.

The assembly halls identified by the Spanish at the time of the Conquest were usually roofed but open-sided, with ample interior space for councils, dances, school classes, and temporary residence of men and boys at festivals. See Schele and Freidel (1990: Chapter 9), for a vignette that starts in such an assembly hall. See Freidel and Sabloff (1984) for a discussion of how archaeologists functionally identify Late Postclassic building remains on Cozumel Island as council houses in conjunction with ethnohistorical accounts. Freidel (1986a) speculates that the large-roomed, multidoored range structures of the Classic and Terminal Classic period, "palaces," may have served these functions as well (see also Harrison 1970 for such arguments as applied to the Central Acropolis "palaces" at Tikal). The Nunnery at Terminal Classic Uxmal in northern Yukatan carries mosaic mat signs alternating with miniature houses on several superior moldings, which together would also read *pop* (mat) *na* (house). These places are such "palaces." The open spaces defined as mat houses at Copan and Waxaktun may be more closely associated with group dancing than with sleeping, storing, and the temporary residential functions found in other *popol naob*.

49. Iconographically, we know that these masks on Structure H-X-Sub-12 are ancestral deities because they are designed with the trilobate scroll element in the neck and lower face zone. These scroll elements mark heads that float in the upper register of early lowland Maya stelae, the position of gods and ancestral spirits when conjured by kings. The reason we think they are ancestral kings and not just gods is that they are adorned with a variant form of the three-jeweled headband that represents, from Late Preclassic times onward, the principal crown of Maya rulership. The fact that the lower jaw is covered by a trilobate element may be significant. The patron of Pax, a Classic-period deity image with the trifurcate scroll in its mouth, is an anthropomorphized expression of *te*, "tree." The outer gate may thus declare two ancestral beings royal world trees.

50. The Maya began constructing formal ballcourts in the Preclassic period. At Late Preclassic Cerros, they built two ballcourts along the main north-south axis of the community (see Chapter 8 and Schele and Freidel 1990: Chapter 3).
51. In March 1993, David Stuart circulated a note to epigraphers in which he laid out his argument that the bent-tree sign we have been reading as te' in actuality read *lakam*, "huge, big" and "banner." His argument rests on the relatively rare occurrence of the phonetic complement *ma* as a suffix and a single occurrence of the prefix *la* with the sign. While his cluster of phonetic complementation is sufficient to support his reading, it was the co-occurrence of the icons associated with this sign to banners and with banner-carrying warriors that convinced us of the reading. The flat middle area of Palenque was called either *lakam ha* or *lakam nab*, "big water" or "big lake." Since this place name is documented in Chiapas (it is the name of one of the principal Lakandon towns), we find its presence at Palenque also convincing evidence.

    The reference to water and lake is also logical in the context of Maya towns. In the rainy season, these wide plastered plazas were indeed, temporarily, covered with water during torrential downpours. Some of the water was channeled off into drinking water reservoirs, valuable and attractive resources during the dry season when surface water was scarce.
52. David Stuart (in a 1987 letter) was the first person to decipher the ancient name of this incredibly sacred mountain. In 1990, he circulated another letter in which he presented his arguments for *lakam* as the reading for the "bent-tree" sign. We accept his evidence and his reading of the name as *Yemal-K'uk'-Lakam-Witz*, "Descending Quetzal Big Mountain." "Big mountain" is a particularly apt description of the tall hill behind the Temple of the Foliated Cross.
53. We have called him Chan-Bahlum since 1973, based on the modern Chol words for his name—"snake-jaguar." However, strong phonetic evidence supports a reading of Kan-Balam, perhaps the eighth-century equivalent of the modern reading. We will retain the more familiar reading in this publication.
54. The texts on the Tablet of the Foliated Cross suggest that K'uk'-Balam, the founder of the lineage, was buried in Descending-Quetzal-Big-Mountain. Ongoing excavation in the pyramid of the Temple of the Cross shows that this temple of Chan-Bahlam was raised on a natural stone hillock at the base of the mountain (Mario Aliphat, personal communication 1992). We suspect that the house of the founder, built in the Early Classic period, underlies Chan-Bahlam's celebration of the rebirth of First Father, the Temple of the Foliated Cross. Like the portal cross shrines of Zinacantan, this pyramid is at the very foot of the mountain that houses the ancestors.
55. The information on the Acropolis contained in this description comes from discussions over the last three years with the members of the Early Acropolis Archaeological Project, especially Barbara Fash, Robert Sharer, David Sedat, Alfonso Morales, Julie Miller, William Fash, Rudi Larios, Ricardo Agurcia, and Richard Williamson (Sharer and others 1990, 1991). Details of the excavations are contained in the projected reports to the Instituto Hondureño de Antropología y Historia and have been presented at a session of the 1991 International Congress of Americanists in New Orleans organized by Ricardo Agurcia and William Fash. Present evidence has confirmed that the earliest levels were built during the time of the founding of the dynasty in the early fifth century and there is growing evidence that they may even belong to predynastic times. Ongoing excavation is revealing construction of a scale and detail unforeseen in previous evaluations of Copan's history but—most intriguing for us—it seems clear to Sharer, Sedat, and the other members of the team of the Early Acropolis Project that many of the most important foci, and perhaps even the shape of the Acropolis, were established during these very early times.
56. When the Maya wished to rebuild or expand on a particular site, they did not raze the existing structure to the ground. They ritually "terminated" the building, defacing or knocking down sculptures on it, to release the power that had built up over its long use as a portal to the Otherworld. Then they would build the next phase of construction over that one, perhaps incorporating part of the old structure into their new design, most likely burying the old under a much larger and more elaborate edifice. It should be understood that a temple was not so much stripped of its power when terminated as "put on hold." Buildings that had served as portals to the Otherworld for centuries were considered very powerful doorways. Through constant use, the membrane between the worlds had become very thin.

Of course, when a center was finally abandoned, due to war, ecological disaster, or other happenstance, all the temples were permanently terminated. Leaving them in operation would not only have been disrespectful to the supernatural beings and gods, but absolutely dangerous, akin to leaving an untended, active nuclear reactor lying around for anyone to wander into. See *A Forest of Kings* (Schele and Freidel 1990) for further information on the ritual termination of buildings.

57. Excavations under the direction of William Fash by Richard Williamson, Joel Palka, and others found examples from three different sets of macaw heads, all buried inside Temple 10L–26, also known as the Temple of the Hieroglyphic Stairs. Strömsvik (1952) found examples from at least three other sets of macaws, and documented at least six different construction phases of the nearby Ballcourt. William Fash and the other members of the team working on the history, restoration, and meaning of the Ballcourt at Copan take this to mean that the symbolism of the Ballcourt has remained the same throughout all its phases, even though the expression of that symbolism adapted to new styles and political purposes throughout the dynasty's history.

58. Elizabeth Newsome (1991) has recently completed a dissertation analyzing the iconography of the stelae in the Great Plaza. She argues persuasively that 18-Rabbit commemorated a series of performances conducted at significant points during the twenty-year k'atun period. These performances by the king related the triumph of the Ancestral Hero Twins over the Lords of Death, and the successful creation of humans at the beginning of the present, or fourth, Creation. She also links these rituals and the arrangement of the stelae to accounts of Creation recorded in the Chilam Balam of Chumayel. As in our own work, Newsome has found extraordinary continuity between the strategies of Classic-period kings at Copan and the mythology, belief, and performance recorded after the Conquest.

    Linda Schele, in addition to her work on the mythology, has found connections between these stelae and other of 18-Rabbit's works to Venus phenomena and movements of the Milky Way. He was linking himself to both astronomical events and creation mythology when he created the architectural and symbolic contexts in which his politics unfolded. This interlocking of political and cosmological frameworks was one of the most successful and widespread of political strategies used by the ancient Maya and their Mesoamerican contemporaries.

    The first big-stone in the cycle, Stela C, celebrates the end of k'atun fourteen on the heliacal rising of the Eveningstar. The east side of the stela shows 18-Rabbit as the Crocodile Tree, and on that night Venus appeared in Sagittarius at the base of the angled crocodile form of the Milky Way.

    Stela B, the culminating monument in the series, records the end of k'atun fifteen, which fell on the maximum elongation of the Eveningstar, with Venus sitting in Virgo, the constellation that shows up on the Hauberg Stela and Tikal Stela 1 as Chak. 18-Rabbit wears the costume of Chak on the stela.

    Stela H and A record dates exactly three k'atuns after the date on Stela 3, which was an elongation of the Morningstar and the twenty-four-haab anniversary of accession of Smoke-Imix-God K, the twelfth king in the succession of Yax-K'uk-Mo', the founder of the lineage. Stela A and H also record a ceremony in which Butz'-Chan, the eleventh in the succession, was conjured up as the bones from his tomb were carved up, perhaps to be taken as relics. The event fell on February 5, the day First Father raised the Wakah-Chan. Nikolai Grube (personal communication 1990) has read the passage recording this event as *susah bakil*, "the bones were cut." On Stela H the king wears the costume of First Father as he acts out the withdrawal of the bones of his own ancestor just as the Hero Twins disinterred the bones of their father.

    Further investigation may very well find more associations of this kind, but it is, even now, apparent to us that the costumes worn by 18-Rabbit and the beings emerging from his Double-headed Serpent Bar are related, at least in part, to the sky on the nights of the events depicted.

59. Inscriptional evidence combined with archaeological data from the excavations of the Copan Acropolis Archaeological Project, directed by William Fash and co-directed by Ricardo Agurcia, suggests that 18-Rabbit commissioned significant work on Temple 26 and twice remodeled or rebuilt the Ballcourt (Schele and Larios 1991; Schele, Grube, and Stuart 1989). Given that he also reworked the Great Plaza, as well as buildings in the East Court, he is considered one of the great builders of Copan's Late Classic history.

60. Excavations by the East Court Archaeological Project under the direction of Robert Sharer and supervised by Alfonso Morales have found at least five earlier versions of Temple 22. The lowest one known to date is called "Chachalaca" in the informal nomenclature of the project. Alfonso Morales (personal communication 1991) has found the remnants of the lowest course of the masks on the northeastern and southwestern corners of Chachalaca. This proves not only that Chachalaca is an earlier version of Temple 22, but that the symbolism of the building remained essentially unchanged throughout its last five or six phases.
61. Schele (1987e) oversaw the restoration of the corner masks after recognizing the nostril and forehead pieces in the great pile of fragments that lay north of the Acropolis in the East Plaza. Stuart (1988b) deciphered the glyphic name of this monster as "mountain" and associated it with the Temple 22 monster. Larios, Fash, and Stuart (n.d.) have also analyzed the building, setting it into the stratigraphic and dynastic history of Late Classic Copan.
62. Wade Collins (1986) first suggested that birds stood on the top of each corner after identifying pairs of bird feet in the rubble associated with Temple 22. His suggestion was confirmed by excavations under the northeast corner, where similar bird feet were found along with fragments from the corner mask (Barbara Fash, personal communication 1990). Since we have only the feet, the identity of the bird cannot be known at this time, but the symbolism of the building suggests it was the Principal Bird Deity, whose name was Itzam-Yeh.
63. Tatiana Proskouriakoff (Trik 1939) reconstructed this mask using information from fallen fragments and the restored walls. Schele (1986a) followed up on her work by identifying more fragments belonging to the door and reconstructing through drawings how it might have appeared.
64. See Chapter 4 for a full discussion of the symbolism of this sacred plate of offering.
65. Stuart (1984, 1988a) identified these lazy-S forms as blood. Subsequently, Ringle (1988), John Carlson (1988), and several other scholars identified the class of symbols to which the lazy-S symbol belonged as *ch'u* and *ch'ulel*, Maya words for "god," "divinity," and the "inner soul." Very recently, Houston and Stuart (1990) and Stone and MacLeod (1991) have independently identified phonetic complements to the glyphic version of the sign as spelling *muy* and *muyal*, words for "cloud" and "cloudy." These identifications make it clear that the lazy-S form is the representation of clouds, but this may also indicate an especially appropriate association with the function of the inner sanctum. Nikolai Grube (circulated in a 1990 letter) deciphered the "fish-in-hand" glyph, which is associated with Vision Serpent scenes, as *tzak*, a word that means both "to exorcise" and "to conjure clouds." We think that the Spanish mistakenly associated the Maya rituals for conjuring beings from the Otherworld and for curing people as the equivalent of their own rituals of exorcism. We find it interesting that this conjuring is associated with "clouds" in exactly the same way as the image on the door of Temple 22's sanctum. And the equivalent scenes from narrative monuments show kings scattering *ch'ul* with their ancestors floating, sometimes in clouds, above them.

It may be significant that there are seven of these S-scrolls arching over the doorway. William Hanks's (1990) informant in Oxkutzkab calls the layered heavenly clouds, *muyal*, and gives the number of such heavens as seven. J. Eric Thompson (1970:195) spoke of thirteen *taz*, "layers" of heaven, and envisioned a stepped model of the Maya sky, six steps to the summit, completed at seven, then six down for a total of thirteen. We think the word and concept in question is *tz'ak* and that it refers to steps but also to nodes (Barrera Vásquez 1980:872), as in knots on a string. As we have seen in Chapter 2, the Celestial Monster is one manifestation of the Milky Way. We think the "nodes" in the heavens possibly are constellations or clusters of stars on the Milky Way. If we are right, then there may be some resonance with modern ritual practice in Yukatan. David Freidel has observed that Don Pablo places thirteen small gourd bowls of precious liquid, corn gruel or honey wine, in little slings hanging from the circular "floating platform of the sky "at the center of his "sky tree" altar. The *xpeten ka'an*, floating platform of the sky, represents the "divine portal," the *u hol gloria* through which blessed magical *itz* flows. As this platform is the sky, so the thirteen gourds may represent the thirteen constellations on the ecliptic. Returning to our hypothesis that the Celestial Monster is the Milky Way in one of its positions in the sky, there is another time of night at which the Milky Way is at the edge of the horizon, forming a semicircular, partly encompassing glow. This is the time we think the sky represents the Black-Transformer, *Ek'-Way*, the major portal to the Otherworld. It may be more than coincidence that both the Classic Maya and the modern Yukatek have this circular symbol of the portal.

Our interpretation of this monster as a cloudy-sky symbol derived from the Milky Way was presented publicly for the first time at the Texas Hieroglyphic Workshop on March 14, 1992. On April 5 of the same year, David Stuart, who had been at the Texas meetings, presented a similar interpretation based on a related, but slightly different set of evidence. He brought out, as had several of the participants at the Texas meetings, that the Cloud-Serpent, Mixcoalt, was the Aztec term for the Milky Way. He added one component that we had not seen—that there is mountain iconography also associated with this Wakah-Muyal-Chan. We note that mountain imagery and glyphs are also associated with another celestial location called Na-Ho-Chan as it is presented on Kerr pot 688.

66. See Sosa 1988: Chapter 5.
67. The kawak signs on the forehead of the *tzuk* heads function as a phonetic complement for the final consonant, *ku*.
68. John Sosa (1986) argued that the Cosmic Monster represents the ecliptic, but it is now clear that it is the Milky Way at sunset on winter solstice.
69. Several pots from the Classic period show these skulls attached to the foreheads of upright Skeletal Maws. This same skull functions as syllabic *xi*, the first syllable of *Xibalba*, which is the Maya name for the Otherworld.
70. In the summer of 1991, David Stuart (personal communication 1991) recognized the first part of this inscription as a first-person quote from 18-Rabbit reading "'My k'atun ended,' he said." Grube and Stuart have also independently recognized a parentage statement naming 18-Rabbit as the child of Smoke-Imix-God K.
71. Based on the presence of these giant mat signs, Barbara Fash (in Fash and Fash 1990) identified Structure 10L–22A as a *popol nah* or "council house." Working with Jeff Stomper and others, she also reassembled a set of glyphs and identified their original position on the building facades. The project epigraphers identified the glyphs as place names and suggested that the figures that sat atop them represented the lords who ruled those sectors. See Note 30 and Barbara Fash et al. (n.d.) for a detailed description of the archaeology and interpretation involved in this fascinating project.
72. Barbara Fash (personal communication 1990–1991) has investigated the function of the Popol Nah in Maya ritual and community structure. She has found that they not only functioned as centers for community consultation, but that dances were taught and performed there. See Chapter 5 for Thomas Gage's sixteenth-century description of this process in a Maya town in Guatemala. Fash has suggested that each division of Copan may have been responsible for teaching and directing particular dances that are reflected in the iconography of the figures associated with each location, and most important, she has identified the long platform above the Venus Stairs and south of the Popol Nah as a dance platform. We believe she is correct in all of her contemplations of the function of the Popol Nah, although we suspect that the audience for the dances in the East Court would have been limited to only the highest officials of the various lineages that composed the Copan polity. We can visualize that the Popol Nah was the origin point for huge pageants that unfolded outward from the house onto the steps and terraces of the north side of the Acropolis both east and west of Structure 26. These pageants would have been visible to vast audiences standing in the great courts lying north of the Acropolis.
73. Barbara Fash (B. Fash n.d., 1989) made the connection between the bat iconography and the closed room of Structure 10L–20 and the myth of the Popol Vuh. She has suggested another connection of Copan's architecture with the Hero Twins story, proposing that the buildings in the East Court functioned at least partially as the backdrop for reenactments of the myth. Her arguments are greatly strengthened by an account of just such a pageant in connection with the inauguration of a sixteenth-century K'eq'chi chief. This account was quoted by Michael Coe (1989) in his reevaluation of the Hero Twins, and is included by us in Chapter 5. The identification of similar features in earlier phases of the East Court suggests that the Hero Twin myth played an important role in Copanec pageant for many generations, if not throughout its entire history.
74. The cruller-eyed jaguar god with Venus signs attached to its head was identified as a Venus God by Grube and Schele (1988). Schele (Schele and Fash 1991; Schele and Larios 1991) has

found that an enormous number of Copan events were timed according to stations of Venus and that this preoccupation was far more extensive and regular than at any other site so far analyzed. The imagery on the Venus Stairs celebrates this Venus God, while a text found in Structure 22A (Schele, Stuart, and Grube 1991) suggests this god was the *way* or "spirit companion" of Yax-Pas and perhaps other Copanec kings.

75. Mary Miller and Stephen Houston (1987) first identified the Hieroglyphic Stairs as the site of Classic Maya ballgames and associated these scenes with the sacrifice of captives as balls. Mary Miller (1988b) went on to identify false ballcourts at Copan (here and behind Temple 11) as exactly this kind of sacrificial place.

76. They are actually a little west of the exact center of the court because they go with the Venus Stairs rather than the full court. The East Court Archaeological Project under the direction of Robert Sharer has found at least one and maybe two earlier versions of these stairs.

77. This centering of the political geography at Chich'en Itza and later at Mayapan was more like the centralized concepts at Teotihuacan, Tula, and Tenochtitlan. The aftereffects appear to be the concept of Tolan, otherwise known as Tula, as the source of political and dynastic legitimacy as it is represented in the Popol Vuh and Annals of the Kaq'chikels of highland Guatemala. Earlier inscriptions from the Maya Classic period acknowledge no single source of legitimacy or group of sources. So we believe this idea of primary sources and the Tolans is a legacy of the Postclassic period. Archaeological interpretations of the Terminal Classic period (A.D. 800–1000) in the northern lowlands are currently dynamic and controversial (Sabloff and Andrews 1986). Scholars of the subject generally agree that there was a significant temporal overlap between the collapsing kingdoms of the southern lowlands and the rising cities of the north in such areas as the Puuk hills region. They also discern some contemporaneity between the Puuk cities of Uxmal, Sayil, K'abah, and Labna, among others, and Chich'en Itza. At present, we see temporal overlap between all of these cities of the north and the collapsing kingdoms of the south. We also follow Robles and Andrews (1985, 1986) in seeing a struggle for hegemony among the cities of the north, an era of war from which Chich'en eventually emerged victorious to dominate its contemporaries. The view that Chich'en Itza ruled in splendid isolation, the sole city of the north after the defeat of its enemies, does not seem viable any longer in light of evidence for overlap. That it was the capital of some regional political organization makes sense of the deliberate attempts to revive its political form in the later city of Mayapan, capital of a regional confederacy.

78. See Schele and Freidel (1990:364–374) for a discussion of the *multepal* government represented in the art of Chich'en Itza. Grube (1991b) has documented that the form of government by council that the Maya called *multepal* was active at Xkalamk'in by 9.15.0.0.0 (A.D. 741). This form of government was operational in northern Yukatan far earlier than we had suspected.

79. Among other contemporary Maya cosmologies, the Tzotzil-speakers of Chiapas regard the solstitial points as the four corners of the world. *Mixik' balamil,* the navel of the universe, lies at the crossing of these two solstitial axes in Zinacanteco thought (Vogt 1985:489).

80. Evon Vogt (1985) argued that the Zinacanteco Maya are hardly aware of the equinoxes, never mind focusing on them. The rattlesnake shadows cast upon the outer edge of the northeast balustrade of the Castillo at Chich'en are indeed a hierophany. As the distinguished astronomer Anthony Aveni puts it: "The Palenque and Chich'en Itza hierophanies are difficult to test with any astronomical accuracy. After all, their symbolism was intended to be purely mythological. These arrangements are so broadly structured that they could never have been intended to serve as the basis for collecting calendric observations of the type we find delineated in the codices; nevertheless, these architectural hierophanies, which find their phenomenal origin in astronomy, provided some of the most powerful religious experiences the common person could witness in the environment of the ceremonial center" (Aveni 1980:-286).

81. The sides of the Venus platform are oriented about 9° east of north or very near magnetic north at Chich'en Itza.

82. See the chapters on warfare and the ballgame for detailed analysis of the imagery and symbolism associated with the great ballcourt.

83. As Alfonso Escobedo pointed out to David Freidel, the heads of these beasts carry the "horns" diagnostic of predatory birds, particularly those of the so-called Mexican Eagle. Their bellies have the scales of snakes, but their bodies are

feathered like birds. The iconographic tension of serpent and bird is age-old in Mesoamerica, registering in Olmec art and persisting through to the present where it appears as the state seal of the modern nation of México. Birds and snakes are animals that live at the upper and lower edges of the ordinary world, obvious candidates for contemplation of the Otherworld beyond.

84. See Freidel (1981) for an elaboration of this argument.
85. See Thompson (1957) for a discussion of this position.
86. See Freidel, Masucci, Jaeger, and Robertson (1991) for a discussion of Preclassic and Classic effigy censer stands and their iconological context in Maya artistic composition.
87. This description of the baptism ceremonies is found in Landa's *Relación de las Cosas de Yucatan* (Tozzer 1941:102–104).
88. Tozzer (1941:145) attributes this translation to Roys.
89. Landa described these rituals, which he must have thought of as pagan bedevilment of the worse kind, in an apology for his harsh treatment of the Yukateks during his rule as bishop of Yukatan. See the notes on pages 68–84 for Tozzer's (1941) discussion of the auto da fé conducted by Landa and the trouble it caused him with Church authorities.
90. These descriptions come from Means (1917) translations of Avendaño's descriptions of his encounters with Kan-Ek'.
91. See Chapter 8 (the ballgame) for a description of Iximche. Like its contemporary cities, the layout of the town incorporates all of the vital components of sacred place—a courtyard, temple-mountains and the divinely made counterparts of the natural world: dance platforms, a ballcourt, and a palace.
92. There are many good discussions of the Caste War. See, for example, Bricker (1981).
93. Nikolai informs us that these books are the descendants of the Books of Chilam Balam. They are still being written by scribes literate in Yukatek, who keep track of the modern history of their villages and the larger world. As with the Chilam Balams, history becomes prophecy—some of which predicts the end of the world at the coming turn of the millennium.
94. A coatimundi, which is relative of the raccoon, is known as a trickster and a clown among modern Yukateks.
95. This same tree ritual was discussed by Redfield (1936) in its form at Tixkakal Guardia and at Sok'otz in northern Belize. He also discusses the role of the coatimundi as a clown and trickster and associates the raising of the *Yax-Che'* with the *Yax-Cheel-Kab* at Tayasal and with similar rituals practiced by the Aztec and other North American cultures. Blaffer (1972) has analyzed a very similar ritual and its associated mythology among the Tzotzil of Chiapas. There the tree is called the *Bolon-Te',* "Jaguar-Tree," which is climbed by people manifesting *H'ik'al,* "Black Men," and by the Jaguar, during the most important festival of the year. At Tixkakal, the young men playing the role of the coatimundi carry stuffed spider-monkey toys up their tree, while the *H'ik'al* and *Bolomtik* at Zinacantan carry stuffed coatimundis and other animals up their tree. Both throw sweets and candies to people below.
96. Several vital performances of the kings of Palenque are in sacred space framed by images of their mothers and fathers. See Schele (1979).

Chapter 4: Maya Souls

1. See Taube (1992b:78) for a discussion of these mirror panels on Late Classic through Early Postclassic period Mesoamerican sculpture. See Schele and Freidel (1990a:394) for a related discussion of the Chich'en Itza mirrors as portals.
2. This is a core thesis in Nancy Farriss's (1984) analysis of the transformation and survival of Maya cosmology through the colonial experience in Yukatan.
3. The comprehensive discussion of *le santoho* as conceived by the Maya of Yalcoba is given by John R. Sosa (1988). We derive our brief summary from this work.
4. See also Hanks (1990:340) for this usage in Oxk'intok', Yukatan.
5. In contrast to other crosses in Yalkoba, these are actually combined images of the crucifix and the World Tree, for the arms curve upward in a half circle. See John R. Sosa (1990) for a discussion of these crosses.

6. See Victoria Bricker (1981) for a systematic analysis of Maya revolts and their cults, both in highland Chiapas and in lowland Yukatan.
7. See López de Gómara (1964:35) for a description of the statue and how it was used.
8. The book is *Unfinished Conversations: Mayas and Foreigners Between Two Wars* (Sullivan 1989) and we strongly recommend it.
9. Nikolai Grube and David Stuart (personal communication in MacLeod 1990a) have identified *sak-a'* as one of the foods recorded in the Primary Standard Sequence on Classic-period pottery.
10. This idea is best expressed in Munro Edmonson's (1971:146) translation:
    And there they found food
    Whence came the flesh
    Of the formed people
    The shaped people
    And water was their blood;
    It became man's blood.
11. See John Sosa (1988).
12. Our descriptions here are drawn primarily from Evon Z. Vogt (1976:18–19).
13. Sixteenth-century Tzotzil (Laughlin 1988:200) has the following glosses: *ch'ulelil* "fate, happiness, luck, mind, soul" and *ch'ul* "holy, spiritual."
14. See Note 59 for a history of the decipherment of the *ch'ul* symbol. The different spelling in Yukatek and Chol results from the linguistic divergence between the two languages. Yukatek *k'* corresponds to *ch'* in Cholan languages.
15. See Barrera Vásquez (1980:299) for the entries on this Yukatek term. The Yukatek term is cognate to a proto-Cholan root **känän*, and its Tzeltal-Tzotzil cognate *kanan* is the word for "guard" and "caretake." Stuart, Houston, and Grube (personal communication 1990) have all suggested that this root is the reading of Proskouriakoff's "captor of" glyph. The problem with this reading is that "guardian" has been reconstructed with a *k* in Yukatekan, Cholan, and Tzotzilan languages, while the title from the Classic-period texts has the glyphs for "snake," "sky," and "four" in free substitution. In Cholan and Tzotzilan, "snake," "sky," and "four" are all *chan*. In order for the "captor of" title of the Classic-period inscriptions to read *kanan* (the word for "guardian"), we must posit that the *k/ch* correspondence was not entirely developed during the Classic period, and there is a large body of evidence supporting this possibility. For example, at Palenque, the words for "snake" and "earth" are spelled phonetically with *ka* rather than *cha* signs. This leaves the possibility open that the "captor of" title was in fact read *kanan*, "the guardian of," and that it is related to the Yukatek protector term.
16. Proskouriakoff (1963:163) identified the wing-shell-sak-ahaw-ik' compound as a glyph for death in her seminal study of the inscriptions of Yaxchilan. Her wing-shell death glyph consists of a bird wing combined with a shell preceding a possessed noun with signs representing the color white conflated with an ahaw sign, and a *na* sign in front of an ik' sign which often has a *li* suffix. Proskouriakoff herself suggested that the death expression "makes some reference to a departing spirit."

    Stuart (in a letter circulated in 1988) found evidence that the wing-shell part of the expressions reads *ch'ay*, "to diminish or die." He found an entry in colonial Tzotzil (Laughlin 1988) for *ch'ay ik'*, "it diminished breath" or "died." Most epigraphers now accept his decipherment as the reading of the verbal component of the expression, but the reading of the possessed noun still eluded us until Nikolai Grube and Werner Nahm found another clue. In a 1990 letter, they identified the ahaw glyph when it is outside the day-sign cartouche as *nik*, "flower." Together with the *sak*, "white," sign, it gives *sak nik*, "white flower," as a word for "soul." Then in 1991, Schele, following a clue from Barbara MacLeod, noticed two examples of the *sak-nik* glyph on the Hieroglyphic Stairs at Copan in which the *ik'* sign is eliminated to leave only *na* and *li* suffixed to the *sak-nik*. In 1992, Nikolai Grube climbed the Hieroglyphic Stairs to check the original step and confirmed the substitution is correctly drawn in Barbara Fash's drawings of the stair. Since the *ik'* sign often has *na* above it even in the context of the day sign, it occurred to Schele that the second half of the word for soul might simply be *-nal*, a suffix meaning something like "born of," "one of the quality of," or "one from." Although we are still collecting evidence to test this idea, it looks promising. We think the word for "soul" was "white-flower-thing."

    Thompson (1970:202) identified the red and white plumeria (frangipani) with the Creator couple in the Lakandon religion. He goes on to cite Robert Bruce's identification of Kakoch as another Creator, who created the waterlily flower from whom all other gods descended. Kakoch is the brother of Hachakyum, who was the first offspring of the plumerias. These associations, combined with Stuart's identification of flowers as one of the principal signals of Maya ahaw status, make *sak-nik*, "white flower," a particularly apt symbol for the soul of human beings.

17. This discovery comes primarily from David Stuart, who circulated a letter in answer to Grube and Nahm's *nik* reading for the ahaw sign. His letter included many of the same interpretations, but based on the amazing variety of flower imagery he had identified with rulers' costumes. Most important, he showed a pattern of substitution where the stamen of the flower was represented with three cylinders and dots in many examples, but by the square-nosed-serpent both with and without jewels in many others. Given this clue, Schele and Mathews recognized a *sak-nik* written with this square-nosed serpent instead of the ahaw sign in the headdress of a figure from Laxtunich. Later, Schele and Barbara Fash recognized the wide distribution of *sak*-square-nosed-serpent motifs at Copan. The final clue came when Barbara Fash was trying to figure out how combinations of *sak* signs were articulated on Structure 22a at Copan, a building she had earlier identified as a *popol nah*, a "mat or council house." At the 1991 meetings of the International Congress of Americanists in New Orleans, Stephen Houston pointed out to us that Barrera Vásquez (1980:570) lists *nikte'il na* as "casa donde se hace junta" and as an alternative to *popol nah*. The white-square-nosed serpents simply mark Structure 22a as a "flower house" as well as a "council-mat house," and thus confirm the identification of the white-ahaw combination that appears so ubiquitously on headdresses and on the tips of tails as white flowers.
18. Stuart and Houston (n.d.) were the first to identify this Matawil glyph as a location in the Otherworld.
19. Weldon Lamb (1983) related these ideas in his unpublished paper on the star lore of the Tzotzil-speaking Maya.
20. George Foster (1944) wrote the definitive article on nawalism.
21. Barbara Tedlock (1982:110) discusses this link between the day on which a child is born and its spirit or nawal.
22. This description is derived from Garret Cook's (1986) fine article on K'iche' theology.
23. Nikolai Grube's new translations of the "k'an" sign as *ol* make it possible that the name here reads *na ol*. We believe the *nawal* reading is the more likely of the two possibilities, but the alternative cannot be eliminated at present.
24. The full name of the town is *Chi Nik'oj Taq'al*, "in the middle of the plain." K'iche' residents call it Chinik' for short, but Spanish speakers have hispanicized that to Chiniqui.
25. John Fox of Baylor University has been studying Utatlan and its history for several decades. He believes the cave was dug by fortune hunters looking for gold under the city during the early nineteenth century.
26. *Ah K'in Mai* was the office of high priest at the last great lowland capital of Mayapan. J.E.S. Thompson (1970) thought *mai* might be a reference to tobacco here, but we think it probably refers to this person's responsibility for recalling the generations of ancestral great souls in the Temple of K'uk'ulkan, the Feathered Serpent who is the Postclassic Vision Serpent.
27. The idea of souls waiting in such a locality might well strike us as something akin to Purgatory in Christianity. In Chapter 6, we give an example of Maya people dancing to temporarily release the souls of their ancestors from such a place, which they liken in that context to a jail. At the time of the Conquest, many Maya regarded the afterlife as an unpleasant sojourn in a dark, damp place like a cave.
28. See Barbara Tedlock (1982:77–82) for a discussion of the different categories of K'iche' shrines in the community of Momostenango.
29. The K'iche' term *warabal ja* is cognate to the Chol-Yukatekan term *wayibal nah*.
30. *Waybil* is the participle of *way*, "to sleep." A house called a *waybil* is a sleeping place just as *warabal* is "sleeping" in K'iche'. Since *way* also means "transformation," a *waybil* can be dormitory or a bedroom—but it can also be a place where one dreams and transforms into the soul companion. Stephen Houston and David Stuart (1989b:11) discussed the glyphic usage of *waybil* as "sleeping place or article," but while they regard the "sleeping house" as a probable reference to places where the *wayob* manifested, they did not report the explicit use of this term in reference to lineage and *cofradía* shrines in modern K'iche' communities. The idea that *waybil* might refer to lineage shrines was proposed by Nikolai Grube and Linda Schele (1990b) in their discussion of the small stone house sculptures bearing this glyphic term at Copan. The hypothesized connection between Classic Maya *waybil* and modern K'iche' *warabal ja* is original to our research.

31. This history is described by Barbara Tedlock (1982:17).
32. See Andrews V and Fash for a description of the archaeology, consolidation, and reconstruction work that reclaimed this very important building from the past.
33. The story of the decipherment of the *way* glyph is one of the really interesting tales in the field. By 1978, Schele (1985a, 1988) had identified a glyph consisting of an ahaw glyph half covered with jaguar pelt that occurred with particular regularity next to the figures of supernatural beings on pottery. She also noted that this glyph like many others has a personified form called a head variant. Moreover, this particular head variant occurred as the head decoration of the main figure on Tikal Stela 31 and as a substitute for the emblem glyph. Based on these occurrences and a set of homophonies in Mayan languages, she proposed that the glyph was read as *balan ahaw,* "hidden lord," as a reference to their status as Otherworldly creatures and *balam ahaw,* "jaguar lord," as a reference to the association of Maya nobles with the jaguar. The weakness of the arguments was that it did not account for the frequent presence of a prefixed *wa* sign and suffixed sign that was subsequently deciphered as *ya.*

In 1989, Houston and Stuart used these two phonetic complements to propose an alternative decipherment of *way.* It turns out that Nikolai Grube had made the same deduction almost simultaneously and that letters relating the arguments from all three epigraphers arrived at Schele's house within a day of each other. Houston and Stuart (1989a) published their results in English, while Grube will publish his ongoing study in German in the near future.

The sum of the argument identifies the reading through the phonetic complements *wa* and *ya,* by taking them to designate the word *way.* Houston and Stuart (1989a:5) provided the following glosses in their analysis:

Yukatek	*way*	"transfigure by enchantment"
	wayak'	"vision in dreams" (Barrera Vásquez 1980:916)
	way	"the familiar that necromancers, witches, and wizards have that is an animal"
	wayasba	"to dream"
	wayak	"prognostication or word of diviners and of dreams" (Martínez Hernandez 1929:888–889)
Lakandon	*äh-way*	"wizard"
	wayäl	"metamorphose" (Bruce 1979:15)
Proto-Cholan	**way*	"to sleep"
	**wayak'*	"dream" (Kaufman and Norman 1984:135)
Chol	*wäy*	"other spirit"
	wäyäl	"sleep"
	wäyibäl	"sleeping place" (Whittaker and Warkentin 1952:114)
Colonial Tzotzil	*way*	"sleep, take lodging"
	wayahel	"witchcraft"
	wayichin	"dream"
	wayahom	"warlock, necromancer" (Laughlin 1988:326–327)
Tzotzil	*wayihel*	"animal transformation of witch, animal companion spirit of witch"
	wayihin	"send animal transformation or animal companion spirit (witch)" (Laughlin 1975:365)

These entries identify the jaguar-spotted ahaw glyph and its head variant as a glyph for "to sleep" and "dream" as well as for the animal companion spirits of human beings, especially those who are adept in the rituals of communication with the Otherworld. *Way* also means the act of transformation that allows adepts to walk the earth in the form of their animal spirit companions. The precise definitions of these animal companion spirits may well have changed in the three millennia of Maya history, but the underlying concepts are the same. The central importance of the concept is now clear, not only to the Maya conception of the supernatural, but also to their understanding of political power and of religious performance.

34. Nikolai Grube has discussed this pattern of distribution with Schele and suggested that the *way* of the Classic period had something in common with the modern Lakandon *onen*, "spirit companions," that are associated with all the members of a single lineage. Nikolai is aware that the match is not exact, but we agree with him that the distribution of *wayob* in the Classic period is constricted in a similar way.

35. In his 1989 letter on *way*, Nikolai Grube made this suggestion based on the 2 Kib 14 Mol passages from the Group of the Cross at Palenque. His deduction was based on two pieces of information derived from the text: (1) the actors of the event are the *way* of the Palenque Triad gods, and (2) the event was the close conjunction of Saturn, Jupiter, and Mars with the full moon. Dütting and Aveni (1982) suggested that this conjunction was conceived to be the reunion of the Palenque Triad with their mother.

36. William Hanks (1990:341) observed while working with a shaman in Oxk'intok', Yukatan: "In the course of prayer, one speaks in order from low to high, to 'the four corners of the earth, the four corners of the sky, the four atmospheres.' . . . *Ah K'iin Tuus*, 'High Priest Deceiver,' the guardian of the west temple in the highest sector of the cosmos, is not addressed in this ceremony, because when he moves, all evil in the world moves also, and the crystals will surely deceive."

37. This god's name was first read phonetically as *k'awil* by David Stuart (1988b). The bifurcate scroll itself is a pervasive motif. Iconographically, it is definitely identified as blood by Stuart 1988 and as *ch'ul*, the basic "soul-force" of the universe, by Ringle (1988) and Carlson (1988). Robiscek (1978) argued in favor of the scrolls as smoke, and some examples of the object in K'awil's forehead are apparently cigars. *K'awil* can mean "green cigar" in Yukatek (Barrera Vásquez 1980:387). Freidel (1985) made the iconographic case for the bifurcate scroll as a general expression of "volatile substances" including smoke, fire, rainwater, and blood. The scroll is the operative element here, attached to celts, cigars, and mirrors in the forehead of K'awil, all of which likely represent objects or actions used to induce the transformative powers of the magical substances.

38. Nikolai Grube has told us that he believes the scroll sign in front of the reclining body of the god is part of a phonetic spelling of *nen*, the word for "mirror." He suggests that the full name was *Nen K'awil*, "Mirror Statue."

39. This mirror sign was read by Grube and Schele (1991) as *tzuk*, the word for "partition" or "province," as in the four partitions of the world. This use of *tzuk* in this name apparently refers to the appearance of four K'awils in the 819-day count, which divides time into four quadrants, each with its own color, direction, and version of K'awil.

40. The second half of this glyph is written with the glyph *yax* which means both "blue-green" and "first" and phonetic *ch'a-ti*. *Ch'at* is the Cholan word for dwarf, an association that makes sense with the childlike proportions of the K'awil's body and the modern associations with a related deity and dwarf. See Note 56 for a discussion of K'oxol, the red dwarf.

41. K'awil, also known as GII of the Palenque Triad, was the third of the three gods born of First Mother. These were GI, born on 1.18.5.3.2, GIII, born four days later on 1.18.5.4.6, and finally GII (K'awil), born on 1.18.5.4.0.

42. Barrera Vásquez (1980:387) discusses the various meanings of *k'awil* in Yukatek.

43. David Stuart (1988b) was the first scholar to link all of these disparate contexts together under the single category of vision rite imagery. He (1988b) also recognized the phonetic spelling of the name of God K (the alphabet designation assigned to the codex version by Schellhas [1904]) in the inscriptions of Chich'en Itza. Much of our interpretation of K'awil and his association with Vision Serpents follows Houston and Stuart's (1989a) study of the *way* glyph and their extension of Stuart's earlier work on Vision Serpents

and the vision rite. We add that Nikolai Grube and Werner Nahm (personal communication 1992) have read the combined name for GII as Nen-K'awil, "Mirror-Statue."

44. Our source is the remarkable study of sixteenth-century Poqom Maya by Susan Miles (1957). She extracted a wealth of cultural information from two sixteenth-century dictionaries, one compiled by Fray Pedro Moran (1720) and the other by Fray Diego Zuñiga (1608).

45. The K'iche'an languages include K'iche', Kaqchikel, and Achi among others. In Kaqchikel, Saenz (1940:174) lists *q'abuil* as "ídolo, estatua, imagen," while *q'abuilaj* is "ofrecer sacrificio a los ídolos; adorar los ídolos." *Q'abuilchajal* is "guardián de los ídolos" and *q'awuiljuyu ru* is "cerro donde sacrifican a los ídolos." The *q'* in the K'iche'an languages corresponds to *k'* in Cholan and Yukatekan languages so that we can be fairly sure that the Poqom term is the same root as the K'iche'an, and that both correspond to *k'awil* of the Classic-period texts.

Dennis Tedlock, translator of the Popol Vuh, analyzed the use of *cabauil* (*q'abawil* in modern orthography) as follows: "*Cabauil* was once a generic Quiché term for images of deities, but the early Dominicans attempted to use it as a translation for Dios (Carmack 1981:318), as if this new and purer meaning could bleach out the word's accumulated pagan stains. But even if the author [of the Popol Vuh] did have the Dominican usage in mind here, he elsewhere uses *cabauil* to refer to wood-and-stone images of the titular deities of the ruling Quiché lineages" (D. Tedlock 1986:78).

Alfred M. Tozzer (1941:143, note 678) provides the following on *k'awil-k'abuil*: "Genet [who did an unpublished translation of Landa into French] has a long note here regarding the name Itzamna Kabuil or Lord Itzamna. He writes that the term *Kabuil*, indicating the gods, fell out of use shortly before the Spanish Conquest." Tozzer goes on to state that the great ethnohistorian Ralph Roys agreed with Genet on this point. Barrera Vásquez (1980:272) includes the following entry: "Itsamna K'abul: . . . 'el Itsamna-que-obra-con-las-manos' . . . *k'abul* posiblemente es un error por k'awil . . . y en Kaqchikel *q'abuil* está definido como estatua or ídolo de bulto or pintado que adoraban (Ms. de Fr. Francisco Barela). . . ."*Ah k'abul* in Yukatek means "artist, maker, knower" (Barrera Vásquez 1980:362). We note also that Houston and Stuart (1989a:7) got very close to this interpretation in their discussion of the Double-headed Serpent Bar in the context of "companion spirits."

46. These statues were found in Burial 195 under the direction of George Guillemin. When the tomb was opened, he and his young assistant, Rudi Larios (1990), found a layer of mud covering the bottom of the tomb. Rather than digging it out, Rudi Larios injected plaster into the negatives, spaces left by rotted objects, and preserved the casts of a wooden ballplayer yoke, a headdress, several gourd and wooden bowls, and these four *k'awilob*.

47. Staff-mounted *k'awil* statues are shown on the piers of House A at Palenque.

48. Elizabeth Newsome (1987) first pointed out this relationship to us and identified the Serpent-Bar deities as eccentric flints.

49. See Dennis Tedlock's (1986) analysis of the hermeneutics of the Popol Vuh.

50. Edmonson (1971:12) used different translations of the names of these three deities: One-Leg Lightning, Dwarf Lightning, and Green Lightning. That different scholars arrived at such different translations is a warning that our work involves personal judgments about the evaluation of bodies of evidence. All text-based analyses are ultimately to be judged by the coherence they lend to larger bodies of evidence. As Dennis Tedlock (1986:79) aptly puts it: "I have already started to shift away from an atomizing approach, in which the text is treated as a collection of artifacts whose provenience must be identified, and in which a stratum containing even a few European artifacts must be sharply segregated from those which do not. The approach I am shifting toward is the hermeneutical one, in which questions of culture history are held in abeyance long enough to get the drift, to hear the tenor, to follow the path, to see the world of the text before us. Comparison is still open to us here, but it is a comparison of tenor and of paths and worlds rather than of artifacts."

51. Michael Coe (1973:116) has long championed the notion that God K and Tezcatlipoca are cognate deities.

52. Dennis Tedlock (1985:365) makes this suggestion in his translation of the Popol Vuh. Karl Taube (1992a) came to the same conclusion in his own analysis of God K.

53. The nineteenth day—Kawak, Chahk, or Chahuk, depending on the language—is the day of lightning and storm (Thompson 1950:87). The same

sign outside the day-sign cartouche was read *ku* phonetically and logographically as *tun* "stone." Because Maya traditionally believe that flint, obsidian, and axheads are made when lightning strikes the ground, they used kawak marking to identify flint axes (and perhaps other kinds of stone). This association of the ax with lightning has long been known and in fact played a significant role in the interpretation of the god Chak as a god of lightning. The most useful studies of modern lightning gods and their relationship to the Classic gods were written by Joanne Spero (1987, 1990). Karl Taube (1992a) used similar data to discuss the associations of lightning with K'awil.

54. See Barbara Tedlock (1982, 1986).
55. Barbara Tedlock (1982:148–149) records "White K'oxol" as the Tzitzimit who protects the cave at Utatlan and helps shamans who come for divinations. Garrett Cook (in a letter sent to us in 1992) gave us the name Q'aqik'oxol, "Red K'oxol," for the actor in the Dance of the Conquest. Clearly the K'oxol were once four in number and associated with the four directions and colors, just as the Chakob and K'awilob of the Classic period were.
56. Karl Taube (1992a) provides an extensive discussion of the relationship between the axes of K'awil and Chak. The connections between these entities are intimate and ancient, dating at least to Early Classic times. With respect to the connections between Chak and K'oxol, Barbara Tedlock (1986:135) says the following: "The stone hatchet carried by the shepherd in the C'oxol role points to the Chakob of pre-Hispanic lowland Maya iconography, who are often associated with such hatchets. . . . Unlike a whip, the C'oxol's hatchet is silent, striking sheet lightning rather than thunderous lightning. To rephrase the various episodes of the narrative in auditory terms, the sounds of log drums and jaguar roars at the beginning later become cracking whips or thunderclaps. Finally, when lightning takes its silent form, there enters the whistling and shawm playing of C'oxol, which moves us closer to the human voice."
57. K'awil, like Chak, had four manifestations associated with the directional partitions of the world, just as the modern dwarf lightnings are associated with the four sides of the K'iche' lakes. *Tzuk* was the Classic-period word for these partitions of the world (Grube and Schele 1991).
58. See Vogt (1976:62) for this discussion and elaboration of Zinacanteco curing rituals.
59. The first clue to the identification of *ch'ul* came at the 1979 Dumbarton Oaks Conference on Human Sacrifice, where Schele (1984) identified the beaded scrolls on the face of the woman on Yaxchilan Lintel 24 as blood. In the following years, Schele and her students at Texas and David Stuart (1984, 1988a) independently identified the beaded material on the face with the material scattered in period-ending scenes throughout the Maya area and with the "water group" set of signs associated with emblem glyphs. Stuart (1988a) vastly expanded the repertoire of images associated with this complex, adding the lazy-S shapes and the floaters of Late Classic images to the inventory. Using the text of Dos Pilas Stela 25 as evidence, he showed that the Maya believed they materialized gods through their bloodletting rites and that they conceived of this action as a kind of "birth." In 1988, John Carlson (1988) saw a photocopy of the original pages from Landa's *Relación* on the calendar and "alphabet," which George Stuart had just acquired from Spain. As Barthel had decades earlier, he recognized that the God C head that occurred in the sign for the month Kumk'u had to read *k'u*. He connected this sign to the word for "god" and "holy." Bill Ringle (1988) had independently made the same connection, and early in 1988, he sent a manuscript outlining the entire argument to George Stuart who published it in his Research Reports. Stuart, Carlson, Ringle, Schele, and many others continued to develop the new insights opened by this decipherment, especially by connecting the *k'u* contexts in the Classic period to the beliefs documented by Evon Vogt and other ethnographers among modern Maya.

Two further clues came in 1990. Schele and Freidel (1990) had taken a new look at the contexts of the "fish-in-hand" glyph and come to the conclusion that it occurred not with scenes of bloodletting, but rather with those showing a Vision Serpent. They suggested that an interpretation of "manifested" or "called up" fit those contexts better than "let blood." Nikolai Grube (in a 1990 letter) confirmed their ideas by finding at Yaxchilan a direct substitute of the "fish-in-hand" glyph for the phonetic spelling *tza-ku*. He found that *tzak* meant both "to grasp in the hand" (in Tzeltal and Tzotzil) and "to conjure" (in Yukatek) and "to follow after" (in Yukatek and Chol). The glyph shows a hand grasping a fish, and since the fish sign was read *ka*, it acts as a phonetic complement for the *tzak* reading. A few of the examples of this verb appear to refer to this action of grasping something, but the greatest majority are clearly "conjured" with special iconographic reference to "conjured clouds." Very soon after the *tzak* decipherment, Stuart and Houston on the one hand, and Andrea Stone and Barbara MacLeod on the other, deciphered the lazy-S sign in which the gods and ancestors float as *muyal*, "clouds."

The pattern of things that can be conjured is also revealing. The verb can be followed by *k'awil* (the word for statue and the spirit in the statue), by *ch'u* and *ch'ul* (the words for "god" and "soul"), and by the proper name of the being who was conjured. Any of these beings could be conjured into clouds, which presumably could be both clouds in the sky and swirls of smoke rising from censers.

Perhaps the most interesting ideas about conjuring came to us from Johannes Wilbert, who has studied the role of tobacco in shamanism throughout South America (Wilbert 1987). In view of the prominence of tobacco in Maya imagery and medicine, he suggested that increased attention be given to the pyschotropic effects of nicotine, the tobacco alkaloid, on the human body. This seemed to him of particular interest because the Maya, like other aboriginal farmers of Mesoamerica, grew *Nicotiana rustica*, a species of very high levels of nicotine content and toxicity.

After reading the manuscript of this book in January 1993, he informed us that several components of the vision experience in Classic-period imagery as well as post-Conquest sources might have been informed by the effects of nicotine intoxication. He sent us the following commentary:

> Seeking nicotine-induced trance states, for instance, South American religious practitioners experience the exchange of life in this world for life in the Otherworld and the sensation of death and rebirth. Due to the biphasic nature of the drug in tobacco, this process of journeying across magic thresholds is experienced as initial nausea, heavy breathing, vomiting, and prostration (symptoms of illness); subsequent tremors, convulsion, or seizure (indicative of agony); and resulting in peripheral paralysis of the respiratory muscle (mimicking death). During this journey toward death, the shaman travels to the Otherworld whence he returns thanks to appropriate dosing and prompt biotransformation of nicotine in the body. If the dosing is wrong, the death journey can become final and permanent.
>
> Nicotine has various effects on the practitioner's sight (e.g., acuteness of vision, night vision), dreaming, and visionary experiences. Vision serpents producing balls of blinding light while imparting instantaneous comprehensive wisdom to the beholder are known to be tobacco-induced experiences. Travels to the dark Otherworld may occur more often than not in a state of tobacco amblyopia, characterized by dimness of vision and color blindness. In this state, the shaman is not blind but moves about freely in an undifferentiated world, shrouded in silvery mist during the day and engulfed in darkness during the night when only white or yellow light may prevail. He may be affected by hemianopsia, i.e., loss of sight in half the visual field. He is unable to recognize the faces of friends he meets, and all people look like corpses to him with yellow and waxy complexions. He himself looks pale and suffers from fatigue, depression, anxiety, and insomnia. Only tobacco abstinence can reverse the effect. Resumed nicotine administration after recovery may bring back the amblyopic state, thus enabling the shaman to move between the worlds of light and darkness.
>
> As in South America, the nawal and shaman-jaguar transformation complex of Mesoamerica may be strongly associated with tobacco consumption. In any case, several effects of the alkaloid on the branches of the human nervous system and the adrenal system allow the shaman not only to imitate the jaguar but to experience his essential identity with the powerful alter ego. Among the tobacco-related characteristics that liken the shaman to the jaguar are night vision, acuteness of vision, wakefulness, raspy voice, furred or rough tongue, and pungent body odor. Most important in connection with nicotine as a transformation agent are the mind-altering effects of nicotine as a liberating agent of epinephrine, norepinephrine, and serotine, among other powerful compounds implicated in the alteration of mood and affective states in humans. In appropriate dosage nicotine produces certain chemical changes that mobilize in the properly enculturated shaman the attack behavior of the were-jaguar, displaying anger, hostility, and sexual aggression.
>
> (Wilbert, personal communication, 1993)

Other ethno-pharmacological effects of tobacco as a hunger depressant, for example, a febrifuge, or analgesic, may provide additional avenues of exploration into Maya religious practice. But to us, Wilbert's suggestions emphasize two important ideas: that Maya imagery of visionary experience is not simply fantasy but that it represents real experience; and that trance experiences were not benign daydreaming but the result of pharmacologically based experimentation with mind-altering drugs like tobacco. These experiences affirmed Maya worldview and the power of religious practitioners who had special access to and familiarity with the Otherworld.

60. See Tozzer (1941:115–121) for an extensive summary of sacrificial practice in Yukatan at the time of the Spanish Conquest.
61. See Barrera Vásquez (1980:680) for entries on *p'a chi*. We understand from Irene Winter that sacrifice for idols in ancient Sumerian society involved similar principles of opening the mouth.
62. This iconographic argument is found in David Stuart (1988a). See also Schele and Miller (1986) for discussion of the art associated with bloodletting.
63. Precolumbian Maya also gave a wide variety of animals in offerings. See Victoria Bricker (1991) on the animal breads of the Maya illustrated in the Dresden Codex.
64. See Schele (1984), Stuart (1988a), and Schele and M. Miller (1986) for discussions of ancient Maya bloodletting practices and the changing perception of Maya autosacrifice on the part of modern experts.
65. See Freidel and Schele (1988a:559) for a discussion of Maya shamanism and the use of these small divination stones.
66. The decipherment of this bloodletting glyph is the result of many different epigraphers' work. Steve Houston and others pointed out examples that have phonetic *ch'a* prefixed to them. He also found examples with phonetic *ba* suffixed to them, thus providing the spelling that led Barbara MacLeod (1991a) to suggest that it read *ch'ab*, "to create," as a metaphor for the act of bloodletting and the conceiving of a child. In the summer of 1991, David Stuart noticed an example on a new text from Copan Temple 22a that has phonetic *ch'a* prefixed and phonetic *ma* suffixed to it. Using that as a key, he and Schele (Schele, Stuart, and Grube 1991) realized that the normal winglike suffix on this sign is actually an early form of *ma*, especially used on the bottom of the initial series introductory glyph, that had become fossilized as part of this bloodletting glyph. Looking up *ch'am*, they found it means "to grasp" and "to harvest." However, in English "harvest" evokes images associated with wheat and other European grains. In the Maya world, to harvest is to pluck the ears of maize from their stalk.
67. These observations were sent to us in a letter dated March 18, 1992.
68. This decipherment was made independently by Houston and Stuart (1990) and by Andrea Stone and Barbara MacLeod in their study of the inscriptions of Nah Tunich (Stone and MacLeod 1991). Stuart (1988a) has shown in his study of the bloodletting complex that supernatural beings were materialized in huge S-scrolls we now know to be clouds. We note also that the Maya word for "conjure," *tzak*, also means "to conjure clouds." Furthermore, the Vision Serpents at Chich'en Itza with scrolls attached to their bodies have long been thought to be the equivalent of the Aztec Mixcoalt, "Cloud Serpent." These new decipherments apparently support that identification in Maya imagery and words. Moreover, the concept is old, because the same kind of "cloud serpent" appears on the Acasaguastlan Pot from the Early Classic period (Schele and Miller 1986:193–194).
69. This serpent is shown on Copan Stela 6, where the text explicitly records that the Waxaklahun-Ubah-Kan was conjured (Schele 1990b). He is associated with the Tlaloc war complex (see Chapter 6) and is the Maya analog of the Feathered Serpent of Teotihuacan that Karl Taube (1992b) identified as a War Serpent.
70. The founder glyph reads *ch'ok te na*, "sprout-tree-house," but here it has another sign attached to it. Schele has taken it to be a *wi* sign, signifying the word for "root." In other words it records the founder as a "sprout-tree-house-root." Grube on the other hand has suggested to Schele that the sign is the glyph form of the bloodletter sitting in the woman's bowl. This would work equally well, for we now know that this bloodletter sign read *ch'am*, "harvest," so that the founder was the "sprout-tree-house-harvest." Both interpretations are entirely appropriate for the context.
71. The earliest Maya Vision Serpents are currently dated to the Late Preclassic period, including a modeled stucco example from Group H, Waxaktun (see Schele and Freidel 1990:Chapter 4) and a black-line graffito on the wall of Structure 5C-2nd at Cerros, in Belize. This graffito shows a tau-toothed god or ancestor in the mouth of the Vision Serpent, the earliest example of this kind of depiction so far. Current work in the Preclassic ruins of the Maya region will no doubt push this concept back even further in time.
72. The Chikchans, who are great fourfold snake deities of the Chorti Maya of eastern Guatemala (see Thompson 1970:262–265 for a summary in English), are likely the modern survivals of the ancient Vision Serpents, especially those depicted with deer horns and ears.

73. Schele (1989) identified this serpent as a *chih-chan*, but at the time she had not seen the examples where an *o* sign precedes the *chi* hand. *Och-kan* was the word for boa constrictor as well as a kind of Vision Serpent.
74. A new book by Kevin Gosner (1992) describes the role that visions and prophecy played in the early rebellions of Chiapas. His translation of Nuñez de la Vega describes a cave as the location of the initiation ceremony. As Thompson noted, the initiate was swallowed by a dragon and passed out the other end transformed into a shaman.
75. In typical fashion for the man, Venancio then grinned and said he had once borrowed an old relative's book of magical incantations and tried to use it to send a letter magically through the air to a friend. He stood at a crossroads at midnight to do this sorcery. The letter never flew, and all he got for his trouble was a rash on his tongue.
76. This raises another thorny problem: did the ancient Maya actually dance around with live snakes? Although we can only evoke depictions from the Classic period as evidence, it seems entirely likely that they did. Dancing with venomous snakes is an anthropologically well-known feature of trancing and vision quest. There is good evidence that the Maya used boa constrictors in some of their activities, but it is just as likely that they danced with fer-de-lances and rattlesnakes as well. Some of the texts that record the conjuring of Vision Serpents may refer to bringing spirit into these living snakes. See Chapter 6 for examples of snake dancing.
77. This ritual is known as "scattering." David Kelley (1976:54–55) suggested the reading *u mal*, "sprinkle," for this action, while David Stuart (1984, 1988a) argued that the stuff being sprinkled is the blood of the visionary. Subsequent decipherments (see Note 59) have shown that the stuff scattered in many examples is *ch'ul*, the "soul-stuff" that resides in human blood. However, Norman Hammond (1981), in summarizing other alternative evidence patterns, argued for the scattering of maize and other substances. We now think that Hammond's arguments are valid and that the category of material scattered includes *ch'ul* in the form of blood, and *itz* in the form of tree resins, beeswax, and other excretions, as well as sustenance in the form of maize gruel and maize dough. Bruce Love (1987) deciphered the word for the droplets as *ch'ah*.
78. Bruce Love (1987) first showed that the signs following the scattering glyph read *ch'ah*, the Yukatek word for "drops" and the Cholan word for "incense." He proposed that the material scattered was drops of tree resin called copal. We accept his reading, adding that *ch'ah* is the Chol verb (Attinasi 1973) for "to fumigate with incense." However, the imagery of the Classic scattering rite clearly shows the beaded matter that emerged when one plunged a stingray spine through the tongue. This matter is clearly blood, although at a larger level, blood was represented as "soul-stuff." Norman Hammond (1981) was also right in supporting the proposition that maize was scattered, for the material placed into the plates and braziers of offering included all these things.

    Bruce Love (1987) suggested that artistic depictions of little scattered droplets signal copal incense, while other more flowing images may depict blood. Freidel (1985) argued that there was a continuous field of iconographic expression incorporating scatterings of droplets, droplet-edged scrolls, and scrolls marked with the dots and stacked dots of liquid (see Schele 1988:300–301, for definition of the dot stacks as liquid). In the case of the codices, which Love uses as his primary field of iconographic reference, scenes that clearly depict water flowing out of jars, urine and vomit flowing out of gods, and smoke or fire flowing during epigraphically identifiable fire-drilling events, and scattering events—all these images employ discrete droplets or dots. Flowing liquids, including blood, are thus demonstrably represented by discrete dots such as those seen in the scattering scenes. Sometimes the blood offering ornamented as *ch'ul* is embodied in a variety of materials, such as bone, jade, spondylus, and volcanic hematite.

    The real challenge is not in distinguishing between the iconographic depictions of blood and other materials, but in differentiating between *ch'ul* and *itz*. The glyph for *itz* is most clearly rendered on Stela C at Quirigua where it is recognizable as a beaded flower with a stamen drooping from its blossom. Sometimes the flower has *ak'bal* in its center and at other times the ahaw glyph Grube and Nahm (in a 1990 letter) identified as *nik*, "flower." Furthermore, this beaded flower is the one depicted on the cantilevered headband of God D and the bird named Itzam-Yeh. *Itz* is the word for the nectar of the flower. *Itz* flowers occupy the background fields of the sarcophagus and the tablets of the Group of the Cross at Palenque and the wall of Burial 48 at Tikal. The identifying elements of *ch'ul*—bone, shell, jade beads, *k'an*, *yax*—share the same domain, so that the environment of sacred actions was filled with both *ch'ul* and *itz*. The beaded flower form of *itz* may also be seen on the chest pectoral so often worn by Maya kings. The attached cylinders and beads represent the stamens of the flower so that the wearing of this form of pectoral may identify its owner as an *itz*-er. Both the *ch'ul* and *itz* sets of imagery partake in overlapping domains of sacredness and the stuff of holiness.

79. This name for the bird is recorded on the Blowgunner's Pot that shows him landing in the World Tree. See Chapter 1, Note 14, for a full discussion of the meaning of *itz* and the god names associated with it.
80. Houston and Stuart (1989a: Note 7) attribute this idea to a personal communication from Karl Taube.
81. Several scholars have worked with the Principal Bird Deity. Bardawil (1976) provides a recent definitional discussion. Andrea Stone (1983) discusses the major iconographic attributes of the bird. She makes the connection between Seven-Macaw of the Classic period and the Principal Bird Deity of the Classic period. Cortés (1986) gives an insightful and comprehensive analysis of the Principal Bird Deity, elaborating the connections between Seven-Macaw and the Principal Bird Deity by structurally analyzing the ambiguities of this being in ways that have inspired our own discussion. Her basic hypothesis is that the bird represented order in nature. Prideful and out of control, it was a sun-pretender, demanding worship from all around it. The Hero Twins brought this misbehaving version of nature back under community control and reestablished the proper order of the world. The king repeats this ordering action through ritual. We concur with her analysis, but add to it the recognition that the Classic-period Itzam-Yeh represented the power of the wizard who manipulates the cosmic sap. He is the symbol of this energy in its cosmic scale, while the god Itzamna was the first wizard to manipulate it. The bird may even be the *way* of the wizard as Taube suggested. Dennis Tedlock (1985:360) identifies Seven-Macaw with the Big Dipper, an identification we believe held also for the Classic period. If the angled configuration of the Milky Way we have identified as the Crocodile Tree also represents this episode, then the Big Dipper behaves in exactly the way described in the Popol Vuh. When the tree is upright, i.e., north to south, the Big Dipper is above the horizon. In fact, on August 13, 1000 B.C., at La Venta, it touches the horizon at sunset. During the Classic period, it was much higher at sunset, but it sank into the horizon at around 343° as the Milky Way turned to its east-west orientation. The Popol Vuh says that after being shot, Seven-Macaw rose up and fell to the ground. This plunge into earth is replicated in the movement of the Big Dipper.

    The bird is one of the earliest images known from lowland Maya iconography, because Richard Hansen (1992) has found a huge pyramid at Nakbe that sports a five-meter-high plaster image of the bird's head. He has dated the pyramid to early Chicanel—400 to 200 B.C. At Copan, the bird was modeled on the side of the temple called Ante and most impressively on the early temple called Rosalila, where he appears with the head of Itzamna in his mouth. The great bird is also modeled in plaster over the doorways leading into the sanctuaries of the Group of the Cross at Palenque. These small inner houses were called *pib na* ("underground house") and *kunil* ("wizard place"). He also perches on top of the skybands that frame the king on the accession monuments of Piedras Negras and on Lintel 3 of Temple 4 at Tikal. In all these contexts, Itzam-Yeh is there to mark the location as a place where *itz* is brought into the world and where the act of *itz*-ing was done.
82. Constance Cortés (1986) identified this scene on Izapa Stela 25.
83. In the Early Classic iconography of incised offering vessels, First Father (GI) wears Itzam-Yeh in his headdress, where it is fused with the Quadripartite God plate. As Nicholas Helmuth suggested to Schele many years ago, the so-called serpent wings on the typical headdress are the bird's wings, and the upper flange of feathers in

the headdress represents his tail. Itzam-Yeh was often worn in the headdress of rulers when they were in other guises as well.

84. Taube and Houston (n.d.) identified the *lak* spelling in the Primary Standard Sequence on plates.

85. This particular version is found on plates used as part of lip-to-lip caches in which one of these huge plates is inverted over another as its lid.

86. Houston and Stuart (1989b) identified the name of this footed plate in the Primary Standard Sequence on footed plates. The name is *hawa(n)te*, meaning "mouth up" and "wide-mouthed."

87. David Stuart (1986a) identified this combination as the glyph for the stone censers at Copan. These large objects are carved with an amazing array of imagery, but the interiors are always in the shape of a large bucket. We suspect that a clay bucket was used as a liner.

88. *Kum* is Yukatek for "olla," a round-bottomed pot for holding liquids. *K'u* is "god." John Carlsen, William Ringle, and Thomas Barthel all independently identified Landa's glyph for the month Kumk'u as a combination of the sign for *ku* and *k'u* in its God C version.

89. These discoveries were made at the 1991 mini-conference on the inscriptions of Palenque sponsored by the John D. Murchison Professorship of the University of Texas. The meetings between Linda Schele, Peter Mathews, David Kelley, and Floyd Lounsbury were held at the Department of Archaeology of the University of Calgary. Peter Mathews made the first connection of the *owal* spelling to the Chol month name, while Schele recognized the connection of the word to the "pot god" reading of the month name in Yukatek. Lounsbury had prepared much of the work on the phonetic reading of the Classic version of the month sign by demonstrating that the upper sign is *o*. See Schele, Mathews, Grube, Lounsbury, and Kelley (1991) for a full discussion of the evidence.

In the fall of 1992, we received a letter from Nikolai Grube suggesting that the *wa* ("maize") sign was not pronounced at all or that it was logographic *ol* in this context. He proposed that the month sign and the noncalendric examples all read *Ol*. Since the month is listed as Ol in several different colonial-period lists for the Cholan and Poqom languages (Miles 1957:744), we have accepted his proposal. After receiving his letter, Houston also sent us evidence that he had independently arrived at the same conclusion.

90. Many of these ideas were worked out with Rebecca Cox (1991), whose thesis on these cache plates provides detailed analysis of the imagery and symbolism.

91. The history of research with the Quadripartite God is a long and rich one. Kubler (1969:33–46) summarized the research up to 1968, citing Maudslay's (1889–1902), Spinden's (1913:53–55), and Seler's (1977) studies of the distribution and meaning of the symbol, while concluding himself that it was associated with fire and burning. Thompson (1970:209–234) identified the double-headed dragon that carries the Quadripartite God as a second head as the "Iguana-House"-Itzamna god, while Merle Robertson (1974) concluded that the symbol was composed of four design units and that it was a badge of rulership. Schele (1976:17–18; Freidel and Schele 1989b; Schele and M. Miller 1986; Schele and Freidel 1990) took the *k'in* sign on the forehead element to identify the god as the sun, and proposed that the double-headed dragon was a cosmic symbol combining Venus as the crocodile front head and the sun as the rear head.

The most important insight came from John Justeson's (n.d.) realization that the Classic-period glyph for "east" combines a "bowl" with a *k'in* infix with the *k'in* glyph to write *lak'in*. Schele realized that Justeson's plate (*lak*) sign was identical to the forehead element of the Quadripartite God. This was the key, for one such sacrificial plate (Schele and M. Miller 1986: 194) has a painting showing a Vision Serpent emerging from the same kind of plate. Clearly the plate both held offerings and allowed things to emerge from the Otherworld, the action exactly shown on the Yaxchilan vision lintels.

Houston and Stuart (1989b) added additional interpretive data by identifying the "kimi" substitution for the crossed-bands knot in the three-part symbol normally shown in the plate as a substitution for *way*, "animal spirit companion" and "to transform." Schele later realized that both it and its substitution with crossed-bands inside was the bib part of a special necklace worn by *wayob*, very probably as a means of phonetically identifying who and what the creature is.

Joralemon (1974) had earlier identified the central element of this three-part symbol as a stingray spine and instrument of bloodletting.

The shell is more difficult to interpret, but we think its meaning may come from its central role in the depiction of *ch'ul* and in the consistent presence of spondylus shell in cached offerings of all sorts. It simply may refer to the spiritual force transformed through bloodletting. All these combined lines of evidence lead us to believe that the Quadripartite God is simply the bowl of offering, the *ol* that leads to the Otherworld, represented as a god. The lower head in this interpretation is the same skeletal head that appears on ahaw and other signs as a head-variant marker.

92. This same skeletal head appears regularly as the personification of a special set of glyphs, including the ahaw sign that reads *nik*, the inverted ahaw that reads *la*, and several others. In those contexts, the skull does not add any specific phonetic or semantic value to the glyph. Schele believes the same is true of the Quadripartite God plate. The skull represents the personification of the power in the plate, but it does not contribute additional semantic or phonetic value to the symbol.

93. Ruth Krochock and Constance Cortés suggested this interpretation to Schele in a seminar paper written for a 1982 seminar. Houston and Stuart (1989a:7) followed the same interpretative tack, but associated the imagery with the appearance of the mother's *way*. In other words, they suggested that the child was born from the *way* of the mother, which is the serpent. This may also be the proper interpretation for the parentage metaphor on the Palace Tablet at Palenque.

    Nikolai Grube provided another important piece of information to our understanding of the Maya conception of the portal pot. Twice on Copan stone censers, the month Kumk'u is written phonetically as *haw*. Nikolai connected this spelling to a Yukatek word for "to open a road."

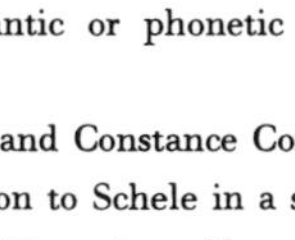

13 Kumk'u spelled *hawal*

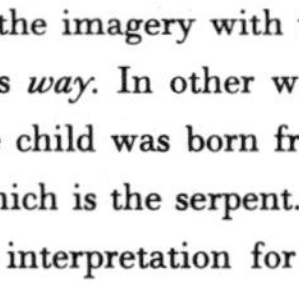

och bih, "he entered the road"

    Specifically, Barrera Vásquez (1980:187) has "*haw* abrir camino, partir tierras; *hawah* abrir camino o apartar la gente, yerba o rama que embarazan el camino para que haya paso" ("to open a road, to break open the land; to open a road or to part people, weeds, or bush that embraces the road so there is a passage"). In this context, the road is the path to the Otherworld. This is the same concept, *och bih* ("he entered the road"), that is illustrated on Pakal's sarcophagus lid at Palenque. In that extraordinary scene, Pakal enters the road by falling down the Milky Way through the gullet of the White-Bone-Snake atop the deified plate of sacrifice. Sacrificial victims were depicted in plates because they too were entering the road through its portal. This concept still exists in Yukatan, for Hanks (1990:299) says that the action of circling the altar in a counterclockwise order is described as *he'ik beel*, "open the road," and *k'axik mesa*, "binding the altar."

94. In other anecdotes throughout the book we have called the Maya priest-shamans with whom we have dealt by their first names with the honorific *Don* affixed. This usage is to acknowledge the friendship Don Pablo has with David Freidel and Don Lucas has with Duncan Earle. Somehow, Sebastián Panjoj falls into a different category in our minds for we went to him as stranger supplicants for a divination. We did not know him as a friend and most people of Chichicastenango call him the Principal Panjoj or the Ah Q'ij Panjoj. We will use his last name or call him by his title as we do with our own medical doctors in normal conversation. This is the way we thought of him and the way we referred to him as our encounter with him unfolded. *Ah Q'ij* means "he of the sun" or "he of days." It is the title used for calendar priests who make divinations using the calendar and *tzite* beans.

95. See Barbara Tedlock's (1982) for a full explanation of the meanings and procedures involved in this kind of divining.

96. Our third author asked, when she read this, why taking on the responsibility is dangerous for an Ah Q'ij and why he would do it for a couple of gringa strangers. The answer is that he believes that sickness, bad luck, and other soul dangers to himself and his family result if he does not perform his responsibilities correctly, diligently, and with proper attention. He risks the danger—even for gringa strangers—because he has been called to the responsibility. If he does not answer the

call, his soul also suffers. Foreigners who visit Maya Ah K'in may not take the divinations seriously, but a good Ah Q'ij or chuchkahaw always does. That is why outsiders should always tread this ground with serious intent and respect—or simply not enter the ground at all.

97. Today, diviners in the highlands receive many requests from visiting tourists for divinations. They cannot take a lot of money from these rich visitors, but they are now trying to organize a cooperative of diviners that could be paid. This communal money would then be used to support young aspirants who want to learn the skills of divining, which takes over a year of concentrated study, or to help diviners who become sick or face other kinds of problems.

Chapter 5: Ensouling the World and Raising the Tree

1. When Federico Fahsen translated this anecdote for Pakal in the summer of 1992, Pakal confirmed that my impression had been correct. His mother and his colleagues had invented a ceremony that seemed appropriate to them.

2. The recognition of these dedication texts began in 1986 with the identification of proper names of the stelae at Copan. At the time, Schele was drawing the inscription on the rear of Altar U, when she asked David Stuart to check the details of a glyph we later learned reads *k'inich*, "sun-eyed." In a flash of insight, he realized that this glyph and the adjacent one were the proper name of the altar, which has a sun-eyed monster with a throne sign on its other side. During the next several weeks, Stuart and Schele (Stuart 1986a, 1986b; Schele and Stuart 1986) worked through the inscriptions of Copan and Quirigua, identifying more proper names and associating them with aspects of the iconography of each monument. They also found the proper names of houses, as well as identifying a pattern of glyphs that accompanied these proper names.

   At the same time, Stuart and Grube (Stuart 1986c; Grube 1991a) had been independently working on the Primary Standard Sequence from pottery, unraveling many of its mysteries and associating it with similar kinds of texts in the inscriptions. Following clues in their work, Ruth Krochock (1988, 1991) identified texts recording the dedication of lintels at Chich'en Itza; Schele (1987g, 1991f; Schele and Freidel 1990) identified similar phrases at Palenque; and Stuart and Grube (n.d.) continued their own investigation of proper names and dedication phrases in the inscriptions. Grube (1990a; personal communication 1987) deciphered the verb for erecting stelae as *tz'ap* "to place in the ground," while David Stuart (in a 1989 letter circulated to epigraphers) read one of the principal verbs for house dedication as *och butz'*, "entered smoke (or incense)."

   Barbara MacLeod (1990b) has suggested that the God N verb is *hoy*, which in the derived form *hoy-bes* means "to use for the first time" and "to bless a church" in Yukatek. More appropriate, in Chorti (Wisdom n.d.), *hoyi* is "make fitting, make proper, make satisfactory," while in Chol and Tzotzil, it has the meaning of "circumambulate."

   Another of the dedication verbs features the glyph of a plate with shallow flaring walls. This glyph for "plate" was first read as *lak* by Justeson (n.d.), as we saw in Chapter 4, and exactly this kind of plate was a portal and the principal container for offerings. In fact, the Maya used two of them laid lip to lip, i.e., one upside down over the other, to deposit offerings inside pits under the floors of their buildings during these dedication rituals. The action, therefore, seems to be the placing of this sort of cache.

   Another important component of these dedication texts, the elbow sign from Glyph B of the lunar series, was read as *u k'aba*, "its name," independently by Nikolai Grube in late 1986 and Judy Maxwell in early 1987. Subsequent research into the nature and meaning of the Primary Standard Sequence and other kinds of dedication texts have shown these dedications to be one of the most widely distributed and important of all inscriptional patterns.

3. Dedicatory sacrifices are recorded in inscriptions, such as that on the Hieroglyphic Stairs at Palenque. However, sacrificial burials, sometimes with dismembered bodies of the victims or their severed heads, are known from excavations of Maya sites from as early as the Late Preclassic period. Pakal's dedicatory text of the Hieroglyphic Stairs suggests that many of the victims were war captives. His own tomb was sealed with the dismemberment sacrifice of at least five peo-

ple ranging in age from young to old. But another burial sacrifice at Palenque, the tomb under Temple 18a, had male and female adults and children.

4. The analysis of these dates is summarized and presented in Mary Miller's (1986a:27–38) study of the Bonampak' murals. We accept her interpretation, but decipherments made since the publication of her study have given us new information. The child's name still is not readable, but we can affirm that he was made an *ahaw* and that he is named in the *ichan* relationship with Shield-Jaguar II of Yaxchilan—in other words, that Shield-Jaguar was the brother of the boy's mother. Lounsbury (in Miller 1986a:30) pointed out that the first date in the inscription is 148 days (an eclipse interval) after a total solar eclipse that was visible at Bonampak' on July 21, A.D. 790. This eclipse is recorded at the nearby site of Santa Elena Poco Winik. The Bonampak' date fell on the next eclipse interval after the Poco Winik date, but the eclipse was not visible anywhere else in the Maya area.

   Before consecrating the building containing the murals, they waited for the Eveningstar to become visible for the first time after its long disappearance at inferior conjunction. The remnants of the event include the *och butz'* event, the proper name of the house, *u k'aba* ("its name"), *yotot* "the house of," and Chan-Muwan's name phrase. We recognize that this passage might record the dedication of another house at Bonampak', but we are assuming that the reference is to this house. In other words, the ritual activated the structure itself—although the painting of the murals required many more years to complete, or so we assume since the images depict events that took place up to a year later. Since two dates and events are recorded in this text, we presume the earlier event, the presentation of the heir, is depicted in the west, south, east sections of the upper register—and that the second event is depicted in the entire lower register and north wall.
5. See Mary Miller's (1986a:69–92) very detailed descriptions and analysis for a deeper understanding of this extraordinary scene.
6. Stephen Houston (1985) first identified this scene as a feather dance, and Nikolai Grube (1990b, 1992) deciphered the verb as *ak'otah* "to dance." Here the *ak'ot* glyph followed the T1.757 auxiliary verb, *u bah*, which translates as "he goes doing something." Following an earlier suggestion by Peter Mathews, Grube also identified the prepositional phrase following *ak'ot* as a reference to an object held or some other part of costume from which the name of the dance was taken. In this example, the verbal phrase reads *u bah ti ak'ot ti k'uk',* "he goes dancing with feathers."
7. See Mary Miller (1986a:79–94, 1988) for a more detailed study of this scene and the musicians.
8. The gourd rattles in the Bonampak' murals are identical to rattles used by contemporary shamans among the Warao of Venezuela as reported by Johannes Wilbert (1973). In particular, the Rattle of the Ruffled Feather, is the *axis mundi* of these people and is "awakened" to facilitate direct communication with the gods. It seems most likely that the ancient Maya felt similarly about these gourd rattles. In particular, an elaborate incised vessel called the Deletaille Tripod vase (Helmuth 1988) depicts a seated lord with two such rattles in his hands, upon a great Vision Serpent that is in the "Raised-up-Sky-Place." In this Bonampak' scene, then, the "musicians" are not just playing music, they are participating in the conjuring of powerful supernaturals into the festival arena.
9. Mary Miller (1986:85–87) discussed this young lord in light of the god impersonators and their appearance in the ballgame scenes at Yaxchilan. She accepted Thompson's original suggestion that this person is the Maize God, although none of his expected attributes are present. We also feel this is the most likely interpretation, although it is also possible he represents some other character in Maya myth.
10. See Chapter 8 for a full discussion of this Yaxchilan ballcourt dedication and the meaning of the ballgame. At the 1991 Texas Meetings on Maya Hieroglyphic Writing, Sandy Bardslay (personal communication 1991) suggested that the Six-Stair-Place shown on Step VI, VII, and VIII of the ballgame panels and named in the text on Step VI was Hieroglyphic Stairs 1, which reused an old ancestral text to create a stairway of six steps in front of Structure 5. Since Bird-Jaguar carefully reset other Early Classic lintels in Structure 12 and 22 on or around his day of accession, we find her suggestion a highly probable one.
11. These Six-Stair-Places are discussed in Chapter 8 and in Grube and Schele (1990a).
12. David Stuart (1986a) first identified a version of this term on the huge stone incensarios at Copan, which were called *sak lak tun.* Thus, we are reasonably sure that the incensarios of Copan were

stone versions of the lip-to-lip caches and that both served to receive offering and to open portals so that the gods could receive them. Today, the Lakandon called their own offering censers *u läk k'uh,* "god plate," in a parallel usage. *Lak* means "plate" in both Cholan and Yukatekan languages, while *sak* is the word for "white" with extended meanings of "clear" and "human-made." The actual form of the *sak lak tun* at Copan, however, is that of a tall cylinder covered with stubby spikes. We know this because the phonetic expression for the incensario substitutes directly for an image of that incensario in the texts. This form is very ancient in Mesoamerica, extending at least into the Late Preclassic period in the southern Highlands of the Maya region. The natural referent for this form is the spiked trunk of trees, a form of natural defense employed by species of the ceiba, the World Tree, among others. The notion of *sak* as "clear" resonates with modern Yukatekan Maya notions of *zuhuy,* "untainted" "pure" in ritual materials.

Images of the designs of several of these cache plates were published by Crocker (1985:230–231) and by Schele and M. Miller (1986: 195, Pl. 175). The known corpus of incised cache plates has been discussed by Rebecca Cox (1991). The argument here is that the sun-marked sacrificial plate is incised as decoration on some *sak-lak* vessels, hence we surmise that in its physical form it was also a *sak-lak* vessel, one of the kind used in lip-to-lip offerings. It certainly existed as a physical object category. The sun-marked plate of sacrifice is depicted, for example, in the hand of Pakal on the Tablet of the Sun and again in the hand of Chan-Bahlam on the outer panel from the Temple of the Sun. In shape, the actual *sak-lak* vessels vary from shallow, out-flaring walled plates to relatively deep, vertical-walled buckets.

13. Dan Potter (1982 and personal communication 1988) described to Linda Schele the moment he fitted the broken blade back onto the core and discovered the second perfect blade above fitted next to it. He suggested the sequence of events to Schele in discussions of this and other caches at Kolha.
14. The cache vessel rested over a deposit of red pigment. Traces of blood have been detected on the blade inside the cache plate (Hester 1983:4–5).
15. See Ricketson and Ricketson (1937:Pls. 21, 23) for photographs of these caches in place.
16. The best descriptions of these are found in Coggins (1976), Shook (1958), and for full details as well as archaeological contexts, W. R. Coe (1990).
17. For the Maya, many precious materials like jade served multiple purposes, including money, state treasure exemplifying the power of government officials, and the material lens of supernatural power (Schele and Freidel 1990:92–93; Freidel 1986).
18. See Schele (1987c, 1990c) and Schele and Morales (1990) for a discussion of the jades in these caches. See Morley (1920), Gann (1926), and Longyear (1952) for photographs and detailed descriptions of the offerings found in earlier caches.
19. See Freidel and Schele (1988a) and Schele and Freidel (1990:121–122) for a discussion of this cache and its significance in royal ritual.
20. Archaeologists called these kind of jades "bib heads." They are associated with the Late Preclassic crown of kingship.
21. Marisela Ayala (n.d.) was the first to associate the bundle in the offering cache from the Lost World Group at Tikal with bundles represented in Classic-period imagery. Schele (Schele and J. Miller 1983) discussed the distribution of bundle imagery in Classic-period art (the reading proposed in the study was not correct, but the discussion of bundles is useful), while Robertson (1972) has speculated on the contents of bundles at Yaxchilan. Stross (1988) deciphered the glyph that often appears on these bundles as *ikatz,* the Tzeltalan word for "bundle" and "burden of office," while Barbara MacLeod (in a letter to David Stuart dated March 23, 1989) identified another collocation for a period-ending phrase as *tzohp,* the Chorti (Wisdom n.d.:730) word for "bundle, collection, cluster." *Yax tz'op* is especially associated with three-tun endings and on Stela 10, *tz'op* is followed by *chanal,* "heavenly," and *kabal,* "earthly." These bundles (Fig. 2:6) also appear in representations of the seven gods at Creation.
22. Peter Harrison (1970, 1989) first realized that this cache vessel identified Structure 5D-57 as Jaguar-Paw's residence. See Schele and Freidel (1990:464) for a discussion of this palace and its relationship to later history.
23. Gann (1926) described just this sort of mercury offering in the cache found in the cruciform vault below Stela 7. We suspect the lowest cruciform may once have been the resting place of Stela 24, which was broken and reused as part of the offering under Stela 7.
24. These properties of mirrors are discussed in

Schele and J. Miller (1983) and Taube (n.d.).

25. In 1990, Freidel saw a set of cache vessels from the site of Comalcalco in Tabasco in the local museum that still contain tiny spangles of these materials, all cut to a uniform size appropriate to scattering with sacrificial blood into the vessel.
26. Tozzer's (1941:110–114, 159–160) notes on Landa's descriptions of the renewal of the stone idols and the rituals that fed them are a detailed source on this subject. He assembles related descriptions from many different sources at the time of the conquest. Blood was drawn from all parts of the body, especially the tongue and genitals, and the idols were anointed with this blood.
27. Landa (Tozzer 1941:160) says that a cedar tree was always used, while Tozzer pointed out that the word for cedar, *k'u che',* meant "god tree."
28. Statuary from Olmec times on was dressed in cloth and other accoutrements. Today the saints in many Indian churches undergo similar redressing rituals until as many as fifty years of clothing are layered on. Chip Morris (personal communication 1986) has used these layers of saints' clothing as a record of lost and forgotten weaving techniques and patterns in the Chiapas highlands.
29. The description of the Lakandon god pots and the rituals associated with them are described by Tozzer (1907), Robert Bruce (1975, 1979, n.d.; Perera and Bruce 1982), and McGee (1990). Other information comes from conversations with Robert Bruce, who participated in a course on Maya hieroglyphic writing held at Universidad Nacional Autónoma de México in the summer of 1979. At that time he presented Linda Schele with a copy of his detailed description of the god-pot renewal ceremony in which he participated in 1970. Titled "Death and Rebirth of the Gods," the manuscript describes the forty-five-day-long ritual in great detail—including his own thoughts and feelings as it unfolded. Robert Bruce had the permission of the Lakandon of Naha to record the ritual in writing and photographs, but he has always believed he was entrusted with special knowledge so that it might be preserved. He also inherited a special responsibility to consider the intentions and feelings of the Lakandon in not exposing them to the curiosity of the Western world without good reason. Since he has not published the account, we will respect his reticence with deep gratitude for being given access to this extraordinary document. We will draw on its information only when his descriptions overlap with accounts and interpretations published by other observers. Schele heard in March of 1992 that old Chank'in made the god pots again during 1991, a little over twenty years after the last ceremony.
30. See McGee (1990:51–52) for a definition of the meaning of the god pots. He quotes an unpublished dissertation by Davis (1978:73) for the body-part associations we incorporate here. Bruce's (n.d.) account confirms the detail of the associations, although he also makes it clear that part of waking up the pot is to give it the proper body parts.
31. Both Bruce (n.d.) and McGee (1990:46) describe this dye made from the same tree (*Bixa orellana*) that yields the spice called "achiote" by Spanish-speaking Mexicans.
32. The descriptions in this paragraph come from McGee (1990:55).
33. His published description of the experience is in Perera and Bruce (1982:30–32).
34. McGee (1990:52) quotes Davis (1978:74), who says the Lakandon liken these stones to radios "by means of these transmitters placed inside the god pots, a man's supplications . . . and offerings reach the gods in heaven." McGee reports they are called *k'anche',* the benches where the gods sit after they arrive to take their offerings.
35. For a description of this tree-column, see Means (1917) and Schele and Freidel (1990:398).
36. These descriptions were gathered by Evon Vogt (n.d.) in his study of crosses in Mesoamerica and their Precolumbian roots. He drew heavily on Roys (1967) for Yukatekan data.
37. Our accounts are drawn from Victoria Bricker's *Indian Christ, Indian King* (Bricker 1981:102–109), which combines many accounts and sources into her analysis of the events of the Caste War of Yukatan.
38. In her discussion of the Cruzob, Bricker (1981:-106–113) presents evidence that the crosses "spoke" through ventriloquists and other such means. The tendency of modern Westerners is to dismiss such behavior as deliberate subterfuge. We suspect that the Maya would say that the people who voiced or wrote the messages were merely the instruments used by the Cross to communicate its oracles. She (Bricker 1981:106) analyzes the name of Manuel Nauat, the ventriloquist who was killed during the enemy raid on Chan Santa Cruz, as follows: "The term *nauatlato* means 'interpreter' in Nahualt. . . . it was applied to Indian translators during the Co-

lonial period. . . . Is it a coincidence that the surname of the interpreter for the Cross was Nauat (*naguat*), or was it an assumed name?" MacDuff Everton (1991:213) photographed a Cruzob Maya named Teodócio Nahuat Canche at Xoken, Yukatan. People in the area today still carry the name "translator."

39. Vogt (n.d.) describes these crosses and the colors used when they are painted. If these crosses are conceptually trees, then they are *yax-te',* the ceiba tree that is the living embodiment of the World Tree.

40. Vogt (1970:14–16) says, "Not only must the 'doorway' be readied with the flowers, but incense must also be burning in a censer for effective communication with the supernaturals to take place. The censer is called a 'place of burning embers' and the incense that is burned is of two types: balls of resin and chips of wood. Both types are burned together in the censer, and the incense must be burning at the cross before the ritual party arrives and begins the prayers. . . . With the flowers in place and incense burning, the participants must also be readied for communication, with the supernaturals and the most important way to prepare is to drink liquor." Finally, music is added to the ceremony from a flute and drum group or a violin, harp, and guitar group. "With flowers on the crosses, incense burning vigorously in a censer, liquor being consumed by the participants, and music being played by musicians, the stage is set for efficient communication with the supernaturals through prayers and offerings, typically candles, which are regarded as tortillas and meat for the gods. As the candles burn down, it is believed that the 'souls' of the candles are providing the necessary sustenance, along with the 'souls' of the black chickens, for the supernaturals who will be so pleased with these offerings that they will reciprocate by restoring the 'inner soul' of a patient, by sending rain for a thirsty maize crop, or by eliminating any number of evils and setting things right for the Zinacantecos."

41. Vogt (1976:6–7, n.d.) identifies the geranium as a symbol of the domesticated world of the community and the pine as a symbol of the wild forest outside the control of human beings.

42. Vogt (1970:6) says these Father-Mothers are the most important deities in Zinacanteco life. They are honored and receive prayer and offerings. They are the remote ancestors of the Zinacantecos who were ordered to take up residence in the sacred mountains by the Four-Corner Gods. They are conceived of as elderly Zinacantecos who live eternally in the mountains, surveying the activities of their descendants and awaiting offerings. They provide the ideal patterns for human life. This includes advice on everything, from the best way to grow maize to building houses, weaving, ritual, marriage, and coping with life in general. Vogt says, ". . . the ancestors are not only repositories of important social rules and cultural beliefs that compose the Zinacanteco way of life, they are the active and jealous guardians of the culture."

43. Crosses are not the only objects with an inner soul in the view of the modern Maya—the scepters they use in pageant and other important ritual objects are power focusers and living beings. In the world of the Zinacantecos (see Vogt n.d., 1970), maize, beans, squash, salt, houses, the fire in the hearth of the house, the saints who reside in the Catholic church, musical instruments played in ceremonies, the Ancestral Gods in the mountains, and domesticated animals, all have inner souls. The staffs of office used in Zinacanteco and Chamula ceremonies, like the cross, are treated with special reverence. There are two types of scepters used—one with a silver head that is retained and passed on from one officeholder to another, and the other that is made by each officeholder as he takes the burden of office. In Chamula, scribes wash the silver-headed batons four times a year, cleaning the head with salt and the baton with warm water. A second washing is finished with water and laurel leaves, and then the water is served to the official in order of their rank. In a third cleaning, the batons are rubbed with salt and a brush of manzanilla leaves which fall into rum. After giving the rum to the same officials, the scribe gives them the batons, which are kissed "so that the baton will accompany you on the road." This washing removes not just the physical dirt and sweat from the previous user, but it rids the staff of his administrative errors. Drinking the water and rum from the cleanings renews the sacred power that comes to them from the Ancestral Gods, through the staffs. The staffs are infallible, so that error comes from the humans that wield them, not from their "inner souls."

In some towns, the staffs of office cannot touch the ground and they are marked and decorated in different ways. In other towns, they are carefully wrapped in cloth to protect them from smoke and incensed once a week by the official's wife. Many Maya believe that these staffs of office were once owned by the Ancestral Gods. Each has an "inner soul" given by the Ancestral God, so that a newly acquired staff must be baptized to lock the soul in and protect it against soul loss. The baton derives its power from the supernatural world "and serves as a reminder to all Zinacantecos of the authority wielded by its current possessor" (Vogt n.d.). See also Ruth Carlsen and Eachus (1977) for a discussion of souls in objects in highland Guatemala.

44. Redfield (1936), Bricker (1981), Carlsen and Prechtel (1991), and Vogt (1976, n.d.) have all made the connection between the World Tree images of the Classic period and the crosses used by the modern Maya. Vogt (n.d.) has extended his study to include many different groups in Mesoamerica. He summarizes the use and meaning of these crosses as follows:

> (1) Cosmologically, the cross is the "world tree," the axis mundi, to the Maya in Yucatan (where it is usually painted green) and in Chiapas (where it is often painted blue or green and decorated with pine tree tops). The Maya classify blue and green as the same color and associate it with the center of the world. Among the Yaqui the cross marks the centers of towns, has strong associations with trees, and is sometimes painted blue.
>
> But the cross is also the quincunx, the Mesoamerican cosmos with its four cardinal directions and its center (its navel of the universe). Linn (1976: 96) stresses this symbolism in Chamula, and Hunt (1977: 197) concludes that the quincunx symbol was replaced by the cross all over Mesoamerica.
>
> (2) As an aspect of the animistic world view of the Indians, the cross is personified, it has an inner soul, and it is "clothed," with huipils in Yucatan, with ritual plants in Chiapas, and on occasion with skirts and necklaces among the Yaqui. The marked contrast: the cross is female in Yucatan and among the Yaqui, where it is called "Our Mother"; it is male in Chiapas where it is "Our Father."
>
> (3) Crosses are "Ancestors" and/or are used to communicate with the Ancestors or with ancestral power. In Chiapas they symbolize "doorways" to the realm of the Ancestral Gods who live inside the hills and mountains and/or represent Ancestors themselves, as the Classic Maya stelae depict rulers or royal ancestors. In Yucatan they have become sacred "santos" and many have been "talking crosses" who communicated messages from the gods, just as Precolumbian idols spoke to worshipers at shrines. . . .
>
> (4) Like Ancestors, crosses protect, guard, or "embrace" human settlements and people from dangers and evils. The Zinacantecos have a concept of "embracing" that includes the way in which parents embrace children, godparents embrace godchildren, ritualists at weddings embrace the bride and groom and create a new bond, shamans embrace patients in a ritual that gathers up parts of lost souls, and Ancestral Gods embrace and care for animal spirit companions of all Zinacantecos in a supernatural corral inside a sacred mountain (Vogt 1970: 105–106). In an important sense, the crosses around Zinacanteco settlements also physically "embrace" the community.
>
> (5) Crosses mark important boundaries: between Nature and Culture, using ritual plants as metaphors as in Chiapas, between towns as among the Yaqui, and at the edges of towns as in Yucatan where the entrance crosses replaced the aboriginal piles of stone.
>
> (6) Finally, crosses have become collective representations of social and political identity. They are symbols of crucial social and political units in the Indian societies, as the Yaqui town crosses symbolize the unity of the town council. The system of cross shrines in Zinacantan provides an accurate map, virtually a mirror, of the social structure of the community: the household cross which symbolizes the unity of the domestic family, the lineage crosses representing the local patrilineage, the waterhole crosses symbolizing the waterhole group, the cross shrines of the hamlet ancestors, and finally the cross shrines in the settlement and at the foot and on the summits of the mountains around Zinacantan Center where all of Zinacantan comes to pray to its tribal Ancestral Gods (Vogt 1970:100–101; Zimbalist 1966). The family, lineage, and village crosses of the Cruzob Maya serve similar functions (Villa Rojas 1945:98).

45. Merle Robertson's (1985a: Fig. 31–33) reproductions of the paintings on the west outer wall of House E at Palenque show row upon row of flowers, some with birds, maize, and deer hanging in the blossom. We wonder if these images might not be related to the Atiteko Creation tree.

46. *Cofradía* is the word used for the organizations which oversee the rituals, pageants, and maintenance of traditional customs among the highland Maya.

Chapter 6: Dancing Across the Abyss

1. The American College Dictionary glosses *pageant* as "1. an elaborate public spectacle, whether processional or at some fitting spot, illustrative of the history of a place, institution, or other subject. 2. a costumed procession, masque, allegorical tableau or the like, in public or social festivities."

2. Gary Gossen is one of many professional anthropologists who have worked with the people

of Chamula over the last three decades, most of them in conjunction with the Harvard Chiapas Project of Evon Z. Vogt. Gossen (1986) has provided an excellent analysis of the Chamula Festival of Games from which our facts are derived. He writes with insight and authority about the Chamula vision of the world (especially Gossen 1974). Relevant works on this Chamula Festival include Linn (1976, 1982), and Wilson (1974).

3. The exact figure given by Culbert and his colleagues (Culbert, Kosakowsky, Fry, and Haviland 1990:117) is "in excess" of 425,000 for the minimal Tikal state at the height of its Classic power.

4. The history of this decipherment is complicated and constitutes an interesting example of how many different researchers find parts of an answer to finally make a whole. The path of discovery began in 1978 when Linda Schele, Nicholas Hopkins, and Kathryn Josserand (Josserand et al., 1985) identified auxiliary verb constructions in the inscriptions. In these expressions, a general verb, such as *bah*, "to go," is followed by a preposition, *ti*, and a verbal noun to form expressions like "he goes dancing." This original study correctly identified one auxiliary verb, *bah*, but the other turned out to be a glyph for "to dance."

   Both of these verbs appear with *ti* expressions—that is, *ti* plus a verbal noun that specified the particular action. Peter Mathews (personal communication 1978) offered a different interpretation for at least some of these expressions—that these *ti* constructions referred to the objects being held by actors in the accompanying scenes, rather than the action they are doing. Several unsatisfactory decipherments were offered for the verb during the ensuing years, but it was not until Stephen Houston (1985) identified a feather dance in the murals of Bonampak' that a break began to appear. In 1990, Schele received photographs of a set of lintels that had been looted from the Yaxchilan region over twenty years ago. One of these lintels had a scene of Bird-Jaguar dancing with a snake in his hands. The verb for this action was the one in question, but the prepositional phrase recorded in this example was *ti chan chan*, "with a sky snake." Here there was no question—the prepositional phrase referred to the object being held in the king's hand. After writing preliminary commentaries on these new lintels, Schele sent copies to Nikolai Grube. He made the next and final leap of imagination by realizing that the verb had to refer to Bird-Jaguar's action: in other words, to his dancing. Using the phonetic complements that had long been known to go with the verb—*a* or *ya* in front and *ta* behind—he looked for the words that meant "to dance." He (Grube 1990b, 1992) found that this verb is *ok'ot* in Yukatek, but *ak'ot* or *ak'ut* in all of the Cholan languages, Tzeltal, and Tzotzil. Not only does the reading fit the contexts with amazing accuracy, but the phonetics support the reading.

5. The first documented image of a dancer was published by Gann (1918:138, Pls. 26–28) from Yalloch, a site in the Naranjo region near the border between Guatemala and Belize. Two more examples were published in the Merwin and Valliant (1932: Pls. 29 and 30) in the report of the excavations of Holmul. Perhaps the most dramatic example of the dance position was excavated at Altar de Sacrificios and reported by R.E.W. Adams (1963). The first scholars to specifically identify the heel-raised position as dance were Michael Coe and Elizabeth Benson (1966:-16) in their analysis of the Dumbarton Oaks panel from the Palenque region. Working from Adams's commentary and Coe and Benson's work, Kubler (1969:13, 17, 25) extended the identification of dance to include the Bonampak' murals, several stelae at Quirigua, and the piers of House D at Palenque. Schele (1988 [presented in 1980]) reevaluated the imagery on the Altar Vase and associated it with Tablet 14 and the Dumbarton Oaks tablet at Palenque.

6. The ideas in this chapter stem from many sources, especially the published and unpublished works of ethnographers who are working with modern Maya communities. We would like to acknowledge in addition long and fruitful conversations about the dance and the nature of performance with Kent Reilly, who has lost his soul to the Olmec, with Nikolai Grube, who not only deciphered the glyph but has worked with the modern Cruzob Maya and shared his insights about both modern and ancient dance with us, and to Matthew Looper and Kathryn Reese, graduate students who have been deeply involved in studying the different aspects of performance and the dance.

7. Victoria Bricker (1981) provides an excellent analysis of the manner in which modern Maya festivals thematically connect the central concern of ethnic conflict (violent definitions of we-they) to a powerful statement of past resistance and future purpose. This brilliant thesis is a fundamental insight into Maya political discourse with

significant applicability to the rhetoric of the Classic Maya kings. Gary Gossen (1986) underscores the cosmic bridging between the primordial world (pre-Christian, but also the time of gods and demons) and the present world in the Chamula Festival of Games. Again, this particular theme is fundamental to royal rhetoric of the Classic period, unequivocally documented in the exegesis on dynasty given by Chan-Bahlam in the Group of the Cross at Palenque (Schele and Freidel 1990:216–261).

8. David Freidel remembers the first time he worked with Maya, mapping the ruin of La Aguada at the northeastern tip of Cozumel Island. One man in his fifties spoke of how his father and mother had been born slaves, but they had been liberated and had become schoolteachers. Slave is the term Maya used for themselves when they were attached to haciendas in the northern lowlands, but in the last century, some Maya were literally sold into slavery to work on Cuba.
9. Evon Vogt (1976:177) cites Edmund Leach (1961) to the effect that "formality" and "masquerade" are paired opposites and modes for moving in and out of the sacred time of ritual performance. That is, ritual episodes that begin with one can end with the other, and large festival cycles can be framed by episodes of the same sort. Vogt illustrates this dynamic in Zinacantan with the great Year-End, New-Year cycle. Christmas rituals begin with the formal building of the crèches, the posada rites, the birth of the Christ children (older and younger brothers) and then progress into a masquerade of "fools of aggression." At the end of this great festival phase of Zinacanteco life in January, the Festival of San Sebastián sees the sequence reversed. The beginning masked pageant is spectacularly inversive of daily life and a recollection of the past, mythical and historical. The Festival ends with the deeply moving and formal transfer of responsibility for sacred articles and official roles from one group of cargo holders to the next. Perhaps it is more than coincidence that the Bonampak' dance scenes open with the Masque of the Monsters in Room 1 and close with a grotesque, paunchy masked character being brought into the final dance on a raised litter in Room 3. These punctuating masquerades are preceded by the formal display of the little heir in Room 1 and followed at the very end by a denouement of formally witnessing noble lords acknowledging the legitimacy of this incipient transfer of power. These are, of course, the central matters of the murals.
10. Spirit possession is still a feature of modern Maya ritual, although not necessarily in the context of masked dance. Garrett Cook gives the following description of K'iche'an ritual: "The aj mesas (he of the table) hold fiestas at their houses for the alcaldes on New Year's Day (March 3), and visit them at the four mountains on behalf of clients. The aj nagual mesas have tables in their houses, and usually a cross and often an image of San Antonio, or some other saint. At midnight they call the naguales of the four mountains to come to their table where they answer questions for paying customers. The mesa in San Bartolo had a large room with about twenty chairs around the walls and a table against the fourth wall. The clients all took seats and the lights were put out. It was quiet. Then there was a noise under the rafters where the ears of corn were hung, like lots of rats, a loud rustling. Then there was a thump on the table, then another, and a third and a fourth. Then there were four voices. I don't remember if they all spoke at once or one at a time. 'We are Tamancu, Pipil, Socop and Iquilaja.' One of the voices was an old man's voice, there were four distinct voices. Everyone was really scared. Then we were allowed to ask our questions [anonymous Momostekan narrator]" (Cook 1986:145).
11. As in the case of modern Maya ritual, there are Classic scenes that appear to be performances in the closed and intimate space of interior courtyards or within buildings, while other scenes, like those of the Bonampak' murals or the Stairway of Temple 33 at Yaxchilan, are definitely public. The pulsing between the space of the home and the space of the center, by way of processions that embrace the totality of the landscape into ritual action, is a fundamental and pervasive feature of contemporary Maya life (Vogt 1976), and the Maya of the ethnohistorically and archaeologically known period of European contact (Tozzer 1941; M. Coe 1965a; Freidel 1981; D. Chase 1985; Freidel and Sabloff 1984). That the Classic Maya royalty performed this way, in public as well as private space, is not simply a matter of extrapolation: they declare in writing that some performances occur "in the plaza" in ways that must refer to public spaces (Schele and Grube 1990a). Merle Robertson (1978) gives a wonderful example of how architectural design confirms the public nature of supernatural transformation. On the piers of the Temple of the Inscriptions at Palenque,

Chan-Bahlam depicts himself as a child being held by parents and ancestors as he transforms into K'awil. The piers were designed to be viewed from the plaza in front of the temple from a distance of twenty to one hundred meters.

12. The interpretation of this vase has long been debated. R.E.W. Adams (1963, 1971), who excavated it, took the name phrases to identify historical individuals, including Bird-Jaguar of Yaxchilan, who had come to Altar de Sacrificios to participate in the funeral rituals of the woman buried within the tomb. Following Coe (1973, 1978) in his many critical discoveries about the interpretation of pottery images, Schele (1988) reevaluated the imagery as representing gods and deities in the Otherworld. Her reasoning was based on the nonnatural representations of the figures, the appearance of similar images on other pots clearly representing supernatural events, the appearance of a jaguar-spotted ahaw glyph in their names, and the recurrence of emblem glyphs in the names of many different god names. She interpreted the jaguar-spotted ahaw glyph as *balam ahaw*, "hidden lord," a reference to their status as Otherworld beings. This interpretation assumed that the images represented either afterlife or otherworldly events. That interpretation was suddenly and convincingly changed when Houston and Stuart (1989a) and Grube (1989) independently read the jaguar-spotted ahaw glyph as *way*, the word for "sorcerer," "animal companion or co-essence," and "to transform" (as in shamanistic ritual). The beings on the Altar de Sacrificios vase became clearly identified as the *wayob* or "animal companions" of lords of different historical cities. Since the identification of these lords includes only emblem glyphs or locational titles, they are never named personally. The images are the *ch'ul-ahawob* of Tikal, Yaxchilan, Kalak'mul, or whatever site transformed into their *wayob*. Adams, then, may have been right in his assumption that the vase represents living people, but his proposed identifications of the figures as specific people was incorrect. Schele was partially correct in identifying the context as supernatural and the beings as having supernatural attributes, but she was incorrect in her assertion that the scene takes place in the Afterlife and that the dance corresponds to the Popol Vuh dances.
13. This is the figure Adams identified as Bird-Jaguar. The name is instead composed of a color sign that Grube (1991a) has taken to be *yax* combined with the head of a jaguar with the scrolls of the Pax patron in its mouth. These scrolls function as *te*, also the reading of the Pax patron head, when they are placed in the mouths of portrait forms of the numbers.
14. The first sign of the name was identified by Nikolai Grube in a letter sent to Schele in 1990 as *bu*. Nahm and Grube (Grube 1991a) have suggested the second sign is *buch* and given this name the reading Buchte-Chan, "Sitting-Snake."
15. The name includes *ch'ak* (Orejel 1990; Grube personal communication 1988), the glyph for "decapitate," combined with the reflexive *ba* in the form of the *ba* rodent. In turn, the rodent is conflated with the portrait head of God A', a human-faced god with *kimi*, "death," marks on his face. Many of these anthropomorphic *wayob* have God A' in their names, but each seems to be distinguished from the others by the glyphs preceding the God A' glyph, just as the many Chak names are distinguished by additional glyphs. Many of these God A' name complements derive from the action the *way* is engaged in. This is the "Self-decapitating God A'."
16. Nikolai Grube, in a letter circulated in November 1990, identified the phonetic spelling of the fish-in-hand glyph as *tzak*, a word meaning "to grasp" and "to exorcise." This hand is the same, but it is drawn grasping the death's-head shown in the arms of the *way*.
17. The nearby text identifies fire as the *way* of the named lord.
18. The use of dance as a trance inducer is documented around the world. Bushmen's trance dancing, for example, has been associated with the 20,000-year-old tradition of their rock painting (Lewis-Williams 1986). Many scholars believe that trance dancing also explains some of the imagery in Paleolithic cave painting in Europe and rock painting in Texas and the Southwest. Dance is a vital component of North American Indian ritual, again with trance states associated with its performance. Amazonian groups still use dance in association with trance state so that trance dancing on an individual and communal level can be understood as a vital component of the worldview shared by all Native American people. For the Maya it became institutionalized as a vital component of state ritual and the performance of the king and his nobles.
19. This is one of two pairs of glyphs first observed on the Tablet of the Sun at Palenque. Several scholars have speculated as to the meaning of

these glyphs, but until Stuart and Houston (1991) discovered how place names are registered in the writing system, the speculations were far off the mark. This glyph is now known to designate a location in the supernatural world, and many epigraphers suspect that both members of the paired opposition are locations. See Kubler (1977) for the most complete published study of the distribution of this pair of glyphs. The location glyph on the Tzum monument consists of the signs *wuk*, *ek'*, *k'an*, and *nal* or "seven," "black," "yellow," and the suffix that marks place names. It reads "Seven-Black-Yellow-Place."

20. Freidel witnessed the dancing of the Alféreces and other officials at the Fiesta of San Lorenzo in Teklum, Zinacantan Center, in August of 1973. While most of the participants sat on benches built into the front of the Church of San Lorenzo next to the doorway, small numbers of men would get up and dance slow and simple steps accompanied by rattles. Their ritual advisers danced with them to make sure that they got the steps right. The point of the exercise seemed not so much to entertain the populace, most of which was milling around and engrossed in conversation (like the witnessing lords on the Bonampak' murals), as it was to show devotion to San Lorenzo and transfer of official responsibility.
21. Vogt (n.d.) has commented on the sacred nature of these objects we think of as inanimate.
22. In most of its appearances, the verb "dance" is followed by a prepositional phrase that Peter Mathews long ago suggested was a reference to the objects held in the hand by the dancers. Following a clue found by Stephen Houston (1985), Nikolai Grube has suggested that the dances were named for the objects held by the dancers. The flapstaff is named with the phonetic sequence *ha-sa-wa ka'an*, or *hasaw ka'an*. We have no idea what this means beyond its being the name of that particular object.
23. The name of this scepter was deciphered in 1991 based on a substitution pattern for the bat head in the glyphic name first recognized by Nikolai Grube and Werner Nahm of Germany. Grube noted that the bat substitutes for phonetic *xa* in several contexts at Yaxchilan. Since it also occurs in spellings using consonant + *u* syllables, he reasoned that it has the phonetic value *xu*. Based on this, he suggested that the name for this scepter and the Copan Emblem Glyph, which is spelled the same way, included the syllables *xu-ku-pi* and/or *xu-ku-pu*. Matthew Looper (1991a) found an entry in Attinasi's (1973) vocabulary of Chol listing *xukpi* as bird. Further research associated this general bird term specifically with the motmot, which is also called *xwukip* and *xwukpik*. Looper's reading both accounts for the scepter and works well with Copan, for the imagery and place names of that kingdom are especially associated with birds of all sorts.
24. Brian Stross (1988) identified the word for this bundle as *ikatz*.
25. See Vogt (n.d.) and Gossen (1986) for discussion of modern bundling practices.
26. A Xerox copy of a photograph of this lintel and five others from the Yaxchilan region were sent anonymously to Linda Schele in 1990. Apparently these lintels had been looted during the 1960s and belonged to a state of Maine collector named Palmer until he died in the early eighties. Some Xeroxes of photographs were seen by several researchers just after his death, but we had lost track of the lintels and access to their study until these copies arrived in 1990. Schele designated this unknown site as R in the designation system started by Thompson (1962) in his catalog of glyphs. Q, the last letter used in that system, designates the snake site, which many think is Kalak'mul.
27. In *Forest of Kings*, we used "kahal" as the reading of this title with the warning that it probably would be replaced with a more correct reading in the near future. That reading has now been produced based on phonetic substitution evidence accumulated by Stuart (1988b), Grube and Nahm (1991), and Schele (1991b). The title reads *sahal*, a term that Grube (personal communication 1990) suggested derived from *sah*, "to fear," as a description of a subordinate as "one who gives fear." Schele and others thought this an improbable interpretation until Stuart (personal communication 1991) pointed out an analogous title used for Aztec subordinates. Grube's interpretation is very probably the correct one. See "Bird-Jaguar and the Kahalob" in *A Forest of Kings* (Schele and Freidel 1990) for a detailed examination of the relationship between the king and his subordinate lords.
28. The snake can be identified as a boa constrictor because of the size and the tooth configuration shown in its mouth. The only other snake of this size and shape in the Yaxchilan region is the "Cuatro Narizes," also known as the *nauyaca* lo-

cally and *batrops atrox* to modern scientists. This very dangerous snake is a pit viper with prominent front fangs, which are not shown on the snakes held in this dance.

29. The identification of this person is difficult because no glyphic texts survive to name her. Merle Robertson (1985b) and Schele have taken this second figure to be a woman, based on an early identification of this net costume with women (J. Miller 1974). At one stage of this writing, we had taken the net suit to be an exclusive reference to the Holmul Dancer (Hellmuth in Taube 1985). However, it is now clear that both the Maize God as the Holmul Dancer and the Moon Goddess wear this costume, based, we think, on their identities as First Father and First Mother of the gods. The net costume, therefore, seems to mark the progenitor gods and those who portray them in ritual. Unfortunately, this co-designation makes the identification of the gender of figures in the absence of glyphic evidence extremely difficult. This figure can be a woman, perhaps the wife or mother of the dancer, in the guise of the Moon Goddess, or a male in the guise of the Maize God.
30. Francis Robiscek and Donald Hales (1981), following Michael Coe (1973), proposed many important parallels between the Popol Vuh and painted pottery scenes from the Classic period. They discuss these resurrection scenes as a fragment of what they hypothesized was a "ceramic codex" in Codex Style vessels, an aggregate of scenes that together conveyed a coherent, if complex, mythical message. While the details of the content and function of such an aggregate ceramic assemblage continue to change with analysis, the hypothesis of a coherent overarching message is proving to be a very fruitful one. Michael Coe (1989) has updated his interpretation of Hero Twin imagery from the Classic period and the critical scenes associated with the myth.
31. This image was called the Holmul Dancer after the second and third major images of him published in the excavation report from Holmul (Merwin and Valliant 1932). Hellmuth (cited in Taube 1985) first put together the imagery of the Holmul Dancer in all of his guises, but it was Taube (1985) who realized that this complex was the Maize God identified by earlier scholars and that the Maize God was the Classic-period prototype of One-Hunahpu and Seven-Hunahpu, the first pair of Twins in the Popol Vuh myth. Matthew Looper (1991b) has also made a detailed study of dance and pageant especially those associated with the Maize God and Chak. We have had long and fruitful discussions with him concerning the meaning of these dances and the pageants associated with them.
32. Nikolai Grube, in a letter sent to Barbara MacLeod in January 1991, identified *Ah Mas,* a title that often appears in the names of dwarves, as a word for *duende.* The Spanish word *duende* refers to the spirits, like imps, goblins, elves in Western European mythology, who live in fields, caves, and mountains. These *duendes* can be malevolent, but mostly they are mischievous. Apparently, the ancient Maya identified dwarves as *duendes* who helped the Maize Gods in their dancing and humans in the ritual activities that led to trance and communication with the Otherworld. Pickands (1986) provides a valuable review of the etymology of Maya words for "dwarf" and the role of dwarves in Maya mythology. Among other things, Pickands associates dwarves with experimental, counterfeit forms of human beings, "mud men" from a previous Creation. First Father as the Maize Lord is of the previous Creation. Here we likely have another elusive fragment of the larger epic myth cycle of which the Popol Vuh is a late redaction.
33. Contemporary Zinacanteco Maya call the lower land of death *K'atin Bak,* "place warmed by bones" (Vogt 1976:13).
34. In earlier studies, Schele (Schele and Miller 1986; Schele 1988) has interpreted the iconography of the sarcophagus with the presumption that the metaphor of death and resurrection underlying the image was based on the story of the second set of Twins—that is, that the king falls as either Hunahpu or Xbalanke to face the trials of Xibalba. After winning, he danced out of Xibalba to become a revered ancestor. With the new understanding of the Maize God and his iconography, however, Pakal's image on the sarcophagus can be identified by his clothing as the Maize God. He thus falls as one of the first set of Twins. Schele and Dorie Reents-Budet (in Everton 1991) have recognized this association independently, while many others, Schele, Coe, Reents-Budet, and we suspect Taube, Stuart, and others, have followed Karl Taube's identification of the Holmul Dancer as the Maize God and independently identified the scene of the Maize God rising from the cracking turtle carapace as the resurrection image. Michael Coe's (1989) reevaluation of the Hero Twins imagery first

published this identification and placed it within the larger complex of the myth cycle. Elizabeth Newsome (1987) first found it in the monumental imagery of the Classic-period stelae by realizing that the great turtle altar of Copan Stela A combines with the image of the king on the west side of the stela to show the dead ruler standing beside the double-headed turtle. Seeing both together and overlapping evokes the image of the Maize God rising from the cracked shell. Finally, this turtle-carapace resurrection image was carved on the top of the Altar at the base of the Hieroglyphic Stairs at Copan (Schele 1990a), where it rests symbolically on the palette of the open jaw of a Vision Serpent from which the history on the stairs emerges. The Maize God is there replaced by the Maize Tree itself, which is twisted with the cords of the ecliptic from Na-Ho-Chan, the "First-Five-Place," and flanked by the Hero Twins. An altar and ritual practitioners sit in its crown. See D. Tedlock's (1985:289–298) notes to these passages in the Popol Vuh for the subtleties of meaning built into the words describing this resurrection event.

Given that Pakal falls into the Otherworld as the Maize God, the imagery of apotheosis and rebirth must be reevaluated in this light. Chan-Bahlam wears sun imagery in his headdress and a skeletal image of the cruller-eyed jaguar on his belt. He is dancing as Yax-Balam, the Hero Twin identified with the sun and perhaps the jaguar-pawed dancer on the Altar de Sacrificios vase. The resurrection image then represents the moment after the defeat of the Lords of Death when the children put their father back together and bring him back to life.

35. Nikolai Grube (personal communication 1990–1992) has suggested that the root is *ip*, "force" or "power." Combined with the number nine, which can have the meaning of "many," it may mean something like "many empowerings." In 1992, Ben Leaf called our attention to an entry in the Barrera Vásquez dictionary in Yukatek of *ip* as "mortecina, animal que se murió" ("dead of natural causes, an animal that has died"). In his reading, it would be "nine-dyings." We are not yet sure of either reading because in both the known examples, the verb has affixes that go only on verbs that have to do with position in or shape of space. Moreover, on Tablet 14 at Palenque, the verb is followed by a phrase naming the White-Bone-Snake as the nawal of K'awil. Here the verb should be some action like conjuring. At Copan, objects acted upon are the bones of a long-dead ancestor. The verb should refer to some action appropriate to both contexts, such as "bring out."

36. In the next phrase, the text also talks of "seating" a *lakam tun*, or "huge stone." Two of the elite tombs excavated in the Acropolis of Copan have been rectangular chambers sealed by a set of huge flat stones. These stones may be the *lakam tun* referred to in the text.

37. This way of interpreting Stela H is supported by two sculptures found on the huge platform that sits on the east side of the Great Plaza. This platform is irregular in shape with a row of houses sitting along its northern side. The interiors of these house are lined with wide stone benches built against three of the walls. A steam bath for purification sat in the middle of this row of buildings and a high pyramid closed the east side of the platform. Its eastern and southern sides dropped down steps into the Great Plaza and the Middle Plaza. In 1895, Gordon (1896:2, 24) discovered two stones, each two feet nine inches high, fifteen feet in front of the northern row of buildings and about a hundred and fifty feet from the Great Plaza. That just about centers them on the northern buildings. They represent the Maize God in his dance position with knees

flexed and arm extended. This platform may well have been the place where the accoutrements of dance were kept, dance was taught, or dancers prepared themselves for performances in the adjacent courts.

38. In his reevaluation of the Hero Twins imagery and First Father, Coe (1989) identified this scene as the gathering of Seven-Hunahpu's head and clothing from the Ballcourt of Sacrifice.

39. The head has been repainted to look like a snake, but the tail is clearly that of a fish and it has water marks along its back. It is a fish monster,

perhaps evoking the memory of the Hero Twins in their catfish aspect before they came from the River of Xibalba.

40. Brian Stross (1988) first read the glyph on the bundle as *ikatz*, a word that still survives in modern Tzeltal as "bundle."

41. The Popol Vuh does not mention a woman in this part of the story so we have no way of determining with certainty the role this goddess played in the Classic-period version of the myth. However, in the Popol Vuh version the Heroes set up a corn plant in the middle of their grandmother and mother's house in the human world. When they are defeated and die in the Otherworld, the plant dies; but when they are alive and triumphant, the plant lives again. The Maize Lord in one of the forms identified by Taube explicitly represents the tasseling of green corn, a critical moment in the life cycle of the plant when it forms edible food. The corn will not tassel without adequate and timely watering.

42. The distance number is written as 5.18.4.7.8.-13.18 in the Maya numerical system.

43. This ruler has appeared under the name Hok', K'an-Xul, and K'an-Hok'-Xul in previous studies of Palenque's dynastic history. New decipherments support a reading of K'an-Hok'-Chitam, "Precious-Leaving-Peccary" or "Precious-Hanging-Peccary." *Hok'* means both "to leave" and "to hang something."

44. The text describes three events: the taking of the K'awil scepter by First Mother, the conjuring of the White-Bone-Snake as the *way* of K'awil by the enigmatic god called Bolon Yokte, and finally, back in human history, Chan-Bahlam's entering yet another supernatural place.

The first event is written *ch'am k'awil*, "it was taken, k'awil." This expression occurs in scenes in which people hold the serpent-footed K'awil scepter while they are dancing or involved in other rituals. The Moon Goddess is here written with her portrait head and the moon sign in a glyphic combination that echoes her representation on Classic-period pottery and on eccentric obsidians in stelae caches at Tikal. In both these contexts, the Moon Goddess sits inside a moon sign. Dennis Tedlock (1985:328) associates the mother of the Hero Twins with the moon, and in the Palenque versions of the Creation mythology, the mother of the Palenque Triad is also associated with the moon by virtue of an astronomical alignment on 9.12.18.5.16 2 Kib 14 Mol (July 23, A.D. 690) in which the moon was in close juxtaposition with Jupiter, Saturn, and Mars. This alignment was apparently taken to be the reassembling of the Palenque Triad, revealed in the form of their *wayob*—the planets—with their mother. See Schele and Freidel (1990:473–474) for a discussion of this alignment and the history of discovery that led to its identification. While it is possible, or even probable, that the Moon Goddess operates here as the mother of the gods, we also note that there were several different moon goddesses, perhaps associated with different phases of the moon.

The second event reads *bolonipnah*. We have found no specific match, but Cholti, a now extinct language related to modern Chol and Chorti, has *bolomac* listed as "nagual de choles" ("wizard, sorcerer, or animal companion of the Chols"). This entry appears to refer to a noun, but it may well be derived from a verb, such as the *bolonipnah* above or the *boloniplah* form recorded on Stela A of Copan. The fact that a particular Vision Serpent is named as the *way* (nagual) of K'awil in the next several glyphs supports the possible association. The actor is Bolon Yokte, a deity identified by Thompson (1950:56) in the Dresden and Paris codices and in the Books of Chilam Balam. The name has also been identified in several inscriptions at Palenque, but unfortunately no image can yet be associated with the name, which in the Classic system seems to have been Bolon-Okte-K'u. We know what the god was called, but we don't know what he looked like.

The final event occurred exactly three haab and one tzolk'in after Chan-Bahlam's death. Floyd Lounsbury (personal communication 1976; 1991) first discovered the numerological components of this distance number, as well as associating the last date with a stationary position of Jupiter. Chan-Bahlam used exactly this Jupiter station throughout his life to time critical events. That a Jupiter hierophany was chosen as his apotheosis date is in keeping with this pattern.

The text records that he entered a supernatural place of some sort and that he made this visit by the authority of a set of gods, all named the *ch'ul* or divine beings of the deceased king. This group included GI and GIII of the Palenque Triad, the Goddess of the Number Two, and the Hero Twin Hun-Ahaw. This Hun-Ahaw god name also appears in a list of gods on the Tablet of the Foliated Cross and the Palace Tablet, and in a reference on the middle panel of the Temple of Inscriptions. Floyd Lounsbury (personal communication 1976) associated this calendric name with the god whose birth on 1 Ahaw 8.5.0 before 4 Ahaw 8 Kumk'u, the Creation date, is recorded on the Tablet of the Cross. He suggested that the Hun-Ahaw name refers to the calendric name of this god as a way of distinguishing him from his firstborn son, who had the same portrait name. Since we now know this god was named Hun-Nal-Ye and that he was the Maize God, Lounsbury's earlier suggestion is less likely. Hun-Ahaw is, however, the documented name for the Headband Twin who wears large black spots on his arms and legs (Schele 1987b). We think this Hun-Ahaw name most likely refers to that god, who was one of the Hero Twins.

This clause seems related to the one accompanying the dressing scene on the pot in Fig. 6:18. There the text reads *Oxlahun Ok Waxak Zip och ha Hun-Nal, Wak-Ahaw, uti Kab Ch'ul Kab Ahaw,* "13 Ok 8 Zip, he entered the water place, Hun-Nal Six-Lord, it happened in the Earth, Holy Earth Lord." The date 13 Ok also appears on Temple 14, but more important is the event recorded as "he entered" the same watery place that is shown in the lower register of the Temple 14 scene. On the pot, a female goddess, presumably the mother of the Maize God, gives her revived, dancing son part of this clothing. Chan-Bahlam's mother gives her son K'awil in the analog of this mythical action.

Earlier interpretations of this scene (Schele 1988; Schele and Miller 1986:272–277) were made without the decipherment of the verb describing the last event. Based on the association of the dance scene with the Popol Vuh story, Schele suggested that this scene shows Chan-Bahlam dancing out of Xibalba after defeating the Lords of Death. We still agree with the main thrust of that argument, but now we know he is entering rather than leaving a supernatural place. This place is not found in any other text, but we suspect it may be the location where the *ch'ulelob* of Palenque royal ancestors resided and helped their descendants. We also caution that ethnographic descriptions make very clear that contemporary ancestors reside in specific mountains and not in generic Maya places. This place may simply be a supernatural location where only Chan-Bahlam and his mother acted by the authority of the gods. The glyph at D7 reads *yeteh,* meaning "the work of" or "the authority of." In this context it appears to specify that this "entering" by Chan-Bahlam was done through their work or authority in the sense that they protected or helped him.

45. The Five-Flower-Valley location is also recorded on Copan Stela J in a phrase recording the actions of gods fourteen bak'tuns before 9.0.8.0.0 (Schele and Grube 1990b). It is one of several known locations in the supernatural world where sacred events are recorded as having occurred. Other locations in this category include *matawil, ho-na-chan,* the Black-Transformer, and "six-shell-in-hand."

    Ch'a means "lying-down." When *ch'a* combines with *k'ak' nab,* the term for "ocean" and the "Primordial Sea," we take it to describe the sea in a state of "lying-downness" before it had been separated from the sky. If we are right, this is the sea in its chaotic state before the moment of Creation.

    The last glyph appears in the main text in direct substitution for *xaman,* "north." The "white" prefix is consistent with the north direction, while the suffix *-nal* often goes with locational nouns, but we do not know the reading of the main sign.

46. This watery surface is marked with "stacked dots," one consistent symbol of liquid in Maya art; the surface in Temple 14 is likewise marked with stacked dots. Freidel (1985) suggests that this liquid refers not only to water but also to fire, smoke, blood, and other volatile substances. This hypothesis is now commensurate with the general category of *ch'ul,* "holiness," which encompasses many examples of such "watery" stuff, particularly the so-called "water-group prefix" in the title *ch'ul ahaw,* "holy lord" (Ringle 1988). However, in the "dance across (or in) the waters" scenes, there are usually pictures of waterlilies or phonetic elements that read *nab* inside the water band. The word *nab,* "waterlily," is homophonous with *nab* meaning "the sea," "lakes," and "canals" (Schele 1988).

47. Robiscek and Hales (1981:89–92) illustrate this complex of skulls, waterlilies, and emergent dancers as Codex Fragment 5, "The Resurrection Codex."

48. In the summer of 1992, Nikolai Grube (personal communication) identified this scene with a Dos Pilas text recording the phrase, *lok'i ti k'an tok kimi,* "he emerged from the Yellow-Torch-Death-Skull." He pointed out that the carapace in both its glyphic and iconic forms has a *k'an* sign. The skull is the *kimi* part of the name, and

the torch, *tok*, emerges from the summit of the skull. He suggested that *k'an tok kimi* is the proper name of the cleft. We suspect that this name also applies to the K'an-cross Waterlily Monster at the base of the Na-Te-K'an tree on the Tablet of the Foliated Cross.

49. The iconography and meaning of these tablets have been the focus of Schele's research both alone and with collaborators for close to twenty years (Schele 1974, 1976, 1985a, 1987g, 1992; Schele and M. Miller 1986; Schele and Freidel 1990a, 1990b, 1991). Much of these interpretations results from these studies and from the collaborative research between Floyd Lounsbury, Peter Mathews, Merle Robertson, David Kelley, and Schele, conducted at the mini-conferences at Dumbarton Oaks between 1974 and 1978.

50. *Pib nail*, "underground house," is the term used for the small sanctuary inside the large temples in the Group of the Cross (Schele 1987g; Stuart 1987:38). In the summer of 1991, Schele, Mathews, Lounsbury, and Kelley discovered that the alternative name for this little house is *kunul*. Barrera Vásquez (1980:352) glosses the Yukatek word *kun* as "enchant, bewitch, conjure" so that a *kunil* is a "place of conjuring."

51. The Palenque Triad refers to a group of three gods, GI, GII, and GIII, so dubbed by Heinrich Berlin (1963), who first recognized them as a unit of related gods in Palenque's inscriptions. These three gods are the offspring of the First Mother and First Father.

52. Schele (1991b) has shown the name of this god reads phonetically as *sak hunal*, a combination of "white" and *hunal*, the word for both "eternal" and "bark paper." The god is the personification of the white cloth headband worn by kings from Olmec times onward (Freidel and Schele 1988a; Fields 1989, 1991).

53. Stuart and Houston (1991) identified *matawil* as a supernatural location based on these texts and its appearance on pots. Grube (personal communication 1990) has shown that the "births" of the Triad gods as originally read by Kelley (1965) read *huli matawil*, "arrived in matawil," so that the births of these gods are characterized as their arrival in this location. Furthermore, their mother is specifically identified as *ch'ul matawil ahäw*, "holy matawil lord." We have no idea what *matawil* meant, although Attinasi (1973:296) reads *mulawil* as "world, existence," and *yambü mulawil* as "death (the other world)."

54. Carlsen and Prechtel (1991) say that the Maya of Santiago Atitlan call the seed of maize *muk*, "interred one," and *joolooma*, "little skulls." Attinasi (1973:243) glosses the Chol word *bük* as "interior, seed, and spirit" and *bük'el* as "bone, long pit of fruit." *Büktal* is "flesh" and *bük'en* is "shock, fear, fall, ghost, spirit which can possess." Here are all the concepts gathered into one, for the seeds of maize and fruit are "bones" and "flesh" as well as "spirit." The human heads on the Na-Te'-K'an have little bones attached to their noses as signals that they have spirit and generative power. The head in the maize, being drawn into the K'an-Hub-Matawil, lacks this bone because it is passing into the Otherworld before the child comes to regenerate it through sacrifice.

55. See Tedlock's (1985:251) penetrating discussion of Creation.

56. First Mother, who was the Moon Goddess, and First Father, who was the Maize God, both wear a net overskirt as one of their principal diagnostic features.

57. Carlsen and Prechtel (1991) describe this tree at the center of the earth. Life was created when it flowered and came to fruition, but the fruit falling to the ground and sprouting is the next generation. The old tree provides shelter and nourishment for the new generation, until it is replaced by the new growth. The stump of that original tree is the "Father/Mother" of everything.

58. Landa (Tozzer 1941:135–148) describes the New Year's celebration in great detail. Since there were 260 days in the tzolk'in calendar and 365 days in the haab, the twenty day names circulated evenly through the year with a remainder of five days at the end of the year. For example, if the first day of the year 1 Pop hits a Kawak this year, it will fall five days later on K'an the next year. The days that can correspond to 1 Pop are called the Year-Bearers and there were elaborate ceremonies in Yukatan to change the Year-Bearers and the gods associated with them. The Yukatekan system has Kawak, K'an, Muluk, and Ix as the Year-Bearers, while the Classic-period calendar had Ben, Etz'nab, Ak'bal, and Lamat. The Madrid Codex has New Year's pages using the Yukatek system, while the Dresden Codex has the Classic-period days. See Coe (1965a) for a discussion of the implications of the

New Year's ceremonies on Maya political and social structure, and see Thompson (1950:124–128) for a discussion of the mechanism involved.

59. Coe (1965a) suggests that the offices within the community circulated through the four quadrants of the town as the responsibility for these ceremonies shifted from year to year. Following these suggestions, Lounsbury has associated the pattern of the New Year's pages in the Dresden Codex with a different kind of social structure involving a community divided into upper and lower moieties. He presented these ideas at the Dumbarton Oaks conference on "Mesoamerican Sites and World Views" in October 1976, and recorded them in two letters to Peter Mathews dated Spring 1977 and November 22, 1977.
60. Before the conquest, they killed a turkey.
61. Here we have another magical object, like the staves and scepters of the Classic kings, that becomes animated through dance. The standards, while intimately linked to great lords, were objects of such immense power and spirit that they transcended the role of official insignia and manifested as gods who watched over entire communities and polities. Even in Landa's description, the Maya are still carrying these *kanhel* (embodiments) of terrifying beings. During the Classic period, festival celebrations providing sacrifice and devotion to the souls of the standards prepared these staffs for their great responsibilities as war monsters and battle beasts. In this way, they became repositories of the spiritual and political center. As such, they were taken to the borders of the kingdom, the frontier of social and cultural order, to confront the enemies of the community.
62. Landa's accounts of life among the Yukatek Maya is full of descriptions of dance, ranging from play among children and adults to great sacrificial rituals. If his observations can be taken as any indication, dance was a vital part of Maya life and ritual. Victoria Bricker (1973) discussed these fire-walking rituals in Yukatan and placed them in the larger context of Maya pageant and ritual.
63. Coe (1989) draws this account from Estrada Monroy's (1979:168–174) publication of the document describing the event.
64. Thomas Gage was born in 1603 and died in Jamaica in 1656. An account of his travels titled *The English-American his Travail by Sea and Land: or A new Survey of the West-Indias, containing A Jornall of Three thousand and Three hundred Miles with the Main Land of America*, was published in London in 1648. A modern version published by the University of Oklahoma Press was edited by J. Eric Thompson, who excised Gage's vitriolic comments about Roman Catholics. The book had been written after Gage converted to Puritanism, and in the dangerous atmosphere of Cromwellian England he sought to prove the profoundness of his conversion by ridiculing his former co-religionists. His account of life in the Spanish colonial world is valuable because it offers a personal description from an outsider.
65. These descriptions are in Gage (1958:243–247).
66. *Chicha* is a home-brewed alcoholic beverage.
67. Victoria Bricker (1981) has brilliantly shown how the history of the Conquest and subsequent ethnic conflicts are institutionalized in the dance-drama of Indians throughout Mesoamerica.
68. Bricker (1981:133–137) has linked a series of these events to prophecies read out in the Ritual of Games at Chamula and in similar dance dramas at Zinacantan and Chenalho in Chiapas. These dance-dramas also recall subsequent confrontations that occurred in later centuries.
69. These descriptions are drawn from Bricker (1981:138–140).
70. Victoria Bricker (1981:139) thinks the White Heads may represent the Indian side of the conflict with Cortés and the Aztec side in the Precolumbian expansion of their empire that had affected the Tzotzils before the Spanish arrived. If she is right about identifying the trilobed nose-pieces in their headgear as referring to Tlaloc, then they may hark back to that very ancient war complex. See Schele and Freidel (1990:130–215) for a detailed discussion of this war complex among the ancient Maya.
71. See Chapter 5 on the emblems of warfare for a discussion of the Waxaklahun-Ubah-Kan War Serpent.
72. Lakandons are a Yukatekan-speaking group of Indians living in the lowlands of Chiapas. According to Bricker (1981:141), Zinacanteco Maya from the highlands took part in 1559 and 1696 expeditions against the Lakandon.
73. Bricker (1981:148–149) made the connection between the death of San Sebastián, tied to a tree and pierced by arrows, and the arrow sacrifice documented in Precolumbian times. See Taube (1988a) for a full discussion of this sacrifice and its meaning among the ancient Maya.
74. Bricker (1981:148) quotes this from Chinchilla

Aguilar (1953:290–291) and she connects this dance to the martyrdom of San Sebastián.

75. Bricker (1981:150) quotes this description from field notes taken by Munro Edmonson in 1962 at San Juan Ixcoy.
76. Gary Gossen (1986) gives a detailed analysis of this part of Carnival and the Ritual of Games that is a part of it. Our description draws from his work, but relates the run as Linda Schele witnessed it in 1992. Also see Chapter 2 for a description of the ceremonies at Calvarios and Chapter 4 for a discussion of the banners.

Chapter 7: Flint-Shields and Battle Beasts

1. Ross Hassig (1988), in his lucid and comprehensive account of Aztec warfare, took the general view that the mystical or spiritual dimensions of Mexica military behavior were subordinate to practical tactical and strategic considerations. However, he noted that the Aztec priests brought images of gods to the battlefield and that they appealed to the supernatural beings for aid before battle and honored them afterward in the wake of victory. Specifically concerning the battle standards, Hassig made several somewhat contradictory observations. He proposed (1988:96–97) that the battle standards (*cuachpantli*) functioned pragmatically as signaling devices to show troops where their units were in the fray and whether they were advancing or retreating. Each major unit had such a standard attached firmly to the back of a warrior, and when the wearer was wounded or killed, another in the unit picked up the standard. While Hassig noted that the Spanish thought that the Aztecs regarded capture of the standard by the enemy as an evil omen, he suggested that such captures caused tactical confusion and disrupted command and control. For this reason, he inferred that the capture of a unit's standard precipitated retreat. This signaling function seems reasonable and it is commensurate with descriptions of coeval Maya battle standards provided by Landa (Tozzer 1941). However, Hassig himself observed the tactical signaling function in battlefield maneuver was only one of several functions served by the Aztec standard: "When a unit standard was taken, the entire unit fled, and if it was the standard of a general or of the king, the entire army retreated." (Hassig 1988:58). In this second case, the issue is strategic rather than tactical; capture of the standard of a king or general signaled a pivotal moment of defeat in an engagement. The strategic importance of the standard was underscored by Hassig in a note (1988:293) concerning the Tlaxcalans, where the standard (presumably the principal standard of the army) was carried in the rear and brought forward after the battle was over for everyone to see. Hassig discounted the source (López de Gómara) as secondhand and hence unreliable. However, Freidel has analyzed (Freidel and Scarborough 1982; see also Simpson 1966) Gómara's account of the battle of Cintla (cornfield) on the Gulf Coast, in which Cortés engaged Maya warriors. This account includes unique details of the cultivated "ridged" fields which could only have derived from accurate eyewitness observation. We, therefore, take his descriptions of the battle standard to be equally viable. The battle standard was of great strategic as well as tactical importance, and its capture signaled defeat. The argument we present to the effect that the standard embodied a spiritual and magical power is circumstantial in the Aztec case and based upon the Maya precedents we discuss in this chapter.

   Ross Hassig (1988), as noted above, has recently written a comprehensive social history of Mexica military organization and the Imperial expansion it allowed. This detailed treatment belies any notion that the Mexica were susceptible to easy loss of battle discipline. On the other hand, Hassig is skeptical of the centrality of belief systems to military performance and, unlike previous historians, passes over the battle of Otumba without comment.
2. Cortés (Pagden 1971:142) himself put it this way: ". . . after we struck camp in the morning and traveled on a league and a half, there came to meet us such a multitude of Indians that the fields all around us were so full of them that nothing else could be seen. We could hardly distinguish between ourselves and them, so fiercely and so closely did they fight with us. Certainly we believed that this was our last day, for the Indians were very strong and we could resist but feebly, as we were exhausted and nearly all of us wounded and weak from hunger."

3. A colleague of ours, Mario Aliphat F. (personal communication to David Freidel, November 1990), suggested that Malinche, the Maya-Mexican woman who accompanied Cortés, may well have understood the power of capturing the standard and have advised this tactic.
4. This great War Serpent, which is sometimes called the Mosaic Monster, has been associated with warfare in Maya imagery since captive iconography was identified by Proskouriakoff (1963–1964). At first, the appearance of this complex with its mosaic image of the War Serpent was considered to signal foreign influence, if not domination, of the lowland Maya. Pasztory (1974), Kubler (1976), Coggins (1976, 1979a, 1979b, 1983), and Stone (1989) have all interpreted the appearances of this iconography as evidence for strong Teotihuacan influence. Schele (1984a, n.d.) began to doubt this interpretation when she found the imagery consistently associated with texts and images that were fully Maya in their historical content. Other researchers, including Mark Parsons (1985, 1986, 1987) and Villela (1990), analyzing the iconography of Teotihuacan found an abundance of bloodletting and sacrificial imagery that had been interpreted in other ways or ignored by other scholars. At the same time, excavations at Teotihuacan (Cabrera, Saburo, and Cowgill 1988, 1991; Sugiyama 1989; López Austin, López Luján, and Sugiyama 1991) revealed a strong militaristic component associated with the Temple of Quetzalcoatl that led Karl Taube (1992b) to identify the War Serpent of Teotihuacan with the famous Feathered Serpent imagery. Specifically, Taube identifies the mask of the Mosaic Monster on that temple as a helmet representing the war god being carried on the back of the Feathered Serpent. Bringing the case back to the Mexica, Taube further argues persuasively that this mosaic serpent is the predecessor of *Xiutecutli*, the Fire Serpent of the Aztecs, who was the spear and weapon of *Huitzilopochtli*, the Aztec god of war. Taube's War Serpent is the Mosaic Monster of the Classic Maya. Our interpretation (Schele and Freidel 1990:159–164) proposes the introduction of this War Serpent-Tlaloc war complex by Teotihuacan trader-emissaries during the latter part of Cycle 8 (that is, after A.D. 250). We further suggest that the successful use of this War Serpent-Tlaloc complex in the Waxaktun-Tikal wars resulted in its first major celebration by Maya kings, an endorsement that led to rapid and pervasive adoption by the Maya. The Mosaic Monster, which was called *Waxaklahun-Ubah-Kan* by the Maya, was, in our view, a borrowed symbol and deity that the Maya made over into one of their major symbols of war. To them, the War Serpent was a supernatural being who influenced its outcome. We think this adoption of the War Serpent by the ancient Maya is an early example of the religious and political syncretism so prevalent in the reaction of their descendants to the arrival of the Spanish and the imposition of European ideas of faith and politics upon their world.
5. Schele first identified *Waxaklahun-Ubah-Kan* as the name of the Mosaic Monster in a letter to Houston and Stuart dated October 30, 1989, based on the text of Stela 6 at Copan. She developed the identification and its implications further in the 1990 Workshop on Maya Hieroglyphic Writing (Schele 1990h), in a Copan Note (Schele 1990b), and in her presentation at the 1990 Maya Meetings at the University Museum of the University of Pennsylvania. The name *Waxaklahun-Ubah-Kan* as a name for the War Serpent continues to resist decipherment. *Waxaklahun* is "eighteen" and *kan* (or its Cholan cognate *chan*) is "snake." *Ubah* is our main problem. In Yukatek (Barrera Vásquez 1980:896) glosses *u'bah* as "sentir generalmente y el sentido e instinto natural ("to feel, perceive, sense in general and feelings and natural instinct"). Attinasi (1973:328) lists *ub* and derived forms such as *ubil* and *ubin* as "sense, feel, perceive, be." The beastie, thus, may be something like "Eighteen Feelings Snake," but we simply have no idea how this translation relates to the concepts of war involved in this image. The other possibility is that rather than being one word, *ubah* is two, with the *u* being a third person possessive pronoun. *Bah* is "gopher" in both Yukatekan and Cholan languages, but *bah* is also "nail" in the Cholan languages as well as in Tzotzil and Tzeltal. The name, thus, could also be "Eighteen [are] his Gophers Snake" or "Eighteen [are] his Nails [perforators?] Snake." The last possibility seems best in keeping with a war context, but there is no reason to believe that the name of this beastie has to be related to war.
6. See "A War of Conquest: Tikal Against Waxaktun" and "Star Wars in the Seventh Century" in *A Forest of Kings* (Schele and Freidel 1990: 130–215).
7. Juan Pedro Laporte excavated both Group 6C-

XVI and the Lost World Group. Our descriptions are based on his detailed reports (Laporte 1988; Laporte and Vega de Zea 1988) and discussions with him over the years. Today this ruin appears to be nothing more than a straight-edged rise in the ground under the dense forest that covers the surface. Juan Pedro cut through a thick plaster cap that sealed it away from view and dug over three kilometers of tunnels to rediscover its lost secrets. This amazing piece of archaeology revealed a complex of courts, buildings, and platforms that had been rebuilt many times before it was abandoned and sealed sometime in the sixth century. Group 6C-XVI was established around A.D. 250, at about the same time that the Holy Lords of Tikal began to refurbish the Lost World pyramid complex with architectural styles imported from Teotihuacan. Many of the monuments in both complexes reveal important information about the War Serpent, Waxaklahun-Ubah-Kan, the imagery of Tlaloc-Venus warfare, and other customs adopted from the Teotihuacanos.

8. This object has been called a ballcourt marker because of its resemblance to similar objects found at Ventilla near Teotihuacan (Aveleyra Arroyo 1963a, 1963b). One of the lower panels in the murals of Tepantitla, furthermore, shows just this kind of object at opposite ends of a field with stick-bearing ballplayers, and other similar objects are known at other sites, including Kaminaljuyu in highland Guatemala (Pasztory 1972:448). At the 1990 Maya Hieroglyphic Workshop, Freidel and several other people realized that this "ballcourt marker" also matched the shape of Maya battle banners. The Teotihuacan ballcourt markers may even be the effigies of captured banners used to mark the ends of the court.

9. The text suggests that the fourth successor to head the lineage was actually the ahaw who went to war. Ch'amak, who may have been his son, erected the standard to commemorate the most glorious event in his lineage's history. William Haviland (personal communication to Linda Schele 1990) noted that we have no archaeological evidence to demonstrate that the standard was ever planted on top of the little shrine; as discussed by Juan Pedro Laporte (1988), it was found buried in a cache with a plaster portrait mask next to the shrine in the context of a termination ritual burying this feature. However, as Federico Fahsen first observed, the text itself includes a logographic expression (glyph block H1) depicting the banner on top of the shrine.

10. As a Vision Serpent, the Waxaklahun-Ubah-Kan manifested the companion spirits or *way* of important warriors, including founders of various lineages. The text on the Tikal effigy battle standard records that Ch'amak, the head of the lineage at the time the standard was dedicated, wielded both the *tok'-pakal* ("flint-shield," see Note 15 for an explanation of the reading) and the Waxaklahun-Ubah-Kan in those dedication rites. Like the War Serpent, the *tok'-pakal* derives from a symbol borrowed from Teotihuacan, and like the War Serpent it denotes war and the warrior. As we shall see, Stela 31 of Tikal shows Stormy-Sky at the moment of his accession as he holds up a *tok'-pakal* headdress for his people to see. As the living embodiment of the *tok'-pakal*, he and all other Tikal rulers acceded into an office called Mah-Chak-Te' or perhaps, Mak-Te'. It is not an accident that the Waxaklahun-Ubah-Kan often includes "West Mah-Chak-Te' " in its name. The text also records that the conqueror of Waxaktun, his grandfather Great-Jaguar-Paw, was also the *tok'-pakal*, written with a spearthrower over the phonetic spelling *kuh*, which was the name of the owl (Schele and Freidel 1990:156–158). Stormy-Sky and all subsequent Tikal kings thus took the *tok'-pakal* as a title descending from the great conqueror, Great-Jaguar-Paw.

Stormy-Sky's act of manifesting the *tok'-pakal* by wearing this crown parallels the flanking portrait of his father Curl-Snout, who wears Waxaklahun-Ubah-Kan as a headdress on one side and a mosaic battle helmet on the other. He holds the spearthrower and shield in his hand, displaying his status as the Mah-Chak-Te', after the death of Smoking-Frog, the brother of the conqueror. On Stela 31, he has been conjured from the Otherworld to pass on these objects of office to his son, just as Pakal passes on the *tok'-pakal* to his sons, Chan-Bahlam and K'an-Hok'-Chitam, on the Tablet of the Sun and the Palace Tablet at Palenque.

The relationships between the *tok'-pakal*, War Serpent, and the Tlaloc-war complex are also clear in these contexts. Like the *tok'-pakal*, the Waxaklahun-Ubah-Kan and Tlaloc have various physical embodiments—magical objects, helmets, masks, staves, among others. Great idols of the Waxaklahun-Ubah-Kan were certainly fashioned at Tikal. While there are several other important supernaturals deployed in Maya warfare, as we discuss in this chapter, these three seem to form something of a triad—not unusual for Maya theology—a constellation of beings that together presided over the warfare ritual complex that the Maya adopted from Teotihuacanos.

11. This cut-shell motif occurs in the art of Teotihuacan in aquatic, and later, military, iconographic contexts. Recently, Ellen T. Baird (1989) has extensively reviewed this symbol and its occurrences in the epiclassic (A.D. 800–1000) art of Teotihuacan, Cacaxtla in Tlaxcala, and in art of the Mixtec region. Baird (1989:116) hypothesized that this cut shell signals Venus in its war-related contexts and that it evolved into the Central Mexican Venus sign. Although this symbol may very well have had such associations at Teotihuacan, this hypothesis cannot be used to identify Venus apparitions as part of the warfare-ritual complex introduced along with Tlaloc and the Waxaklahun-Ubah-Kan. Ironically, as Baird pointed out, it is the association of this half-star, or its whole-star expression, with explicitly military and Venus-related contexts in the Maya lowlands that lends credence to its identification with Venus in the Mexican highlands. To argue that it is Venus at Teotihuacan because the symbol is a military one in the Maya lowlands and because the Classic Maya do associate Venus with warfare, by means of a very distinct symbol, is not tenable. Nevertheless, the occurrence of the cut shell as the insignia of this standard does reinforce our interpretation of it as a military instrument. In another valuable recent discussion of the star-cut-shell element and its variations, Virginia Miller (1989) underscored the military associations of the symbol at Chich'en Itza and elsewhere in the Terminal Classic period. Miller made several other pertinent points. The star element is associated with writhing serpents at Chich'en Itza in border designs such as the Mercado dais. Human faces emerge from half-stars carved on reliefs from Chich'en Itza. Finally, some Chich'en Itza warriors appear to be wearing a giant version of the star element as a skirt in a manner that clearly represents a symbolic state rather than a piece of clothing. In the Upper Temple of the Jaguars murals at Chich'en Itza, a star-skirted warrior stands next to a seated lord who is inside a great spiked disk. Clemency Coggins (Coggins and Shane 1984:162) suggested that this is a Venus star-warrior engaged with the Sun Disk-Maya Lord. Later in this chapter we shall show that this Sun Disk is, in fact, a late Maya expression of the *tok'-pakal.* Wrapping warriors in great snakes, feathered or otherwise decorated but always rattlesnakes, constituted the other expression of mystical or spiritual state on the battlefield in the Chich'en Itza murals. Evidently, warriors could manifest the being symbolized by the cut-shell-star just as they could manifest the spirit serpents, the most important of which was the Waxaklahun-Ubah-Kan.

12. Hasso von Winning (1987:65) attributed the first identification of this symbol with the War Tlaloc to Beyer (1922), who called it an "emblem of Tlaloc." He further cited Seler, who suggested it represented the footprint of a jaguar, and Caso, who associated the lower part with the lip of Tlaloc. Von Winning went beyond the simple association by showing that the symbol was particularly used on standards with a feather-fringed disk atop a shaft which were carried by warrior figures.

13. We first noticed the direct association of this symbol combination through Hasso von Winning's (1987:93–96, Figs. 7–9) treatment of Tlaloc B (the War Tlaloc) in his comparative study of Teotihuacan iconography. He had long ago recognized that this owl-weapon medallion functioned as a symbol of war (von Winning 1948) for the people of Teotihuacan and appeared with figurines of warriors. James Langley (1986:65) briefly reviewed this medallion complex and Clara Millon (1988:124) reaffirmed its designation as a war emblem.

14. This imagery is especially associated with the ritual war complex in the Patio of Atetelco, which is the fullest expression of the iconography that is directly related to the Tlaloc-Venus war complex among the Maya. The most complete discussions of this complex known to us are Oakland (1982), Parsons (1985, 1986), and Schele (n.d.).

15. At Teotihuacan, the warrior figures in the Atetelco sacrifice-war complex wear rings around

their eyes and carry javelins, spearthrowers, and small round or rectangular shields as signals of their status. We cannot prove that these weapons became the metaphor for "war" at Teotihuacan because we do not know the language they spoke, and hence we cannot prove any hypotheses requiring control of language. However, in the Classic Maya context, where we do control language through deciphered texts, we can be sure that the two elements of the metaphor, the flint blade (*tok'*) and the shield (*pakal*), enjoy the kind of autonomy required by the Aztec metaphor for war as described by Fran Karttunen in the note immediately below (for example, we have the phrase "his flint, his shield" as a direct substitution for "his flint-shield"). Furthermore, the consistent presence of these images at Teotihuacan, and of the owl with war imagery there, certainly supports the profound association with war that could have transferred to the Maya with the Tlaloc-war complex.

16. We asked Fran Karttunen, a highly respected Nahuatl specialist, about this metaphor and she provided the following commentary in February 1991:

> The Nahuatl difrasismo/metaphor for "war" you are looking for is *mill chimalli* "arrow shield" rather than *atlatl chimalli* "atlatl shield." No need to comb Olmos; it's right in Molina, identified as a metaphor. It appears numbers of times in the *Cantares mexicanos.* Another apparent difrasismo is *chimalli tlacochtli* "shield dart." It appears several times in the *Cantares* but not in Molina. Bierhorst thinks it is a metaphor for "warrior," and that would make sense, since a common way of constructing a difrasismo was to name an entity by two of its characteristics. Hence, "skirt blouse" for a woman, "turtle rabbit" for an armadillo (i.e., shell and long ears). The Molina stamp of approval on *mill chimalli* is incontrovertible.
>
> By the way, these are true difrasismos. *Mill* does not modify *chimalli.* It's not an "arrow-shield" or an "arrow-kind-of-a-shield." That would be *michimalli.* The arrow and the shield have equal status. *Mill chimalli* is like saying, "it [war] is an arrow, it is a shield." Likewise, *chimalli tlacochtli* would mean "He [a warrior] has a shield, he has a dart."

Mill can be translated as either "arrow" or "arrows," *tlacochtli* as either "dart" or "darts," and *chimalli* as either "shield" or "shields," since Nahuatl makes no distinction between singular and plural for inanimate entities. Hence the metaphor for war could also be translated as "It is arrows, it is shields" and for a warrior as "He has a shield, he has darts."

17. In *A Forest of Kings,* we identified the scene on Stela 31 as a representation of the last recorded action in the text—a bloodletting by the king as the javelin-shield person. Since we published that interpretation, we have learned a great deal about the knot-in-hand accession phrase that appears in this text and at Palenque. Barbara MacLeod and Schele (Schele 1991b) realized that the central act of accession it records was the tying on of the royal headdress—usually the Jester God, whose name was Sak-Hunal. Several scenes of this headdress ritual show an attendant, who is sometimes an official and other times a dead parent or god, lifting up the headdress to show it to an audience before it is put on the king's head. On Stela 31, Stormy-Sky holds up his *tok'-pakal* in exactly this action. Schele and Villela (1992) have also associated the Tikal accession with similar depictions of accession at Palenque.

18. In our discussion of the Tikal-Waxaktun war and the inscriptions associated with it, we (Schele and Freidel 1990:156, 449–450) showed that the spearthrower-owl and spearthrower-shield are simply phonetic substitutions for each other. In a substitution first explained by Peter Mathews, the shield is phonetically marked as *kuh,* a term known in both Chol and Yukatek as "owl of omen." We also showed how the *kuh* owl from the early Tikal complex was soon replaced by indigenous Maya term for "shield," *pakal.* The spearthrower was quickly replaced by the principal Maya weapon of war, the flint-headed spear called *tok'.* Thus, the spearthrower-owl-shield symbol was quickly replaced by "flint-shield," *tok'-pakal* (Schele 1990b), which became the emblem of war for the Maya.

19. Houston (1983) identified a phonetic substitution, *to-k'(a)* for the flint blade, thus confirming its reading as *tok'.* Many intriguing problems still surround the relationship between objects and powers in Classic Maya reality. We know for a fact that the Maya created beautiful chipped stone objects, especially of honey-dark chert and obsidian but also in all cryptocrystalline materials. Archaeologists call these objects eccentric flints—"eccentric" because they do not conform to obviously useful shapes like choppers, knives, spearheads or javelin tips. As we shall see, the category of *tok'* depicted in Maya art includes not only spearheads, but evidently also such eccentric chipped-stone objects. Archaeological excavated examples depict ab-

stract shapes, animals such as snakes, and multiarmed shapes with human profiles sporting the smoking celt of K'awil. The eccentric flints are a very ancient tradition among the lowland Maya, extending back at least into the Late Preclassic period (400 B.C.–A.D. 200) and probably into the Middle Preclassic (1000–400 B.C.). They precede the advent of the Tlaloc-Venus ritual complex of war by many centuries. The javelin-spearhead-owl complex at Teotihuacan, which provides the antecedent for the Maya war symbol *tok'-pakal*, does not employ obvious eccentric forms of chipped stone. We believe, therefore, that the Maya absorption and adaptation of the Teotihuacan concept integrated a venerable Maya category of magical objects associated with war and sacrifice with this powerful new formula and rationale for conquest. This syncretism may well account for the rapid substitution of the atlatl and its feathered javelin by the Maya stabbing spear as the *tok'* component in the symbol. In light of archaeological context, we can say that some *tok'-pakalob* were likely to have been heirlooms passed down through families. Because the retrieved ones ended up in cached offerings and burials, however, we know that Maya stoneworkers were fashioning new *tok'-pakalob* for noble patrons throughout the Classic period. These stoneworkers were no doubt the same people who created the weapons of war used by these nobles from the same materials.

20. That the effigy battle standard erected in Group 6C-XVI was a war emblem, rather than a ballcourt marker, is supported by the final phrase of the text as well as by the appearance of the emblem of war in the Teotihuacan tradition. It relates directly to a similar construct at the conclusion of a long dynastic history recorded on a set of Early Classic lintels at Yaxchilan. Both of these terminal phrases have *u ch'a way*, "received it, the soul companion."

    The last phrase on the Group 6C-XVI standard begins with *u ch'a*, several unrecognizable signs, and in the very last glyph block, a hand holding a spearthrower over the Tlaloc-marked shield that appears on the side of Stela 31. A similar clause at Yaxchilan Lintel 35 has *u ch'a* preceding the glyph for *way* and the portrait heads of two gods—one named O-Chak and the other with a variant of GIII. *U ch'a* means "to take, to carry, to use, to appropriate, to receive" in Yukatek (Barrera Vásquez 1980:119). *Way* is the term for "spirit or animal companion." This seems to be a final dedicatory phrase recording that it (the inscription and its history) was received by the *way* of these two important gods. The battle standard has the same *u ch'a ski-?? way* followed by a phrase concluding with the spearthrower-shield combination. We think this may record that the *way* of this war emblem and the Waxaklahun-Ubah-Kan beast it conjures received the inscription on the standard.

    Although we can't be absolutely sure, the logical thing being received in these phrases is the text of sacred words itself, received as the Maya gods and ancestors still receive prayers, explanations, and admonitions from human beings. At Yaxchilan, the *wayob* who received the words carved on the lintels were two gods, one named O-Chak and the other shown as a cruller-eyed jaguar god. We shall meet both of them again. On the Tikal standard, the being who received the sacred words was either the spirit of Tlaloc-Venus war itself as housed in the spearthrower-shield talisman, or it was the divine monster who accompanied the spirit of this talisman, Waxaklahun-Ubah-Kan. The dedication phrase of the standard identifies its proper names as the Tlaloc emblem and the spearthrower-owl signs, the insignia of the upper medallion. But the final text phrase identifies it as part of a larger category of magical objects called spearthrower-shield and later *tok'-pakal*.

21. Chris Jones made this connection during the 1990 Texas Workshop on Maya Hieroglyphic Writing. This recognition led us to realize that the spearthrower-shield combination corresponded to the flint-shield war symbol. Pasztory (1974) first demonstrated the relationship between the Stela 31 image and her Tlaloc B at Teotihuacan.

22. The dance in the lower register of Room 1 has been taken to be part of the heir-designation rite recorded in the upper register (M. Miller 1986a), but new decipherments now allow the identification of the last event in the main text as the dedication of a house. Presumably, the dance corresponded to this ritual. Landa described the use of such feathered banners in the rites of K'uk'ulkan at Mani in Yukatan: ". . . they regarded him as a god, and fixed a time for him in which they should celebrate a festival to him as such, and this was celebrated throughout all the land until the destruction of Mayapan. After this (city) was destroyed, it was celebrated only in the

province of Mani, and the other provinces, in recognition of what they owed to K'uk'ulkan, presented, one year and another, to Mani four and sometimes five magnificent banners of feathers . . . they placed the banners on top of the temple" (Tozzer 1941:157–158; parens original). We shall note later that K'uk'ulkan was, in all likelihood, a transformation of the Waxaklahun-Ubah-Kan. Tozzer, in a note, offers further evidence of a direct connection between processional banners and battle banners: "At Zaci or Zaquivae, the native name for Valladolid, there was, as already noted, an idol to the god Aczaquivae and 'every four years they had a certain combat among the natives, some against others, over whom they carried a banner which they had of a regular military type and hoisted on that high hill (temple) in the midst of (the town)' " (Tozzer 1941:157; parens original).

The third room shows the sacrificial rites that sanctified the events of Room 1—the heir-designation and house dedication. Again the central dancers are flanked by standard-bearers. The two lords on the right lead the musicians in the great dance, and the two on the left hold their standards high in the air and stride toward the dancers. These last two figures may represent a victorious warrior and his captive, soon to be sacrificed. Mary Miller (1986a:144) pointed out that the second of these two figures wears the same green rope as the sacrificial victim in the center of the scene. She suggested both figures as the next victims, while we see the first as the captor leading his captive toward his fate.

23. Landa described this scene in his sixteenth-century discussion of Maya warfare in Yukatan: ". . . Guided by a tall banner they went out in great silence from the town and thus they marched to attack their enemies, with loud cries and with great cruelties, when they fell upon them unprepared" (Tozzer 1941:123). See note 1 for discussion of the tactical use of the Aztec battle standard as a signaling device. Note that while the Aztecs attached the standards to the backs of warriors, Maya warriors carried their standards. There are a variety of attached elements in the Maya repertoire as well, generally called backracks, and the dancers in Room 3 of the Bonampak' painted temple have banners attached to their backs. The Teotihuacan warriors wear their great war medallions as pectorals, and there are Late Classic Maya kings who similarly display the owl on their chests as part of the Tlaloc-Venus ritual complex.

24. The sacred banners of Chamula and the saints and sacred objects of other modern Maya people are kept in wooden boxes very much like the Bonampak' box.

25. Mary Miller (personal communication 1989) thinks the same Knot-eye-Jaguar may have been captured by a king of Piedras Negras. Later, we shall see that Hasaw-Ka'an-K'awil of Tikal is deeply concerned with the standards of his city, one example of which was likely lost with his father when that hapless lord was captured by Flint-Sky-God K of Dos Pilas.

26. The text of the Temple of the Sun also records the heir-designation of Chan-Bahlam, when he was displayed from the pyramid as the enterer (Schele 1985b). Five days later on summer solstice, he became the sun in the company of GI, who was probably his father in the guise of the god. At Bonampak', this same heir-designation ritual is associated with war and the taking of captives for sacrifice (M. Miller 1986a) so that we presume that Chan-Bahlam's heir-designation was also associated with warfare conducted by Pakal. We are considerably more convinced now than when we wrote *A Forest of Kings* (Schele and Freidel 1990: 470–471) that the smaller figure in the three scenes is in fact Pakal, rather than Chan-Bahlam as a child. The shell under him on the Tablet of the Foliated Cross is identified glyphically as K'an-Hub-Matawil, "Precious-Shell-Matawil," which Stuart and Houston (1991) have identified as a supernatural location in the Otherworld from which the First Mother, First Father, and their children came. On the Tablet of the Cross, he is standing on a 9 glyph, which usually pairs with a *Wuk-Ek'-K'anal* glyph. We now know these two glyphs also record supernatural locations in the Otherworld. Although we still do not understand the being supporting him on the Tablet of the Sun, we take his location in the Otherworld on the other two tablets to locate him in supernatural space. This location is consistent with his identification as a dead ancestor rather than as a yet-to-be crowned child.

27. Zhang He (1990) first associated these Palenque thrones and their counterparts at Copan with images of bone thrones shown in pottery scenes.

28. This drum-major headdress is the Late Classic analog of the domed mosaic headdresses worn by Curl-Snout on the side of Stela 31 and by Venus-

Tlaloc warriors throughout the Classic period. Schele and Villela (1992) have explored these associations with the accession compounds used at Palenque for putting on this headdress.

29. See *A Forest of Kings* (Schele and Freidel 1990: 262–305) for a full recounting of her story.
30. The text next to his head reads *u bah nawal wi-ch'ok-te'-na* or "it goes the transformation (spirit) companion of the root-sprout-tree-house." An alternative reading of the *nawal* collocation as *na ol* may refer to the coming out through the heart of the beastie we are seeing. The initial verb may also be a combination of the *ba* rodent head and a tree sign with numbers on it. We still do not know how to read that particular verb, but it seems to be associated with bloodletting contexts.
31. The god's name appears here, in an exactly parallel passage on Lintel 42, and on Lintel 14 of Yaxchilan. Barbara MacLeod (1991c) suggested that the full name of this god may be Ol Chak and that this name refers to the *ol* portal in the turtle carapace of the Maize God's resurrections.
32. Recognition that Stela 6 did not depict a king came from long-term conversations between David Stuart, Nikolai Grube, and Schele. With the recognition that the Waxaklahun-Ubah-Kan phrase on the side of the monument was not the name of a person but a Vision Serpent, it became clear that the protagonist of this monument was not a king (Schele 1990b).
33. Archaeological and epigraphic evidence suggests that Tikal lost a war with Caracol and its allies ending in A.D. 562, another ending in A.D. 637, and a final one with Dos Pilas and its allies in A.D. 679. A newly excavated stairs at Dos Pilas records that the *tok'-pakal* of his father, Shield-Skull, was brought down by Flint-Sky-God K (Houston 1993).
34. Temple 1 may have been built after his death by his son, Ruler B, but he surely oversaw the design of the imagery that would be carved on the lintels of that building.
35. We know this is the Waxaklahun-Ubah-Kan because the name is written in the block following the now-destroyed verb and preceding the name of the king (Schele 1990).
36. We think this jaguar god may be related to a being called *Nupul-Balam*, "War Jaguar," who is named the *way* of the lord of Tikal on the famous polychrome vase from Altar de Sacrificios. Fig. 5:18). Ben Leaf was the first one to notice the shell earflare of the God Chak in the name of this great beast, who is apparently also shown on the Initial Series pot from Waxaktun. There he holds a bound lip-to-lip cache vessel.
37. This place is identified by an emblem glyph with a snake head as its main sign. Peter Mathews called it Site Q in an early study of the many looted panels associated with that emblem glyph. Evidence supporting its identification as Kalak'mul is still debatable, with Peter Mathews and Nikolai Grube especially doubtful that it can be that kingdom. We will use Kalak'mul for the purposes of this discussion, while warning that the identification may well change in the years ahead. A newly discovered hieroglyphic stairway from Dos Pilas, which has been deciphered by Stephen Houston and David Stuart, records Flint-Sky-God K as the vassal lord of the king of Site Q, wherever it was. Lords from this kingdom also participated in two earlier wars against Tikal and its next king died at Hasaw-Ka'an-K'awil's hands.
38. Nikolai Grube, in a letter sent to the authors in February 1990, demonstrated the phonetic reading of the verb as *hub*, which means among other things "bajar" or "to put something down." To conquer in the Maya idiom was to put down the *tok'-pakal* of your enemy. The *tok'-pakal* was a class of objects including the War Serpent and the feather-fringed standards, and perhaps the capacity to make war itself.
39. Again the text records this quite directly as *u tzak k'ul tu ch'am ti yak'il*, "he conjured the god by his harvesting from his tongue." In 1991 David Stuart recognized an example of the lancet glyph on a text from Copan Structure 22a. This example has *ch'a* and *ma* as phonetic complements spelling the word it represented. *Ch'am* means "to grasp or receive," but it is also glossed in Chol and Chorti as "to harvest," as when one plucks an ear of corn from its stalk. We now know that the Maya thought of bloodletting as the act of "harvesting" human sustenance. The same expression was used to record the relationship of a child to his parents as their "harvest" (Schele, Stuart, and Grube 1991).
40. The date in the Maya system was 9.12.6.16.17 11 Kaban 11 Zotz', which curiously enough was exactly one k'atun after a date at Palenque when a person of the same name, Nu-Bak-Chak, visited Palenque and participated in house dedication rites in which captives were sacrificed. We do not know how the two events are related, although

we suspect the Tikal king may have been trying to establish an alliance with Pakal. In any case, the fact that he was sacrificed exactly one k'atun later seems unlikely to have been coincidental.

41. This Kalak'mul king was probably the father of the man Hasaw-Ka'an-K'awil captured and killed.
42. The king's name in Mayan would have been Balah-Ka'an-K'awil.
43. The text on this lintel records a star war against an unknown site that took place on 9.15.12.11.13 7 Ben 1 Pop (February 8, A.D. 744). On the same day, presumably for the battle itself, he did something (it's the same verb as Lintel 3 of Temple 1) with the litter, which was named *Nik Pilip K'in-Balam Ek' Hun u k'ul Yax-Muy-Ka'an-Chak, Sak Chuwen,* "Flower ???? Sun-Jaguar, the Black-Headband, the god of First-Cloud-Sky-Chak, the White-Artisan." The cruller-eyed, jaguar-eared god, who is also the god of number seven and the patron of the month Wo, is named K'in-Balam and he was the god and palanquin of a Naranjo lord captured in battle. An important new insight into this event has been provided by Simon Martin, a young British epigrapher, who attended the 1991 Workshop on Maya Hieroglyphic Writing at Texas where he impressed everyone with the astuteness of his knowledge of the Site Q inscriptions. During the summer of 1991 he extended his studies to include Naranjo and Tikal. He (S. Martin n.d.) identified a Tikal attack in which a Naranjo lord was captured and brought back to the city. The captured palanquin shown on Lintel 2, which has the Naranjo emblem glyph on its base, was taken from this unfortunate captive. Martin went on to identify this very captive as the man who struggles at Ruler B's feet on Stela 5.
44. Hassig (1988) details the manner in which Aztec armies signaled conquest of a town by burning its main temple and either destroying or taking captive the images of the principal gods of the community.
45. Proskouriakoff (1963-64) identified the glyph in this phrase as "captor of" based on its association with known captives. We have long known that it read *u kan,* but it was not until 1990 that its full meaning became apparent. A panel depicting the first bloodletting rite of Dos Pilas Ruler 3's son was discovered by the Dos Pilas project in 1990 (Palka 1990; Houston 1993). The man who helps the youngster perforate his penis is titled *u kan ch'ok* where he is clearly not a captor of the boy. David Stuart, Stephen Houston, and Nikolai Grube realized from this and other independent evidence that the glyph corresponded to the term *kan,* which means "to guard" in Yukatek. The captor is the guardian of his captive and of the trophies such as the battle litters that are taken in battle. This may explain why particular names, like Tah-Mo', "Torch-Macaw," recur in these guardian phrases so often.
46. David Stuart has long been uncomfortable with the *ka* value that had been proposed for the dotted-comb sign by Lounsbury in an 1977 analysis of the Yax-Pas, "New Dawn," name at Copan. By 1988, Stuart (1988b) had found this sign substituting in the glyph for elder brother, *sukun* or *sakun.* Since *pas* was a Cholan word for "dawn," he felt *sa* was the more likely reading. In 1990, Grube and Nahm (1990) found an independent set of substitutions that confirm the *sa* value and Schele (1991b) found yet another set. The evidence is now very strong that the dotted-comb sign had the value *sa.* Combined with the moon sign, which read *ha,* we now know his actual name was Hasaw-Ka'an-K'awil.
47. The precise relationship between the *tok'-pakal* and the *hasaw-ka'an* remains elusive. The Waxaklahun-Ubah-Kan at Tikal was conjured in conjunction with a decorated staff that likely constituted a *hasaw-ka'an.* The seated king here carries the bunched javelins and shield emblematic of the *tok'-pakal* and mentions this talisman in his defeat of his enemy, Jaguar-Paw, on the adjacent lintel. Clearly the Classic Maya employed several kinds of objects in the conjuring of supernaturals for war. So far, the *tok'-pakal* is the precedent-setting object for hegemonic war of territorial expansion among these people. Perhaps the Maya already had a principle of battle talisman in place into which they incorporated the *tok'-pakal,* but this remains to be seen with future investigation of the issue. Present evidence indicates that the principle expanded and elaborated after the initial success of Tikal against Waxaktun.

    Grube (1991a) associated the *hasaw-ka'an* "flapstaffs" with battle banners and quotes Villagutierre Soto-Mayor's description of their use by the Itza Maya:

    > They had two other idols which they adored as gods of battle: one they called *pakoc* [*pak'ok*], and the other, *hexunchan.* They carried them when they went to fight the Chinamitas, their mortal frontier enemies, and

when they were going into battle they burned *copal*, and when they performed some valiant action their idols, whom they consulted, gave them answers, and in the mitotes and dances they spoke to them and danced with them.

(Villagutierre 1983:303)

Following a clue from David Stuart, Grube went on to associate the *hasaw-ka'an* banners with a war banner called *tlauhquecholpamilt*, "red flamingo feather banner," that appears in the Lienzo de Tlaxcala. It was part of the war dress of Xipe-Totec in the aspect of the sky.

48. As it happens, David was substituting on a Far Horizons tour for Schele, who was unable to come on the trip. Mike and Harry Parker and Sibyl Masquelier, among many other great people on the tour, shared this moment, as did Arq. Guadelupe Belmonte, Mexican government archaeologist who kindly allowed Freidel to examine the Tonina facade closely. Mary Dell Lucas, owner of Far Horizons, and photographer Barry Norris were not on the scene, but were very helpful in thinking it through.
49. In his letter of 1989 on the *way* glyph and its related concepts, Nikolai Grube suggested that the title present in the names of Chan-Bahlam, K'an-Hok'-Chitam, and Bahlam-K'uk' reads not *bakel balam-ahaw* as Schele (1985a) had earlier proposed, but instead *Bakle(l) Way*, "Bony Soul Companion." He further suggested that all the members of the royal family of Palenque—or perhaps only those who became kings of Palenque—had one of the great dancing skeletal gods so frequently shown in Maya pottery scenes as their *way* or "soul companion."
50. In the inscriptions of the Hieroglyphic Stairs centered between their six portraits, they are recorded in the *yitah* "sibling" relationship on the day House C was dedicated (Schele n.d.). Six men from different places became brothers by dying together in these dedication rites. Perhaps as interesting, the same inscription refers to a person named Nu-Bak-Chak, otherwise nicknamed Shield-Skull, who was the father of Hasaw-Ka'an-K'awil of Tikal. Furthermore, the west panel of the Temple of Inscriptions records that he "arrived" (*huli*) on 9.11.6.16.17, exactly one k'atun earlier than the date that his *tok'-pakal* was put down by the king of Dos Pilas.
51. New phonetic evidence accumulated since this king was named has confirmed the *hok'* value of the bundle glyph. The animal in the name has a flat nose, small ear, and fur or hair along its jaw. *Xul* was originally used because it is a neutral term for "animal" in Q'eqchi as well as the name of the month written with a similar head, but the consistent presence of *ma* as a phonetic complement identifies it as a word ending in *-am*. Only one animal name fits these conditions—*chitam* or "peccary." This king's actual name was K'an-Hok'-Chitam.
52. We suspect that the need of captives for the dedication of the new north facade of the Palace sent K'an-Hok'-Chitam against Tonina, because it is the last major project we know of archaeologically before his capture and death at the hands of the Tonina king.
53. It turns out that the last glyph in this phrase is Pia Ahaw, so that the human soul companion of this terrible being is identified as a lord of the unknown place of Pia. Interestingly, this same glyph shows up in titles and as a toponym in the inscriptions of Pomona (called Pakab in ancient times), the first kingdom to the east of Palenque. One of the six sacrificial victims brothered for the dedication of House C was also a Pia Ahaw (Fig. 7:23). We wonder if Pia was not a province in the kingdom of Pomona and if Pia was not associated in some way with Tonina. We wish also to acknowledge that Stephen Houston and David Stuart on the one hand and Nikolai Grube on the other independently identified these place names and identified the people associated with them. Looper and Schele (1991) have furthermore recognized an earlier war between Pia and Palenque in which the lords of Pia penetrated into Palenque and damaged the central area of the city called Lakam-Ha.

In January 1993, Schele had the opportunity to visit Tonina with yet another tour group. While there she examined the stucco mural with great care. The "turtle-head" sign has a dotted circle attached to its forehead. This sign is problematic because a dotted circle is phonetic *mo*. The glyph for macaw, which are called *mo'* in most Maya eyes, have this dotted circle around the eye as a phonetic complement. In fact, the heads for macaws and turtles are drawn in the same way with only this dotted eye to distinguish the bird from the turtle. Thus, it is possible that the glyph we read as *pi-a* is, instead, *Pi-mo*. If the later reading is the correct one, then the location cannot be associated with any presently-known place.

54. The netting pattern also recalls the background to the Atetelco murals, which depict the Tlaloc-war complex in greatest detail at Teotihuacan.
55. Arthur Demarest, Stephen Houston, and their colleagues are investigating this problem in the field (Demarest and Houston 1990).
56. The severed and bleeding head of the jaguar is a well-established sacrificial image on stucco and painted decorated lowland temples already in the Late Preclassic period, centuries before the visible impact of Teotihuacan conventions on such buildings. This image remains central to Maya war iconography through the Classic period, as expressed for example on the Tablet of the Sun at Palenque. Examples of Preclassic bleeding-mouthed jaguars are known from Cerros in Belize and Tikal in Guatemala. Other examples of the jaguar at Cerros, bearing the sun glyph on their cheeks, and associated with the mat symbol of royal office, show that this beast represented majestic power as well as sacrifice (Schele and Freidel 1990: Chapters 3, 4, 6).
57. David Freidel's student from the People's Republic of China, Binsheng Wang, has written a persuasive paper to the effect that the Maya Late Classic warfare bears a much stronger resemblance to the Chinese "Spring and Autumn" periods of warfare between shifting alliances of many capitals than to the "Warring States" phase between rival imperial aspirants who had already consolidated substantial territorial control.
58. None of the previous interpretations of the imagery of Teotihuacan have ever identified a figure who could represent a ruler. As a result, most interpretations of the political and social organization of Teotihuacan have assumed some kind of anonymous rulership in which the public representation of political power was not an important component of Teotihuacan's system. This began to change in a 1991 seminar at the University of Texas, when Nancy Deffenbach presented an analysis of the murals of Atetelco showing that the figures in the net surrounding the entry portal into the compound are categorically different from the others in the same mural. These human figures wear the nosepiece and headdress of the female Creatrix in the Tlalocan murals of the Tepantitla complex. Deffenbach also showed that the net pattern on this wall is also different in that it lacks the nawal animals from the other walls and is marked with blood symbols. When Schele saw her analysis, she immediately recognized that this reproduced the pattern by which the Maya identified their rulers—that is, by surrounding them with emblems drawn directly from the central cosmic imagery of the World Tree. The Tepantitla murals represent the Teotihuacan analog of the same imagery. By analogy to the Maya case, human beings who wear symbols drawn from that imagery may depict rulers who wield that power on earth. We suggest that these human figures on the Atetelco entry wall represent the ruler of Teotihuacan, and we have since noticed many other images showing humans wearing the same emblems as they hold scepters of various sorts. If we are correct in identifying these representations as those of the ruler of Teotihuacan, the person of the king may not have been as anonymous as we had previously thought.
59. See Schele and Freidel (1990: Chapter 10) for an account of the collapse and the century that followed it. In the last decade, it has become increasingly clear that the collapse was neither as universal nor as widespread as had been previously thought. In the Copan valley, for instance, the outlying groups were occupied for a century or more after the end of the central government. The same may have been true of the central Peten area. Moreover, northern Belize and northern Yukatan never collapsed at all. Sites like Lamanai (Pendergast 1990) and Santa Rita (D. Chase and A. Chase 1986) were never abandoned and have continuous habitation for three thousand years.
60. K'uk'ulkan, the Feathered Serpent of Chich'en Itza, was also associated with another snaked god, Hapay Kan, translated "Great Sucking Snake," who was mentioned in the Books of Chilam Balam. Ralph Roys, noted ethnohistorian of these chronicles, offers this translation and points out that Hapay Kan refers to an individual who was, according to the accounts, taken from Chich'en Itza to Uxmal and pierced by an arrow. Before that, however, he states: ". . . the son (or possibly the sons) of 'Holy Izamal' was given in tribute (or perhaps under tribute) to feed and nourish Hapay Can." (Roys 1967:179). Hapay Kan was at Chich'en Itza. Roys then supplies evidence that Hapay Kan does refer to a spirit among the modern Lakandon and that Feathered Serpents preside over sacrifice in scenes at Chich'en Itza. The argument is clearly speculative, but the point remains that the gods of

Chich'en Itza served all of that city's human lords and not just its king.

61. Victoria Bricker (1981:138–144) notes the presence of two categories of character, the Feathered Serpent and the White Heads (Tlaloc-masked) in the modern Maya festival of San Sebastián in Zinacantan. Seeing the Precolumbian antecedents of these characters in K'uk'ulkan-Quetzalcoatl and Tlaloc, Bricker cites ethnohistorical literature that not only documents the contact-period relationship between Tlaloc and Quetzalcoatl as rain and wind respectively, but also links both to warfare and sacrifice (Bricker 1981:146–149). The mask of the White Heads is a three-pronged element placed over the upper face (see the mask of Tlaloc emerging from the lower head of the Waxaklahun-Ubah-Kan on Lintel 25 of Yaxchilan for a remarkably similar element and placement). We argue that the Waxaklahun-Ubah-Kan is the Classic Maya name for the Mesoamerican Mosaic Serpent, the Classic Teotihuacan version of which Taube (1992b) persuasively suggests is a war deity. As a helmet, this god is carried on the frieze of the Temple of Quetzalcoatl by the undulating body of the Feathered Serpent. In sum, Bricker's associations for these characters among the Maya can be extended a thousand years before the Spanish Conquest.
62. *Cintla* is the Nahuatl word for "cornfield." This battle is described in detail by Lopéz de Gómara, biographer of Cortés, who likely had access to eyewitness accounts (López de Gómara 1964).
63. Our description of this encounter is drawn from Victoria Bricker's (1981:31–41) brilliant analysis of these events from both the Spanish and Indian sides.
64. Bricker (1981) in a translation from Recinos (1957), uses the entire description from *Títulos de la casa Ixquin-Nahaib,* an important K'iche' historical document. A second document called the *Título K'oyoi*, translated and interpreted by Robert Carmack (1973:265:345), is now in the Robert Garett Collection of the Princeton University Library. Written at Utatlan, the capital of the K'iche', by K'oyoi rulers from Quetzaltenango, it provides details of the battle and the preparation for it that are not found in any other sources. Our account combines these two sources.
65. This is described in the *Título K'oyoi* (Carmack 1973:302).
66. Alvarado was a man of arrogant carriage and very blond features, which led to the Aztecs calling him *Tonatiuh,* the Sun (Recinos and Goetz 1953:119). *Tunadiú* is the Maya transformation of the Nahuatl nickname.
67. Recinos (1953:119) makes the day 1 Ganel (Q'anel) equal to the day February 20, 1524, in the Julian calendar. Using the 584283 correlation constant, February 20 fell on 11.15.4.1.8 12 Kimi 14 Sak. 1 Q'anel, which is equal to 1 Lamat in the Yukatekan system, fell two days later on February 22, 1524, or 11.15.4.1.8 1 Lamat 16 Sak.
68. Susan Webster (1992) has just finished a study of the confraternities of Seville in Spain during the period of the Conquest and the first two centuries of the colonial period. Her work has contributed enormously to our understanding of the process of adaptation and imposition that went on shortly after the Conquest. Seville was the official port in Spain from which all ships, people, and products came and went from Spain to the New World. Moreover, the confraternities (*cofradías* in Spanish) of Seville proliferated exactly during the period between 1510 and 1700 and became the most powerful and numerous in Spain. They provided refuge from the Inquisition and the civilian government of Seville to the thousands of foreigners who flooded into the city to take advantage of the wealth there or to wait for transport to the New World. Neither institution liked the foreigners and the confraternities offered a safe way for them to display piety and to make political allies. Most important to the development of parallel organizations in the Americas, the confraternities of Seville were independent of both the state and the civil authority. Their primary ritual functions were processions in which members carried great litters bearing images of Christ, the Virgin, and a few saints. Members whipped themselves publicly in acts of penitence and piety that rivaled the Maya penchant for self-inflicted bloodshed. Webster recounts how the penitents would often deliberately spray beautiful women with their blood.

    According to Webster, many of the statues were thought to perform miracles. They were also made with movable arms and heads, and were painted and dressed to present the most realistic experience possible. Both the statues and the processions enjoyed great popularity among the general populace. Some of these statues were carried into battle by the king of Spain as early as the fourteenth century. Most important for us, the confraternities were brought to the Americas both by civilians and soldiers and by the mendicant orders, especially the Franciscans and the Dominicans. There, the institution of the *cofradía* provided a very special refuge for the Maya and other native people by providing a shelter for preserving the old mythology and pageant under the thin disguise of Christianity. As in Seville, the *cofradías* began in the church but remained largely independent of it. Today in Guatemala, members of the *cofradías* are called *costumbristas*. They are perceived as the conservators of the old ways and beliefs in the face of the spread of evangelical Christianity and the pressure to assimilate into the dominant culture of the Ladino. The processional statues were very likely carried by Cortés's soldiers wherever they went into battle. These are the angels and white virgins the K'iche' Maya saw.

69. Our description draws on the discussion of this *santo* in Momostenango by Garrett Cook (1986:149). The native name of Santiago is Nima Chumil, Morningstar. Significantly, the festival of Santiago in Momostenango is the occasion for the reenactment of the Conquest Dance in which Tekum is defeated by Alvarado (Cook 1986:152).
70. The lords of Chich'en Itza undoubtedly had foreign allies, but current evidence suggests that they were Mayan-speaking people. See Schele and Freidel (1990: Chapter 9) for a summary discussion of this issue.
71. Gary Gossen provided the following description of the banners of Jesus carried by the Pasiones at the Chamula Festival of Games, a major modern Maya festival: "The Pasiónes are the custodians of, sponsors of, and actors in the cult to the head of Sun/Christ, who is the principal deity honored at the festival. The head of the Sun/Christ is represented by four metal flagpole tips. Each of three sets (one for each barrio) of flagpole tips is kept in a sacred wooden chest, which is brought out during the festival and, during the rest of the year, is maintained in the outgoing Pasión's home in an elaborate chamber-like shrine. . . .

    ". . . whereas the Pasión is in charge of the head of the Sun/Christ (the flagpole tips and ribbon), the Flower is in charge of the body (the banners) and the skeleton (the poles). The four banners (flowered cotton cloth on various colored backgrounds) are kept carefully folded in a wooden chest, receiving a regular 20-day flower renewal ritual. . . . The flagpoles are kept wrapped in white cotton, lying horizontally behind the chest that contains the banners" (Gossen 1986:232–233). Vogt (n.d.) observed that ribbons directly substitute for feathers in highland Maya staff decoration.
72. Arthur Demarest (1978) discussed this subject under the rubric "situational ethics."
73. Rani Alexander (1993) has included the Hacienda Ketelak and its surrounding settlement in her Ph.D. dissertation on the historical archaeology of this part of Yukatan for the Department of Anthropology, University of New México.

Chapter 8: Gaming with the Gods

1. The Centro de Investigaciones Regionales de Mesoamérica, located in Antigua, Guatemala, provided this laboratory for workshops on hieroglyphic writing given in 1987, 1989, 1990, and 1992.
2. The three workshops—1987, 1989, and 1990—were organized by Nora England of the University of Iowa and Martín Chacach Cutzal, the former director of the Proyeto Lingüistico Francisco Marroquín. Steve Eliot of CIRMA provided the facilities, support for the workshops, and helped translate into Spanish the English-language workbook from the hieroglyphic workshop Schele gives each year at the University of Texas at Austin.
3. In 1990, Mareika Sattler pointed out the plaque on the side of the ruined building, identifying it as the place where *The Conquest of New Spain* was written.
4. Some of the names at Palenque were best-guess translations that have since been refined by new and continuing decipherments.
5. James Fox (1984) had identified the *pi* sign in a numerical classifier spelled *pis* in the inscriptions of the Casa Colorada at Chich'en Itza, and David Stuart (1987) had identified the dotted-maize sign as *tzi*. In the fall of 1989, Stuart, Schele, MacLeod, and several other epigraphers had put these two readings together and independently found the Yukatek entry of *pitz* as "play ball."
6. Since Schele learned to speak Spanish in the field without formal instruction, she speaks it under-

standably, but ungrammatically. Her Spanish is generously salted with naive constructions and literal translations of English syntax and logic into Spanish.

7. Both Q'eqchi and Q'anhobal are more closely related to Cholan and Yukatekan, the principal languages of the inscriptions, than Kaqchikel or K'iche'. Of all of the languages spoken by participants of the workshop, these two are the ones that would be expected to have closely related vocabularies as well as words that had survived from the Classic period.
8. Many other authors have also written on the Mesoamerican ballgame and tried to make sense of the images and myths associated with it. These include Clune (1963), Knauth (1961), Stern (1948), Krickenberg (1948), Pasztory (1972), Cohodas (1975), Gillespie (1991), and Schele and Freidel (1991). See Kowalski and Fash (1991) for a very useful summary of past interpretations of the game and its mythology. See two new volumes edited by Scarborough and Wilcox (1991) and by Van Bussel et al. (1991) for the most recent investigations and interpretations of the ballgame tradition.
9. Coe and Diehl (1980:340–343) identified Monument 32 as a kneeling ballplayer and suggested that the once-movable arms were manipulated to match different postures in the game. Many scholars suspected that the famous Olmec heads represent rulers in the gear of ballplayers, but this is still a much-debated interpretation.
10. Ignacio Bernal (1968, 1973) reported these sculptures and identified them as images of ballplayers. See Leyenaar and Parsons (1988) for a very useful history of ballgame scholarship and a full summary of past and present thought about the Mesoamerican ballgame.
11. Leyenaar and Parson (1988:25–34) summarized the evidence for Late Preclassic ballgames in the Pacific coast and highland area of Guatemala and Chiapas. They propose portable markers like those at Ventilla at Teotihuacan were used in openfield courts and they suggest that the "trophy head" cult, large heads with closed eyes and puffy features at Monte Alban, may represent sacrificial victims from a ballgame cult.
12. In the ordinary procedure, Freidel and Scarborough had devised a typology of building groups describing the variation in size and complexity of supposed household facilities at Cerros. The Structure 50 group constituted the most elaborate proposed residential complex in the community before it revealed its true purpose. Freidel and Scarborough might be forgiven being so off-center in their surface analysis. At the time of the research, there were no reported Preclassic ballcourts in the Maya lowlands and very few Early Classic ones (Scarborough et al. 1982). It is a good example of archaeological expectations being driven by the established record of what is known rather than by what is potentially knowable. The evidence for Preclassic ballplaying from other parts of Mesoamerica should have encouraged the latter view.
13. This ballcourt was found by Scarborough and Mitchell. The other ballcourt was discovered by Carr.
14. Taladoire (1981) and Leyenaar and Parsons (1988:34–39) summarize the different kinds of ballcourts in Mesoamerica, where and how many of each type have been found in the various cultural areas.
15. Ted Leyenaar (1978) and Leyenaar and Parsons (1988) document a game that is still played in Sinaloa that resembles closely surviving images of the Precolumbian Maya game. He describes a very complicated scoring system that does not advance in a simple accumulation of points. There is a special emphasis on numbers we know to have been sacred to the ancient peoples, but there is no way to tell if the ancient Maya used the same kind of scoring system. The March 1990 issue of the *National Geographic* magazine shows one of the teams from Sinaloa playing their game in the ancient ballcourt at Xochicalco. Used to playing on a flat field, they reported that it was easier to keep the ball in play against the angled benches.
16. The entire quote from Durán is as follows:

    " . . . to speak of the ball game of which a description is offered according to that promised in the chapter head, and the drawing shows. It was a game of much recreation to them and enjoyment specially for those who took it as pastime and entertainment, among which were some who played it with such dexterity and skill that they during one hour succeeded in not stopping the flight of the ball from one end to the other without missing a single hit with their buttocks, not being allowed to reach it with hands, nor feet, nor with the calf of their legs, nor with their arms. They were so clever both those of one side and those of the other in not allowing the ball to stop that it was marvelous—for if to see those of our country (Spain) play ball with their hands gives us such pleasure and surprise, then seeing skill and speed with which some of them play, how much more must we praise those who with such skill and dexterity and elegance play it with buttocks and with knees, counting it a foul to touch the ball with hands or any other part of the body except the two said parts, the buttocks or the knees. And with the practice there were such skillful and excellent players, that they not only were held in high esteem, but the rulers made them gifts and lodged them in their houses and courts, and they were honored with special insignia.

"Many times have I seen this being played and in order to satisfy myself of how much it enchanted the old ones to copy the ancient, but as the best (most important) was lacking, which was the enclosure inside of which they played and the holes (rings) through which they drove and passed the ball, and over which the combat and dispute was to held, it was soft now as compared to how it was in the time of their infidelity and differed as much the live from the painted. And so that we may understand the way it was done and enjoy the art and dexterity which this game was played it should be understood that in all the cities and towns which had some renown and standing and serious authority, as well among the citizens as among the rulers (of which they made much) and in order not to be inferior, one to the other, they built ball-courts enclosed with fine walls, and well ornamented, with the whole floor inside very smooth and covered with mortar, with many paintings of effigies of idols and devils to whom the game was dedicated and whom the players had for protectors in that sport.

" . . . and at the ends of the court they had a quantity of players on guard and to defend against the ball entering there, with the principal players in the middle to face the ball and the opponents. The game was played just as they fought, i. e., they battled in distinct units. In the center of this enclosure were placed two stones in the wall one opposite the other; these two (stones) had a hole in the center, which was encircled by an idol representing the god of the game. He had a face like that of a monkey, whose feast, as we see in the calendar, was celebrated once every year. That we may understand the purpose which these stones served it should be known that the stone on one side served that those of one party could drive the ball through the hole which was in the stone; and the one on the other side served the other party and either of these who first drove his ball through (the hole in the stone) won the prize. Those rings also served them as cord, because straight in front of them (*derecho*) on the ground was a black or green line made with a certain herb, which for reason of superstition had to be made with a particular herb and with no other. All the time the ball had to pass this line in which case they did not lose, because even when the ball came rolling on the ground, as they had hit it with the seat or knee as soon as it passed the line, though only two fingers distance, then it was not a default, but counted as such if it did not pass. He who hit the ball through said hole in the stone was surrounded by all, and they honored him and they gave him a certain special prize of feathers or loin clouts, a thing they valued highly, though the honor was what he appreciated most and most highly esteemed, because they practically honored him as a man who in special battle of even sides had conquered, and ended the dispute.

"All those who entered this game, played with leathers placed over their loin-clouts and they always wore some trousers of deerskin to protect the thighs which they all the time were scraping against the ground. They wore gloves in order not to hurt their hands, as they continuously were steadying and supporting themselves on the ground. . . . A great multitude of nobles and gentlemen took part, and they played with such content and joy, changing now some and later others, from time to time, in order all to enjoy the pleasure and so content that the sun would go down before they knew of it. Some of them were carried dead out of the place and the reason was that as they ran, tired and out of breath, after the ball from one end to the other, they would see the ball come in the air and in order to reach it first before others it would rebound on the pit of the stomach or in the hollow, so that they fell to the ground out of breath, and some of them died instantly, because of their ambition to reach the ball before anybody else. Some were so outstanding in playing this game and made so many elegant moves in it that it was worth seeing and I will specially relate one which I saw done by Indians who had practiced it, and it was that they used a curious thrust or hit, when seeing the ball in the air, at the time it was coming to ground. They were so quick in that moment to hit with their knees or seats that they returned the ball with an extraordinary velocity. With these thrusts they suffered great damage on the knees or on the thighs, with the result that those who for smartness often used them, got their haunches so mangled that they had those places cut with a small knife and extracted blood which the blows of the ball had gathered."

17. Maudslay (1889–1902:III, Pl. 29) shows one of the huge ballcourt markers from Chich'en Itza next to a Maya workman. The size of the center hole is large enough for a Maya ball to have passed through, although it must have been a rare event indeed. Ring markers mounted in the center of the vertical walls on both sides of the playing alley were thought to be Mexican in style, but the earliest and most frequent examples occur in northern Yukatan.
18. Schele and Miller (1986:256) included a figurine of a ballplayer holding one of these handstones in *The Blood of Kings* exhibition, while an actual handstone carved in the form of a skull was found in Tomb B/2 from Caracol (Clancy et al. 1985:189).
19. The retrieval of this yoke is one of the great stories in Maya archaeology. When George Guillemin (1968) opened Burial 195, he found the bottom part of the tomb filled with sediment and the remnants of several wooden objects that had rotted away. A very young Rudi Larios (1990) argued for permission to try to recover the impressions of these objects by injecting plaster into the holes. The result was the miraculous recovery of several statues of the god K'awil, wooden bowls, and a corner of the yoke. To everyone's wonderment the plaster casts made by Larios also recovered a fine layer of plaster and paint that had once adhered to the rotted objects. The casts were covered with the blue-painted and black-line details on the K'awil statues and the extraordinarily fine painting from the bowls. The cast of the yoke fragment had the parallel groves that are frequently shown on the yokes of Maya ballplayers.
20. Leyenaar (1978:70–71) reports that modern players wear a leather strap low on their hips to keep their buttocks from spreading when they slide along the ground. So great is the protection given by this belt that players will wear it when all other gear is discarded. The Precolumbian players very probably wore something just like it.
21. Tedlock (1985:354) analyzes the name *puzbal chaah* as *puzbal*, "place of sacrifice," and *chaah*.
22. Tedlock (1985:338) gives the K'iche' *nim xol carchah* as the name of the ballcourt. According to his interpretation, *nim xol* is "abyss" and Carchah is a possible reference to San Pedro Carchá, a town well to the east of the K'iche' region, the direction from which the K'iche' migrated into their present homeland. He also suggests that these Twins are responsible for the Morningstar, which is a phenomenon also associated with the east. For our purposes, it is the concept of chasm or abyss that is important.
23. Barbara Fash (n.d.) has identified Temple 20 of Copan with its huge, over-sexed bats as the Bat House of the Popol Vuh myth.
24. Seven-Hunahpu, the uncle of the Twins, is specifically named, but they call him their father. Since the story does not tell us what happened to One-Hunahpu, it is possible that both One-Hunahpu and Seven-Hunahpu remained in the Ballcourt to be worshiped by human beings as the symbol of resurrection and rebirth.
25. Michael Coe in his public lecture at the opening of the exhibition, *The Blood of Kings*, in Fort Worth, Texas, May 1986, first drew this analogy between *The Iliad* and *The Odyssey* and the story of the Hero Twins. He published his views on this subject in Justin Kerr's *Maya Vase Book* (Coe 1989).
26. See Schele and Freidel (1990: Chapter 10) for a discussion of the idea that the attempts by Classic Maya kingdoms to master each other politically by means of conquest warfare ended in the general disaster of the ninth-century collapse. Arthur Demarest espouses a similar interpretation of the causes of the Maya collapse.
27. Mary Miller (Schele and M. Miller 1986:256–257) first identified the imagery of this scene with Tlaloc warfare and the sacrifice of defeated enemies. The text on the ballcourt bench appears to read "he ties on the *hun chak-te*, 'the headdress of Chak-te,' " which should refer to the Tlaloc-war regalia the victor wears. The loser is identified as an *itzin*, "younger brother," of the person named in the text. Presumably his value as a captive lay in his kinship to the older brother. The ballgame took place during a fifteen-tun period-ending, perhaps 9.12.15.0.0. The actor is named with the same "4-White-Jaguar" glyph that identifies an actor on Altar 5 at Tikal that Simon Martin (n.d.) has recently associated with

the Kalak'mul-Site Q kingdom. The game took place at Ox-Nab-Tunich, a toponym already identified by Stuart and Houston (1991) in the inscriptions of Kalak'mul. See Schele and Freidel (1990: Chapter 5) for a discussion of the Kalak'-mul-Tikal-Dos Pilas wars.

28. The west side of Stela J at Copan (Schele and Grube 1990b) records the end of fourteen bak'-tuns before the historical date of 9.0.18.0.0. The mythological date fell in the past creation when the gods were the actors.

29. The main sign is a hand with a jade bauble hanging from its first finger. It has the number five, which is phonetic *ho*, prefixed to it and phonetic *ma* and signs act as phonetic complements for the hand, thus giving the reading *hom*, or "chasm." In January 1993, Schele checked the original stone at Yaxchilan and confirmed that the suffix is a *ma* sign (T74 in the Thompson number system).

30. Michael Coe (1965b:18) identified this shape in Relief 1 at Chalcatzingo as a cave in which a person sat and clouds emerged.

31. *Wak* means "six" and *eb* is "step" so that the name of this place is a straightforward description of the stairway. However, as we have seen before, *wak* also means "raised-up, stood-up" as in the Wakah-Chan, the "Raised-up-Sky" established by First Father at the beginning of this Creation. The Wakah-Chan contrasts to Ch'a-Chan, "Lying-down-Sky," which was the state of space and time before the act of Creation. Given the Maya penchant for cabalistic wordplay, we must bear in mind the possibility that this place of ballgame sacrifice was also understood as a "Raised-up-Stair" in the context of other "six" or "raised-up" places established at the time of the Creation. This interpretation makes it yet another member of the "Six" places—that is, the Wakah-Chan ("Raised-up-Sky) and Wak-Nabnal ("Raised-up-Sea-Place"), the location painted on an Early Classic tomb wall at Río Azul in northeastern Peten.

32. The badly eroded text on this structure records the accessions of the first ten kings of Yaxchilan, as well as visits by important dignitaries from other kingdoms. This stairway was once surmounted by a temple decorated with lintels recording this same dynastic history. Bird-Jaguar moved the lintels to another building, but left the stairway and its dynastic history intact to function as his Six-Stair-Place of sacrifice. This *Wak-Eb* stairway lies equidistant between two ballcourts (Structures 14 and 67), and while we can't prove it, we think this is where the dedication rituals described on the ballgame panel occurred.

    To our knowledge, David Stuart (personal communication 1985) was the first epigrapher to realize the contents of this badly eroded inscription. Schele and Mathews (1991) went on to discuss these accessions in connection with royal visits and connected them with the Early Classic lintels recording the same kings and visitors. It is clear that the stairway once went with the building housing these lintels.

33. M. Miller and Houston (1987) first identified the imagery of hieroglyphic stairs in ballgame scenes and connected these scenes to sacrificial iconography. Mary Miller (1986b and 1988b) went on to show that the reviewing stand on the south side of Temple 11 and on the west side of the East Court were false ballcourts and the location of the kinds of ritual shown in scenes at other sites. Six-staired places as locations of ballgame sacrifices were first discussed by Grube and Schele (1990a) in conjunction with the Temple 11 false ballcourt at Copan.

34. The dates given for these games are 13 Manik' 5 Pax, 9 K'an 12 Xul, and 1 Ahaw 13 Xul. Unfortunately, the distance numbers linking these dates—5.19.0.17 and 3.8.10.14.11—do not work arithmetically. In addition, they are not linked to a period-ending or other anchor date, so that we do not know when in time they occurred. The distance numbers, however, cover a range of almost fifteen hundred years and extend beyond the range of Yaxchilan's dynastic history. We suspect the dates on the left side of the panel occur in the third Creation and lay the mythological foundation of the historical ballgame that is the subject of the scenes.

35. Schele noticed in 1985 that this ax glyph appeared with self-decapitation scenes on pottery. Taking this clue, Jorge Orejel (1990) proposed reading *ch'ak*, "to decapitate," which was later independently confirmed by Nikolai Grube. The *ba* suffix is normally a reflexive so that "decapitated himself" might also be the intended reading here, although if we are right in our analysis of the meaning (Schele and Freidel 1991), the sacrifices refer to the original Popol Vuh myth in which the Heroes both decapitated themselves in dance and were decapitated by the Lords of Death.

36. David Stuart has suggested that the "successor"

glyph in the ordinal construction simply reads *-tal*, a suffix used for ordinal constructions in Palenque's inscriptions. Here it occurs with *u na*, a construction meaning "first" (Schele 1978). The glyph following "first" reads *ahal*, which we (Schele and Freidel 1991) have previously interpreted as "manifested." However, Jorge Orejel (personal communication 1989) found an entry in the Moran dictionary of Cholti listing *ahal* as "vencer (to defeat, to conquer, to vanquish)." Grube (personal communication 1990) found the same entry independently and believes that the "vanquish" meaning fits the overall context much better. We will use the "vanquish" or "conquer" meaning in the paraphrase, but the "manifest" meaning was clearly intended in some contexts. We think it possible that both meanings were intended here.

37. The date in the Maya calendar was 9.15.13.6.9 3 Muluk 17 Mak. The long count, however, is written in a very rare form using eight cycles above the 400-tun cycle as part of the notation. These higher cycles are all set at thirteen because long counts record the elapsed time since a zero date which was written at Koba as 13.13.13.13.13.13.-13.13.13.13.13.13.13.13.13.13.13.13.13.13.0.0.0.0 4 Ahaw 8 Kumk'u. When a complete cycle accumulates in this system, the number with that particular unit changes to one. This Yaxchilan date simply records that the units above the bak'tun were still set at thirteen when the ballcourt was dedicated.

38. In our original analysis of this passage, we (Schele and Freidel 1991) suggested a decipherment of *Ox-Ahal-Em*, "Thrice-manifested descent." Last year, however, Nikolai Grube found good evidence that the final sign in the second glyph is *bu* rather than *mu* as we had taken it to be. It turns out that the *bu* and *mu* sign are very close graphically and distinguished (in many but not all instances) by the presence of cross-hatched circles on the inner scroll of the *bu* sign. Grube combined the new *bu* reading with the winal frog *e* sign to get *eb*, the word for "step." Almost simultaneously a new dedication step was discovered at Copan with exactly this spelling in an unequivocal context so that Grube's ideas were confirmed (Grube and Schele 1990c; Schele and Mathews 1990; Morales, Miller, and Schele 1990).

39. The names for these stairs were first identified by Grube and Schele (1990c).

40. Schele (1979a) and Schele and M. Miller (1986:-177) has suggested that some narratives employ the device of showing sequential events in a single rite on several different lintels, while using the text to imply that this same sequence of actions occurred on different dates.

41. There are very good political reasons for Bird-Jaguar to feature his paternal grandmother in the context of ballgame sacrifice with its connotations of war. The male kin of this woman were also allied by marriage to Bird-Jaguar's father's senior wife, Lady K'abal-Xok. Lady K'abal-Xok, by Maya tradition, should have borne the heir to Shield-Jaguar's throne (see Schele and Freidel 1990: Chapter 7). Bird-Jaguar was child to a junior wife of royal blood from the city of Kalak'-mul. Bird-Jaguar defeated rival claimants and acceded to the throne some ten years after his father's death. The mention of his grandmother here suggests that her male kin sided with Bird-Jaguar and against their other in-laws, the men of Lady K'abal-Xok's lineage.

42. Panel 11 shows another woman throwing a ball toward a stairway so that play may have been restarted during the game following some sort of delay.

43. This glyph has been read as *sahal*, a title derived from the root for "to fear." These sahals were nobles of various ranks who served the kings in various functions, including as war chiefs, administrators, and governors of towns. There were many sahals at any one site, and of these, one was singled out to be the "first," *ba*, of his kind. K'an-Tok was the *ba sahal* of Yaxchilan. Other ranks and titles also carried this "first" designation. For example, the heir was the *ba ch'ok*, "first sprout." We also have the "first ahaw," "first sculptors," and "first scribes" (Schele 1990g).

44. Mary Miller (1986:84–88) first commented on the similarity of the masks worn in these two contexts. She also suggested that the heavy belts of the Bonampak' dancers may identify them as ballplayers. Following her lead, Schele and Freidel (1991) associated these scenes with maize rituals and the Popol Vuh myth of First Father and the quest for maize for the creation of human beings in the last and finally successful creation.

45. On the day of this house dedication, October 21, 744, Venus was near maximum elongation as Morningstar, 46.325° and at maximum altitude. Since both Coe (1973:93) and Tedlock (1985:353) associated One-Hunahpu and Seven-Hunahpu, the Popol Vuh versions of the Maize God, with

Venus as Morningstar and Eveningstar, the position of Venus on this day may be relevant. If Bird-Jaguar is in the role of Morningstar, then the dwarves may represent the constellation in which Venus appeared. Lounsbury (M. Miller 1986a:47–49) has shown that images were used to represent constellations in the sky on a first appearance of Morningstar. Moreover, the constellation represented on the Hauberg Stela, on Tikal Stela 1, and the Xultun stelae with the small manikins all show Virgo as the god Chak. These left dwarf wears the shell earflare of Chak and thus may represent Virgo. If so, the other dwarf may represent the region of the ecliptic we call Leo.

46. The text reads 1 Imix 9 te Ch'en *yalah* (it was thrown), *u chan Balamnal* (the guardian of the jaguar), *u k'aba* ([was] its name), *Bolon-nab* (the ball). The text continues to record that this happened at Tamarindito and that the actor was the same Dos Pilas king. Here is direct hieroglyphic evidence that captives were bound up and used as balls (Grube and Schele 1990a).

47. The first major excavation and the restoration of the final phase of the Ballcourt was done by Edwin Shook and others under the direction of Gustav Strömsvik (1952) between 1937 and 1941. Charles Cheek (1983) in his excavations of the Great Plaza trenched all around the Ballcourt and associated its phases with important floors in the Plaza. Finally, during the Copan Mosaics Project and the ongoing Copan Acropolis Archaeological Project under the direction of William Fash, tunneling into Structure 10L-26 has revealed many earlier structures that had been "killed" and then sealed before new constructions began over them. In the fill separating these construction phases, archaeologists have found at least three different styles of macaw heads. The presence of these carefully cached sculptures demonstrates the continued association of the Ballcourt with Structure 10L-26 in all its manifestations and confirms the fact that the iconography of the Ballcourt remained the same throughout its history.

48. Strömsvik (1952) reports finding two macaw heads lying on the ground near Mound 7 across the court from the Ballcourt, while examples of at least two other sets of macaw heads have been found inside the fill of Structure 10L-26, deposited apparently as offerings when various phases of the building were terminated. The construction history of Temple 26 has been reported by W. Fash, Williamson, and others (n.d.) and by William Fash (1991:81–87).

49. The reassembly of these macaws was the result of an extraordinary project conducted by William and Barbara Fash and Jeff Kowalski in the Copan Mosaics Project (Kowalski and Fash 1991). In 1986 they began a project designed to test methodology and techniques that could be used to reconstruct the iconography of fallen Copan buildings using the thousands of sculptural fragments. Since Strömsvik had left the fragmentary sculpture from the Ballcourt buildings in known piles, they began by photographing, drawing, and cataloging all known fragments. Using minimum counts of the elements and a huge sandbox for experimental reconstructions, they were able to determine that there were sixteen birds in all and to produce best guesses as to their original compositions (see Kowalski and Fash 1991:62). Barbara Fash had earlier discovered that some of the birds had triangular tenons signaling that they went on the corners of the buildings, and that the height of the birds required that they go between the doors. As an experiment in 1990, Rudi Larios placed one of Barbara's reconstructed birds back into the eastern building. He found that the birds had to be set at a slight angle, while Barbara continued to refine her fits as they went back into the wall. This experimental work also confirmed that Strömsvik's reconstruction of the vaulting was at least two courses too short. Each building then had eight macaws distributed around the upper half of the outer walls.

Baudez (1984), Schele and Miller (1986:251–253), Kowalski and Fash (1991) have all discussed the meaning of the markers from Ballcourt A-IIb and agree on the cosmological intent of the imagery and the direct reference to the Popol Vuh myth. Much of our own interpretation is based on these earlier analyses.

50. Kowalski and Fash (1991:64–65) identified the macaws as Seven-Macaw, the prideful macaw in the Popol Vuh. We entirely agree with their identification, but add that the bird on the Ballcourt at Copan is shown before his defeat. MacLeod (1991d) pointed out that a macaw was also present in a Creation-related "Ritual of the Angels" in the book of Chilam Balam of Chumayel. "Then resounded the eight thousand k'atuns at the word of the first stone of grace, at the first ornamented stone of grace. It was the macaw that warmed it well behind the acantun. Who was born when our father descended? Thou

knowest. There was born the first macaw who cast the stones behind the acantun. . . ." The Ballcourt macaw at Copan appears to be related to these birds present in the past cosmos or at the Creation of the present one.

51. Andrea Stone (1983:217) connected this story with the imagery of the Principal Bird Deity of the Popol Vuh myth. Constance Cortés (1986) further suggested that "the adventures of the Hero Twins are attempts at reconciling the imbalances caused by overextended powers. In the process, the Twins come to represent a blueprint of the 'correct' behavior to be emulated by future generations of Maya nobility." In this interpretation, Vuqub-Kaqix acts in the Maya version of hubris. The Twins bring nature into correct behavior by defeating him.
52. 18-Rabbit is a nickname that came from a misreading of the T757 glyph. As David Stuart first pointed out, the actual reading is Waxaklahun-Ubah-K'awil. Like Hasaw-Ka'an-K'awil of Tikal, this Copan king was named for one of the great war symbols—the War Serpent Waxaklahun Ubah.
53. Schele, Grube, and Stuart (1989) were able to reconstruct the text once recorded on the raised band running down the center of the west bench. Two dates are recorded there—the accession of 18-Rabbit on 9.13.3.6.8 7 Lamat 1 Mol and the dedication of the Ballcourt on 9.15.6.8.13 10 Ben 16 K'ayab. This dedication date fell on a maximum elongation of the Eveningstar. As we shall see, the inscription on the opposite bench recorded the date 9.12.0.0.0 10 Ahaw 8 Yaxk'in, which was also a maximum elongation date. This earlier date may be associated with Ballcourt A-IIb.
54. David Stuart (personal communication 1988) associated censer fragments found inside the fill between the final phase of Temple 26 and the next phase inward with 18-Rabbit and the date 9.13.8.0.0 (February 2, A.D. 700). In conversations with Stuart, Larios, and Fash, we have tentatively assigned this date to 18-Rabbit's remodeling of Temple 26 and the refurbishment of Ballcourt A-II.
55. The earliest sets with Ballcourt I were made of plaster and did not survive.
56. Mary Miller (in Schele and Miller 1986:251) associated this rope with the Popol Vuh story. She also took the shapes attached to the rope to be a plan-view ballcourt into which the ball will fall when the rat chews through the rope. The altar of Stela 4 is a huge effigy of this tied ball. Leyenaar (personal communication 1986) pointed out to Schele that a Mesoamerican rubber ball will not hold the round without a mold. He suggested that hanging and turning the ball might have served this purpose.
57. David Stuart (personal communication 1985) first pointed out to Schele the contrast of these halfmasks on the ancestor figures in the upper cartouches of Yaxchilan's stelae. The fathers, who sit in k'in-marked cartouches, wear upward-turning, square-nosed face masks, while the mothers, who sit in moon signs, wear this downward-turning face mask. He suggested that the second mark referred to the moon and the former to the sun. While we have not confirmed the sun association, the correspondence of the downturned face mask does correlate very well to moon imagery. Since Hunahpu and Xbalanke rose into the sky where the sun belonged to one and the moon to the other, this moon mask may signal the moon-owner.
58. Schele and Miller (1986:251) first identified the left player as Hun-Ahaw based on the glyphic text next to his face. Schele (1987) went further in showing that Hun-Ahaw was one of the headband twins, who are now recognized as the Classic prototypes of the Hero Twins.
59. Just this kind of sacrifice has been found archaeologically in the sacrificial burials at the Temple of Quetzalcoatl at Teotihuacan (Sugiyama 1989).
60. Barbara Fash has worked diligently and for years with the sculptural fragments excavated from the sides of Structure 10L-26 during the Copan Mosaics Project. See a discussion of her reconstructions and the project of excavation on Structure 26 in Fash (1991:142–151). In a prodigious feat of patience, persistence, inspiration, and the best iconographic and pattern-finding skills working in the field, she has demonstrated that the outer facade of the upper temple was covered with huge front-view Tlalocs. Stuart (1986c) has determined that the inner inscription of the temple recorded the dynastic successors. In this full-figured inscription, most of the glyphs wear Tlaloc symbolism in addition to their normal attributes.
61. Stuart and Schele (1986), Schele (1990d), Schele and Grube (1990), and Grube and Schele (1990) have worked out the decipherment of sections of the stairway inscriptions, although its badly eroded condition and the random order that was used to remount much of it makes a complete reading highly unlikely.

62. Owls, shields, spears, spearthrowers, and captives are represented throughout the iconography of the final phase of Structure 10L-26.
63. Houston and Stuart (1989b:12) show the "kimi" sign in direct substitution for the standard *way*. The sign may, in fact, derive from a special necklace and bib worn by *wayob*. The necklace part, which consists of disembodied eyes, functions as the phonetic sign *u* to give the first sound of *way*. The word for "necklace" is *u*. The bib part of the ornament usually sports a "kimi" sign and the same trileaf motif that characterizes this glyph. We suspect that because the bib and necklace read *way*, it identified its wearer as a *way* of a human being transformed into his *way*. As glyphic signs, the necklace became phonetic *u* and the bib became a logograph for *way*.
64. The Ballcourt at Tikal had markers showing bound captives known to have been taken by Ruler B, the son of Hasaw-Ka'an-K'awil (Ruler A). Tikal's Ballcourt was also associated with the sacrifice of war captives.
65. The text on each of these shields (M31, M52, M65, and M72) begins with "nine" and *te*, which can be a numerical classifier. The next unknown glyph is followed by a name, then *yahaw* "the vassal lord of," *ah pitzil* "the ballplayer," and the emblem glyph. M65 may have a "guardian of the Palenque ahaw" as a record that the king captured someone from that city.
66. The date on the capture panel unfortunately was never completely finished so that only 13 Ak'bal 16 ??? is readable. There are enough details to determine that the month must be Keh, Yax, or Ch'en, but not which color was intended in the superfix. However, only one of the possible combinations occurs within K'an-Hok'-Chitam's reign—9.13.19.13.3 13 Ak'bal 16 Yax or August 30, 711. This would have given him a reign of about nine years, but since a new king did not accede at Palenque until 9.14.10.4.2, we presume that K'an-Hok'-Chitam was kept prisoner for almost eleven years before he was sacrificed.
67. Peter Mathews (n.d.) first deciphered the dates on this ballcourt marker and recognized that the text recorded some posthumous event 260 days after the death of the protagonist. He read the dates as 9.17.4.12.5 8 Chikchan 18 Sak (death of the protagonist), 9.17.5.0.0 6 Ahaw 13 K'ayab (an anchor date), and 9.17.5.7.5 8 Chikchan 13 Xul. Stephen Houston (personal communication 1988) later read the second verb as *tu muknal*, "in his tomb." The separation of dates may refer to a reburial of the bones as we have discussed on Altar 5 of Tikal and the apotheosis of the soul.
68. Marisela Ayala (1988) associated the hole under the central alley marker in the Tonina ballcourt with the consistent representation of just such a hole in ballcourts in the Mixtec and Borgia group codices. In these sources, the ballcourt was also depicted as a gateway to the Otherworld, which played a crucial role in origin and creation mythology. This hole may represent the *Wuk-Ek'-K'anal* part of the ballcourt name at Tonina.
69. Stuart (n.d.) first read this glyph as *ba al* and proposed that it means "firstborn child of woman." Since "child of" term in Mayan languages are differentiated according to the gender of the parents—that is, "man's child" versus "woman's child"—this kinship terms identifies this person in reference to his mother, not his father. We suspect that he is identified as the firstborn child of one of the king's wives, and that this ballcourt marker records his premature death and burial. That he warranted a record in such an honored place suggests that he was the heir to the throne of Tonina.
70. The oval with line of dots usually marks the body of the Cosmic Monster, while the mirror sign in the center has been read by Grube and Schele (1991) as *tzuk*, "partition." It appears to identify the location as a partition inside or on the Cosmic Monster—in other words, in the sacred space of the Otherworld.
71. Edward Kurjack and his colleagues (1991) reviewed the known distribution of ballcourts in the northern Maya lowlands. These scholars noted that masonry courts are rare in the north and date primarily to the Terminal Classic period (A.D. 800–1000) although some Late Classic courts have also been documented. In contrast to this general pattern, Chich'en Itza has thirteen identified ballcourts (Kurjack, Maldonado, and Robertson 1991:150).
72. These are characteristics found in other Terminal Classic-period courts in the northern Maya lowlands. See Kurjack, Maldonado, and Robertson (1991).
73. Freidel (1986a) suggested that the Late Classic and Terminal Classic governments of the northern lowlands lacked the focus on dynastic kings found in the southern Classic lowlands. He hypothesized that the emphasis on interior space in "palaces" registered a focus on the aggregate over the individual. Freidel and Schele (1989b)

and Schele and Freidel (1990:346–376, 393–395) proposed that Chich'en Itza was ruled by a set of siblings under a form of government by council called *multepal* by later Yukateks. Grube (1991b), in a study of inscriptions from other northern Yukatekan sites, found evidence for the same kind of government at least one hundred years earlier at Xkalumk'in. Siblings (Schele 1989; Grube 1991b) are also recorded as ruling at Uxmal so that this form of government by council seems to have been extant in northern Yukatan from the second half of the Late Classic period on. The Itza apparently adopted this form of governance and used it as a basis of empire. Wren and Schmidt (1991; Wren n.d.) reviewed the iconography of the North Temple and provided new restoration drawings of the reliefs. They suggested that these scenes celebrate accession ceremonies of a ruler in ways that parallel later accession rituals at Tenochtitlan in México. We find persuasive Wren and Schmidt's arguments that accession is the subject of the central scenes of the north wall of the North Temple. Confederate government in the northern lowlands of later periods included a "first among equals" called *yax halach winik*, "first true person" among the *halach winikob*, the rulers. Who this person is at Chich'en Itza remains a matter of conjecture, for although he wears a special mosaic spangled shirt in this scene, he has no texts or other diagnostics of a personal sort. No preserved texts accompany this event or any other in the Great Ballcourt.

74. See Maudslay (1889–1902: Pls. 26 and 28) for plans and photographs of the Great Ballcourt at the time of his investigation late in the last century. He provides the base plan of these summit structures as portal arches, stone gates facing the sun's east-west path.

75. This arrangement is somewhat reminiscent of the north and south buildings of the Twin Pyramid complexes of Tikal. While the north building does not hold a stela as it would at Tikal, it does house extensive pictorial reliefs.

76. A. M. Tozzer (1957) systematically investigated the art of Chich'en Itza from the vantage of a Toltec versus Maya ethnic arrangement.

77. This is not to deny the complex and syncretic nature of the regalia worn by lords at Chich'en Itza. In *A Forest of Kings* (Schele and Freidel 1990), we argued that these cosmopolitan lords appealed successfully to all Mesoamerica and had vital alliances with highland Mexican counterparts at cities like Tula. See Kurjack, Maldonado, and Robertson (1991) for further discussion of the Maya iconography of the two competing groups in these reliefs.

78. Several of the players on the Great Ballcourt reliefs wear the Jester God adornment that is a signal of kingship in the southern lowlands until late in Classic history. Toward the collapse period, the Jester God became increasingly available as a status marker for lords other than kings. It is certainly still a statement of lordship at the time of Chich'en Itza. See Freidel (1990:77) for a discussion of the Jester God in the Terminal Classic period.

79. We have no doubt that the disk is a portal. The precise nature of this portal is a matter of debate. We have discussed the *tok'-pakal* blade-shield talisman of war at some length in Chapter 7. In *A Forest of Kings* (Schele and Freidel 1990:503), we reviewed the various arguments concerning the Chich'en Itza Sun Disk and its warrior occupant. The Sun Disk in the northwest mural of the Upper Temple of the Jaguars carries the four square-nosed dragons radiating at four corners. These also can mark the Ancestor Cartouche of southern lowland Classic iconography. Inside the Sun Disk is a warrior seated on the jaguar throne and grasping the throwing stick and bunched spears of Tlaloc-Venus war. We propose that this anonymous figure represents the kingship, elevated at Chich'en Itza to the status of an abstraction, the eternal spiritual king, by the assembly of living lords who together rule the government and imperial state. The syncretic quality of the spiked Sun Disk is an area for further productive investigation of the relationship between the lowland Maya and western Mesoamerica, including the Mexican highlands, in the Terminal Classic period (A.D. 800–1000).

80. During the spring semester, 1991, Schele held a graduate seminar at the University of Texas that focused on warfare and its imagery throughout Mesoamerica. Research projects included the Maya of the southern lowlands, Teotihuacan, Monte Alban, Chich'en Itza, Cacaxtla, the Codex Nutall of the Mixtec, and the Aztec. Most of these areas had multiple researchers working on various aspects of war, but the most remarkable result was the identification of a war and ritual complex that is clearly present with little alteration in the representation of its symbolism throughout Mesoamerica from about A.D. 400 until the conquest. We have discussed much of

the paraphernalia associated with this complex in the chapter on warfare under the rubric of Tlaloc-Venus war. It is clearly present in the imagery of Maya in the southern lowlands and at Chich'en Itza, but it is also extremely important at Teotihuacan, as many researchers have already observed (see Chapter 5 for a full discussion of this research and interpretation). However, what we learned by comparing all the various versions was its presence at Cacaxtla and in the Codex Nutall associated with wars fought as a part of foundation and origin mythology. Following John Pohl's interpretative studies of the Codex Nutall, Rex Koontz showed how the foundation war by the Mixtec against the "original stone people" was required to establish land claims and legitimacy that Eight-Deer used in his own campaign to establish himself at Tilantango. Kathy Reese associated the almost identical iconography of the slaughter scene, the star-chamber imagery, and the jaguar and eagle men murals to a foundation myth recording a terrible war between rival factions that was resolved by alliance to create the Olmeca-Xicalanco. Similar foundation war mythology and iconography are associated with Huizilapochtli of the Aztec. The murals of the Upper Temple of the Jaguars appear to record exactly the same kind of seminal battles in which the Itza established their legitimacy as rulers and holders of dominion over empire. The same imagery is associated with founders on Lintel 25 at Yaxchilan and in Temple 16 and Altar Q of Copan, both of which present dynastic founders in this war complex. The *Ox-Ahal-Ebnal* of Chich'en Itza not only links to the traditions of war of Teotihuacan and the southern Maya lowlands but it participates in the epi-Classic adaptation of the complex to a pan-Mesoamerican expression of origin and divine sanction.

81. This image has always been interpreted as Toltec and Mexicanized, but it, like almost all other imagery at Chich'en Itza, has prototypes in the southern lowlands. There are many birds, usually the Muwan bird screech owl associated with God L, that appear on pottery with human heads in their mouths. The most telling, however, was discovered by Ricardo Agurcia at Copan in 1990 (Fash 1991:101). After clearing the west wall of an older temple completely encased by Temple 16, he found huge stucco relief sculptures flanking the doorway. These are certainly Itzam-Yeh standing front view with its wings extended, its claws before it, and the aged face of Itzamna, God D in Paul Schellhas's nomenclature, in its mouth. Iconographically, this image is clearly one Classic-period precedent for the kind of bird images with human faces emerging from the mouths found at Chich'en Itza and other northern Yukatekan sites a few generations later. Indeed, architecturally the west front facade of the Temple of the Warriors virtually replicates the composition on the west wall of the Itzamna temple at Copan. These magical birds accompany all of the portrayed lords on the columns of the Northwest Colonnade below the Temple of the Warriors. The Itza were greatly feared as magicians and it seems likely to us that their very name derives from *itz.* See *National Geographic* magazine, 98–99, September 1991, for a full-color restoration drawing of the Itzamna temple at Copan.

82. For discussions of the iconography of Maya sacred liquids see Schele (1988), Stuart (1988a), and Freidel (1985). The jar shape depicted on this frieze is not exactly like known Terminal Classic vessels from the area, but it generally conforms to the pedestal vase shape that is common in this period. It is remarkably similar to defined types of censerwares from the Middle and Late Postclassic periods (Smith 1971). The three droplets and four jars are no doubt ritually significant.

83. The roof comb had completely fallen off the building at the time of modern scientific observation. The restoration is based on the discovery at the base of the structure of tenoned elements that clearly constituted part of that decoration. Our illustration and interpretation of the roof comb follows Marquina (1964: Fig. 860), and Tozzer (1957: Fig. 82) illustrates this motif. Maudslay (1888–1889; reprinted 1974: Pl. 32) restores the roof comb with a single *tok'-pakal* element directly over the doorway and then scroll elements around the rest of the roof. This is not a central issue to the interpretation, for in either case the element we describe is certainly part of the roof comb.

84. In addition to the Upper Temple of the Jaguars and the North Temple, the chamber of the Lower Temple of the Jaguars is elaborately decorated with scenes germane to the state rituals of the Great Ballcourt. We have already discussed the Lower Temple of the Jaguars in Chapter 9 of *A Forest of Kings* (Schele and Freidel 1990). The central ritual in that chamber is a bundled offer-

ing given by a Tlaloc-masked lord entwined in a Feathered Serpent to a higher lord seated majestically within a spiked sun-shield. We argued that the person in the sun-shield is an ancestral king elevated to the status of state divinity and that the person below is one of many high officials of the government who rule together as brothers in the name of that dead king. The hypothesis that Chich'en Itza was ruled by a brotherhood of lords is supported in glyphic texts from elsewhere in the city. The notion that there was a first among equals, a superior lord, is hinted at in some texts and declared in the accession iconography of the North Temple as we describe further on in this chapter. The idea that such a superior lord ruled in the name of a dead king, presented in an ancestor cartouche portal, is still under investigation, but it makes sense of the iconography of the city as we understand it. In the ritual sequence of the Great Ballcourt, the Lower Temple of the Jaguars is a more public and accessible display of the lordly accession depicted in the inaccessible North Temple. It is accompanied by tribute processions, military processions, and sacrificial offerings to the sun-cartouched divinity. This temple contains the stone jaguar throne on which that divinity is shown seated. Presumably, his earthly proxy sat upon it before the crowds in the great plaza of the city after the ritual games in the Ballcourt.

At the 1992 Texas Workshop on Hieroglyphic Writing, a team consisting of Linnea Wren, Ruth Krochock, Erik Boots, Lynn Foster, Peter Keeler, Rex Koontz, and Walter Wakefield added additional detail to the Creation association. They suggested that the old male and female gods on the two columns are the Creator Couple. In addition, they pointed out the image of a cleft mountain/turtle carapace with a human figure emerging from either end. They also identified the being emerging from the cleft top as the Maize God and the tree-vine he brings with him as the vine growing from the neck stump of the sacrificial victim in the ballplayer panels.

85. The Chich'en artists were using a tried and true way of ensuring that their message would survive through many generations of use by carving the imagery into the wall permanently. Murals like those of Bonampak' required masters the equal of the original painters to repaint and repair them. Once the painted images were carved into the wall, anyone who knew the rules of color symbolism could repaint them without needing special training or skills as an artist or scribe. Thus, the interior reliefs used at Chich'en were indeed the functional equivalent of the mural tradition, even though we also note that the Itza also painted quite extraordinary murals as well.

86. The artist divided the north wall into three major pictorial fields: two registers on the upper vault wall and a huge single scene on the vertical wall. The main scene is, in turn, sub-divided into three episodes by lining up three rows of people, standing one above the other, as they converge on a central action. These rows of figures do not incorporate the formal device of separate ground lines to separate the registers; nevertheless, each row shows a highly ordered set of people engaged in a different activity.

The major event occurs in the middle row, where a man dressed in a jade-sequined, knee-length shirt appears to accede to an office. His special status is emphasized not only by his unique shirt, but by the fact that he stands, whereas everyone else in his row sits on pillow chairs. The special status of the jade-shirted person is also emphasized by his position on the large grotesque head below his feet. Snaggly teeth dangle from the front of the upper jaw of this monster and smaller rear teeth stud its length. Elsewhere in the reliefs, and on the staircase balustrades, trees grow out of such heads—including a cruciform World Tree on the west wall. The substitution of a person for the tree in this central scene harks back to the Creation mythology of the Classic Maya, and especially to the manifestation and apotheosis of First Father.

Even the Feathered Serpent-enwrapped man facing him sits in his presence. These seated men, like those depicted in seated positions in the Temple of Chak-Mol and the Temple of the Warriors on the other side of the main plaza from the Great Ballcourt, represent the government council in session.

In the lower row immediately below the jade-shirted protagonist, an empty bench draped with the skin of a animal and a ball sit as the focus of attention. In other scenes at Chich'en Itza (including in the upper row of this panel), lords sit on this kind of jaguar bench as demonstrations of their power and rank. Archaeologists found a real example of this stone jaguar bench in the debris of the Lower Temple of the Jaguars and that may be where the scene commemorated here actually unfolded. If this imagery records the accession rites of the ruler of Chich'en, then the ball is connected directly to the seat of power. Warriors armed with spearthrowers and javelins stand four on the left and five on the right of the bench. Among the warriors, fourth on the left, is a man wearing the net skirt of the Maize God, Hun-Nal-Ye, First Father, and he is holding an offering plate in his arm. This connection to the myth of the Hero Twins is reinforced by a blowgunner scene just below the houses in the upper vault registers.

87. This reclining jade-shirted figure also stretches out above the center of the murals in the Upper Temple of the Jaguars. There are a number of features of the iconography and compositional strategies of Chich'en Itza that resemble contemporary art of Tajin, an important city in north-central Veracruz (Wilkerson 1991: Fig. 3:4). One scene on a carved column commemorating an important ruler named 13-Rabbit depicts sacrificial disembowelment and the arranging of the entrails over a wooden rack decorated with maize plants. The effect is that of a tree growing out of the already decapitated victim. In this scene, Wilkerson notes the presence of ballgame gear and suggests that this ritual may immediately follow a ballgame (Wilkerson 1991:53). We also note that Taube (1990) identified this prone figure as an Earth Goddess.

88. Wren (1991) identified the dedication date of the Ballcourt as November 18, A.D. 864 (10.1.15.3.6 11 Kimi 14 Pax). This day was just short of the nadir position of the sun. Four days later, in fact, the sun was at nadir at midnight, while the Gemini-Orion pivot—that is, Three Stones of Creation—was at zenith. As important, the full moon was also directly at zenith position at midnight. Considering that most dedication rituals took several days to complete, we suggest that the recorded dedication date was the beginning of the ceremonies that climaxed on this date. Susan Milbrath (1988) has also commented on the importance of this time of the year at Chich'en and on other important alignments in the Great Ballcourt.

Moreover, we now know that on the day of Creation, the Three Stones of Creation were laid at Na-Ho-Chan. Pictures of this location in the Paris Codex (Fig. 2:32) show twisted cords emerging from the belly of the Maize God like his umbilicus. A pot (Fig. 2:31) depicting Na-Ho-Chan shows these snake cords intertwining through the scene. We think the snakes coming from the belly of the jade-shirted being relates to this same imagery.

89. Divided into two registers, the vault images begin with two men, one with his genitals exposed flanking a huge phallus, itself perforated by two awls. Then the following sequence of scenes unfolds: a warrior stands before two canines; two warriors flank a captive in front of some kind of rocklike structure; a corpse is attended before two houses; and finally two warriors bearing paper-wrapped lances walk past a tree holding a bird with a human head in its mouth. The lower register from left to right has the following scenes: a ritual at a water tank; a vomit ritual involving men and women holding a large plate; a bloodletting ritual on a platform (the seated figure has a plate in his lap and a lancet in his hand); a ritual held before a house and around a tree, and last, blowgunners shooting birds out of a tree. Only a few fragmentary sections of the south wall have been reassembled, but they too show important rituals. In one scene, priests dressed in long skirts decorated with a crossed-bones pattern carry offering bowls to a Vision Serpent, while the other fragment depicts the serpent dance taking place before a low platform with seated people and a dead body. Again we do not understand all of these scenes, but they appear at least partially to involve the purification and bloodletting rituals that prepared practitioners for important ritual performance. These vision and purification rituals appear to parallel the conjuring ritual in the Yaxchilan ballgame sequence.

90. Kurjack, Maldonado, and Robertson (1991) summarized the history of ideas on the Great Ballcourt and credited Stephens (1843) with first recognizing similarities to central Mexican ballcourts and the first expression of the Toltec hypothesis to Charnay (1888). They traced the ideas about the Toltec origin of war and ballcourt iconography that culminated in their most eloquent expression in the work of Tozzer (1957).

91. Again Kurjack and others (1991) summarized the opponents' viewpoints, especially citing Thompson's early opposition and Charles Lincoln's more recent and systematic criticism of the Toltec hypothesis. See Schele and Freidel (1990:346–377) for our own ideas on the matter.

92. The description Landa wrote of the Yukatek Maya, their history, customs, religion, and calendar, is called *Relación de las Cosas de Yucatan* (Tozzer 1941).

93. This vantage does not explain why the northern

lowland Maya relinquished the game as a political theater after the fall of Chich'en Itza but before the Spanish arrived. We can point out that the game as played in masonry courts never had a strong foothold in the northern lowlands. It was principally a phenomenon of the Late and Terminal Classic periods, when the struggles for political hegemony racked the entire lowland region. If we are right that the game as a political instrument was introduced from the southern lowlands during this episode, it may never have been important to local towns or villages who would have had to sustain it for it to be present at the time of the Spanish Conquest.

Afterword

1. My graduate student Tim Albright helped me in my endeavor by doing much of the library work and consulting with Dr. Tadzia after I became engaged in the unfolding of Creation.
2. This is described in Chapter 2.

REFERENCES

Adams, Richard E. W.

1963 A Polychrome Vessel from Altar de Sacrificios, Petén, Guatemala. *Archaeology* 16:90–92.

1971 *The Ceramics of Altar de Sacrificios.* Papers of the Peabody Museum of Archaeology and Ethnology 63(1). Cambridge: Harvard University.

Agurcia, Ricardo F., and William L. Fash, Jr.

1991 Maya Artistry Revisited. *National Geographic* 180(3):94–105.

Alexander, Rani

1993 Colonial Period Archaeology of the Parroquia de Yaxcaba, Yucatan, Mexico: An Ethnohistorical and Site Structural Analysis. A Ph.D. dissertation, University of New Mexico.

Andrews, Anthony P., and Fernando Robles C.

1985 Chichén Itzá and Cobá: An Itzá-Maya Standoff in Early Postclassic Yucatán. In *The Lowland Maya Postclassic,* edited by Arlen F. Chase and Prudence M. Rice, 62–72. Austin: University of Texas Press.

Attinasi, John J.

1973 *Lak T'an: A Grammar of the Chol (Mayan) Word.* Ann Arbor: University Microfilms.

Aulie, H. Wilbur, and Evelyn W. de Aulie

1978 *Diccionario Ch'ol-Español: Español-Ch'ol.* Serie de Vocabulario y Diccionarios Indigenas "Mariano Silva y Aceves" 21. México: Instituto Linguïstico de Verano.

Aveleyra Arroyo de Anda, Luis

1963a *La Estela de la Ventilla.* México: Instituto Nacional de Antropología e Historia.

1963b An Extraordinary Composite Stela from Teotihuacan. *American Antiquity* 29:235–237.

Aveni, Anthony F.

1980 *Skywatchers of Ancient Mexico.* Austin: University of Texas Press.

1992 *The Sky in Maya Literature.* New York: Oxford University Press.

Ayala, Marisela

1988 The Sacred Landscape at Tonina. A paper prepared for an art history seminar at the University of Texas at Austin.

n.d. El bulto ritual de Mundo Perdido, Tikal, y los bultos mayas. A MS in the possession of the authors.

Baird, Ellen T.

1989 Stars and War at Cacaxtla. *In Mesoamerica After the Decline of Teotihuacan, A.D. 700–900,* edited by Richard A. Diehl and Janet C. Berlo, 105–122. Washington D.C.: Dumbarton Oaks Research Library and Collection.

Bardawil, Lawrence

1976 The Principal Bird Deity in Maya Art: An Iconographic Study of Form and Meaning. In *The Art, Iconography, and Dynastic History of Palenque, Part III: Proceedings of the Segunda Mesa Redonda de Palenque,* edited by Merle Greene Robertson, 195–210. Pebble Beach: Robert Louis Stevenson School.

Barrera Vásquez, Alfredo

1980 *Diccionario Maya Cordemex, Maya-Español, Español-Maya.* Mérida: Ediciones Cordemex.

Baudez, Claude F.

1984 Le roi, la balle, et le maïs: Images du jeu de balle Maya. *Journal de la Société des Américanistes* n.s. 70:139–152.

1991 The Cross Pattern at Copán: Form, Rituals and Meanings. In *Sixth Palenque Round Table, 1986,* gen. ed. Merle Greene Robertson, vol. ed. Virginia M. Fields, 81–88. Norman: University of Oklahoma Press.

Berlin, Heinrich

1963 The Palenque Triad. *Journal de la Société des Américanistes,* n.s. 52:91–99. Paris.

Bernal, Ignacio

1968 The Ball Players of Dainzú. *Archaeology* 21:246–251.

1973 Stone Reliefs in the Dainzu Area. In *The Iconography of Middle American Sculpture,* 13–23. New York: The Metropolitan Museum of Art.

Bernal-Garcia, Maria Elena

1988 La Venta's Pyramid: The First Successful Representation of the Mesoamerican Sacred Mountain. A paper prepared for a graduate seminar, Art Department, University of Texas.

Berry, Richard

1987 *Discover the Stars.* New York: Harmony Books.

Beyer, H.

1922 Relaciones entre la civilización teotihuacana y la azteca. In *La población del Valle de Teotihuácan,* edited by M. Gamio.

Blaffer, Sarah

1972 *The Black-man of Zinacantan: A Central American Legend.* Austin: University of Texas Press.

Blom, Frans, and Oliver La Farge

1926–1927 *Tribes and Temples.* Middle American Research Institute Pub.1, 2 vols. New Orleans: Tulane University.

Bowditch, Charles, trans.

1904 *Mexican and Central American Antiquities, Calendar Systems, and History.* Bureau of American Ethnology Bulletin 28. Washington, D. C.: Smithsonian Institution.

References

Brasseur de Bourbourg, Charles Etienne

1864 Relation des choses de Yucatan de Diego de Landa. Coll. Doc. Lang. Indig. de l'Am. Anc. 3. Paris.

Bricker, Victoria

1973 *Ritual Humor in Highland Chiapas.* Austin: University of Texas Press.

1981 *The Indian Christ, the Indian King: The Historical Substrate of Maya Myth and Ritual.* Austin: University of Texas Press.

1983 Directional Glyphs in Maya Inscriptions and Codices. *American Antiquity* 48: 347–353.

1988 A Phonetic Glyph for Zenith: Reply to Closs. *American Antiquity* 53:394–400.

1991 Faunal Offerings in the Dresden Codex. In *Sixth Palenque Round Table, 1986,* gen. ed. Merle Greene Robertson, vol. ed. Virginia M. Fields, 285–292. Norman: University of Oklahoma Press.

Broda, Johanna

1987 The Provenience of the Offerings: Tribute and Cosmovision. In *The Aztec Temple Mayor,* edited by Elizabeth H. Boone, 211–256. Washington, D.C.: Dumbarton Oaks Research Library and Collection.

Bruce, Robert D.

1975 *Lacandon Dream Symbolism. Vol. 1: Dream Symbolism and Interpretation.* México: Ediciones Euroamericanas Klaus Thiele.

1979 *Lacandon Dream Symbolism: Dream Symbolism and Interpretation Among the Lacandon Maya of Chiapas, México.* México: Ediciones Euroamericana, Klaus Thiele.

n.d. Death and Rebirth of the Gods. An unpublished manuscript given to Linda Schele in 1979.

Brunhouse, Robert L.

1975 *Pursuit of the Ancient Maya: Some Archaeologists of Yesterday.* Albuquerque: University of New Mexico Press.

Bunzel, Ruth

1952 *Chichicastenango: A Guatemalan Village.* Seattle: University of Washington Press.

Cabrera Castro, Rubén, Saburo Sugiyama, and George Cowgill

1988 Discoveries at the Feathered Serpent Pyramid. A paper presented at the 1988 Dumbarton Oaks Conference on "Art, Polity, and the City of Teotihuacán."

1991 The Templo de Quetzalcoatl Project at Teotihuacan. *Ancient Mesoamerica* 2: 77–92.

Carlsen, Robert, and Martin Prechtel

1991 The Flowering of the Dead: An Interpretation of Highland Maya Culture. *Man* n.s. 26:23–42.

Carlson, John

1988 The Divine Lord: The Maya God C as the Personification of *k'u,* Divinity, Spirit, and the Soul. A paper presented at the 46th International Congress of Americanists, Amsterdam. Copy provided by author.

Carlson, Ruth, and Francis Eachus

1977 The Kekchi Spirit World. In *Cognitive Studies of Southern Mesoamerica,* edited by Helen Neuenswander and Dean Arnold, 38–65. Dallas: SIL Musuem of Anthropology.

Carmack, Robert

1973 *Quichean Civilization: The Ethnohistoric, Ethnographic, and Archaeological Sources.* Los Angeles: University of California Press.

1981 *The Quiché of Utatlán.* Norman: University of Oklahoma Press.

References

Charnay, Desiré

1888 *Ancient Cities of the New World.* London: Chapman and Hall.

Chase, Arlen F.

1991 Cycles of Time: Caracol and the Maya Realm, with an Appendix on Caracol Altar 21 by Stephen D. Houston. In *Sixth Palenque Round Table, 1986,* gen. ed. Merle Greene Robertson, vol. ed. Virginia M. Fields, 32–42. Norman: University of Oklahoma Press.

Chase, Arlen F., and Diane Z. Chase

1987 *Investigations at the Classic Maya City of Caracol, Belize: 1985–1987.* Pre-Columbian Art Research Institute, Monograph 3. San Francisco: Pre-Columbian Art Research Institute.

Chase, Diane Z.

1985 Ganned but Not Forgotten: Late Postclassic Archaeology and Ritual at Santa Rita, Corozal, Belize. In *The Lowland Maya Postclassic,* edited by Arlen F. Chase and Prudence M. Rice, 104–125. Austin: University of Texas Press.

Chase, Diane Z., and Arlen F. Chase

1986 *Offerings to the Gods: Maya Archaeology at Santa Rita, Corozal.* Orlando: University of Central Florida.

Cheek, Charles

1983 Excavaciones en la Plaza Principal. In *Introducción a la arqueología de Copán, Honduras. Tomo II,* 191–290. Tegucigalpa: Instituto Hondureño de Antropología e Historia.

Chincilla Aguilar, Ernesto

1953 *La inquisición en Guatemala.* Guatemala City: Editorial del Ministerio de Educación Pública.

Clancy, Flora S., Clemency C. Coggins, T. Patrick Culbert, Charles Gallenkamp, Peter D. Harrison, and Jeremy A. Sabloff

1985 *Maya Treasures of an Ancient Civilization,* gen. eds. Charles Gallenkamp and Regina E. Johnson. New York: Harry N.Abrams, Inc., in association with the Albuquerque Museum.

Closs, Michael

1988 A Phonetic Version of the Maya Glyph for North. *American Antiquity* 53:386–393.

Clune, Francis J.

1963 A Functional and Historical Analysis of the Ball Game in Mesoamerica. Ph.D. dissertation, University of California.

Coe, Michael D.

1965a A Model of Ancient Community Structure in the Maya Lowlands. *Southwestern Journal of Anthropology* 21:97–114.

1965b *The Jaguar's Children: Pre-Classic Central Mexico.* New York: The Museum of Primitive Art.

1973 *The Maya Scribe and His World.* New York: The Grolier Club.

1978 *Lords of the Underworld: Masterpieces of Classic Maya Ceramics.* Princeton: The Art Museum, Princeton University.

1989 The Hero Twins: Myth and Image. *The Vase Book: A Corpus of Rollout Photographs of Maya Vases,* Vol. 1, by Justin Kerr, 161–184. New York: Kerr and Associates.

1992 *Breaking the Maya Code.* London: Thames and Hudson

Coe, Michael D., and Elizabeth P. Benson

1966 *Three Maya Relief Panels at Dumbarton Oaks.* Studies in Pre-Columbian Art and Archaeology, no. 2. Washington, D.C.: Dumbarton Oaks Research Collection and Library.

References

Coe, Michael D., and Richard A. Diehl

1980 *In the Land of the Olmec, Vol. 1, The Archaeology of San Lorenzo Tenochtitlán.* Austin: Univeristy of Texas Press.

Coe, William R.

1990 *Excavations in the Great Plaza, North Terrace and North Acropolis of Tikal.* Tikal Report No.14, 5 vols., University Museum Monograph 61. Philadelphia: The University Museum.

Coggins, Clemency C.

1976 *Painting and Drawing Styles at Tikal: An Historical and Iconographic Reconstruction.* Ann Arbor: University Microfilms.

1979a A New Order and the Role of the Calendar: Some Characteristics of the Middle Classic Period at Tikal. In *Maya Archaeology and Ethnohistory,* edited by Norman Hammond and Gordon R. Willey, 38–50. Austin: University of Texas Press.

1979b Teotihuacán at Tikal in the Early Classic Period. *Actes de XLII Congrès International des Américanistes* 8:251–269. Paris.

1983 An Instrument of Expansion: Monte Alban, Teotihuacán, and Tikal. In *Highland-Lowland Interaction in Mesoamerica: Interdisciplinary Approaches,* edited by Arthur G. Miller, 49–68. Washington, D.C.: Dumbarton Oaks Research Library and Collection.

1988 Reply to: A Phonetic Version of the Maya Glyph for North. *American Antiquity* 53:401–402.

Coggins, Clemency C., and Orin C. Shane III

1984 *Cenote of Sacrifice: Maya Treasures from the Sacred Well at Chichen Itza.* Austin: University of Texas Press.

Cohodas, Marvin

1975 The Symbolism and Ritual Function of the Middle Classic Ball Game Cult: The Late Middle Classic Period in Yucatan. In *Middle Classic Mesoamerica: A.D. 400–700,* edited by Esther Pasztory, 85–107. New York: Columbia University Press.

Collins, Wade

1986 Minimum Counts with an Eye Toward Reconstruction of the Corner Masks, Copan Temple 22. Field report for the 1986 season of the Copan Acropolis Archaeological Project.

Cook, Garrett

1986 Quichean Folk Theology and Southern Maya Supernaturalism. In *Symbol and Meaning Beyond the Closed Corporate Community: Essays in Mesoamerican Ideas,* edited by Gary H. Gossen, 139–153. Albany: Institute for Mesoamerican Studies, the University at Albany, State University of New York.

Cortéz, Constance

1986 The Principal Bird Deity in Late Preclassic and Early Classic Maya Art. M.A. thesis, University of Texas at Austin.

Coto, Fray Thomás de

1983 *Vocabulario de la lengua cakchquel v Guatemalteca, nueuamente hecho y recopilado con summo estudio, trauajo y erudición,* edited by René Acuña. México:Universidad Nacional Autónoma de México.

Cox, Rebecca

1991 The Symbolism of Lip to Lip Cache Vessels of the Early Classic Maya. M.A. thesis, University of Texas at Austin.

Culbert, T. Patrick, Laura J. Kosakowsky, Robert E. Fry, and William A. Haviland

1990 The Population of Tikal, Guatemala. In *Precolumbian Population History in the Maya*

Lowlands, edited by T. Patrick Culbert and Don S. Rice, 103–121. Albuquerque: University of New Mexico Press.

Davis, Virginia Dale

1978 Ritual of the Northern Lacandon Maya. Ph.D. dissertation, Tulane University. Ann Arbor: University Microfilms.

Demarest, Arthur A.

1978 Interregional Conflict and "Situational Ethics" in Classic Maya Warfare. In *Codex Wauchope: Festschrift in Honor of Robert Wauchope*, edited by Munro Edmonson et al., 101–111. New Orleans: Tulane University Press.

Demarest, Arthur A., and Stephen D. Houston, eds.

1990 *Proyecto Arqueológico Regional Petexbatun: Informe Preliminar #2, Segunda Temporada 1990.* Vanderbilt University.

Drucker, Philip, Robert F. Heizer, and Robert Squire

1959 *Excavations at La Venta, Tabasco.* Bureau of American Ethnology, Bulletin 170. Washington, D.C.: Smithsonian Institution.

Dunham, Peter

1980 The Maya and the Milky Way: A Descriptive Approach. An unpublished paper provided by the author.

Durán, Fray Diego

1971 *Book of the Gods and Rites of the Ancient Calendar,* trans. Fernando Horcasitas and Doris Heyden. Norman: University of Oklahoma Press.

Dütting, Dieter, and Anthony F. Aveni

1982 The 2 Cib 14 Mol Event in the Palenque Inscriptions. *Zeitschrift für Ethnologie* 107. Branschweig.

Edmonson, Munro

1971 *The Book of Counsel: The Popol Vuh of the Quiche Maya of Guatemala.* Middle American Research Institute, Pub. 35. New Orleans:Tulane University.

1986 *Heaven Born Mérida and Its Destiny: The Book of Chilam Balam of Chumayel.* Austin: University of Texas Press.

Estrada Monroy, Augustín

1979 *El mundo k'echi' de la Vera-Paz.* Guatemala: Editorial del Ejército.

Everton, Macduff

1991 *The Modern Maya: A Culture in Transition.* Albuquerque: University of New Mexico Press.

Farriss, Nancy M.

1984 *Maya Society Under Colonial Rule: The Collective Enterprise of Survival.* Princeton: Princeton University Press.

Fash, Barbara

1989 Late Classic Mosaic Sculpture Themes from the Copan Acropolis. A paper presented at the 88th annual meeting of the American Anthropological Association.

n.d. Temple 20 and the House of Bats. A paper presented at the Seventh Round Table of Palenque, in Palenque, Chiapas, México, June 1989.

Fash, Barbara, William Fash, Sheree Lane, Rudi Larios, Linda Schele, and David Stuart

n.d. Classic Maya Community Houses and Political Evolution: Investigations of Copán Structure 22A. A paper submitted to the *Journal of Field Archaeology*.

References

Fash, William L.

1991 *Scribes, Warriors, and Kings: The City of Copan and the Ancient Maya.* London: Thames and Hudson.

Fash, William L., and Barbara Fash

1990 Scribes, Warriors, and Kings. *Archaeology* 43:26–35.

Fash, William L., Richard V. Williamson, Carlos Rudi Larios, and Joel Palka

n.d. The Hieroglyphic Stairway and Its Ancestors: Investigations of Copan Structure 10L–16. *Ancient Mesoamerica.* (in press)

Fernandez, José

1992 A Stellar City: Utatlan and Orion. A paper presented at the International Symposium, "Time and Astronomy at the Meeting of Two Worlds." Warsaw, Poland.

Fields, Virginia M.

1989 The Origins of Divine Kingship Among the Lowland Classic Maya. Ph.D. dissertation, University of Texas at Austin.

1991 The Iconographic Heritage of the Maya Jester God. In *Sixth Palenque Round Table, 1986,* gen. ed. Merle Greene Robertson, vol. ed. Virginia M. Fields, 167–174. Norman: University of Oklahoma Press.

Foster, George

1944 Nagualism in México and Guatemala. *Acta America* 2:85–103.

Fox, James

1984 The Hieroglyphic Band in the Casa Colorada. A paper presented at the American Anthropological Association, November 17, Denver, Colorado.

Freidel, David A.

1976 The Ix Chel Shrine and Other Temples of Talking Idols. In *A Study of Changing Pre-Columbian Commercial Systems: The 1972–1973 Seasons at Cozumel, Mexico.* Monographs of the Peabody Museum, Number 3, edited by Jeremy A. Sabloff and William L. Rathje, 107–113. Harvard University.

1979 Cultural Areas and Interaction Spheres: Contrasting Approaches to the Emergence of Civilization in the Maya Lowlands. *American Antiquity* 44:6–54.

1981 Continuity and Disjunction: Late Postclassic Settlement Patterns in Northern Yucatán. In *Lowland Maya Settlement Patterns,* edited by Wendy Ashmore, 371–382. A School of American Research Book. Albuquerque: University of New Mexico Press.

1985 Polychrome Facades of the Lowland Maya Preclassic. In *Painted Architecture and Polychrome Monumental Sculpture in Mesoamerica,* edited by Elizabeth H. Boone, 5–30. Washington, D.C.: Dumbarton Oaks Research Library and Collection.

1986a Terminal Classic Lowland Maya: Successes, Failures, and Aftermaths. In *Late Lowland Maya Civilization: Classic to Postclassic,* edited by Jeremy A. Sabloff and E. Wyllys Andrews V, 409–430. A School of American Research Book. Albuquerque: University of New Mexico Press.

1986b Maya Warfare: An Example of Peer Polity Interaction. In *Peer Polity Interaction and the Development of Sociopolitical Complexity,* edited by Colin Renfrew and John F. Cherry, 93–108. Cambridge: Cambridge University Press.

1990 The Jester God: The Beginning and End of a Maya Royal Symbol. In *Vision and Revision in Maya Studies,* edited by Flora S. Clancy and Peter D. Harrison, 67–78. Albuquerque: University of New Mexico Press.

Freidel, David A., Maria Masucci, Susan Jaeger, and Robin A. Robertson

1991 The Bearer, the Burden and the Burnt: The Stacking Principle in the Iconography of the Late Preclassic Maya Lowlands. In *Sixth Palenque Round Table, 1986,* gen. ed. Merle Greene Robertson, vol. ed. Virginia M. Fields, 175–183. Norman: University of Oklahoma Press.

Freidel David A., and Jeremy A. Sabloff

1984 *Cozumel: Late Maya Settlement Patterns.* New York: Academic Press.

Freidel, David A., and Vernon L. Scarborough

1982 Subsistence, Trade and Development of the Coastal Maya. In *Maya Agriculture: Essays in Honor of Dennis E. Puleston,* edited by K. V. Flannery, 131–155. New York: Academic Press.

Freidel, David A., and Linda Schele

1988a Kingship in the Late Preclassic Lowlands: The Instruments and Places of Ritual Power. *American Anthropologist* 90(3):547–567.

1988b Symbol and Power: A History of the Lowland Maya Cosmogram. In *Maya Iconography,* edited by Elizabeth P. Benson and Gillett G. Griffin, 44–93. Princeton: Princeton University Press.

1989a Dead Kings and Living Temples: Dedication and Termination Rituals Among the Ancient Maya. In *Word and Image in Maya Culture: Explorations in Language, Writing, and Representation,* edited by William F. Hanks and Don S. Rice, 233–243. Salt Lake City: University of Utah Press.

1989b Tlaloc War and the Triumph of the Confederacy at Chichen Itza. Paper delivered at the 54th annual meeting of the Society for American Archaeology, April 6, New Orleans.

Gage, Thomas

1958 *Thomas Gage's Travels in the New World.* Edited and with an introduction by J.E.S. Thompson. Norman: University of Oklahoma Press.

Gann, Thomas

1918 *The Maya Indians of Southern Yucatan and Northern Britsh Honduras.* Bureau of American Ethnology Bulletin 64. Washington, D.C.: Smithsonian Institution.

1926 *Ancient Cities and Modern Tribes: Exploration and Adventure in Maya Lands.* London: Duckworth.

Gillespie, Susan D.

1991 Ballgames and Boundaries. In *The Mesoamerican Ballgame,* edited by Vernon L. Scarborough and David R. Wilcox, 317–345. Tucson: University of Arizona Press.

Girard, Rafael

1949 *Los Chortis ante el problema maya: Historia de las culturas indígenas de América, desde su origen hasta hoy.* 5 vols. Mexico, Antigua Libreriá Robredo.

1966 *Los Mayas: Su civilización, su historia, sus vinculaciones continentales.* Mexico: Libro Mexicano.

Gordon, George Byron

1896 *Prehistoric Ruins of Copan. 1896. Preliminary Report on Excavations, 1891–1895,* Memoirs of the Peabody Museum of Archaeology and Enthnology, Vol. I (5). Cambridge: Harvard University.

Gosner, Kevin

1992 *Soldiers of the Virgin: The Moral Economy of a Colonial Maya Rebellion.* Tuscon: University of Arizona Press.

Gossen, Gary H.

1974 *Chamulas in the World of the Sun: Time and Space in a Maya Oral Tradition.* Cambridge: Harvard University Press. New edition 1984, Waveland Press, Prospect Heights, Illinois.

1986 The Chamula Festival of Games: Native Macroanalysis and Social Commentary in a Maya Carnival. In *Symbol and Meaning Beyond the Closed Corporate Community: Essays in Mesoamerican Ideas,* edited by Gary H. Gossen, 227–267. Albany: Institute for Mesoamerican Studies, the University at Albany, State University of New York.

Grove, David C., ed.

1987 *Ancient Chalcatzingo.* Austin: University of Texas Press.

Grube, Nikolai

1989 Commentary on the *way* glyph. A letter dated November 17, with notes written on September 10, sent to the authors and other epigraphers.

1990a Die Errichtun von Stelen: Entzifferung einer Vebhieroglyphie auf Monumenten der klassichen Mayakultur. In *Circumpacifica: Festschrit für Thomas S. Barthel,* edited by Bruno Illius and Matthias Laubscher, 189–215. Frankfurt am Main: Peter Lang.

1990b T516 as a Glyph for *Ak'ot* "Dance." Preliminary comments circulated in a letter dated May 1990.

1991a An Investigation of the Primary Standard Sequence on Classic Maya Ceramics. In *Sixth Palenque Round Table, 1986,* gen. ed. Merle Greene Robertson, vol. ed. Virginia M. Fields, 223–232. Norman: University of Oklahoma Press.

1991b Hieroglyphic Sources for the History of Northwest Yucatan. Revised version of a paper presented at The First Maler Symposium on the Archaeology of Northwest Yucatan, Bonn University, Bonn, West Germany, August 20–25, 1990.

1992 Classic Maya Dance. *Ancient Mesoamerica* 3:201-218.

n.d. The God C-Mirror Sign in Monument Names at Copan. A public lecture presented at the 1989 Texas Meetings on Maya Hieroglyphic Writing.

Grube, Nikolai, and Werner Nahm

1990 A Sign for the Syllable *mi. Research Reports on Ancient Maya Writing 33.* Washington, D.C.: Center for Maya Research.

1991 Signs with the phonetic value *sa.* An unpublished note circulated from Bonn, Germany, in January 1991.

Grube, Nikolai, and Linda Schele

1988 A Venus Title on Copán Stela F. *Copán Note* 41. Copán: Copán Acropolis Archaeological Project and the Instituto Hondureño de Antropología e Historia.

1990a Six-Staired Ballcourts. *Copán Note* 83. Honduras: Copán Mosaics Project and the Instituto Hondureño de Antropología.

1990b Royal Gifts to Subordinate Lords. *Copán Note* 87. Honduras: Copán Acropolis Archaeological Project and the Instituto Hondureño de Antropología.

1990c Two Examples of the Glyph for "Step" from the Hieroglyphic Stairs. *Copán Note* 91. Honduras: Copán Mosaics Project and the Instituto Hondureño de Antropología.

1991 *Tzuk* in the Classic Maya Inscriptions. *Texas Notes on Precolumbian Art, Writing, and Culture* No. 15. Austin: CHAAAC, the Center for the History and Art of Ancient American Culture, Art Department, University of Texas.

Guillemin, Jorge F.

1968 Un "yugo" de madera para el juego de pelota. *Antropología e Historia de Guatemala* 20:25–33.

1977 Urbanism and Hierarchy at Iximche. In *Social Process in Maya Prehistory: Studies in Honour of Sir Eric Thompson,* edited by Norman Hammond, 227–264. London: Academic Press.

References

Hammond, Norman

1981 A Reexamination of Piedras Negras Stela 40. *México* 3:77–79.

Hanks, William F.

1990 *Referential Practice, Language and Lived Space Among the Maya.* Chicago: The University of Chicago Press.

Hansen, Richard D.

1989 Las investigaciones del sitio Nakbe, Petén, Guatemala: Temporada 1989. A paper delivered at the Tercer Simposio del Arqueología Guatemalteca, Guatemala City, July.

1992 *The Archaeology of Ideology: A Study of Maya Preclassic Architectural Sculpture at Nakbe, Guatemala.* Ph.D. dissertation, University of California, Los Angeles.

Harrison, Peter D.

1970 The Central Acropolis, Tikal, Guatemala: A Preliminary Study of the Functions of Its Structural Components During the Late Classic Period. Ph.D. dissertation, University of Pennsylvania.

1989 Architecture and Geometry in the Central Acropolis at Tikal. A paper presented at the Seventh Round Table of Palenque, held in Palenque, Chiapas, México, June.

Hassig, Ross

1988 *Aztec Warfare, Imperial Expansion and Political Control.* Norman: University of Oklahoma Press.

Hatch, Marion P.

1971 An Hypothesis on Olmec Astronomy, with Special Reference to the La Venta Site. In *Papers on Olmec and Maya Archaeology: Contributions of the University of California Archaeological Research Facility* No. 13: 1–64. Berkeley: University of California.

Heizer, Robert F.

1968 New Observations on La Venta. In *Dumbarton Oaks Conference on the Olmec, October 28–29, 1967,* edited by Elizabeth P. Benson, 9–40. Washington: Dumbarton Oaks Research Library and Collection.

Hellmuth, Nicholas M.

1987 *Monster und Menschen in der Maya-Kunst.* Graz: Akademische Druck u. Verlagsanstalt.

1988 Early Maya Iconography on an Incised Cylindrical Tripod. In *Maya Iconography,* edited by Elizabeth P. Benson and Gillett G. Griffin, 152–174. Princeton: Princeton University Press.

Hester, Thomas

1983 *A Preliminary Report on the 1983 Investigations of Colha.* San Antonio: Center for Archaeological Research, University of Texas at San Antonio.

Houston, Stephen D.

1983 A Reading for the Flint-shield Glyph. In *Contribution to Maya Hieroglyphic Decipherment I,* edited by Stephen Houston, 13–25. New Haven: Human Relations Area Files.

1985 A Feather Dance at Bonampak, Chiapas, Mexico. *Journal du Société des Américanistes* n.s. 70:127–138. Paris.

1991 Political Expansion and Collapse in the Petexbatan Region, Guatemala: Epigraphy and Monuments. A paper presented at the 47th International Congress of Americanists, New Orleans.

1993 *Hieroglyphics and History at Dos Pilas: Dynast Politics of the Classic Maya.* Austin: University of Texas Press.

References

Houston, Stephen D., and David Stuart

1989a Folk Classification of Classic Maya Pottery. *American Anthropologist* 91:720–726.

1989b The *Way* Glyph: Evidence for "Co-essences" Among the Classic Maya. *Research Reports on Ancient Maya Writing* 30. Washington, D.C.: Center for Maya Research.

1990 T632 as *Muyal,* "Cloud." *Central Tennessean Notes in Maya Epigraphy,* No. 1. A note circulated by the authors.

Houston, Stephen D., and Karl Taube

n.d. "Name-tagging" in Classic Mayan Script: Examples on Objects of Stone and Ceramic. A unpublished paper provided by the authors.

Hunt, Eva

1977 *Transformation of the Hummingbird: Cultural Roots of a Zinacantan Mythical Poem.* Ithaca: Cornell University Press.

Johnson, Richard, and Michel Quenon

1992 Comments on the Paris Codex Pages 23 and 24: A Maya Zodiac. A report circulated by the authors.

Jones, Christopher

1977 Inauguration Dates of Three Late Classic Rulers of Tikal, Guatemala. *American Antiquity* 42:28–60.

Jones, Grant D.

1989 *Maya Resistance to Spanish Rule: Time and History on a Colonial Frontier.* Albuquerque: University of New Mexico Press.

Jones, Tom

1989 A Review of the Evidence for the *Xaman* Reading. *U Mut Maya II,* edited by Tom and Carolyn Jones.

Joralemon, David

1974 Ritual Blood-Sacrifice Among the Ancient Maya: Part I. *Primera Mesa Redonda de Palenque, Part II,* edited by Merle Greene Robertson, 59–76. Pebble Beach, Calif.: Robert Louis Stevenson School.

1976 The Olmec Dragon: A Study in Pre-Columbian Iconography. In *Origins of Religious Art and Iconography in Preclassic Mesoamerica,* edited by Henry B. Nicholson, 27–72. Los Angeles: UCLA Latin American Center Publications and Ethnic Arts Council of Los Angeles.

Josserand, Kathryn, Linda Schele, and Nicholas A. Hopkins

1985 Auxiliary Verb + *ti* Constructions in the Classic Maya Inscriptions. *Fourth Palenque Round Table, 1980, Vol. VI,* edited by Elizabeth P. Benson, 87–102. San Francisco: Center for Pre-Columbian Art Research.

Justeson, John

n.d. An interpretation of the Classic Maya Hieroglyphic Spelling of "East." Manuscript provided by author.

Kaufman, Terrence S., and William M. Norman

1984 An Outline of Proto-Cholan Phonology, Morphology, and Vocabulary. In *Phoneticism in Mayan Hieroglyphic Writing,* edited by Lyle Campbell and John S. Justeson, 77–167. Albany: Center for Mesoamerican Studies, State University of New York at Albany.

Kelley, David

1965 The Birth of the Gods at Palenque. In *Estudios de Cultura Maya* 5:93–134. México: Universidad Nacional Autónoma de México.

1976 *Deciphering the Maya Script.* Austin: University of Texas Press.

Knauth, Lothar

1961 El juego de pelota y el rito de la decapitación. *Estudios de Cultura Maya* 1:183–198. México: Universidad Nacional Autónoma de México.

Kowalski, Jeff K., and William L. Fash

1991 Symbolism of the Maya Ball Game at Copán: Synthesis and New Aspects. In *Sixth Palenque Round Table, 1986,* gen. ed. Merle Greene Robertson, vol. ed. Virginia M. Fields, 59–67. Norman: University of Oklahoma Press.

Krickenberg, Walter

1948 Das mittelamerikanische Ballspiel und seine religiöse Symbolik. *Paudeuma* 3:118–190. Hamburg.

Krochock, Ruth

1988 The Hieroglyphic Inscriptions and Iconography of Temple of the Four Lintels and Related Monuments, Chichén Itzá, Yukatan, México. M.A. thesis, University of Texas at Austin.

1991 Dedication Ceremonies at Chichén Itzá: The Glyphic Evidence. In *Sixth Palenque Round Table, 1986,* gen. ed. Merle Greene Robertson, vol. ed. Virginia M. Fields, 43–50. Norman: University of Oklahoma Press.

Kubler, George

1969 *Studies in Classic Maya Iconography.* Memoirs of the Connecticut Academy of Arts and Sciences XVIII. Hamden: Archon Books.

1976 The Double-Portrait Lintels at Tikal. *Actas del XXIII Congreso Internacional de Historia del Arte España Entre el Mediterráneo y el Atlántico.* Granada.

1977 Aspects of Classic Maya Rulership on Two Inscribed Vessels. *Studies in Pre-Columbian Art and Archaeology 18.* Washington, D.C.: Dumbarton Oaks Research Library and Collection.

Kurjack, Edward B., Ruben Maldonado C., and Merle Greene Robertson

1991 Ballcourts of the Northern Maya Lowlands. In *The Mesoamerican Ballgame,* edited by Vernon L. Scarborough and David R. Wilcox, 145–159. Tucson: University of Arizona Press.

La Farge, Oliver

1947 *Santa Eulalia: The Religion of a Cuchumatán Indian Town.* Chicago: University of Chicago Press.

La Farge, Oliver, and Douglas Byers

1931 *The Year Bearer's People.* Middle American Research Institute Pub. 3. New Orleans: Tulane University.

Lamb, Weldon

1986 Tzotzil Maya Starlore. A paper read at the 1983 Conference on Ethnoastronomy, Washington, D.C. Copy provided by the author.

Langley, James C.

1986 *Symbolic Notation of Teotihuacan: Elements of Writing in a Mesoamerican Culture of the Classic Period.* BAR International Series 313. Oxford, England.

Laporte Molina, Juan Pedro

1988 Alternativas del Clásico Temprano en la relación Tikal-Teotihuacán: Grupo 6C–XVI, Tikal, Petén, Guatemala. A dissertation for a Doctoral en Antropología, Universidad Nacional Autónoma de México.

Laporte Molina, Juan Pedro, and Lillian Vega de Zea

1988 Aspectos dinásticos para el Clásico Temprano de Mundo Perdido, Tikal. In *Primer Simposio Mundial Sobre Epigrafía Maya*, 127–141. Guatemala: Asociación Tikal.

Larios, Rudi

1990 The Archaeology of Burial 195 at Tikal. A presentation at the 1990 Texas Meetings on Maya Hieroglyphic Writing, featuring the history of Tikal.

1990 The Archaeology of Burial 195 at Tikal. A lecture delivered at the Symposium on the History and Archaeology of Tikal at the 1990 Maya Meetings at the University of Texas.

Larios, Rudi, William Fash, and David Stuart

n.d. Architectural History and Political Symbolism of Temple 22, Copán. A paper presented at the Seventh Round Table of Palenque, in Palenque, Chiapas, México, June 1989.

Laughlin, Robert

1962 El símbolo de la flor en la Región de Zinacantan. *Estudios de Cultura Maya* 2:123–139. México: Universidad Nacional Autónoma de México.

1975 *The Great Tzotzil Dictionary of San Lorenzo Zinacantán.* Smithsonian Contributions to Anthropology 19. Washington, D.C.: Smithsonian Institution Press.

1988 *The Great Tzotzil Dictionary of Santo Domingo Zinacantán.* Smithsonian Contributions to Anthropology 31. Washington, D.C.: Smithsonian Institution Press.

Leach, Edmund

1961 Two Essays Concerning the Symbolic Representation of Time. In *Rethinking Anthropology*, 124–136. London: The Athlone Press.

Lewis-Williams, J.D.

1986 Beyond Style and Portrait: A Comparison of Tanzanian and Southern African Rock Art. In *Contemporary Studies of the Khoisan*, edited by R. Vossen and K. Keuthmann: 93-139. Hamburg: Helmut Buske Verlog.

Leyenaar, Ted J. J.

1978 *Ulama: The Perpetuation in México of the Pre-Spanish Ball Game Ullamaliztli.* Leiden, The Netherlands: Rijkmuseum voor Volkenkunde.

Leyenaar, Ted J. J., and Lee Parsons

1988 *Ulama: The Ballgame of the Mayas and Aztecs.* Leiden: Spruyt, Van Mantgem & De Does bv

Linn, Priscilla Rachun

1976 The Religious Office Holders of Chamula: A Study of Gods, Rituals, and Sacrifice. Ph.D. dissertation, Oxford University.

1982 Chamula Carnival: The "Soul" of Celebration. In *Celebration: Studies in Festivity and Ritual*, edited by Victor Turner, 190–198. Washington, D.C.: Smithsonian Institution Press.

Longyear, John

1952 *Copán Ceramics: A Study of Southeastern Maya Pottery.* Carnegie Institution of Washington Pub. 597. Washington, D.C.

Looper, Mathew

1991a The Name of Copan and of a Dance at Yaxchilan. *Copán Note* 95. Honduras: Copán Acropolis Archaeological Project and the Instituto Hondureño de Antropología.

1991b The Dances of the Classic Maya Deities *Chak* and *Hun Nal Ye.* A master's thesis, Art Department, University of Texas at Austin.

1991c The Peccaries Above and Below Us. *Texas Notes on Precolumbian Art, Writing, and Culture* No. 10. Austin: CHAAAC, the Center for the History and Art of Ancient American Culture, Art Department, University of Texas.

Looper, Mathew, and Linda Schele

1991 A Little Conflict Between Palenque and Pomona. *Texas Notes on Precolumbian Art, Writing, and Culture* No. 11. Austin: CHAAAC, the Center for the History and Art of Ancient American Culture, Art Department, University of Texas.

López Austin, Alfredo, Leonardo López Luján, and Suburo Sugiyama

1991 The Temple of Quetzalcoatl at Teotihuacan: Its Possible Ideological Significance. *Ancient Mesoamerica* 2:93–105.

López de Gómora, Francisco

1964 *Cortés: The Life of the Conqueror by His Secretary,* translated and edited by Lesley Byrd Simpson from the *Istoria de la Conquista de México,* printed in Zaragoza, 1552. Berkeley: University of California Press.

Lounsbury, Floyd

1976 A Rationale for the Initial Date of the Temple of the Cross at Palenque. In *The Art, Iconography, and Dynastic History of Palenque, Part III: Proceedings of the Segunda Mesa Redonda de Palenque,* edited by Merle Greene Robertson, 211–224. Pebble Beach, Calif.: Robert Louis Stevenson School.

1991 A Palenque King and the Planet Jupiter. In *World Archaeoastronomy,* edited by Anthony Aveni, 246–259. Cambridge, England: Cambridge University Press.

Love, Bruce

1987 Glyph T93 and Maya "Hand-Scattering" Events. *Research Reports on Maya Writing* 4 & 5. Washington D.C.: Center for Maya Research.

1991 A Text from the Dresden Pages. In *Sixth Palenque Round Table, 1986,* gen. ed. Merle Greene Robertson, vol. ed. Virginia M. Fields, 293–302. Norman: University of Oklahoma Press.

McGee, R. Jon

1990 *Life, Ritual, and Religion Among the Lacandon Maya.* Belmont, Calif.: Wadsworth Publishing Co.

MacLeod, Barbara

1990a Deciphering the Primary Standard Sequence. A Ph.D. dissertation, University of Texas at Austin.

1990b The God N'Step Set in the Primary Standard Sequence. In *The Maya Vase Book,* Vol. 2: 331–347. New York: Kerr Associates.

1991a Retooling the Lancet: A New Interpretation of the Yaxchilan Bloodletting Phrases. A paper presented at *Visions of Yaxchilan,* the VIIth Symposium of the Texas Meetings on Maya Hieroglyphic Writing.

1991b Some Thoughts on a Possible Hal Reading of T153, the "Crossed Batons" Glyph. *North Austin Hieroglyphic Hunches* 2. A note circulated by the author on February 5, 1991.

1991c The Classic Name for *Cumku* and Other Tales of *Ol. North Austin Hieroglyphic Hunches* 8. A note circulated by the author.

1991d Maya Genesis: The First Steps, 5. *North Austin Hieroglyphic Hunches* 5. A note circulated by the author.

1992 Maker, Modeler, Bearer, Begetter: The Paddlers as *Chan Itz'at.* In *Workbook for the XVIth Maya Hieroglyphic Workshop at Texas, with commentaries on the Group of the Cross at Palenque*: 257–259. Austin: Art Department, University of Texas.

MacLeod, Barbara, and Dennis Puleston

1979 Pathways into Darkness: The Search for the Road to Xibalba. *Tercera Mesa Redonda de Palenque, Vol. IV,* edited by Merle Green Robertson, 71–79. Palenque: Pre-Columbian Art Research, and Monterey: Herald Printers.

Maler, Teobert

1901–1903 *Researches in the Central Portion of the Usumasintla Valley.* Memoirs of the Peabody Museum of American Archaeology and Ethnology, Harvard University, II. Cambridge.

1908–1910 *Explorations of the Upper Usumasintla and Adjacent Region.* Memoirs of the Peabody Museum of American Archaeology and Ethnology, Harvard University, IV. Cambridge.

Marquina, Ignacio

1964 *Arquitectura prehispánica.* México: Instituto Nacional Autónoma de México. Reprint of Instituto Nacional de Antropología e Historia, *Memorias,* I, 1951.

Martínez Don Juan, Guadalupe

1984 *Teopantecuanitlan, Guerrero: Un sitio olmeca.* México: Revista Mexicana.

Martínez Hernandez, Juan

1929 Diccionario de Motul, Maya-Español atruido a Fray Antonio de Ciudad Real y Arte de Lengua Maya por Fray Juan Coronel. Mérida: Talleres de la Copañia Yucateca.

Martin, Simon

n.d. Some Thoughts and Work-in-Progress, Summer 1991. A manuscript circulated by the author.

Mathews, Peter

1979 Notes on the Inscriptions of "Site Q." Unpublished manuscript in the possession of the authors.

n.d. Tonina Dates: 1. A Glyph for the Period of 260 Days? Maya Glyph Notes, No. 8. MS provided by author.

Matos Moctezuma, Eduardo

1987 Symbolism of the Templo Mayor. *The Aztec Temple Mayor,* edited by Elizabeth H. Boone, 185–210. Washington, D.C.: Dumbarton Oaks Research Collection and Library.

Maudslay, Alfred P.

1889–1902 *Archaeology: Biología Centrali-Americana.* Vol. I–IV. London: Dulau and Co. Reprint edition, 1974, Milparton Publishing Corp.

Mayer, Karl Herbert

1991 *Maya Monuments: Sculptures of Unknown Provenance, Supplement 3.* Austria: Verlag von Flemming.

Means, Philip Ainsworth

1917 *History of the Spanish Conquest of Yucatán and of the Itzás.* Papers of Peabody Museum of American Archaeology and Ethnology, Harvard University, Vol. 7. Cambridge: Harvard University.

Mendelson, E. Michael

1956 Religion and World View in Santiago Atitlan. Ph.D dissertation, University of Chicago. Microfilm Collection on American Indian Cultural Anthropology, Series 8, Nos. 52–54. Chicago: University of Chicago Library.

Merwin, Raymond E., and George C. Vaillant

1932 *The Ruins of Holmul, Guatemala.* Memoirs of the Peabody Museum of American Archaeology and Ethnology, Harvard University, Vol. 3, No. 2. Cambridge.

Milbrath, Susan

1988 Astronomical Images and Orientations in the Architecture of Chichen Itza. In *New Directions in American Archaeoastronomy,* edited by Anthony Aveni. BAR International Series 454. Oxford, England.

Miles, Susan

1957 The Sixteenth Century Pokom-Maya: A Documentary Analysis of Social Structure and Archaeological Setting. *Transactions of the American Philosophical Society* 47: 733–781. Philadelphia.

Miller, Arthur G.

1973 *The Mural Painting of Teotihuacan.* Washington, D.C.: Dumbarton Oaks Research Collection and Library.

1974 The Iconography of the Painting in the Temple of the Diving God, Tulum, Quintana Roo, México: The Twisted Cords. In *Mesoamerican Archaeology: New Approaches,* edited by Norman Hammond, 167–186. London: Duckworth.

Miller, Jeffrey

1974 Notes on a Stelae Pair Probably from Calakmul, Campeche, México. In *Primera Mesa Redonda de Palenque, Part I,* edited by Merle Greene Robertson, 149–162. Pebble Beach, Calif.: Robert Louis Stevenson School.

Miller, Mary E.

1986a *The Murals of Bonampak.* Princeton: Princeton University Press.

1986b Copán: Conference with a Perished City. In *City-States of the Maya: Art and Architecture,* edited by E. Benson, 72–109. Denver: Rocky Mountain Institute for Pre-Columbian Studies.

1988a The Boys in the Bonampak Band. In *Maya Iconography,* edited by Elizabeth P. Benson and Gillett G. Griffin, 318–330. Princeton: Princeton University Press.

1988b The Meaning and Function of the Main Acropolis, Copán. In *The Southeast Classic Maya Zone,* edited by Elizabeth H. Boone and Gordon R. Willey, 149–195. Washington, D.C.: Dumbarton Oaks Research Library and Collection.

Miller, Mary E., and Stephen D. Houston

1987 The Classic Maya Ballgame and Its Architectural Setting: A Study in Relations Between Text and Image. *RES* 14, 47–66.

Miller, Virginia E.

1989 Star Warriors at Chichen Itza. In *Word and Image in Maya Culture: Explorations in Language, Writing, and Representation,* edited by William F. Hanks and Don S. Rice, 287–305. Salt Lake City: University of Utah Press.

Millon, Clara

1988 A Reexamination of the Teotihuacan Tassel Headdress Insignia. In *Feathered Serpents and Flowering Trees: Reconstructing the Murals of Teotihuacan,* edited by Kathleen Berrin, 114–134. San Francisco: The Fine Arts Museums of San Francisco.

Moore, Dan

1982 *Xaman* as "North." A note reprinted in 1989 *U Mut Maya II*, edited by Tom and Carolyn Jones.

Morales, Alfonso, Julie Miller, and Linda Schele

1990 The Dedication Stair of "Ante" Structure. *Copán Note* 76. Honduras: Copán Mosaics Project and the Instituto Hondureño de Antropología.

Moran, Fray Pedro

1720 *Arte breve y compendioso de la Lengua Pocomchi de la provincia de la Verapaz, compuesto y ordenado por el Venerable Padre Fray Dionysio de Çuñiga para los principiantes que comiençan a apprender, y traducido en la Lengua Pocoman de Amatitlán.* Photocopy in the Gates Collection. Modern transcription and analysis by Lyle Campbell; manuscript provided by Peter Mathews.

Morley, Sylvanus Griswold

1920 *The Inscriptions at Copán.* The Carnegie Institution of Washington Pub. 219. Washington, D.C.

1938 *The Inscriptions of the Peten.* Carnegie Institution of Washington Pub. 437. Washington, D.C.

1946 *The Ancient Maya.* Stanford: Stanford University Press.

Morrison, Frank, C. W. Clewlow, Jr., and Robert Heizer

1970 Magnetometer Survey of La Venta Pyramid. *Contributions of the University of California Archaeological Research Facility 8,* 1–20. Berkeley: University of California.

Newsome, Elizabeth

1987 Blood Sacrifice, Venus, and the World Tree: The Iconography of 18-Rabbit's Stelae at Copan, Honduras. An unpublished paper written for a graduate seminar in art history, University of Texas at Austin.

1991 The Trees of Paradise and Pillars of the World: Vision Quest and Creation in the Stelae Cycle of 18-Rabbit-God K, Copán, Honduras. Ph.D. dissertation, University of Texas at Austin.

Norman, V. Garth

1973 Izapa Sculpture. Part 1: Album. *Papers of the New World Archaeological Foundation, No. 30.* Provo: New World Archaeological Foundation, Brigham Young University.

Nuñez de la Vega, Fray Francisco

1988 *Constituciones diocesanas de obispado de Chiapa.* Fuentes para el estudio de la cultura Maya, 6. México: Universidad Nacional Autónoma de México.

Oakes, Maud

1951 *Two Crosses of Todos Santos: Survivals of Mayan Religious Ritual.* Bollingen Series 27, Princeton: Princeton University Press.

Oakland, Amy

1982 Teotihuacán: The Blood Complex at Atetelco. A paper prepared for a seminar on the transition from Preclassic to Classic times, held at the University of Texas, 1982. Copy in possession of author.

Orejel, Jorge L.

1990 The "Axe-Comb" Glyph (T333) as *ch'ak. Research Reports on Ancient Maya Writing* 32. Washington, D.C.: Center for Maya Research.

Pagden, Anthony R.

1971 *Hernán Cortés: Letters from Mexico.* New York: Grossman Publishers.

Palka, Joel

1990 Operación DP15: Excavaciones en el Grupo Residencial K4-1. In *Proyecto Arqueológico Regional Petexbatun, Informe Preliminar #2, Segunda Temporada 1990,* edited by Arthur A. Demarest and Stephen D. Houston, 146–165. Nashville: Vanderbilt University.

Parsons, Mark

1985 Three Thematic Complexes in the Art of Teotihuacán. A paper prepared at the University of Texas. Copy in possession of author.

1986 Blood and Death at Atetelco, Teotihuacan. A unpublished paper prepared for a graduate seminar at the University of Texas.

1988 The Iconography of Blood and Sacrifice in the Murals of the White Patio, Atetelco, Teotihuacan. MA thesis, University of Texas at Austin.

Pasztory, Esther

1972 The Historical and Religious Significance of the Middle Classic Ball Game. In *Religión en Mesoamerica, XII Mesa Redonda,* 441–455. México: Sociedad Mexicana de Antropología.

1974 *The Iconography of the Teotihuacán Tlaloc.* Studies in Pre-Columbian Art and Archaeology 15. Washington, D.C.: Dumbarton Oaks Research Library and Collection.

Pendergast, David M.

1990 Up from the Dust: The Central Lowlands Postclassic as Seen from Lamanai and Marco González. In *Vision and Revision in Maya Studies,* edited by Flora S. Clancy and Peter D. Harrison, 169–179. Albuquerque: University of New Mexico Press.

Perera, Victor, and Robert D. Bruce

1982 *The Last Lords of Palenque: The Lacandon Mayas of the Mexican Rain Forest.* Berkeley: University of California Press.

Pickands, Martin

1986 The Hero Myth of Maya Folklore. In *Symbol and Meaning Beyond the Closed Corporate Community: Essays in Mesoamerican Ideas,* edited by Gary H. Gossen, 101–124. Albany: Institute for Mesoamerican Studies, the University at Albany, State University of New York.

Pérez, Juan Pió

1843 Cronologiá antigua de Yucatán y exámen del método con que los indios contaban el tiempo, sacados de varios documentos antiguos. In *Incidents of Travel in Yucatan,* by John L. Stephens, 2 vols., 434–59. New York: Harper.

Potter, Daniel R.

1982 Some Results of the Second Year of Excavation at Operation 2012. In *Archaeology at Colha, Belize: The 1981 Interim Report,* edited by Thomas R. Hester, Harry J. Shafer, and Jack Eaton, 98–122. San Antonio: Center for Archaeological Research, the University of Texas.

Proskouriakoff, Tatiana

1960 Historical implications of a pattern of dates at Piedras Negras, Guatemala. *American Antiquity* 25:454–75.

1961 Lords of the Maya realm. *Expedition* 4(1):14–21.

1963–1964 Historical Data in the Inscriptions of Yaxchilán, Parts I and II. *Estudios de Cultura Maya* 3:149–167 and 4:177–201. México: Universidad Nacional Autónoma de México.

Recinos, Adrián

1950 *Popol Vuh: The Sacred Book of the Ancient Quiche Maya.* Translated by Delia Goetz and S. G. Morley. Norman: University of Oklahoma Press.

1952 *Pedro de Alvarado, conquistador de México y Guatemala.* Mexico City: Fondo de Cultura Económica.

1957a *Crónicas indígenas de Guatemala.* Guatemala: Editorial Universitaria.

1957b Títulos de la casa Ixquin-Nehaib, señora del territorio de Otzoya. In *Crónicas indígenas de Guatemala,* edited by Adrián Recinos, 71–94. Guatemala: Editorial Universitaria.

1957c Historia de los Xpantzay de Tekpan Guatemala. In *Crónicas indígenas de Guatemala,* edited by Adrián Recinos, 121–132. Guatemala:Editorial Universitaria.

Recinos, Adrián, and Delia Goetz

1953 *The Annals of the Cakchiquels.* Norman: University of Oklahoma Press.

Redfield, Robert

1936 The Coati and the Ceiba. *Maya Research,* Vol. 3: 231–243.

Redfield, Robert, and Alfonso Villa Rojas

1934 *Chan Kom: A Maya Village.* Carnegie Institution of Washington Pub. 448. Washington, D.C.

Reilly, Kent

1987 The Ecological Origins of Olmec Symbols of Rulership. M.A. thesis, University of Texas.

1988 Olmec Conceptions of the Sacred Mountain as Underworld Entrance. An unpublished manuscript, Institute of Latin American Studies, Universtiy of Texas.

1989 Enclosed Ritual Space and the Watery Underworld in Formative Period Architecture: New Observations on the Function of La Venta Complex A. A paper presented at the Séptima Mesa Redonda, Palenque, Chiapas, June 1989.

1991a Olmec Iconographic Influences on the Symbols of Maya Rulership: An Examination of Possible Sources. In *Sixth Palenque Round Table, 1986,* gen. ed. Merle Greene Robertson, vol. ed. Virginia M. Fields, 151–174. Norman: University of Oklahoma Press.

1991b Cosmos and Rulership: The Function of Olmec-Style Symbols in Formative Period Mesoamerica. An informal paper prepared for the *First Sibley Family Symposium on World Traditions of Culture and Art: The Symbolism of Kingship: Comparative Strategies Around the World,* April 18–21, University of Texas, Austin.

n.d. *La Venta and the Olmec: The Function of Sacred Geography in the Formative Period Ceremonial Complex.* A Ph.D dissertation, Institute of Latin American Studies, University of Texas at Austin.

Remington, Judith

1977 Current Astronomical Practices Among the Maya. In *Native American Astronomy,* edited by Anthony Aveni, 75–88. Austin: University of Texas Press.

Ricketson, Oliver G., and Edith B. Ricketson

1937 *Uaxactún, Guatemala: Group E 1926–1931.* Carnegie Institution of Washington Pub. 477. Washington, D.C.

Riese, Berthold

1984 Hel Hieroglyphs. In *Phoneticism in Mayan Hieroglyphic Writing,* edited by John Justeson and Lyle Campbell, 263–286. Albany: Institute for Mesoamerican Studies, State University of New York.

Ringle, William

1988 Of Mice and Monkeys: The Value and Meaning of T1016, the God C Hieroglyph. *Research Reports on Ancient Maya Writing* 18. Washington, D.C.: Center for Maya Research.

Robertson, Merle Greene

1972 The Ritual Bundles of Yaxchilán. A paper presented at the symposium on "The Art of Latin America," Tulane University, New Orleans. Copy in possession of author.

1974 The Quadripartite Badge—A Badge of Rulership. *Primera Mesa Redonda de Palenque, Part 1,* edited by Merle Robertson, 129–137. Pebble Beach: Robert Louis Stevenson School.

1978 An Iconographic Approach to the Identity of the Figures on the Piers of the Temple of Inscriptions, Palenque. In *Tercera Mesa Redonda de Palenque, Vol. IV,* edited by Merle Greene Robertson, 129–138. Palenque: Pre-Columbian Art Research; Monterey: Herald Printers.

1985a *The Sculpture of Palenque. Vol. II: The Early Buildings of the Palace and the Wall Painting.* Princeton: Princeton University Press.

1985b The Late Buildings of the Palace. *The Sculpture of Palenque, Vol. III.* Princeton: Princeton University Press.

Robiscek, Francis

1978 *The Smoking Gods, Tobacco in Maya Art, History, and Religion.* Norman: University of Oklahoma Press.

Robiscek, Francis, and Donald Hales

1981 *The Maya Book of the Dead: The Ceramic Codex.* Charlottesville: The University of Virginia Museum. Distributed by the University of Oklahoma Press.

Robiscek, Francis, and Donald Hales

1982 *Maya Ceramic Vases from the Classic Period: The November Collection of Maya Ceramics.* Charlottesville: University Museum of Virginia.

Robles C., Fernando, and Anthony P. Andrews

1986 A Review and Synthesis of Recent Postclassic Archaeology in Northern Yucatán. In *Late Lowland Maya Civilization,* edited by Jeremy A. Sabloff and E. Wyllys Andrews V, 53–98. A School of American Research Book. Albuquerque: University of New Mexico Press.

Roys, Lawrence

1931 The Engineering Knowledge of the Maya. *Contributions to American Archaeology,* Vol. II, No. 6; Carnegie Institution of Washington Pub. No. 436, 27–105. Washington, D.C.

Roys, Ralph L.

1943 *The Indian Background of Colonial Yucatán.* Carnegie Institution of Washington Pub. 548. Washington, D.C.

1957 *The Political Geography of the Yucatán Maya.* Carnegie Institution of Washington Pub. 613. Washington, D.C.

1962 Literary Sources for the History of Mayapán. In *Mayapán, Yucatán, México.* Carnegie Institution of Washington Pub. 619. Washington, D.C.

1965 *Ritual of the Bacabs.* Norman: University of Oklahoma Press.

1967 *The Book of the Chilam Balam of Chumayel.* Norman: University of Oklahoma Press.

Ruppert, Karl, J. Eric S. Thompson, and Tatiana Proskouriakoff

1955 *Bonampak, Chiapas, Mexico.* Carnegie Institution of Washington Pub. 602. Washington D.C.

Sabloff, Jeremy A., and E. Wyllys Andrews, eds.

1986 *Late Lowland Maya Civilization: Classic to Postclassic.* A School of American Research Book. Albuquerque: University of New Mexico Press.

Sabloff, Jeremy A., and William L. Rathje, eds.

1976 *A Study of Changing Pre-Columbian Commercial Systems: The 1972–1973 Seasons at Cozumel, Mexico.* Monographs of the Peabody Museum, Harvard University, No. 3.

Saenz de Santa María, Carmelo

1940 Diccionario Cakchiquel-Español. Guatemala.

References

Scarborough, Vernon L., Beverly Mitchum, H. Sorayya Carr, and David A. Freidel

1982 Two Late Preclassic Ballcourts at the Lowland Maya Center of Cerros, Northern Belize. *Journal of Field Archaeology* 9:21–34.

Scarborough, Vernon L., and David R. Wilcox, eds.

1991 *The Mesoamerican Ballgame.* Tucson: University of Arizona Press.

Schele, Linda

1974 Observations on the Cross Motif at Palenque. In *Primera Mesa Redonda de Palenque, Part I,* edited by Merle Greene Robertson, 41–62. Pebble Beach: Robert Louis Stevenson School.

1976 Accession Iconography of Chan-Bahlum in the Group of the Cross at Palenque. *The Art, Iconography, and Dynastic History of Palenque, Part III. Proceedings of the Segunda Mesa Redonda de Palenque,* edited by Merle Greene Robertson, 9–34. Pebble Beach, Calif.: Robert Louis Stevenson School.

1978 A Preliminary Commentary on the Tablets of the Temple of Inscriptions at Palenque, Chiapas. A preliminary study in preparation for a publication of the Miniconferences held at Dumbarton Oaks between 1974 and 1976.

1979a A Preliminary Study of Pictorial Devices Used in Maya Narrative Art. An unpublished paper prepared for a graduate readings course at the University of Texas.

1979b Genealogical Documentation in the Tri-Figure Panels at Palenque. *Tercera Mesa Redonda de Palenque, Vol. IV,* edited by Merle Greene Robertson, 41–70. Palenque: Pre-Columbian Art Research, and Monterey: Herald Printers.

1982 *Maya Glyphs: The Verbs.* Austin: University of Texas Press.

1984a Human Sacrifice Among the Classic Maya. In *Ritual Human Sacrifice in Mesoamerica,* edited by Elizabeth Boone, 7–49. Washington, D.C.: Dumbarton Oaks Research Library and Collection.

1984b *Notebook for the Maya Hieroglyphic Workshop at Austin, Texas.*

1985a Balan-Ahau: A Possible Reading of the Tikal Emblem Glyph and a Title at Palenque. *Fourth Round Table of Palenque, 1980, Vol. 6,* gen. ed. Merle Greene Robertson, vol. ed. Elizabeth Benson, 59–65. San Francisco: Pre-Columbian Art Research Institute.

1985b Some Suggested Readings of the Event and Office of Heir-Designate at Palenque. *Phoneticism in Mayan Hieroglyphic Writing,* 287–307. Albany: Institute of Mesoamerican Studies, State University of New York at Albany.

1986a Interim Report on the Iconography of the Architectural Sculpture of Temple 22 (from the 1986 field season). *Copán Note* 19. Copán Mosaics Project and the Instituto Hondureño de Antropología e Historia.

1986b The Tlaloc Heresy: Cultural Interaction and Social History. A paper given at "Maya Art and Civilization: The New Dynamics," a symposium sponsored by the Kimbell Art Museum, Fort Worth, May.

1987a The Figures on the Central Marker of Ballcourt IIa at Copan. *Copan Note* 13. Copán: Copán Mosaics Project and the Instituto Hondureño de Antropología e Historia.

1987b The Reviewing Stand of Temple 11. *Copan Note* 32. Copán Mosaics Project and the Instituto Hondureño de Antropología e Historia. Copán, Honduras.

1987c A Cached Jade from Temple 26 and the World Tree. *Copan Note* 34. Copán Mosaics Project and the Instituto Hondureño de Antropología e Historia. Copán, Honduras.

1987d New Fits on the North Panel of the West Door of Temple 11. *Copan Note* 38. Copán Mosaics Project and the Instituto Hondureño de Antropología e Historia. Copán, Honduras.

1987e Report on the Reconstruction of the Corner Masks of Temple 22 During the 1987 Field Season. Copán: Copán Acropolis Archaeological Project and the Instituto Hondureño de Antropología e Historia. Copán, Honduras.

1987f The Figures on the Central Marker of Ballcourt IIa at Copan. *Copan Note* 35. Copán: Copán Mosaics Project and the Instituto Hondureño de Antropología e Historia.

1987g *Notebook for the Maya Hieroglyphic Writing Workshop at Texas, with Commentaries on the Texts from the Group of the Cross.* Austin: Institute of Latin American Studies, University of Texas.

1988 The Xibalba Shuffle: A Dance After Death. In *Maya Iconography,* edited by Elizabeth Benson and Gillett Griffin, 294–317. Princeton: Princeton University Press.

1989 Brotherhood in Ancient Maya Kingship. A paper presented at the SUNY, Albany, conference on "New Interpretation of Maya Writing and Iconography," held October 21–22.

1990a The Glyph for "Hole" and the Skeletal Maw of the Underworld. *Copán Note* 71. Honduras: Copán Mosaics Project and the Instituto Hondureño de Antropología.

1990b Further Comments on Stela 6. *Copán Note* 73. Honduras: Copán Mosaics Project and the Instituto Hondureño de Antropología.

1990c Speculations from an Epigrapher on Things Archaeological. *Copán Note* 80. Honduras: Copán Mosaics Project and the Instituto Hondureño de Antropología.

1990d Lounsbury's Contrived Numbers and Two 8 Eb Dates at Copan. *Copán Note* 81. Honduras: Copán Mosaics Project and the Instituto Hondureño de Antropología.

1990e A Possible Death Statement for 18-Rabbit. *Copan Note* 90. Honduras: Copán Mosaics Project and the Instituto Hondureño de Antropología.

1990f House Names and Dedications at Palenque. In *Vision and Revision in Maya Studies,* edited by Flora S. Clancy and Peter D. Harrison, 143–157. Albuquerque: University of New México Press.

1990g Ba as "First" in Classic Period Titles. *Texas Notes on Precolumbian Art, Writing, and Culture* No. 5. Austin: CHAAAC, the Center for the History and Art of Ancient American Culture, Art Department, University of Texas.

1990h The Proceedings of the Maya Hieroglypic Workshop, March 10–11, 1990, with Commentary on the Inscriptions of Tikal, transcribed by Phil Wanyerka. Unpublished transcription available from the transcriber.

1991a The Workshop on Maya Hieroglyphic Writing, with Commentary on the History of Bird-Jaguar, transcribed by Phil Wanyerka. Austin: Art Department, University of Texas, and distributed by the transcriber.

1991b *Workbook for the 1991 Workshop on Maya Hieroglyphic Writing, with Commentary on the Inscriptions of Bird-Jaguar of Yaxchilan.* Austin: Art Department, University of Texas.

1992 *Workbook for the XVIth Maya Hieroglyphic Workshop at Texas, with Commentaries on the Group of the Cross at Palenque.* Austin: Art Department, University of Texas.

1992a Religion und Weltsicht. *Die Welt de Maya,* 215–238. Mainz am Rhein: Verlag Philipp Von Zabern.

n.d.a Some Thoughts on the Inscriptions of House C at Palenque. A paper presented at the Seventh Round Table of Palenque, held in Palenque, Chiapas, 1989.

Schele, Linda, and Barbara Fash

1991 Venus and the Reign of Smoke-Monkey. *Copán Note* 100. Honduras: Copan Acropolis Archaeological Project and the Instituto Hondureño de Antropología e Historia.

Schele, Linda, and David A. Freidel

1990a *A Forest of Kings: The Untold Story of the Ancient Maya.* New York: William Morrow and Company, Inc.

1990b The Maya Message: Time, Text, and Image. In *Art and Communication,* edited by Dan Eban. Jerusalem: The Israel Museum.

1991 The Courts of Creation: Ballcourts, Ballgames, and Portals to the Maya Otherworld. In *The Mesoamerican Ballgame,* edited by Vernon L. Scarborough and David R. Wilcox, 289–315. Tucson: University of Arizona Press.

Schele, Linda, and Nikolai Grube

1990a The Glyph for Plaza or Court. *Copan Note* 86. Honduras: Copán Acropolis Archaeological Project and the Instituto Hondureño de Antropología.

1990b A Suggested Reading Order for the West Side of Stela A. *Copán Note* 88. Honduras: Copán Mosaics Project and the Instituto Hondureño de Antropología.

1990c Building, Court, and Mountain Names in the Text of the Hieroglyphic Stairs. *Copán Note* 92. Honduras: Copán Mosaics Project and the Instituto Hondureño de Antropología.

Schele, Linda, Nikolai Grube, and David Stuart

1989 The Date of Dedication of Ballcourt III at Copán. *Copán Note* 59. Honduras: Copán Acropolis Archaeological Project and the Instituto Hondureño de Antropología e Historia.

Schele, Linda, and Rudi Larios

1991 Some Venus Dates on the Hieroglyphic Stairs at Copán. *Copán Note* 99. Copán, Honduras: Copán Acropolis Archaeological Project and the Instituto Hondureño de Antropología e Historia.

Schele, Linda, and Peter Mathews

1990 A Proposed Decipherment for Portions of Resbolon Stair 1. *Texas Notes on Precolumbian Art, Writing, and Culture* No. 3. Austin: CHAAAC, the Center for the History and Art of Ancient American Culture, Art Department, University of Texas.

1991 Royal Visits Along the Usumacinta. In *Classic Maya Political History: Archaeological and Hieroglyphic Evidence,* edited by T. P. Culbert, 226–253. A School of American Research Book. Cambridge: Cambridge University Press.

Schele, Linda, Peter Mathews, Nikolai Grube, Floyd Lounsbury, and David Kelley

1991 New Readings of Glyphs for the Month Kumk'u and Their Implications. *Texas Notes on Precolumbian Art, Writing, and Culture* No. 15. Austin: CHAAAC, the Center for the History and Art of Ancient American Culture, Art Department, University of Texas.

Schele, Linda, Peter Mathews, and Floyd Lounsbury

1990 The Nal Suffix at Palenque and Elsewhere. *Texas Notes on Precolumbian Art, Writing, and Culture* No. 6. Austin: CHAAAC, the Center for the History and Art of Ancient American Culture, Art Department, University of Texas.

Schele, Linda, and Jeffrey H. Miller

1983 The Mirror, the Rabbit, and the Bundle: Accession Expressions from the Classic Maya Inscriptions. *Studies in Pre-Columbian Art & Archaeology* no. 25. Washington, D.C.: Dumbarton Oaks Research Library and Collection.

Schele, Linda, and Mary Ellen Miller

1986 *The Blood of Kings: Dynasty and Ritual in Maya Art.* New York: George Braziller, Inc., in association with the Kimbell Art Museum, Fort Worth.

Schele, Linda, and Alfonso Morales

1990 Some Thoughts on Two Jade Pendants from the Termination Cache of "Ante" Pyramid at Copán. *Copán Note* 79. Honduras: Copán Mosaics Project and the Instituto Hondureño de Antropología.

Schele, Linda, and David Stuart

1986 Te-tun as the Glyph for "Stela." *Copán Note* 1. Copán: Copán Mosaics Project and the Instituto Hondureño de Antropología e Historia.

Schele, Linda, David Stuart, and Nikolai Grube

1991 A Commentary on the Inscriptions of Structure 10L–22A at Copan. *Copán Note* 98. Honduras: Copán Acropolis Archaeological Project and the Instituto Hondureño de Antropología.

Schele, Linda, and Khristaan Villela

1992 Some New Ideas About the T713-757 "Accession" Phrases. *Texas Notes on Precolumbian*

Art, Writing, and Culture No. 27. Austin: CHAAAC, the Center for the History and Art of Ancient American Culture, Art Department, University of Texas.

Schellhas, Paul

1904 *Representation of Deities of the Maya Manuscripts.* Papers of the Peabody Museum of American Archaeology and Ethnology, Harvard University 4(1). Cambridge.

Scholes, Franz V., and Elenor B. Adams

1960 Relación que en el Consejo Real de las Indias hizo sobre la pacificación y población de las provincias del Manché y Lacandón, el Licenciado Antonio de León Pinelo. In *Relaciones Historico-Descriptivas de la Verapaz, el Manche y Lacandon, en Guatemala,* 253–272. Guatemala: Editoria Universitaria, Universidad de San Carlos de Guatemala.

Scholes, Franz V., and Ralph Roys

1948 *The Maya Chontal Indians of Acalan-Tixchel.* Carnegie Institution of Washington Pub. 560. Washington, D.C.

Seler, Eduard

1902–1923 *Gesammelte Abhandlungen zür Amerikanischen Sprach-und Alterthumskunde von Eduard Seler.* Berlin: Asher.

1977 *Observations and Studies in the Ruins of Palenque (Beobachtungen und Studien in den Ruinen von Palenque),* translated by Gisela Morgner, edited by Thomas Bartman and George Kubler. Pebble Beach:Robert Louis Stevenson School.

1990 *Collected Works in Mesoamerican Linguistics and Archaeology,* 5 vols., supervised by Charles Bowditch, edited by J. Eric S. Thomspon and Francis B. Richardson. Labyrinthos: Culver City, Calif.

Sharer, Robert, David Sedat, and Alfonso Morales

1990 Investigaciones en el Patio Oriental de la Acropolis de Copan: Temporada de 1990. P.A.A.C., Ruinas de Copán, 1 de junio, 1990.

Sharer, Robert, David Sedat, Alfonso Morales, Julia Miller, Loa Traxler, Andrew Weiss, and Luis Reina

1991 Investigaciones del Programa de la Acropolis Temprana de Copan: Temporada de 1991. P.A.A.C., Ruinas de Copán, 11 de mayo, 1991.

Shook, Edwin M.

1958 The Temple of the Red Stela. *Expedition* 1(1):26–33.

Simpson, Lesley B., trans. and ed.

1966 *Cortés: The Life of the Conqueror by His Secretary Francisco Lopéz de Gómara.* Berkeley and Los Angeles: University of California Press.

Smith, Robert E.

1971 The Pottery of Mayapan, Including Studies of Ceramic Material from Uxmal, Kabah, and Chichen Itza. *Papers of the Peabody Museum of Archaeology and Ethnology,* vol. 66. Cambridge: Harvard University.

Sólis Alcalá, Ermilo

1949 *Códice Pérez,* translated into Spanish by Ermilo Sólis Alcalá. Mérida: Inprenta Oriente.

Sosa, John R.

1988 The Maya Sky, the Maya World: A Symbolic Analysis of Yucatec Maya Cosmology. Ph.D. dissertation, State University of New York at Albany.

1990 Cosmological, Symbolic, and Cultural Complexity Among the Contemporary Maya of

Yucatan. In *World Archaeoastronomy*, edited by Anthony Aveni, 130–142. Cambridge: Cambridge University Press.

Spero, Joanne

1987 Lightning Men and Water Serpents: A Comparison of Mayan and Mixe-Zoquean Beliefs. M.A. thesis, The University of Texas at Austin.

1990 Beyond Rainstorms: The Kawak as an Ancestor, Warrior, and Patron of Witchcraft. In *Sixth Palenque Round Table, 1986*, gen. ed. Merle Green Robertson, vol. ed. Virginia Fields, 184–193 Norman: University of Oklahoma Press.

Spinden, Herbert J.

1913 *A Study of Maya Art, Its Subject Matter and Historical Development.* Memoirs of the Peabody Museum of American Archaeology and Ethnology, Harvard University, VI. Cambridge.

Stephens, John L.

1843 *Incidents of Travels in Yucatán, Vols. I and II.* New York: Harper and Brothers. Reprint: New York: Dover Publications, 1963.

Stern, Theodore

1948 The Rubber Ball Games of the Americas. *Monographs of the American Ethnological Society, no. 17.* New York.

Stone, Andrea J.

1983 *The Zoomorphs of Quirigua, Guatemala.* A Ph.D dissertation, University of Texas at Austin. Ann Arbor: University Microfilms.

1989 Disconnection, Foreign Insignia, and Political Expansion: Teotihuacan and the Warrior Stelae of Piedras Negras. In *Mesoamerica After the Decline of Teotihuacan, A.D. 700–900*, edited by Richard Diehl and Janet C. Berlo. Washington, D.C.: Dumbarton Oaks Research Library and Collection.

Stone, Andrea, and Barbara MacLeod

n.d. A Study of the Painting of Naj Tunich Cave, Guatemala. A study in preparation.

Strömsvik, Gustav

1952 The Ball Courts at Copán. *Contributions to American Anthropology and History* 55:185–222. Washington, D.C.: Carnegie Institution of Washington.

Stross, Brian

1988 The Burden of Office: A Reading. *Mexicon* 10: 118–121.

Stuart, David

1984 Blood Symbolism in Maya Iconography. *RES* 7/8, 6–20.

1986a A Glyph for "Stone Incensario." *Copán Note* 1. Copán: Copán Mosaics Project and the Instituto Hondureño de Antropología e Historia.

1986b The Hieroglyphic Name of Altar U. *Copán Note* 4. Copán: Copán Mosaics Project and the Instituto Hondureño de Antropología e Historia.

1986c Thoughts on the Temple Inscription from Structure 26. *Copán Note* 11. Copán: Copán Mosaics Project and the Instituto Hondureño de Antropología e Historia.

1986d The "Lu-bat" Glyph and Its Bearing on the Primary Standard Sequence. A paper presented at the "Primer Simposio Mundial Sobre Epigrafía Maya," a conference held in Guatemala City in August 1986.

1987 Ten Phonetic Syllables. *Research Reports on Ancient Maya Writing* 14. Washington, D.C.: Center for Maya Research.

1988a Blood Symbolism in Maya Iconography. In *Maya Iconography*, edited by Elizabeth P. Benson and Gillett G. Griffin, 175–221. Princeton: Princeton University Press.

1988b *Maya Glyph Observations* No. 2. An unpublished commentary prepared for Dr. William Fash in the Copan Acropolis Project.

n.d. Kinship Terms in Mayan Inscriptions. A paper prepared for "The Language of Maya Hieroglyphs," a conference held at the University of California at Santa Barbara, February 1989.

Stuart, David, and Nikolai Grube

n.d. Observations on the Mayan Nomenclature of Stone Monuments, Architecture, and Certain Portable Objects. A manuscript in preparation.

Stuart, David, and Stephen Houston

1991 Classic Maya Place Names. An unpublished paper circulated by the authors in March 1991, submitted and accepted for publication by Dumbarton Oaks.

Stuart, David, and Linda Schele

1986 Interim Report on the Hieroglyphic Stair of Structure 26. *Copán Note* 17. Copán, Honduras: Copán Mosaics Project and the Instituto Hondureño de Antropología e Historia.

Stuart, David, Linda Schele, and Nikolai Grube

1989a A Mention of 18-Rabbit on the Temple 11 Reviewing Stand. *Copán Note* 62. Honduras: Copán Mosaics Project and the Instituto Hondureño de Antropología.

1989b A Commentary on the Restoration and Reading of the Glyphic Panels from Temple 11. *Copán Note* 64. Honduras: Copán Mosaics Project and the Instituto Hondureño de Antropología.

Stuart, George E.

1992 Quest for Decipherment: A Historical and Biographical Survey of Maya Hieroglyphic Investigation. In *New Theories on the Ancient Maya*, edited by Elin C. Danien and Robert Sharer: 1–65: Philadelphia: University Museum, University of Pennsylvania.

Sugiyama, Saburo

1989 Burials Dedicated to the Old Temple of Quetzalcoatl at Teotihuacán, México. *American Antiquity* 54:85–106.

Sullivan, Paul

1989 *Unfinished Conversations: Mayas and Foreigners Between Two Wars.* New York: Alfred A. Knopf.

Taladoire, Eric

1981 Les terrains de jeu de balle (mesoamérique et sud-oest des Etats-Unis). *Etudes Mesoaméricaines* Série II:4. Mission Archaeologique et Ethnologique Française au Mexique.

Taube, Karl A.

1985 The Classic Maya Maize God: A Reappraisal. In *Fifth Palenque Round Table, 1983, Vol. VII*, gen. ed. Merle Greene Robertson; vol. ed. Virginia M. Fields, 171–181. San Francisco: The Pre-Columbian Art Research Institute.

1986 The Teotihuacan Cave of Origin: The Iconography and Architecture of Emergence Mythology in Mesoamerica and the American Southwest. *RES*: 51–82.

1988a A Study of Classic Maya Scaffold Sacrifice. In *Maya Iconography*, edited by Elizabeth Benson and Gillett Griffin, 331–351. Princeton: Princeton University Press.

1988b The Ancient Yucatec New Year Festival: The Liminal Period in Maya Ritual and Cosmology. A Ph.D dissertation, Yale University.

1989 *Itzam Cab Ain*: Caimans, Cosmology, and Calendrics in Postclassic Yucatán. *Research Reports on Ancient Maya Writing* 26 & 27. Washington D.C.: Center for Maya Research.

1990 The Iconography of Toltec Period Chichen Itza. A paper presented at the First Maler

Symposium on the Archaeology of Northwest Yucatan. Seminar für Völkerkunde, Universität Bonn.

1992a *The Major Gods of Ancient Yucatan: Schellhas Revisited.* Studies in Pre-Columbian Art and Archaeology 82. Washington, D.C.: Dumbarton Oaks Research Library and Collection.

1992b The Temple of Quetzalcoatl and the Cult of Sacred War at Teotihuacan. *RES* 21:53–87.

n.d.a The Iconography of Mirrors at Classic Teotihuacan. A paper presented at the 1988 Dumbarton Oaks Symposium on Art, Polity, and the City of Teotihuacan.

Tedlock, Barbara

1982 *Time and the Highland Maya.* Albuquerque: University of New Mexico Press.

1985 Hawks, Meteorology, and Astronomy in Quiché Maya Agriculture. *Archaeoastronomy* 8:80–88.

1986 On a Mountain in the Dark: Encounters with the K'iche' Maya Culture Hero. In *Symbol and Meaning Beyond the Closed Community: Essays in Mesoamerican Ideas,* edited by Gary H. Gossen, 125–138. Albany: Institute for Mesoamerican Studies, The University at Albany, State University of New York.

1992 The Road of Light: Theory and Practice of Mayan Skywatching. In *The Sky in Mayan Literature,* edited by Anthony Aveni. New York: Oxford University Press.

Tedlock, Dennis

1985 *Popol Vuh: The Definitive Edition of the Mayan Book of the Dawn of Life and the Glories of God and Kings.* New York: Simon and Schuster.

1986 Creation in the Popol Vuh: A Hermneutical Approach. In *Symbol and Meaning Beyond the Closed Community: Essays in Mesoamerican Ideas,* edited by Gary H. Gossen, 77–82. Albany: Institute for Mesoamerican Studies, The University at Albany, State University of New York.

1992 Myths, Maths, and the Problem of Correlation in Mayan Books. In *The Sky in Mayan Literature,* edited by Anthony Aveni. New York: Oxford University Press.

n.d. The Sowing and Dawning of All the Sky-Earth: Sun, Moon, Stars, and Maize in the Popol Vuh. *Museum Monographs.* Philadelphia: University Museum (in press).

Thompson, J. Eric

1930 Ethnology of the Mayas of Southern and Central British Honduras. *Field Museum of Natural History Pub. 274, Anthropological Series Vol. XVII, No.2.* Chicago.

1950 Maya Hieroglyphic Writing: An Introduction. *Carnegie Institution of Washington Pub. 589.* Washington, D.C.

1957 Deities Portrayed on Censers at Mayapan. *Carnegie Institution of Washington Current Reports,* no. 2, 599–632.

1962 *A Catalog of Maya Hieroglyphics.* Norman: University of Oklahoma Press.

1970 *Maya History and Religion.* Norman: University of Oklahoma Press.

Tozzer, Alfred

1907 *A Comparative Study of the Mayas and the Lacandones.* New York: Archaeological Institute of America, The Macmillan Co.

1941 *Landa's Relación de las Cosas de Yucatán: A Translation.* Papers of the Peabody Museum of American Archaeology and Ethnology, Harvard University, Vol. XVIII. Reprinted with permission of the original publishers by Kraus Reprint Corporation, New York, 1966.

1957 *Chichén Itzá and Its Cenote of Sacrifice: A Comparative Study of Contemporaneous Maya and Toltec.* Memoirs of the Peabody Museum of Archaeology and Ethnology, Harvard University, XI and XII. Cambridge.

Trik, Aubrey S.

1939 Temple XXII at Copan. *Contributions to American Anthropology and History 27.* Washington, D.C.: Carnegie Institute of Washington.

References

Valdés, Juan Antonio

1987 El Grupo H de Uaxactun: Evidencias de un centro de poder durante el Preclásico. Paper presented at the II Coloquio Internacional de Mayistas, Campeche, Mexico, August.

Van Bussel, Gerard W., Paul L. F. van Dongen, and Ted J. J. Leyenaar, eds.

1991 *The Mesoamerican Ballgame,* 81–90. Leiden: Rijksmuseum voor Volkenkende.

Villagutierre Soto-Mayor, Don Juan de

1983 *History of the Conquest of the Province of the Itza,* translated by Brother Robert D. Wood, edited and notes by Frank E. Comparato. Culver City, Calif.: Labyrinthos.

Villa Rojas, Alfonso

1945 *The Maya of East Central Quintana Roo. Carnegie Institution of Washington* Pub. 559. Washington, D.C.

Villela, Kristaan

1990 The Iconography of Auto-sacrifice at Teotihuacan. Honors thesis submitted to the Department of Latin American Studies, Yale University.

Vogt, Evon Z.

1970 The Zinacantecos of Mexico: A Modern Maya Way of Life. *Case Studies in Cultural Anthropology.* New York: Holt, Rinehart and Winston.

1976 *Tortillas for the Gods: A Symbolic Analysis of Zincanteco Rituals.* Cambridge: Harvard University Press.

1985 Cardinal Directions and Ceremonial Circuits in Mayan and Southwestern Cosmology. *National Geographic Society Research Reports* 21:487–496.

1986 How the Yucatec Survived the Conquest. *Annual Reviews in Anthropology* 13(1):37–41.

n.d. Indian Crosses and Scepters: The Results of Circumscribed Spanish-Indian Interactions in Mesoamerica. In *Interethnic Encounters: Discourse and Practice in the New World,* edited by J. Jorge Klor de Alva, Gary H. Gossen, Manuel Gutiérrez-Estevez, and Miguel León-Portilla. Albany: Institute for Mesoamerican Studies, The University at Albany, and State University of New York Press (forthcoming).

von Winning, Hasso

1948 The Teotihuacan Owl and Weapon Symbol and Its Association with "Serpent Head X" at Kaminaljuyu. *American Antiquity* 14:129–132.

Wauchope, Robert

1938 Modern Maya Houses: A Study of Their Significance. *Carnegie Institution of Washington Pub. 502.* Washington, D.C.

Wauchope, Robert, ed.

1973–1975 *Handbook of Middle American Indians,* Vols. 13–15. Austin: University of Texas Press.

Webster, Susan Verdi

1992 The Processional Sculpture of Penitential Confraternities in Early Modern Seville. A Ph.D dissertation, University of Texas at Austin.

Whittaker, Arabelle, and Viola Warkentin

1952 The Supernatural World of Curing. *The Ethnology of Middle America,* edited by Sol Tax, 119–134. New York: Macmillan Company.

Wilbert, Johannes

1973 The Calabash of the Ruffled Feathers. *Arts Canada* nos. 184–187:90–93.

1987 *Tobacco and Shamanism in South America.* New Haven: Yale University Press.

Wilkerson, S. Jeffery K.

1991 And Then They Were Sacrificed: The Ritual Ballgame of Northeastern Mesoamerica Through Time and Space. In *The Mesoamerican Ballgame*, edited by Vernon L. Scarborough and David R. Wilcox, 45–71. Tucson: University of Arizona Press.

Wilson, George C.

1974 *Crazy February: Death and Life in the Mayan Highlands of Mexico.* Berkeley: University of Califoria Press, Reprint Edition.

Wisdom, Charles

1940 *The Chorti Indians of Guatemala.* Chicago: University of Chicago Press.

n.d. Materials on the Chorti Languages. Collection of Manuscripts of the Middle American Cultural Anthropology, Fifth Series, No. 20. Microfilm, University of Chicago.

Wren, Linnea H.

1991 The Great Ball Court Stone at Chichén Itzá. *Sixth Palenque Round Table, 1986*, gen. ed. Merle Greene Robertson, vol. ed. Virginia Field, 51–59. Norman: University of Oklahoma Press.

n.d. Ceremonialism in the Reliefs of the North Temple, Chichen Itza. A paper presented at the Seventh Round Table of Palenque, June 1989.

Wren, Linnea H., Ruth Krochock, Erik Boot, Lynn Foster, Peter Keeler, Rex Koontz, Walter Wakefield

1992 Maya Creation and Recreation: The Great Ball Court at Chichén Itzá. A manuscript prepared after the 1992 Advanced Seminar of the Texas Meetings on Maya Hieroglyphic Writing.

Wren, Linnea H., and Peter D. Schmidt

1991 Elite Interaction During the Terminal Classic Period: New Evidence from Chich'en Itza. In *Classic Maya Political History: Archaeological and Hieroglyphic Evidence*, edited by T. Patrick Culbert, 199–225. Cambridge: A School of American Research Book, Cambridge University Press.

Zhang, He

1990 Bone-Thrones in Classic Maya Pottery Scenes. A paper prepared for a graduate seminar at the University of Texas.

Zimbalist, Michelle

1966 La Granadilla: un modelo de la estructura social Zinacanteca. In *Los Zinacantecos: Un pueblo tzotzil de los altos de Chiapas*, edited by Evon Vogt: 275–297. Colección de Antropología Social 7. México: Instituto Nacional Indígenista.

Zuñiga, Fray Diego de

1608 *Diccionario Pocomchi—Castellano y Pocomchi de San Cristóbal Cohcoh.* Copy of the photographic copy in the Gates Collection, Tulane University.

INDEX

Page numbers beginning with 405 refer to notes.

Index

Index

Index

Index

Index

Index

Index

Index

Index